LOUGH NEAGH

MONOGRAPHIAE BIOLOGICAE

VOLUME 69

Series Editors

H.J. Dumont and M.J.A. Werger

The titles published in this series are listed at the end of this volume.

Lough Neagh

The Ecology of a Multipurpose Water Resource

Edited by

R.B. WOOD and R.V. SMITH

KLUWER ACADEMIC PUBLISHERS
DORDRECHT / BOSTON / LONDON

Library of Congress Cataloging-in-Publication Data

Lough Neagh : the ecology of a multipurpose water resource / edited by R.B. Wood and R.V. Smith.
p. cm. -- (Monographiae biologicae ; v. 69)
Includes bibliographical references and index.
ISBN 0-7923-2112-X (alk. paper)
1. Lake ecology--Northern Ireland--Neagh, Lough. 2. Limnology--Northern Ireland--Neagh, Lough. 3. Hydrology--Northern Ireland--Neagh, Lough. 4. Neagh, Lough (Northern Ireland) I. Wood, R. B. II. Smith, R. V. III. Series.
QP1.P37 vol. 69
[QH143.5]
574 s--dc20
[574.5'26322'09416] 92-44827

ISBN 0-7923-2112-X

Published by Kluwer Academic Publishers.
P.O. Box 17, 3300 AA Dordrecht. The Netherlands.

Kluwer Academic Publishers incorporates
the publishing programmes of
D. Reidel, Martinus Nijhoff, Dr W. Junk and MTP Press.

Sold and distributed in the U.S.A. and Canada
by Kluwer Academic Publishers
101 Philip Drive, Norwell, MA 02061, U.S.A.

In all other countries, sold and distributed
by Kluwer Academic Publishers Group,
P.O. Box 322, 3300 AH Dordrecht, The Netherlands.

Printed on acid-free paper

Printed in the Netherlands

Table of contents

List of Contributors

ANDREW, T.E. Freshwater Laboratory, University of Ulster, Traad Point, Ballyronan, Magherafelt, Co Derry, Northern Ireland, BT45 6LR.

BATTARBEE, R.W. Environmental Change Research Centre, Department of Geography, University College London, 26 Bedford Way, London WC1H OAP, UK. (Formerly, Freshwater Laboratory, University of Ulster)

BRIGGS, M. Formerly, Freshwater Laboratory, University of Ulster, Traad Point, Ballyronan, Magherafelt, Co Derry, Northern Ireland, BT45 6LR.

CARTER, C. Freshwater Laboratory, University of Ulster, Traad Point, Ballyronan, Magherafelt, Co Derry, Northern Ireland, BT45 6LR.

CARTER, R.W.G. Department of Environmental Studies, University of Ulster, Cromore Road, Coleraine, Co Derry, Northern Ireland, BT52 1SA

CROZIER, W.W. Aquatic Sciences Research Division, Department of Agriculture for Northern Ireland, New Forge Lane, Belfast, Northern Ireland, BT9 5PX.

DAVIDSON, R.D. Lough Neagh Nature Reserves Warden, Environment Service, Department of the Environment for Northern Ireland, Visitor Centre, Oxford Island, Craigavon, Northern Ireland, BT66 6NJ.

FERGUSON, A. School of Biology and Biochemistry, The Queen's University of Belfast, Belfast, Northern Ireland, BT7 1NN.

FITZSIMONS, A.G. Aquatic Sciences Research Division, Department of Agriculture for Northern Ireland, New Forge Lane, Belfast, Northern Ireland, BT9 5PX.

FLOWER, R.J. Environmental Change Research Centre, Department of

Geography, University College London, 26 Bedford Way, London, WC1H OAP, UK. (Formerly, Freshwater Laboratory, University of Ulster)

FOY, R.H. Aquatic Sciences Research Division, Department of Agriculture for Northern Ireland, New Forge Lane, Belfast, Northern Ireland, BT9 5PX.

FURPHY, J.S. Countryside and Wildlife, Environment Service, Department of the Environment for Northern Ireland, Calvert House, Castle Place, Belfast, Northern Ireland, BT1 1FY.

GIBSON, C.E. Aquatic Sciences Research Division, Department of Agriculture for Northern Ireland, New Forge Lane, Belfast, Northern Ireland, BT9 5PX.

JEWSON, D.H. Freshwater Laboratory, University of Ulster, Traad Point, Ballyronan, Magherafelt, Co Derry, Northern Ireland, BT45 6LR.

JONES, R.I. Division of Biological Sciences, Institute of Environmental and Biological Sciences, University of Lancaster, Lancaster, LA1 4YQ, UK. (Formerly, Freshwater Laboratory, University of Ulster)

KENNEDY, G.J.A. Aquatic Sciences Research Division, Department of Agriculture for Northern Ireland, New Forge Lane, Belfast, Northern Ireland, BT9 5PX.

MACFADYEN, A. Freshwater Laboratory, University of Ulster, Traad Point, Ballyronan, Magherafelt, Co Derry, Northern Ireland, BT45 6LR.

MITCHEL, N.C. Institute of Irish Studies, The Queen's University of Belfast, Belfast, Northern Ireland, BT7 1NN.

MONTGOMERY, C.R. Freshwater Laboratory, University of Ulster, Traad Point, Ballyronan, Magherafelt, Co Derry, Northern Ireland, BT45 6LR. Present address: Ministry of Agriculture, Fisheries and Food, Nobel House, 17 Smith Square, London SW1P 3JR.

MURPHY, P.M. Freshwater Laboratory, University of Ulster, Traad Point, Ballyronan, Magherafelt, Co Derry, Northern Ireland, BT45 6LR. Present address: European Commission, 200 Rue de la Loi, 1049 Brussels, Belgium.

QUINN, J.P. Freshwater Laboratory, University of Ulster, Traad Point, Ballyronan, Magherafelt, Co Derry, Northern Ireland, BT45 6LR. Present address: School of Biology and Biochemistry, The Queen's University of Belfast, Belfast, Northern Ireland, BT7 1NN.

RIPPEY, B.H.R.T. Freshwater Laboratory, University of Ulster, Traad Point, Ballyronan, Magherafelt, Co Derry, Northern Ireland, BT45 6LR.

SMITH, R.V. Aquatic Sciences Research Division, Department of Agriculture for Northern Ireland, New Forge Lane, Belfast, Northern Ireland, BT9 5PX.

STEWART, D.A. Formerly, Biometrics Division, Department of Agriculture for Northern Ireland, New Forge Lane, Belfast, Northern Ireland, BT9 5PX.

TOBIN, C.M. Freshwater Laboratory, University of Ulster, Traad Point, Ballyronan, Magherafelt, Co Derry, Northern Ireland, BT45 6LR. Present address: School of Applied Sciences, University of Wolverhampton, Wulferuna Street, Wolverhampton, WV1 1FB, UK.

VICKERS, K.U. Formerly, Aquatic Sciences Research Division, Department of Agriculture for Northern Ireland, New Forge Lane, Belfast, Northern Ireland, BT9 5PX.

WARD, R.D. Department of Human Sciences, University of Technology, Loughborough, Leicester, LE11 3TU, UK.

WILSON, J.P.F. Freshwater Laboratory, University of Ulster, Traad Point, Ballyronan, Magherafelt, Co Derry, Northern Ireland, BT45 6LR. Present address: Le Floriana, 406 chemin de St-Claude, 06600 Antibes, France.

WINFIELD, I.J. Freshwater Laboratory, University of Ulster, Traad Point, Ballyronan, Magherafelt, Co Derry, Northern Ireland, BT45 6LR. Present address: Institute of Freshwater Ecology, Ferry House, Far Sawrey, Ambleside, Cumbria, LA22 OLP.

WOOD, R.B. Freshwater Laboratory, University of Ulster, Traad Point, Ballyronan, Magherafelt, Co Derry, Northern Ireland, BT45 6LR.

WOODMAN, P.C. Department of Archaeology, University College Cork, Lee Maltings, Prospect Row, Eire. Formerly: Department of Antiquities, Ulster Museum, Botanic Gardens, Belfast BT9 5AB.

WOODWARD, E. Formerly, Freshwater Laboratory, University of Ulster, Traad Point, Ballyronan, Magherafelt, Co Derry, Northern Ireland, BT45 6LR.

1. Introduction: Lough Neagh as a multipurpose resource

R.B. WOOD

1.1. Background

"... but, like Aesop's bundle of sticks, these scattered records, lying loose and disconnected, are here gathered together and tied in a bundle". So Marshall (1934) began the Foreword to his essay 'Lough Neagh in Legend and History' and, given Ireland's long history of story-telling, it is perhaps appropriate that myth and fairy-tale should start this gathering of Lough Neagh's scientific sticks. Being both Irish, and Ireland's largest lake, Lough Neagh predictably boasts a mythical origin, indeed (and equally predictably) two rival origins, on the grand scale. In one, a lid left off a magic well allowed its waters to inundate Eochaid's camp and form 'loch nEchach' (= loch Neagh) resulting in the still current story of a submerged city, and granting Eochaid's daughter, Liban, 50% of her wish to become a salmon (*Salmo salar* L.) and thus the first mermaid. Her dog became the first otter (*Lutra lutra* L.). Today otters are occasionally seen; mermaids seem locally absent. Giraldus Cambrensis (1186) authenticated this origin, embellishing it with overtones of Sodom and Gomorah, although it must be recorded that Giraldus Cambrensis has been described as "a zealous if rather credulous observer ...".

The second legend, on an appropriate Gargantuan scale, has it that the Irish giant Fion MacCuil routed a Scottish rival who fled and outran the victor. Hurling a gigantic handful of Irish rock at the Scot, Fion MacCuil over-compensated for his inadequacy as a middle distance runner by setting a new record in the shot-putt. The rock became the Isle of Man in the Irish Sea and the hole left filled to become Lough Neagh (see Fig. 1.1).

In the chapters which follow we provide a somewhat less colourful gathering of the sticks of what is known about Lough Neagh, its origins, geological and human history followed by the interacting complex of factors described under the broad heading of ecology. This we try to set against the backcloth of the lough as a valuable, multipurpose resource.

While Giraldus Cambrensis (1186) reported fairly accurately the dimensions of Lough Neagh (Chapter 2), intensive study had to wait approximately

R.B. Wood and R.V. Smith (eds.), Lough Neagh, 1–9.

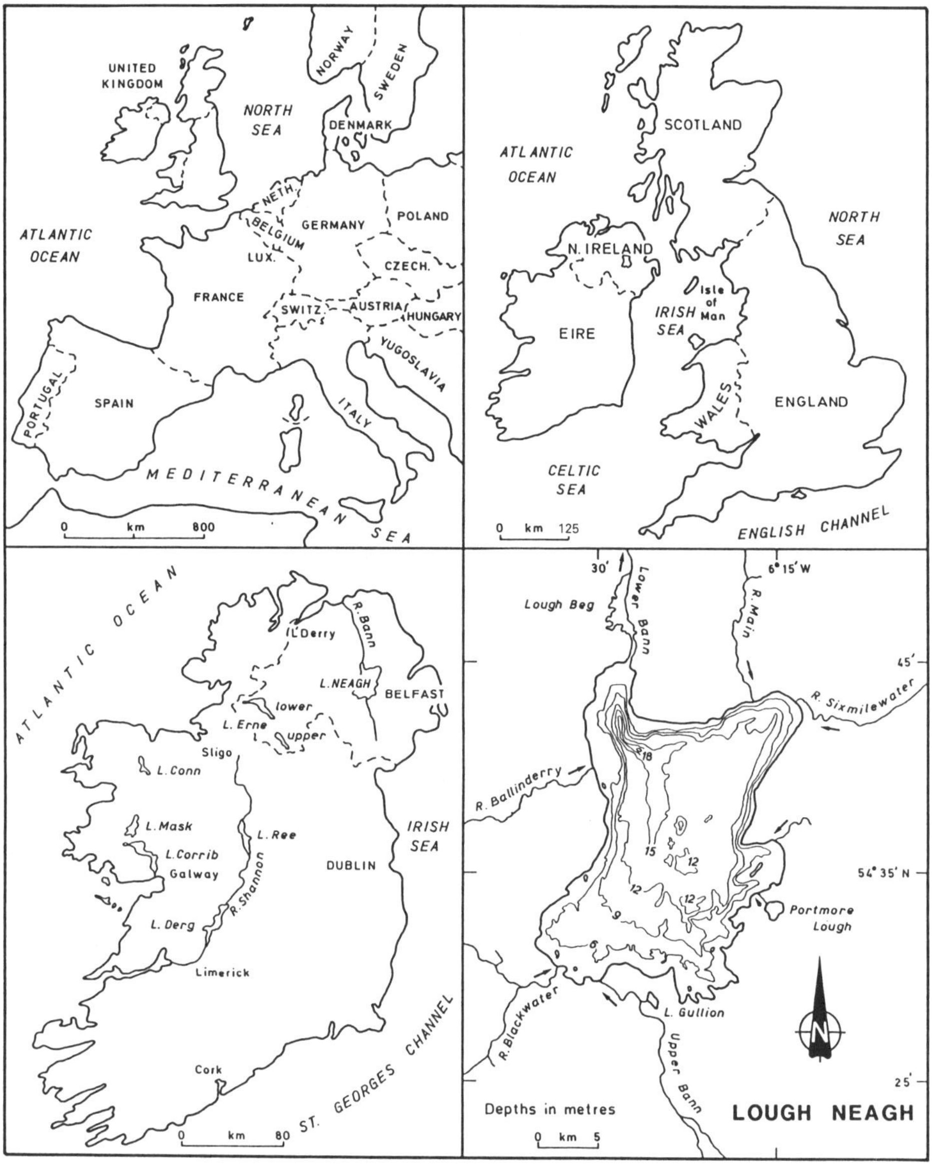

Figure 1.1. The location of Lough Neagh.

another 800 years. Individual chapters provide references to early work, but the vast bulk, upwards of 300 scientific papers, has been written since 1967. That year saw two institutional innovations which changed Lough Neagh from being among the least to probably the most studied body of water in

Ireland. The fisheries of Lough Neagh had long been exploited and occasionally fought over (Chapters 8 and 18); increasing urbanisation and industrialisation had highlighted in the 1960s the inadequate knowledge of Lough Neagh as a potential water source; agricultural productivity was rapidly increasing and land drainage and river improvements were in hand. To the increasing awareness of the resource potential of Lough Neagh were added, in 1967, perhaps 10,000 tons dry weight of *Anabaena flos-aquae*. Lough Neagh had produced a considerable, media-noticed algal bloom which killed eels in keep boxes, clogged filtration plants and was aesthetically unpleasing. Two responsible government departments, Agriculture (DA(NI); fisheries and drainage) and Development – later Environment (DoE(NI); water quality, water supply and effluent disposal) agreed that DA(NI) should establish the Freshwater Biological Investigation Unit (FBIU) to seek out the causes and remedies for the unwelcome bloom.

Predating the bloom by a year or so, the embryonic New University of Ulster (NUU) had decided to specialise in freshwater studies, and chose Lough Neagh as its prime study, building a lough-side laboratory permanently staffed. The increasing output of research, indeed this monograph, marks the past twenty years of productive cooperation principally between these two laboratories together with others, for example in the Water Service of DoE(NI), the Water Quality Branch of the (now) Department of Economic Development (DED(NI)), the Fisheries Research Laboratory DA(NI), and the Queen's University of Belfast.

That a range of disciplines (from archaeology to optics, by way of civil engineering, geography, farming and palynology, in addition to the expected chemistry and biology) is represented in the monograph should come as no surprise (Wood, 1980, 1989). Management requires knowledge, not merely of the immediate exploitation, like catching fish or drinking water or watching birds, but of the interaction of these and the processes which the lake, its catchment, its climate and its use encompass.

1.2. The resource

As shown in Fig. 1.1., Lough Neagh is situated close to the Atlantic seaboard of Europe. Its overall mild, damp climate results from its nearness to the relatively warm surface waters of the North Atlantic drift and its lying in the path of depressions and frontal systems tracking from the South West. These result in rapid changes in weather, predominance of cloud cover (average 5.8 octas and only 3.5 hr d^{-1} average bright sun), mild night temperatures, and precipitation (only rarely as snow or hail) averaging 880 mm yr^{-1} at Toomebridge, and up to 1600 mm yr^{-1} on the modest high ground of the Sperrin Mountains, Antrim Plateau and Mourne Mountains (peak heights 683 m, 554 m and 850 m, respectively) which bound the catchment (see Fig.

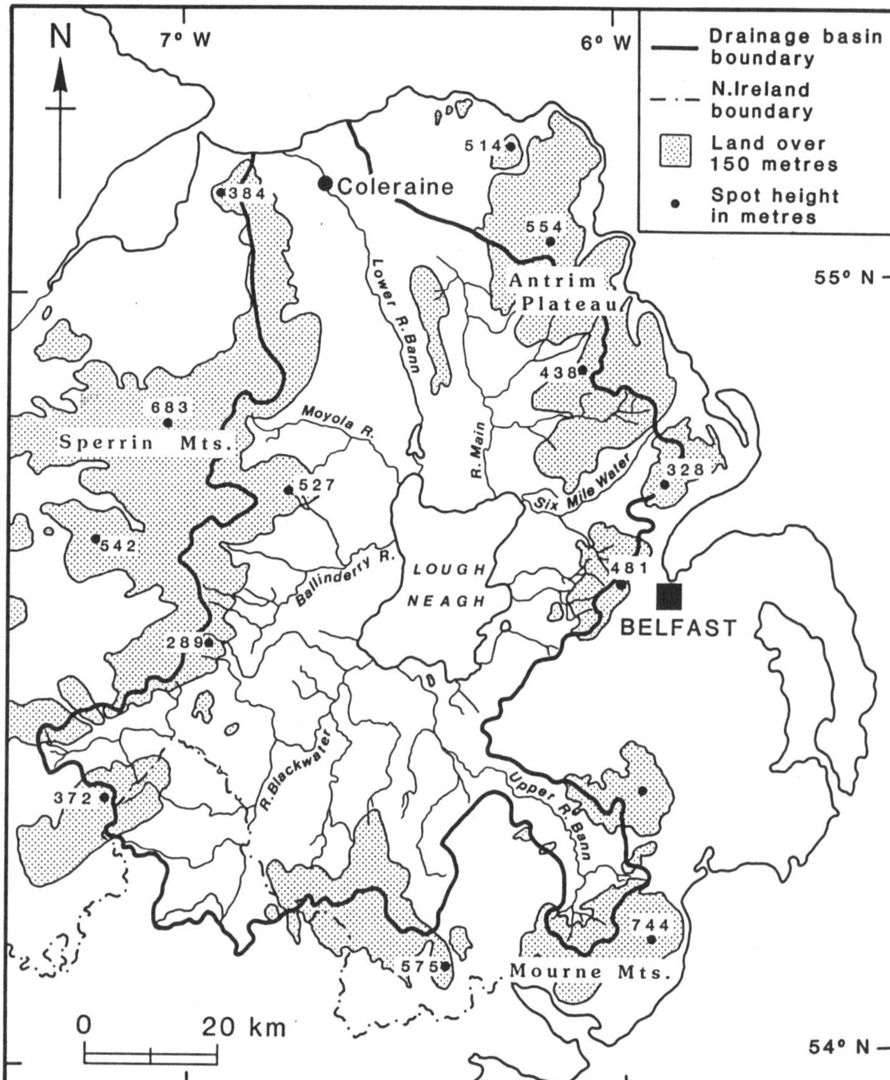

Figure 1.2. Relief map of the Lough Neagh catchments.

1.2). Extremes of temperature (e.g. at Armagh +30.6° C July 1934 average, −11.1° C January 1963 average) and rainfall/drought (with an average of approximately 200 days per year having >0.2 mm rainfall) are rare. Even within year seasonality is muted. A survey of N. Ireland's climate is given in Betts (1982) and points of specific interest are dealt with in the chapters which follow.

The bland climate has enabled highly productive agriculture to develop and the Lough Neagh catchment is essentially improved grassland bounded by higher peatlands with comparatively low human populations, little industry of its own, and up-wind of the great industries of the rest of Europe.

Chapter 6 introduces and Chapter 18 describes in some detail the fisheries of Lough Neagh. In socio-economic terms the total commercial fisheries, principally for eels, provide employment for about 500 men in an area of chronic unemployment. The total financial turnover may be upwards of £3 million per year. To this must be added the recreational and tourist value of largely salmon, trout and cyprinid fishing offered by perhaps 1000 km of Lough Neagh's principal rivers and their minor tributaries. It is not possible to distinguish from the rest of Northern Ireland the attraction which the Lough Neagh system has for anglers but, even allowing for the fully justified popularity of the Erne and Foyle systems, Lough Neagh and its rivers generate a significant percentage of the total angling effort as judged by the licences issued which, for 1983, are shown in Table 1.1 (see also Wood and Smith, 1986). Some 12,000 residents of Northern Ireland are members of organised angling clubs and a good many others, less organised, fish regularly. The recreational and tourist value of the Lough Neagh angling resource must be high, and would probably be higher in the absence of civil unrest. Certainly, the annual variation in 15-day non-resident licences issued since 1969 shows a strong inverse correlation with the intensity of local violence (Anon, 1984).

Table 1.1. Angling licences issued in Northern Ireland in 1983 (data from the Fisheries Conservancy Board (FCB) Anon 1984). Values in parentheses indicate non-residents i.e. tourists

Game fishing: season	9113	(27)
Game fishing: 15-day	305	(148)
Coarse fishing: season	1118	(117)
Coarse fishing: 15-day	2053	(1658)

Chapter 18 includes an example of the effects of water supply on fisheries; specifically of water discharge on the movement of eels and salmon in the R. Lower Bann. To satisfy the fishermen the lough needs to be regulated, with higher levels at the start of summer to sustain salmon movement upstream (May to August) and silver eel migration downstream (September to November) as well as summer demands for increased water supply. But agricultural land is poorly drained (Chapter 2) and better drainage calls for lowered, not raised, lough levels. Even nesting wildfowl have their preferences for water levels (Chapter 20). Added to which, in Northern Ireland, heavy rain in summer and dry periods in winter are by no means impossible.

The fact that Lough Neagh is a huge reservoir (total volume 3.45×10^6 Ml, Chapter 2) when compared with demand for water abstraction, currently some 124 Ml d^{-1} net export from the catchment, does not obviate the need, recognised by among others the Lough Neagh Working Group (Bell, 1971), that the impact of one use upon another must be reconciled.

The Lough Neagh Working Group concluded that, *properly managed*, Lough Neagh might permanently export for water supply up to 400 Ml d^{-1} without adversely affecting other users. 'Other users' continue to question even quite small increases in abstraction, not least because of the fishing vs. farming conflict of interest.

Scenically Lough Neagh may have fairly low amenity value with few mountains and islands to break up the great expanse of open water, but for bird watchers (and gun clubs) (Chapter 20) and those who appreciate fens and bogs (Chapter 19) it has much attraction. Relatively few sailors and bathers use it, though the broad, non-tidal waters and good winds offer remarkable sailing, and the Lower R. Bann is navigable to the sea. Imaginative development of marinas, canoe stretches, riverside walks and picnic sites, of nature reserves and birdhides by such departments as DED(NI), DA(NI) and DoE(NI) has made Lough Neagh a considerable, if somewhat local, asset.

1.3. The deteriorating resource

Perhaps the most critical interaction between one use and all others concerns effluent disposal, summarised early in the recent intensive period of study by one of the Lough Neagh Working Group's conclusions: "The continuing critical deterioration in L. Neagh due to eutrophication must be arrested or reversed". The situation in 1967 had brought the lough very much to the public notice; British Enkalon's man-made fibre plant recently attracted to Northern Ireland by, among other things, a plentiful supply of good fresh water was in danger of temporary closure because its filtration plant could not cope and in subsequent years the poor quality of the water led to DoE(NI) choosing to abstract water from the lower R. Moyola rather than the open lough, and to propose further ecologically dubious impoundments in the Mourne Mountains rather than risk the supply from Lough Neagh (Dennison and Sweetnam, 1979). The water abstraction works at Dunore Point with a design capacity of 134 ML d^{-1} frequently lost 30% of output during algal blooms, occasionally as much as 60%. Extra filter cleaning and pre-chlorination kept supplies flowing but at high cost.

The biological implications for the lough were equally worrying. Lough Neagh is shallow and exposed to strong winds, it rarely stratifies and even then not for long periods. Yet such was its accumulation of algal biomass (see Chapter 9) that respiration and decomposition put demands on the dissolved oxygen which photosynthesis and turbulence could not always satisfy. Anoxia, albeit transient at depth, affected nutrient release from the sediments (Stevens and Gibson, 1977; Rippey, 1977 and Chapter 9) and must have affected bottom feeders among the fish. Carter in Chapter 16 discusses the effect of these conditions on the chironomid fauna and its distribution.

The enormous nuisance swarms of midges were, and still are, by no means negligible in considerations of amenity and development.

Long-term surveillance produced the evidence summarised in Fig. 1.2. where whole lake integration of dissolved oxygen, allowing for the relatively small volumes of deep O_2-depleted water showed that full saturation might not be reached for months on end in spite of the lough's climate and morphometry.

Individual profiles (see Chapter 9) at times showed extreme O_2-depletion at depth, and *in situ* studies by Gibson (Chapter 9) in the open lough showed that anoxia in the bottom 1 m could develop in only 4–5 days when wind fell below 5 knots (~9 km.p.h.). Laboratory measurements of O_2-consumption by sediments and deep water confirmed the general concern that a week or so of calm conditions could hardly avoid resulting in fish deaths on a spectacular scale. Much of the study of nutrients (Chapters 8 and 9), algal populations (Chapter 10) and primary production (Chapter 11) had its origin in this concern for the over-enriched state of Lough Neagh, after the first, fairly superficial, analysis of the problem (Wood and Gibson, 1973).

Although, because of enrichment, the quality of Lough Neagh open water has given cause for grave concern over the past 20 years, the rivers in the catchment are generally of good quality. All those of sufficient size have salmon runs and over 400 km of the principal rivers meet the salmonid standard set by Directive 78/695/EEC and a further 107 km are designated cyprinid waters as is Lough Neagh itself in spite of its supporting a commercially exploited trout population (see Chapter 18, Part C). There are reaches of rivers below towns and downstream of particular enterprises which are unsatisfactory but DoE (NI) statistics (Anon 1981) recorded for Northern Ireland as a whole (and there is little reason to suppose Lough Neagh rivers are significantly different) some 85% of all rivers to be Class Ia – unpolluted, or Class Ib – high quality (National Water Council classification where e.g. BOD < 3 and $< 5\,\mathrm{mg\,l^{-1}}$ respectively occur for at least 95% of the time). A more recent survey (Anon, 1985a) shows little change; class 3 (poor) has risen from 2.7% to 3.9% and Class 1A (Good) has fallen from 23.3% to 16.3% but the combined class 1A + 1B has risen very marginally. The 1983 pollution statistics (Anon, 1985b) for the whole of Northern Ireland and from which it is not possible to separate Lough Neagh data, show a total of 1515 incidents of which the Industrial and Sewage categories together contribute 498, and Agricultural incidents number 777. In the latter the relatively recent switch to silage making is reflected in the sharp peak of incidents in June and July. Within the 1983 statistics involving fish kills and investigated by the Fisheries Conservancy Board (Anon, 1984), 673 incidents occurred of which 335 were attributed to silage and a further 129 to general farm pollution. While considerable effort is expended to minimise this pollution, the effects on angling, fish farming and on river abstractions of drinking water are considerable, and to some extent the high percentage of Class I

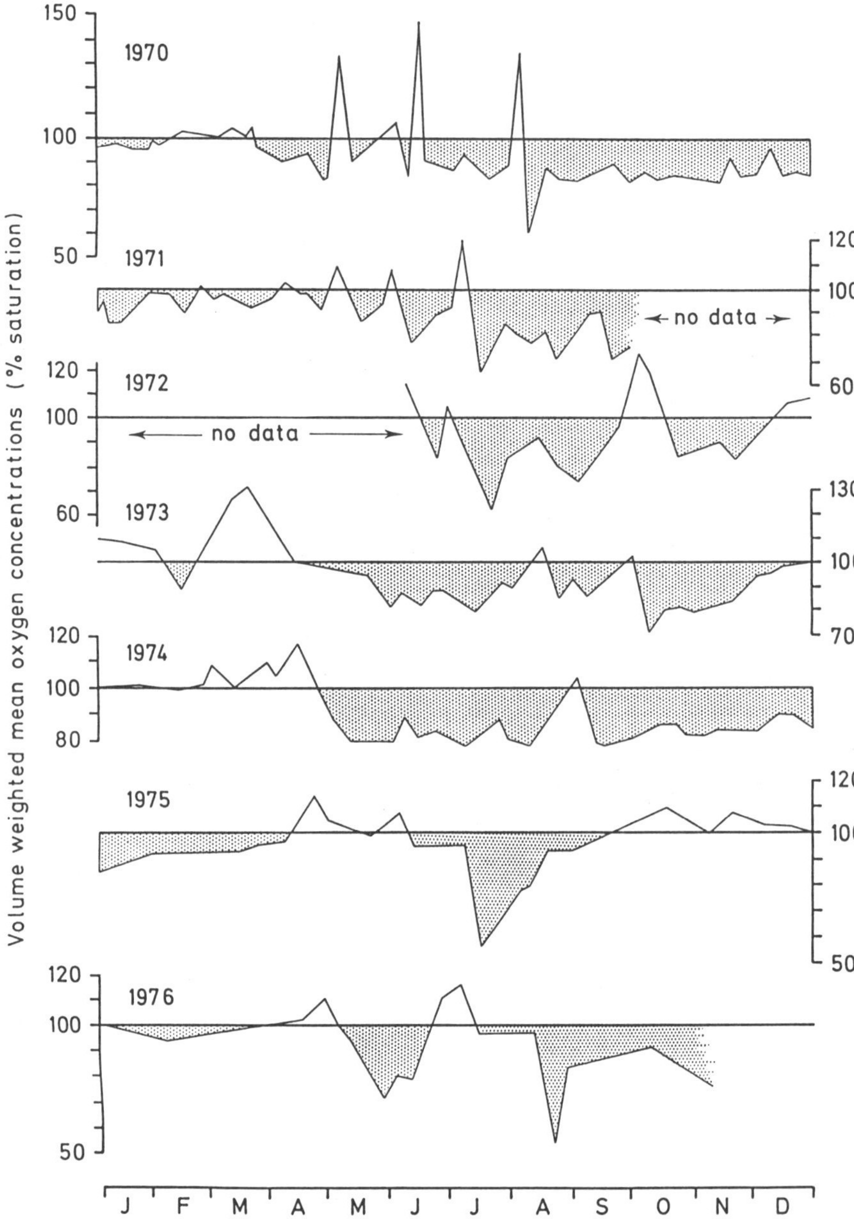

Figure 1.3. Volume-weighted mean oxygen saturation at a NW station 1970–1976.

rivers does not take into account catastrophic conditions of relative short duration.

1.4. Summary

It is against these human values, amongst them man's need for productive labour and satisfying recreation, for creature comforts like food, piped water and waste disposal, and not least his need to exercise both his innate curiosity in general and his trusteeship of this remarkable resource in particular, that we offer this monograph.

1.5. References

Anon (1981) Northern Ireland Water Statistics 1980. Department of the Environment for Northern Ireland.

Anon (1984) Annual Report of the Fisheries Conservancy Board for Northern Ireland 1983, 43 pp.

Anon (1985a) River quality in Northern Ireland. Dept. of the Environment, Environmental Protection Division. HMSO Belfast, 16 pp.

Anon (1985b) Northern Ireland Water Statistics 1983. Department of the Environment for Northern Ireland.

Bell, P., ed. (1971) Lough Neagh Working Group Advisory Report Vols. I and II. Government of Northern Ireland.

Betts, N.L. (1982) Climate. In: J. G. Cruickshank and D.N. Wilcock (eds.), Northern Ireland Environment and Natural Resources. QUB & NUU x-294 pp.

Cambrensis, Giraldus (1186) Topography of Ireland. Translated by Thomas Forester 1863. Publ. H.G. Bohn, London.

Dennison, M. and Sweetnam R.J.N. (1979) The Kinnehala Scheme: Public Enquiry. Water Appeals Commission, Belfast.

Marshall, J.J. (1934) Lough Neagh in Legend and History. Tyrone Printing Co. Dungannon, 81 pp.

Rippey, B. (1977) The behaviour of phosphorus and silicon in undisturbed cores of Lough Neagh sediments, pp. 348–353. In: H.L. Golterman (ed.), Interactions between sediments and freshwaters. Junk/Pudoc.

Stevens, R.J. and Gibson, C.E. (1977) Sediment release of phosphorus in Lough Neagh, Northern Ireland, pp. 343–347. In: H.L. Golterman (ed.), Interactions between sediments and freshwaters. Junk/Pudoc.

Wood, R.B. (1980) Pollution of Lough Neagh: a multidisciplinary approach. Analytical Proceedings 17: 369–374.

Wood, R. B. (1982) Lakes. In: J.G. Cruickshank and D.N. Wilcock (eds.), Northern Ireland Environment and Natural Resources. QUB & NUU x-294 pp.

Wood, R.B. (1989) Lough Neagh: Managing to stay resourceful. Went Memorial lecture Royal Dublin Society. Occ. Papers in Irish Sci. and Technol. No. 6, pp. 24.

Wood, R.B. and Gibson, C.E. (1973) Eutrophication and Lough Neagh. Water Res. 7: 173–187.

Wood, R.B. and Smith, R.V. (1986) Lough Neagh: 100 years of man's influence on the water quality of a major inland fishery. In: W.W. Crozier and P.M. Johnston (eds.), Proc. 17th Inst. Fisheries Management Conference, N. Ireland, pp. 111–126.

2. Geology, hydrology and land-use of Lough Neagh and its catchment

R.W.G. CARTER

2.1. Introduction

The aims of this chapter are twofold, to provide a background for the remainder of this volume on Lough Neagh, and to cover some of those aspects of the physical environment e.g. geological history, lake sedimentation and drainage, which would not otherwise be examined but which are germane to an understanding of the Lough Neagh ecosystem. In defence of the first aim it must be emphasised that many of the topics are dealt with more fully in other chapters, so that the inadequate interpretations here are merely signposts to better things. The second aim is to provide a reasonably balanced and integrated account of environmental conditions both in and around Lough Neagh. The somewhat 'fuller' sections are designed to supply basic information for the later, more specialised accounts to refer to. Throughout a regional approach has been adopted, mainly for the sake of brevity.

2.2. Geology

2.2.1. *Pre-tertiary and tertiary eras*

Geologically the Lough Neagh Basin is a downfaulted and downwarped depression (map Fig. 2.1) formed as the result of numerous tectonic episodes, dating back at least as far as the Caledonian earth movements 400 Ma ago. These movements imparted a pervasive NE–SW basement structure to much of NW Europe, which has controlled many subsequent events. As early as the Palaeozoic era, the Lough Neagh Basin was acting as a depositional graben (Anderton *et al.*, 1978; Holland, 1981). This general role as a sediment sink has remained a conspicuous element in the history of the Basin since that time. Evidence of infilling is provided by the deposition of shelf-marginal Carboniferous sediments (Charlesworth, 1963; Wilson, 1972) and

R.B. Wood and R.V. Smith (eds.), Lough Neagh, 11–33.

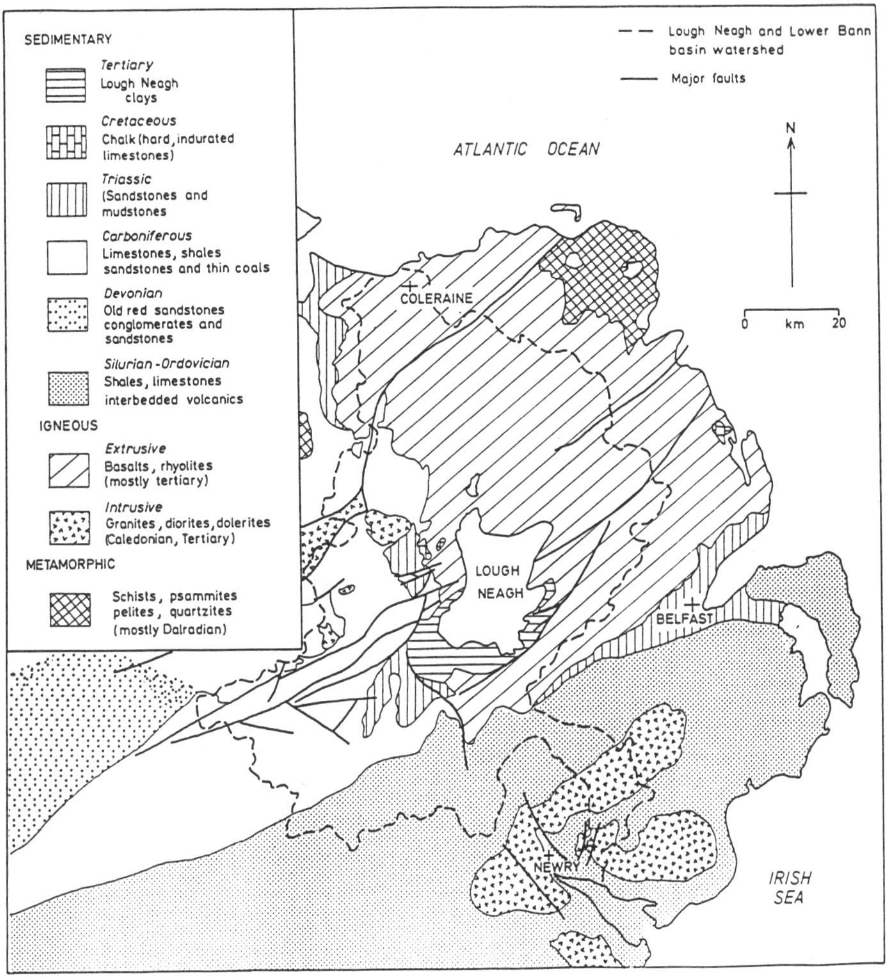

Figure 2.1. Simplified solid geology map of the Lough Neagh catchment and surrounding areas in Northern Ireland. The lough itself lies within a downfaulted basin, active since at least the carboniferous era.

marine transgressions in both the early Jurassic (190 Ma) (Fowler and Robbie, 1961) and late Cretaceous (65 Ma) (George, 1967) (Table 2.1).

In the early Cainozoic (65 Ma) initial rifting and sea-floor spreading of the Atlantic ocean presaged a comparatively short geological period of intense, but episodic, volcanic and plutonic activity in NW Britain. In NE Ireland, sub-aerial volcanic eruptions leading to extensive cone and plateau sheet basalt extrusions, followed by dyke injections (mainly along a NW–SE lineament) and the emplacement of acid-igneous plutons in SE Ulster and N Leinster (Preston, 1981), are all testimony to this period of crustal extension.

Table 2.1. Summary of stratigraphical episodes in N.E. Ireland

Era	Period		Epoch	Radiometric age (Ma)	Geological events in the Lough Neagh Basin
CAINOZOIC	Quaternary		Holocene	10,000 a	Present Lough Neagh formed
			Pleistocene	1.8 Ma	glaciations, drainage disruption
	TERTIARY	NEOGENE	Piocene	5	
			Miocene		'Normal' landform development
		PALAEOGENE	Oligocene	22.5	Infilling by river and lake sediments
			Eocene	37.5 53.5	Crustal downwarping of Lough Neagh Basin Vulcancity ceases
			Palaeocene	65	Intense volcanic activity in N.W. Europe Basalt flows in N.E, Ireland intrusive plutonic granites
System				Radiometric age (Ma)	
CRETACEOUS				65 135	Chalk sea in N.E. Ireland
JURASSIC				195	Subsidence and marine infilling of Lough Neagh Basin
TRIASSIC				235	Worldwide transgression

Consequential faulting and flexuring of the crust seems to have led to an accentuation of Lough Neagh Basin structure. Wilkinson *et al.* (1980) suggest that initially the basin was 'open' (possibly to the SE) as there was sediment throughput rather than accumulation, but by the late Oligocene (20 Ma) the basin had become closed which led to the deposition of the Lough Neagh Clay Formation.

The Lough Neagh Clay Formation marks the existence of a proto-Lough Neagh, forming under geological conditions not unlike those of today. This early Lough (see Fig. 2.2) covered at least 550 km^2 spreading across the southern end of the present lake, westwards into Co. Tyrone and eastwards in SE Co. Antrim. Although well known for over a century (Hardman, 1875;

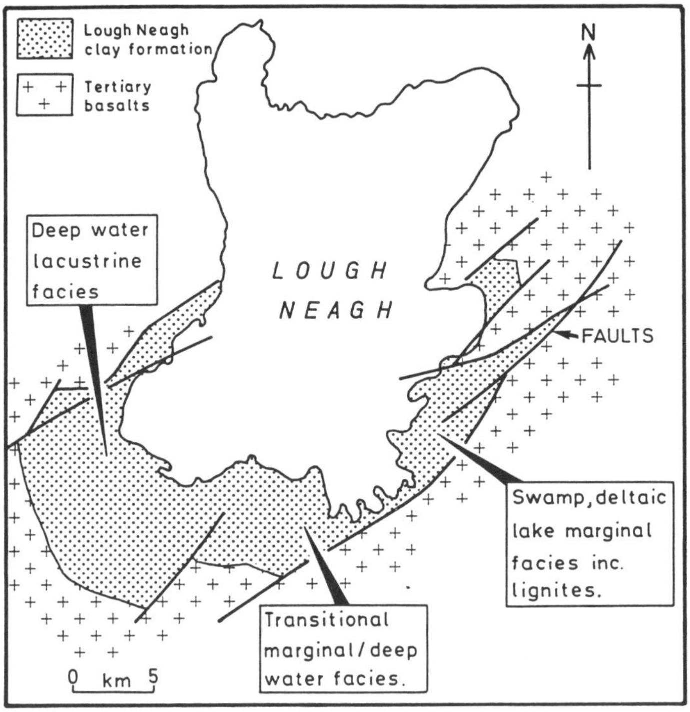

Figure 2.2. The Lough Neagh Clay Formation indicates the extent of the Tertiary precursor of the modern lake. The Tertiary lake comprised shallow eastern and western shelves and a deepwater central NE–SW trough. Allogenic organic deposits on the shallow shelf have formed thick lignites, which are now being considered for commercial exploitation (based on Wilkinson *et al.*, 1980).

Wright, 1924), the clays are poorly exposed, and both stratigraphical and facies information is drawn from a scatter of bore-holes, mainly sunk to assess the commercial lignite and ball-clay potential of the deposits. The formation comprises interbedded clays, sands and lignites plus occasional gravel stringers, particularly near the base. Fowler and Robbie (1961) show that sediment was largely derived from the Carboniferous rocks and regoliths to the west although all types of material are present to some degree. Lateral impersistency of beds and marker horizons precludes simple stratigraphical correlation, although Wilkinson *et al.* (1980) have concluded that the palaeo-environment consisted of a series of coalescing lake marginal deltas in the south-west and west, infilling a central deep-water lake with a low-lying eastern shore (Fig 2.3). Sediment influx led to further subsidence and faulting of the basin, although Wilson (1972) points out that lake level must have risen faster than subsidence as the lacustrine members transgressively overstep the original lake margins and bounding faults. The maximum thickness of the

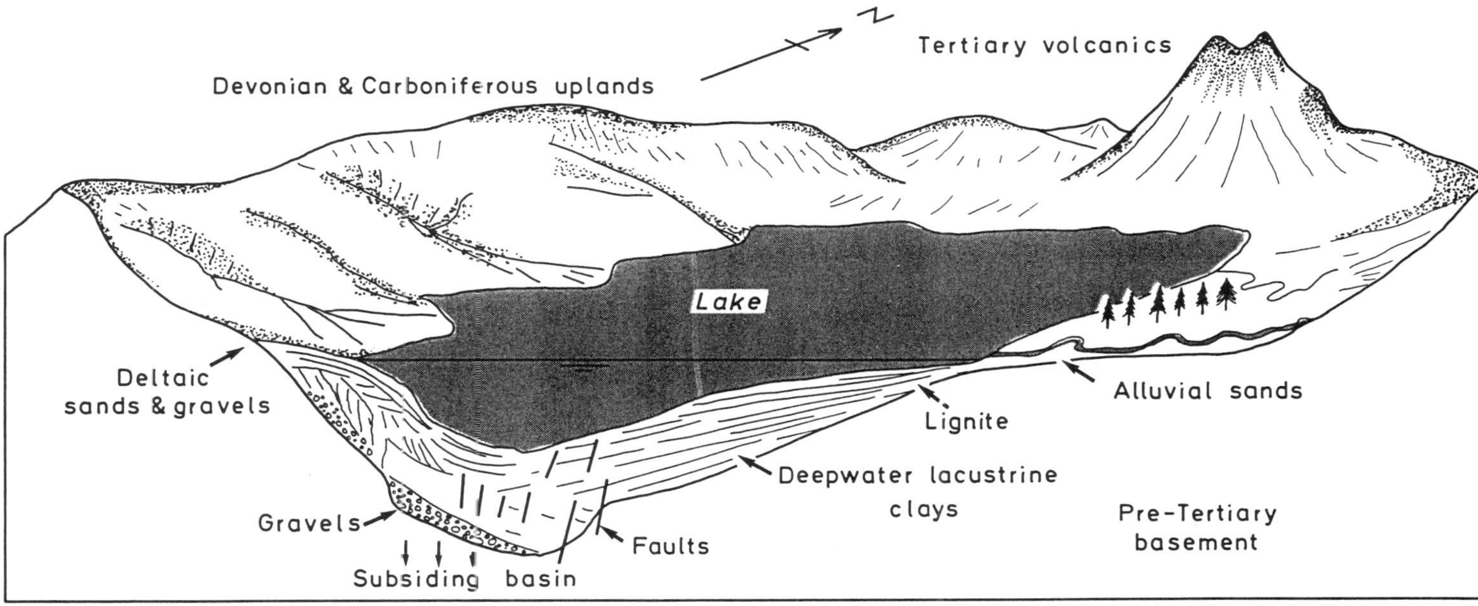

Figure 2.3. A hypothetic block diagram of Tertiary 'Lough Neagh'. Sediment influxes from the peripheral highlands formed alluvial and swampy deltaic fans along the shoreline. It is not known where the lake outlet was, but it probably drained to the south or west.

formation is 550 m near Washing Bay, in Co. Tyrone. The deposits are palynologically important (Watts, 1963, 1970; Boulter, 1980; Wilkinson *et al.*, 1980) containing abundant plant fossils, derived from lake marginal swamps and woodlands. Terrestrial facies are indicated by allochonothous lignite beds, up to 22 m thick near Crumlin and the presence of alluvial flood plain deposits. Westward the lignite beds split into numerous thin horizons as they become mixed with lacustrine deposits. At the present time plans for the commercial exploitation of the lignite are well advanced (Newbould, 1982). The initial proposal concerns an area of 6 km^2 to the north and east of Crumlin, where 105×10^6 tonnes of lignite could be mined by open cast methods. Ultimately it may be feasible to recover a further 350×10^6 tonnes from under the lake bed. This would probably involve the temporary sealing off of the shallow bay to the east of Ram's Island between Gartree Point and Tunny Point by a clay bund. The lignite resource may be used in an integrated electric power generation scheme. Additionally, commercially recoverable lignite deposits have now been located on the west shore of the Lough near Ardboe, and further north in the Bann Valley, near Ballymoney.

It is worth noting, in passing, that the much publicised petrification properties of the present Lough Neagh, are due to the presence of eroded Oligocene silicified lignite fragments in the modern sediments reported by, among others, Barton (1751). Although the correct interpretation was put forward by Scoular in 1837, the mythological origin is still accepted by many local people!

The processes leading to the cessation of lake sedimentation and abandonment of the Oligocene Lough Neagh remain a mystery. Mitchell (1980) postulates that younger Neogene material may have been removed by erosion but there is no evidence of this. Continuing surface warping (George, 1967; Davies, 1970; Mitchel, 1981) coupled to watershed lowering may have allowed the lake to drain and the basin to revert to one in which sediment throughput was the norm. It is just possible that land and sea-level changes allowed marine incursion of the Basin, and contributed to the formation of high-level planation surfaces, so beloved by denudational chronologists in the 1950s and early 1960s (Proudfoot, 1954; George 1967; Davies and Stephens, 1978, p. 95). Although there are many compelling arguments in favour of such surfaces (see Ollier, 1981, p. 147 *et seq.*) the lack of associated deposits (in Ireland) rather devalues their significance.

It is probably because the surface veneer of later sediments of glacial origin present so many interesting problems that serious investigation of the Tertiary landscape has been ignored. Tertiary landforms in Ireland remain, to quote Davies (1970) 'enigmatic', although progress, excellently summarised by Mitchell (1980), is being made. In the Lough Neagh Basin there is obviously considerable scope for further valuable studies of Tertiary geomorphology and sedimentology.

Table 2.2. Glacial and post-glacial successions in Central Ulster (after Colhoun, 1971a; Stephens *et al.*, 1975; and Dardis, 1980)

			Lough Neagh Basin	Co. Tyrone	N.W. Co. Antrim N. Co. Derry N.E. Co. Donegal
Holocene	Littletonian		Moyola river terraces Soil and slope development Bann Valley channel erosion	Soil and slope development	Coastal sea-level changes – dunes and beach ridges formed Estuarine clays deposited
Pleistocene	Midlandian	late	Bann Valley channel erosion	Moneymore interstadial(?)	Armoy glacial readvance
		middle	Early Lough Neagh?	Curglassen retreat phase (Early Lough Neagh?)	Lough Neagh basinal ice; North Channel ice
		early	Lough Neagh basinal glaciation	Central Tyrone glaciation	
	Munsterian		Interglacial – not yet recorded in Ireland		
			Central Ireland basinal glaciations		Scottish ice sheet and local ice caps

2.2.2. *Quaternary era*

2.2.2.1. *The Pleistocene* (*180,000 BP to 10,000 BP*)

The surface of the Lough Neagh basin is largely mantled by thick glaciogenic formations of Pleistocene age. The somewhat chaotic, present day drainage network has developed on this Pleistocene cover. Both soils and land use are, to a considerable extent, determined by the distribution and character of the glacial deposits.

Over the last 200,000 years Northern Ireland has been subjected to a number of glacial episodes. Only the most recent of these, the Midlandian (see Mitchell *et al.* (1973) for a correlation of British and Irish Quaternary stages), is extensively represented in terms of sediments and landforms. Colhoun (1970, 1971a) estimates that the Midlandian stage in Central Ulster lasted from 35,000 BP to 13,000 BP, and obliterated most of the evidence for earlier glacial events. Even though glacial deposits abound, and surface exposures are reasonably commonplace, confusion exists over their regional correlation and interpretation. Intra-regional facies variation is considerable, and many glacial geomorphologists and sedimentologists have presented alternative versions of local stratigraphic successions, with little cross correlation, or chronological framework. Table 2.2, taken largely from Stephens *et al.* (1975) and Dardis (1980), represents probably the best summary currently available of the Pleistocene succession in central Ulster.

Early workers (Portlock, 1843; Kilroe, 1888, 1913; Charlesworth, 1924;

1939) tended to stress that the glaciations were due to the extension of highland ice caps, either from Donegal or NW Scotland, overrunning central Ulster. However more recent interpretations (Synge and Stephens, 1960; Colhoun, 1971a and b; Stephens *et al.*, 1975), have relegated mountain ice caps to a subordinate role and emphasised instead the accumulation of basinal ice masses. Certainly in prolonged cold conditions, intermontane depressions like the Lough Neagh basin, the Irish Sea basin to the east or the Erne basin to the west, stand far more chance of developing thick snow cover, which engenders ice sheets. Although valley glaciers may contribute to local ice masses, it is accumulation in these basins which dominates the regional patterns of glaciation.

Throughout the Pleistocene, we can envisage the periodic development of a Lough Neagh ice sheet. Under gravitational stresses within the ice the mass would have spread out in centrifugal fashion. Exact directions of ice movement would have been determined, initially at least, by the pre-existing topography. However as the ice mass thickened topographic control would have diminished. Colhoun (1971a) showed how Lough Neagh ice in the Midlandian was fully capable of crossing the western basalt plateau and the Sperrin watershed at up to +600 m O.D. and moving into the Foyle valley to the N.W. From such evidence we can estimate the ice thickness to have been at least 600–800 m during the maximum glaciation. To the north, south-east and south-west the Lough Neagh ice came into contact with other ice sheets. These movements and associations are shown in Fig. 2.4.

Beneath the main ice mass erosional scouring is likely to have accentuated the central Lough Neagh depression, while at the margins deposition would have dominated. In the Bann valley and all the associated lowlands to the south and west of the present Lough, depositional glacial landforms abound, including moraines, eskers, kame terraces and drumlins. In many places the Midlandian ice sheet simply reworked existing material; for example many of the drumlins which are so characteristic of the Ulster landscape, were formed through the rheological deformation of pre-Midlandian lodgement till.

The coincidence of distal ice sheets around the basin provide intriguing pro- and en-glacial depositional environments. On many occasions there must have been vast volumes of melt water with no obvious drainage routes. Undoubtedly supra-glacial channels relieved much of the melt water pressure, but nonetheless large ephemeral lakes must have been common. The idea of extensive pro-glacial lakes throughout central Ulster is not new, as both Kilroe (1913) and later Charlesworth (1924, 1939) recognised glacio-lacustrine sediments and put forward, often complex, explanations to account for them. To the south and south-west of Lough Neagh, particularly in the upper Ballinderry valley, there are impressive, high-altitude deltaic structures that could only have been formed at the edges of ice-bounded lakes (Dardis, 1980). More recently, Dardis (1986a, 1986b) has postulated the existence of a large (at least 600 km^2) late-glacial Lough Neagh, trapped between two ice

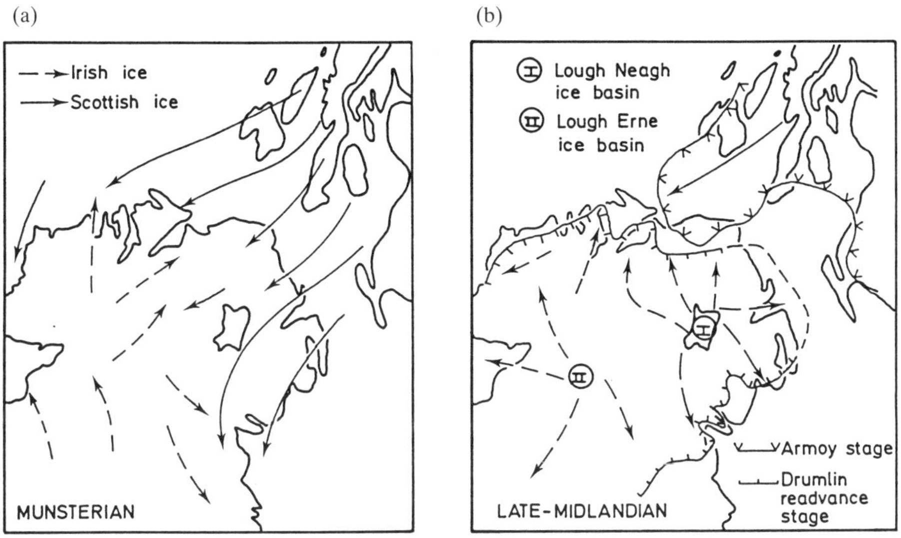

Figure 2.4. Two patterns of Quaternary ice movement across the north of Ireland. (a) shows the pattern during the Munsterian (>50,000 years BP) during which a Scottish ice sheet covered north-east Ireland, abutting an indigenous Irish ice-sheet to the south and west of Lough Neagh. (b) shows the Lough Neagh basin with an ice centre formed during the later-Midlandian (22,000–18,000 years BP). The later invasion of Scottish ice along the north coast of Ulster probably blocked the lower Bann valley causing flooding throughout the upper Bann and Lough Neagh lowlands (Figure follows Stephens *et al.*, 1975 and Mitchell, 1976).

sheets, one to the south of the present lake and one near the north coast of Ireland. This lake, which would have reached the present-day 60 m contour, may have existed for several hundred years, before gradually being drawn down as the ice fronts receded, and drainage lines reopened. Although there is some morphological and sedimentological evidence for this lake, particularly near Moneymore in Co. Londonderry and the town of Antrim, the general lack of high level lake shorelines, lake and lake-margin sediments, and the unexpectedly widespread preservation of purportedly 'submerged' landscape throw doubts upon the validity of Dardis' ideas.

The eventual decay and dissolution of the ice sheets after 18,000 BP was probably unspectacular, as evidence (Colhoun, 1971a; Creighton, 1974) points to gradual melting *in situ* rather than measurable retreat. The lower Bann valley, and the Moyola valley, to the west of Magherafelt, both contain sandur-type facies including numerous 'dead-ice' structures – kettle holes etc. Obviously the late-glacial amelioration of the climate was very slow and, to judge by the existence of small readvance moraines (Colhoun, 1971a) contained several reversals of the general warming trend. Withdrawal of the Lough Neagh ice from the upper Bann valley allowed a final incursion around 14,500 BP, of ice into Ulster. This stage – the Armoy Stage – is marked by an extensive push moraine (Creighton, 1974; Shaw and Carter, 1980) around Ballymoney and Coleraine. The presence of ice in the northern Bann valley obviously impeded drainage from the Lough Neagh Basin, and may have caused extensive flooding. Evidence of late-Midlandian glacio-fluvial and glacio-lacustrine deposition abounds throughout west Co. Antrim, east Co. Tyrone and south Co. Londonderry. Around Tobermore and Cookstown thick deposits of sand and gravel are commercially important (Cameron and Calvert-Harrison, 1977). Dardis (1986a and b) argues that small delta complexes around Moneymore, Cookstown and Antrim may all be associated with a single lake with a level about +60 m O.D. – see above. Varved clays, which are commonly encountered throughout the Bann valley (Kilroe, 1907; Creighton, 1974; Carter *et al.*, 1977) may be related to the same lake, although there is no conclusive evidence that this is the case.

Temperatures throughout Northern Ireland in the late-Midlandian would have remained low. Colhoun (1971b) has recorded intense periglacial activity in and around the Sperrin mountains dating from this period. Mitchell (1976, 1981) shows that open tundra vegetation would have prevailed (mosses, grasses, sedges, dwarf willow) until 13,000 BP after which juniper became dominant until 12,000 BP (Woodgrange Interstadial). Between 12,000 and 10,500 BP climate deteriorated rapidly and open tundra returned (Naganagan Stadial stage). Giant Deer, reindeer, bears, wolves and Arctic foxes are likely to have lived in Central Ulster through the late-Midlandian. Numerous fossil remains of Quaternary age have been found in and around Lough Neagh (Doughty, 1966; James, 1982; McCabe *et al.*, 1987).

2.2.2.2. *The Holocene (10,000 BP to present day)*

The Holocene or Littletonian Stage is inevitably important in the development of the present Lough Neagh. Aspects of this development are dealt with by Woodman and Mitchel (Chapter 6) and Battarbee and Flower (Chapter 7). During the Littletonian the drainage pattern would have matured, the Lough level become reasonably stabilised and the sediment budget established. The early Littletonian saw a rapid improvement in the Irish climate (Mitchell, 1976) which by about 7,000–6,000 BP had become slightly warmer (1–2 °C), and possibly drier, than today. Development of modern soils probably dates from the early Littletonian, in conjunction with the evolution of the Irish woodlands. Jessen (1949), Smith (1961, 1970), Mitchell (1976) and others have unravelled the Littletonian floristic succession in central Ulster and established a reference framework of regional pollen assemblage zones. The early beech forest gave way to hazel and pine, then oak, elm and alder. Much of the forest was modified by early agricultural practices which are dealt with in Chapter 6 by Woodman and Mitchel.

Drainage of Lough Neagh would have been affected by sea-level. The altitudinal difference, assuming a reasonably constant horizontal distance, between the lake level and the marine base level would have a profound influence on the discharge regime of the lough. At present the Lower Bann falls only 15 m over the 64 km between Toome and Coleraine. Carter (1982) has described the late- and post-glacial fluctuations in sea-level around the coast of Northern Ireland which have been controlled by the balance between local isostatic delevelling and world-wide eustatic factors. In north-east Ireland sea-level stood at *c.* +20 m O.D. 18,000 years ago, but then fell to *c.* −30 m O.D. by 11,000 B.P. The early Littletonian climatic improvement saw a rapid rise in sea-level ($8\,mm\,y^{-1}$) up to, and perhaps slightly above, the present level around 6,500 and 5,000 B.P. The high, late-Midlandian sea-levels imply that there was potential for substantial marine incursion into the Lough Neagh basin. However no marine or glacio-marine deposits have ever been found, and we must assume that ice blocked the way. As sea-level fell so discharge from Lough Neagh created an incised channel in the lower Bann valley. The sides of this channel are still visible between Kilrea and Coleraine. The weirs at Portna and Movanagher are probably testaments to the lower base-level causing thalweg rejuvenation. Marine transgressions and regressions during the Littletonian stage were responsible for flooding and infilling the Bann estuary (Battarbee *et al.*, 1984; Hamilton and Carter, 1983) and would also have affected discharge from Lough Neagh. The stream rapids at the Salmon Leap, 2 km south of Coleraine at the present tidal limit of the river, are probably a legacy of these mid-Littletonian variations in discharge regime.

The water level in Lough Neagh during the Littletonian is not known. The lake had obviously shrunk in area from the possible +60 m O.D. Midlandian level by the opening of the Littletonian stage 10,000 years ago, although neither morphological nor sedimentological evidence of the draw-down stages has been found. Throughout the Littletonian the unregulated lake level was probably between +15 and +20 m O.D. Mitchell (1976, p. 85–p. 103) suggests, on the basis of circumstantial archaeologic and climatic evidence (see Chapter 6), that the lake was somewhat lower, and thus smaller, around 7,000 BP, subsequently rising in the mid-Littletonian to deposit the thick, laminated sequences of diatomite up to 1 m thick north of Toome Bay (see Battarbee and Flower, Chapter 7). Smith (1981) has pointed out the paradoxical nature of the diatomite deposition at around +12 to +13 m O.D. contemporaneous with a supposedly low lake level (O'Sullivan *et al.*, 1973) between 5,000 and 4,000 BP. Smith suggests that a lacuna in the pollen records of both Mitchell (1955) from Toome Bay and O'Sullivan *et al.* (1973) from Lough Neagh during the mid-Littletonian (6,000–4,500 BP) may represent river erosion of, rather than non-deposition of, lake sediments. However as Smith (p. 256) says 'the whole situation is clearly ill-understood and discussion has perhaps led us unreasonably far into the realms of speculation'. Peat cropping-out on the lake shore at Toome was considered by Jessen (1949) to be of Zone VIb age (about 8,000 BP on a ^{14}C chronology). This outcrop may indicate a period of relatively low lake levels in the early Littletonian. Unregulated, seasonal water level fluctuations were responsible for periodic flooding around the shores of Lough Neagh; McCutcheon (1980,

p. 71) records that as late as the mid-19th century some 5,500 ha were seriously affected by annual inundations. Historical lake level variations are dealt with in Chapter 3, Section 3.3. Littletonian changes in lake level may have led to the formation of river terraces, particularly obvious along the Moyola valley around Castledawson.

2.3. Hydrology, soils and land-use

2.3.1. *The drainage pattern and hydrological regimes*

Davies and Stephens (1978, p. 111 *et seq.*) outline three main stages of drainage evolution within the Lough Neagh basin (Fig. 2.5). First, a centripetal drainage pattern was established on the newly-formed Tertiary basalt surface; second, this pattern gradually became superimposed in places onto the underlying Mesozoic and Palaeozoic rocks and third the pattern suffered dismemberment through competing, adventitious streams. The best example of the latter is provided by the River Lagan, which has, in eroding along the Triassic sandstone outcrop, destroyed the basic centripetal network to the south-east of the lake. In addition, minor glacial diversions of the drainage are common.

Six major rivers, the Blackwater, the Upper Bann, the Six Mile Water, the Main, the Moyola and the Ballinderry, flow into Lough Neagh. The eastern and southern shores are drained by a number of small streams, including the Dunore, Crumlin, Glenavy, Closet and Flush. The Lower Bann provides the exit (Fig. 2.5). The total catchment area draining into Lough Neagh is 4,453 km^2 about 43% of Northern Ireland, plus a small area of the Republic of Ireland.

Average precipitation (mostly rain of frontal, cyclonic origin) in the catchment is 1,095 mm y^{-1} (Betts, 1982) of which approximately 60% contributes to streamflow. Rainfall is well distributed throughout the year, with a slight winter (November–February) maximum. Within the basin, rainfall varies from *c.* 2,000 mm yr^{-1} over the western highlands to 750–900 mm yr^{-1} in the lower, south-east. Annual variations rarely exceed ±20%.

Under these circumstances of equable annual precipitation it is somewhat surprising that streamflow variations, based on the Q_{95}/Q_5 ratio (Table 2.3), are considerable. Two reasons may be put forward to explain this; one, that the decade (1970–1980) on which the majority of records are based was one of climatic extremes (Wilcock, 1982) and two, that arterial drainage works have almost totally altered the normal discharge regime, emphasising extreme flows (see section 2.3.3.4 below).

Streamflow statistics for Northern Ireland are based on a network of 34 rated gauges (Wilcock, 1977a) and catchment run-off patterns are related to rainfall, surface geology, topography and vegetation. These factors appear to be fairly standard throughout the Lough Neagh basin (Wilcock, 1982).

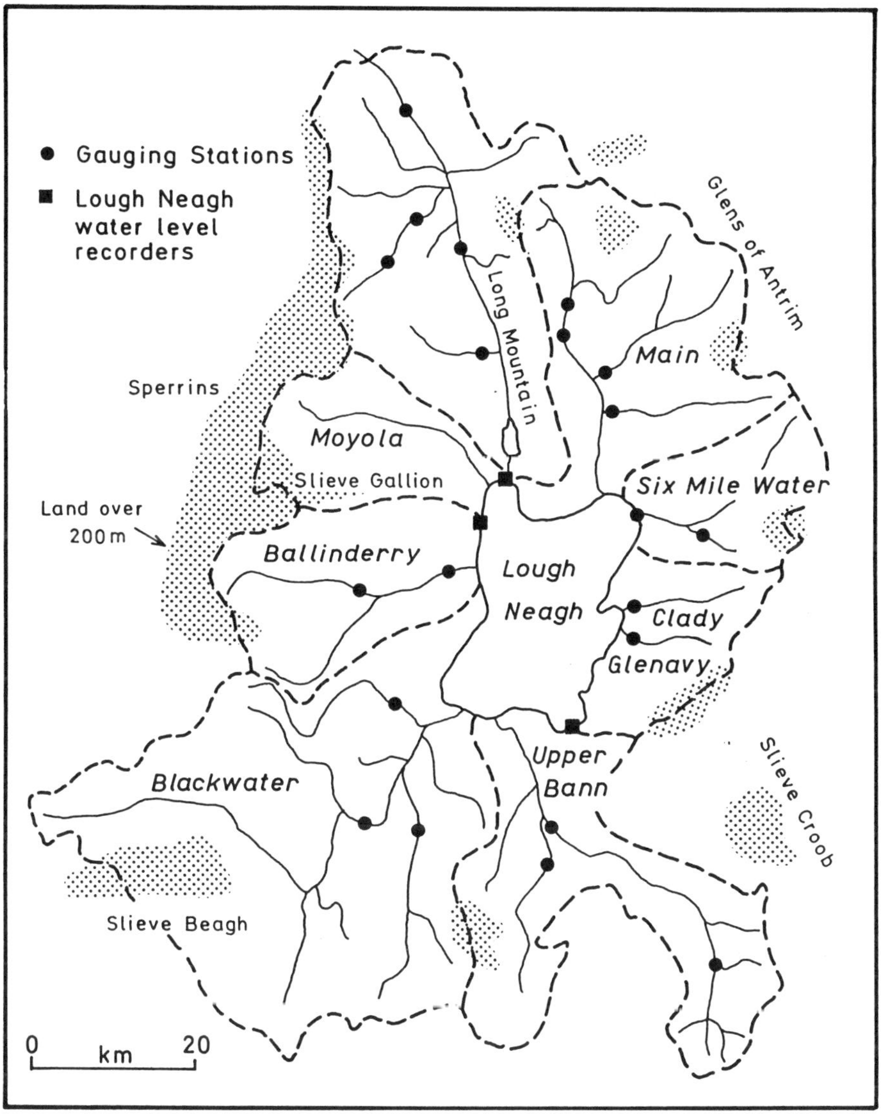

Figure 2.5. The stream network and sub-catchments of the Lough Neagh basin.

Stream management is preoccupied with extreme flows as these affect flooding, abstraction, pollution dispersal and dilution and fishing. High flows are more frequent in winter when saturated soil conditions cause rapid run-off, augmented by groundwater seepage. In summer a higher proportion of precipitation enters groundwater storage or is involved in evapotranspiration processes, so that low flows are more likely. This seasonal distribution of

Table 2.3. Streamflow and catchment statistics for rivers draining into Lough Neagh

River	Catchment area km^2	Catchment area above gauging site km^2	Average precipitation mm yr^{-1}	Drainage density km km^{-2}	Mean annual flood $m^3 s^{-1}$ $\bar{Q}$ curves	Average daily flow $m^3 s^{-1}$ curves	Stream variability Q_{95}/Q_5	Flow conversion factor*
Main	709	647	1211	1.1	51.3	5.3	20.9	1.046
Six Mile Water	301	277	1068	1.3	88.5	5.0	25.7	1.076
Blackwater	1481	951	1043	2.7	77.3	13.9	60.3	1.515
Upper Bann	661	335	1016	1.9	120.1	5.3	55.4	1.821
Ballinderry	430	419	1135	2.7	118.4	7.3	nd	1.012
Moyola	312	306	1295	2.1	88.4	6.3	nd	1.020
Direct drainage	559	nd	nd	nd	nd	nd	nd	nd
Total	4453							

*Conversion factor allows rated discharges to be converted into approximate lake inflows.
nd = no data.
Sources: Smith (1977) and Hanna (1982).

high and low flows was faithfully reflected in the volume of water held in Lough Neagh, prior to regulation of the Lower Bann at Toome. Even today sudden heavy rainfall can cause a rapid 20–30 cm rise in lake level. Average annual flows in the six major inflowing rivers are given in Chapter 5, Table 5.2.

2.3.2. *Groundwater*

Hydrogeologic conditions within the Lough Neagh basin vary according to the permeability and porosity of the rocks. The best aquifers are likely to occur in the permeable Permo-triassic sandstones to the south and south-west of the lake, where well yields of *c.* 1.3 to 4.5 $Ml\ d^{-1}$ are to be expected (Bell, 1971). Potential yields from the widespread basalt and limestone rocks are generally low ($< 0.5\ Ml\ d^{-1}$), except where local jointing and faulting increases porosity. The Lough Neagh Clay Formation is an aquaclude. Superficial deposits, especially Pleistocene fluvio-glacial gravels and sands, are often highly permeable, with transmissabilities exceeding 1.5 $Ml\ d^{-1}$. These deposits are often recharged directly from surface flows. High quality, groundwater supplies, in commercial quantities (usually $> 1.0\ Ml\ d^{-1}\ well^{-1}$), are obtained from the superficial aquifers beneath the Upper Bann at Portadown and the River Main south of Ballymena.

2.3.3. *Vegetation, soils and land-use*

2.3.3.1. *Natural vegetation*

Smith (1970) and Mitchell (1976) have summarised changes to the natural Irish vegetation that occurred during the Littletonian stage. Initially these changes were largely adjustments to varying climate, but since *c.* 4,000 BP man's activities have played an increasingly important role. A fluctuating, but steadily increasing, agricultural influence gradually altered the natural ecosystems, which today survive only by accident or in inaccessible, inhospitable spots. Spread of grasses, cereals and weeds parallelled the decline of the native trees; birch and hazel scrub replaced oak, elm and ash woodlands; raised bogs expanded in the wetter hollows. Active soil erosion, especially after the introduction of iron coulters and mouldboards on ploughs around 2,000 BP, provided minerals and nutrients perhaps increasing rates of both bog productivity and accumulation. Lowland raised bogs dominated by sphagnum mosses are common throughout the Lough Neagh basin, especially around Portadown and along the Lower Bann, Main and Moyola valleys. Increasing intensity of peat cutting, especially in the nineteenth century removed a number of bogs altogether, leaving either birch, willow or alder scrub on the marginal cut-over (as in the Birches Peatland Park, northwest of Craigavon), or rushy pasture on the underlying gley soils, or, as in the Brackagh Moss National Nature Reserve, south of Portadown (see Chapter 19) has caused the bog vegetation to revert to fen and fen carr.

Along the southeast edge of Lough Neagh, between the pre-regulation summer and winter high watermarks, extensive areas of open fen developed (Duff, 1930; Small, 1931; White, 1932). When the lake level was both lowered and stabilised by 1930–1942 Shepherd Scheme (see Chapter 3, Section 3.4.2) many of these fens were colonised by willow and alder. Further lake lowering in 1959, plus agricultural reclamation means that little fen now remains. At higher altitudes, normally above +300 m O.D., the destruction of forests after 4,000 BP (which had extended up to +600 m O.D.) led to deteriorating soil conditions accompanied by water-logging (due to iron pan formation) and the onset of blanket peat development. Typical species include heathers, rushes, *Nardus*, *Molinia* and cotton grass (*Eriophorum*). For the current situation see Chapter 19. In addition to being extensively 'cut-over' for fuel, almost all the upland peat is suffering serious erosion (Tomlinson, 1981).

The Plantation of Ulster by immigrants from Scotland during the 17th Century added to a steadily increasing natural population and resulted in the rapid, but final expansion of agriculture into marginal hill areas. Contemporaneously, the remaining native woodlands were exploited for charcoal, used in iron smelting, and staves for the barrel making industry (McCracken, 1971).

2.3.3.2. *Soils*

Regional descriptions of soil types and their associated pedogenic factors are to be found in Kilroe (1907) and Cruickshank (1970) and Gardiner and Radford (1980).

Soil associations in the Lough Neagh catchment have evolved slowly over the last 14,000 years in concert with the biogeographical and climatological changes outlined earlier. Glaciogenic parent materials dominate, although all the geological formations are important contributors to pedogenesis. Two phases of accelerated weathering may be recognised. First, intense physical weathering occurred during the late-Midlandian under periglacial freeze/thaw conditions. Second, enhanced chemical weathering of the regolith has accompanied the agricultural changes of the last 4,000 years. Where clay- and silt-sized particles form more than 45% of the soil, waterlogging or gleying occurs, even on steep slopes. Much of the land to the west of Lough Neagh, underlain by either the Carboniferous shales and/or Pleistocene till, is gleyed. The soils are characterised by perched water tables, base saturation often less than 10%, high organic content (peaty gleys) and a mottled appearance. pH varies from 4.0 to 5.5. Often trampling or 'poaching' tends to accentuate gleying, and many fields support rushes (*Juncus* spp.).

On the freer draining Ordovician sandstones in west Co. Tyrone and the Pleistocene sands and gravels in the Moyola, Main and Upper Bann valleys, brown earths or even podzols have developed. The latter have well-developed A and B horizons and are low in exchangeable cations due to surface leaching. Fossil podzols often underlie blanket peat.

Soils to the east and north of Lough Neagh, plus a small area to the west around Moneymore and Magherafelt, are formed either on Tertiary basalt or on basalt-derived boulder clays (soil associations No. 3, 17 and 41 in Gardiner and Radford (1980)). These give rise to unusual soils which have attracted a good deal of attention from pedologists (Brown, 1954; Smith, 1957; McConaghy and McAleese, 1957a and b; McAleese and Mitchell, 1958a, b and c; McAleese and McConaghy, 1958). The soils are characterised by high cation exchange capacities (up to 50 milli-equivalent 100 g^{-1} (Curtin and Smillie, 1981a)), high exchangeable calcium and magnesium (up to 40% Mg of the total cations in some soils (Cruickshank, 1970)), abundant sesquioxides of iron and aluminium and a cloddy or massive structure. The clay component is dominated by hydroscopically expanding vermiculite which causes severe shrinkage in droughts although all species of clay minerals may occur according to soil status (McAleese and Mitchell, 1958a; Curtin and Smillie, 1981b). Soils of basalt origin have a high base status, pH 6.5–7.0, require no liming, but are generally deficient in phosphates.

Throughout the basin, management is an important pedogenic factor. Invariably lowland gleys are 'improved', in terms of production, by artificial drainage (see Section 2.3.3.4) and application of fertilizers (Wilcock, 1979). In the upper Moyola valley, Cruickshank (1978) has shown how marginal gley soils are improved by initial application of lime and superphosphate followed by annual treatments of NPK fertiliser, but that, after only a short period of neglect, the soil returns to its former condition.

2.3.3.3. *Land-use*

Land utilisation surveys were conducted in Northern Ireland in 1938 and 1956 (Symonds, 1963). More recent information is available from the Annual General Reports of the Department of Agriculture (Northern Ireland) (see Table 2.4). Changes in land-use from 1971 to 1979 are outlined in Foy *et al.* (1982, Table 3) for each of the six major Lough Neagh catchments. Over 80% of the land may be classified as agricultural, nine-tenths of this as permanent pasture. Much of the land is of marginal quality requiring treatment before it can be profitably farmed. In the Sperrin Mountains, Cruickshank and Cruickshank (1977) have described the changing pattern of upland 'improvements' and the slow drift toward neglect and abandonment of enclosed fields. There is little doubt that recent farming practices have concentrated on intensifying the use of better land, at the expense of more marginal tracts.

Throughout the Lough Neagh basin, beef and milk production are the traditional farm enterprise, although attractive subsidies have diversified this pattern. Only 10% of the land is given over to arable crops, although this percentage increases on the flat land around the lake. Crops include potatoes, cereals (mainly barley, but also oats and winter wheat) and fodders (mainly kale). To the south and southwest of Lough Neagh, between the Blackwater and Upper Bann, there is a long established fruit-growing industry, concen-

Table 2.4. Land use in the river catchments of the Lough Neagh Basin

	Arable			Pasture			
River	Cereals	Potatoes	Other crops	Grass	Rough grazing	Woods and other land	Total (Drainage basin area)
Six Mile Water	16.9 = km^2	2.7	0.6	212.8	57.0	11.4	301.4
	(5.6) = %	(0.9)	(0.2)	(70.6)	(18.9)	(3.8)	(100.0)
Upper Bann	50.9	13.2	10.6	506.6	52.8	26.4	660.6
	(7.7)	(2.0)	(1.6)	(76.7)	(8.0)	(4.0)	(100.0)
Main	46.8	22.7	0.7	416.5	184.4	37.6	709.2
	(6.6)	(3.2)	(0.1)	(58.8)	(26.0)	(5.3)	(100.0)
Moyola	27.5	9.4	0.3	193.8	65.3	16.2	312.6
	(8.8)	(3.0)	(0.1)	(62.0)	(20.9)	(5.2)	(100.0)
Blackwater	60.7	17.8	19.2	1206.4	115.5	60.7	1480.3
	(4.1)	(1.2)	(1.3)	(81.5)	(7.8)	(4.1)	(100.0)
Ballinderry	24.9	5.6	0.4	294.1	83.0	22.4	430.1
	(6.3)	(1.3)	(0.1)	(67.8)	(19.3)	(5.2)	(100.0)
Direct Drainage	35.2	5.6	6.1	408.2	81.0	22.9	559.2
	(6.3)	(1.0)	(1.1)	(73.0)	(14.5)	(1.1)	(100.0)
Totals	262.9	77.0	37.9	3238.4	639.0	197.6	4453.4
	(5.9)	(1.7)	(0.8)	(72.8)	(14.4)	(4.4)	(100.0)

Source: Smith (1977).

trating mainly on apples. After several years of decline attempts are being made to revive this industry, although market prospects remain uncertain.

The best economic use of upland areas has long been considered to be afforestation (Wilcock, 1978) although this depends on site factors and a number of economic assumptions. In 1947 the Babington Agricultural Inquiry Committee recommended that 61,000 ha of forest should be planted in Northern Ireland by the year 2000. Approximately 53,000 ha of mainly (98%) coniferous forest had been planted by 1980 towards this goal, much in the higher western parts of the Lough Neagh catchment. Use of fertilizers to promote tree growth at altitude has led to deteriorating water quality in some upland reservoirs (Gibson, 1976). In addition to providing timber, the forests are widely used for recreation, with over one million visitors in 1980 throughout Northern Ireland (Rainey, 1982).

2.3.3.4. *Land drainage*

As will be seen throughout this volume, land drainage is of great importance to the well-being of Lough Neagh as it affects secular variations in both water quality and quantity. Wilcock (1979) has calculated that half the agricultural land in Northern Ireland has already benefitted directly from artificial land drainage schemes, a proportion five times greater than any other U.K. region. The generally flat terrain together with the impeded natural drainage, and pervasive soil gleying has for a long time made land drainage an attractive proposition to Ulster landowners. As early as 1842 a national coordinated effort was made to improve drainage (Common, 1970), although it is only in recent years that large-scale, integrated arterial schemes have been undertaken. A Drainage Council under the guidance of the Department of Agriculture for Northern Ireland was set up by the Drainage Act of 1947. This Council, reorganised in 1973, has overseen the enormous expansion of 'improved' watercourses. Wilcock (1979) estimates that, by 1976, 1,878 km of 'main' rivers and 3,759 km of 'minor' rivers had been 'improved' at a cost of over £50 million (1976 prices). Field drainage, mainly by mole drains, is the responsibility of landowners, who receive Government subsidies of between 50 and 70% of total costs.

The benefits and detriments of land drainage are hotly contested. Shoard (1980, p. 96) has argued that "there is little doubt that the cost to the community (as opposed to the lucky farmers) of the drainage schemes currently going ahead . . . would outweigh the benefits by a substantial margin". She feels that the real detriments to landscape and wildlife are totally under-assessed by the criteria currently employed in drainage cost/benefit analysis. Convery (1981) and Bruton and Convery (1982) note that cost benefit analysis (CBA) from the State's viewpoint often gives answers quite different from CBA from the individual farmer's view. While many plans for the early arterial drainage schemes in Northern Ireland went through unopposed, the ramifications of the River Main drainage scheme were widely debated in a Public Inquiry (Hutton, 1972) and certain aspects of the scheme were subsequently modified. The effects of drainage on streamflow vary. Burke (1975) claims that in blanket peat catchments, typical of those in the higher parts of the Lough Neagh basin, a more equable stream regime results with lower floods and more consistent low flows. Wilcock (1977a and b) monitored changes in 15 km^2 of the upper Agivey catchment (a left bank tributary of the lower Bann, typical of many Lough Neagh catchments) following arterial drainage. He concluded that after twelve to eighteen months of groundwater storage withdrawal, replenishment occurred, and within four years pre-drainage conditions had been re-established, channel maintenance having been negligible.

There is no doubt that better soil moisture conditions encourage farmers to increase their use of fertilizers, which may be reflected in both river and lake water quality (Smith, 1977, Wilcock, 1979).

2.4. References

Anderton, R., Bridges, P.H., Leeder, M.R. and Sellwood, B.W. (1978) A dynamic stratigraphy of the British Isles. Allen & Unwin, London, 301 pp.

Barton R. (1751) Lectures in natural philosophy, designed to be a foundation, for reasoning pertinently upon the petrifications, gems, crystals and sanative quality of Lough Neagh in Ireland; and intended to be an introduction to the natural history of several counties contiguous to that lake, particularly the County of Ardmagh. George and Alexander Ewing, Dublin.

Battarbee, R.W., Scaife, R. and Phethean, S.J. (1984) Palaeoecological evidence for sea-level change in the Bann Estuary in the early Mesolithic. In: P.C. Woodman (ed.), Excavations at Mountsandel. Arch. Res. Monographs, No. 2, HMSO, Belfast, pp. 211–220.

Bell, P.D. (1971) Lough Neagh Working Group Advisory Report Vol. 1. Government of Northern Ireland, 93 pp.

Betts, N. (1982) Climate. In: J.G. Cruickshank and D.N. Wilcock (eds.), Northern Ireland: environment and natural resources. The Queen's University, Belfast, and the New University of Ulster, pp. 9–42.

Boulter, M.C. (1980) Irish Tertiary plant fossils in a European context. J. Earth Sci. R. Dubl. Soc. 3: 1–12.

Brown, W.O. (1954) Some soil formations of the basaltic region of north-east Ireland. Ir. Nat. J. 11: 120–132.

Bruton, R. and Convery, J. (1982) Land drainage policy in Ireland. Economic and Social Research Institute, Dublin, 91 pp.

Burke, W. (1975) Effect of drainage on the hydrology of blanket bog. Ir. J. agric. Res. 14: 145–162.

Cameron, I.B. (2nd edition by Calvert-Harrison, E.N.) (1977) Sources of aggregate in Northern Ireland. Rep. Inst. Geol. Sci. 77/1, 21 pp.

Carter, R.W.G. (1982) Sea-level changes in Northern Ireland. Proc. Geol. Assoc. 93: 7–23.

Carter, R.W.G., Fisher, G.C. and Roberts, J.C. (1977) Carbonate nodules from a pro-glacial lacustrine environment at Greys Town, Co. Londonderry. Ir. Nat. J. 18: 2–4.

Charlesworth, J.K. (1924) The glacial geology of the north-west of Ireland. Proc. R. Ir. Acad. 36B: 174–314.

Charlesworth, J.K. (1939) Some observations on the glaciation of north-east Ireland. Proc. R. Ir. Acad. 45B: 255–295.

Charlesworth, J.K. (1963) The historical geology of Ireland. Oliver and Boyd, Edinburgh, 565 pp.

Colhoun, E.A. (1970) On the nature of the glaciations and final deglaciation of the Sperrin Mountains and adjacent areas in the north of Ireland. Ir. Geogr. 6: 162–185.

Colhoun, E.A. (1971a) The glacial stratigraphy of the Sperrin Mountains and its relation to the glacial stratigraphy of north-west Ireland. Proc. R. Ir. Acad. 71B: 37–52.

Colhoun, E.A. (1971b) Late Weichselian periglacial phenomena of the Sperrin Mountains, Northern Ireland. Proc. R. Ir. Acad. 71B: 53–71.

Common, R. (1970) Land drainage and water use in Ireland. In N. Stephens and R.E. Glasscock (eds.), Irish Geographical Studies, Belfast, Queen's University, pp. 342–359.

Convery, F. (1981) The economics of drainage: a preliminary analysis. Conference preprint, Dublin, 30 pp.

Creighton, J.R. (1974) A study of the late Pleistocene geomorphology of north central Ulster. Unpublished Ph.D. Thesis, The Queen's University, Belfast.

Cruickshank, J.G. (1970) Soils and pedogenesis in the North of Ireland. In: N. Stephens and R.E. Glasscock (eds.), Irish Geographical Studies, The Queen's University, Belfast, pp. 89–104.

Cruickshank, J.G. (1978) Soil properties and management levels of marginal hill land in the Sperrin Mountains, Co. Tyrone and Co. Londonderry. Ir. J. Agric. Res. 17: 303–314.

Cruickshank, J.G. and Cruickshank, M.M. (1977) A study of neglected agricultural land in the Sperrin Mountains, Northern Ireland. Ir. Geogr. 9: 36–43.

Curtin, D. and Smillie, G.W. (1981a) Contribution of the sand and silt fractions to the cation exchange capacities of some Irish soils. J. Earth Sci. Roy. Dub. Soc. 4: 17–20.

Curtin, D. and Smillie, G.W. (1981b) Composition and origin of Smectite in soils derived from basalt in Northern Ireland. Clays. Clay Mineral 29: 277–284.

Dardis, G.F. (1980) The Quaternary sediments of Central Ulster. In: K.J. Edwards (ed.), County Tyrone, Northern Ireland, IQUA Field Guide, No. 3, pp. 5–29.

Dardis, G.F. (1986a) Late-Pleistocene glacial lakes in South-central Ulster, Northern Ireland. Ir. J. Earth Sci. 7, 133–144.

Dardis, G.F. (1986b) Fossil ice and sand wedges in South-central Ulster, Northern Ireland. Ir. Geogr. 19: 51–58.

Davies, G.L.H. (1970) The enigma of the Irish Tertiary. In: N. Stephens and R. E. Glasscock (eds.), Irish Geographical Studies. The Queen's University Belfast, pp. 1–16.
Davies, G.L.H. and Stephens, N. (1978) The geomorphology of the British Isles – Ireland. London, University Paperbacks, 250 pp.
Doughty, P.S. (1966) Giant deer remains from Lough Neagh. Ir. Nat. J. 15: 187.
Duff, M. (1930) The ecology of the Moss Lane region, Lough Neagh. Proc. R. Ir. Acad. 39B: 477–496.
Fowler, A. and Robbie, J.A. (1961) Geology of the country around Dungannon. Mem. Geol. Surv. N.I.
Foy, R.H., Smith, R.V. and Stevens, R.J. (1982) Identification of factors affecting nitrogen and phosphorus loadings to Lough Neagh. J. Environmental Management 15: 109–129.
Gardiner M.J. and Radford T. (1980) Soil associations of Ireland and their land use potential. An Foras Taluntais, Dublin, 142 pp.
George, T.N. (1967) Landform and structure in Ulster. Scott. J. Geol. 3: 413–448.
Gibson, C.E. (1976) An investigation into the effects of forestry plantations on the water quality of upland reservoirs in Northern Ireland. Wat. Res. 10: 995–998.
Hamilton, A.C. and Carter, R.W.G. (1983) A Littletonian moss-bed from interdune deposits near Articlave, Co. Londonderry. Ir. Nat. J. 21: 73–75.
Hanna, J. (1982) Analysis of extreme events in the streamflow record of Northern Ireland. Unpublished M. Phil. Thesis, The New University of Ulster, Coleraine; 104 pp.
Hardman, E.T. (1875) On the age and mode of formation of Lough Neagh, Ireland; with notes on the physical geography and geology of the surrounding country. J. R. Geol. Soc. Irl. 14: 170–198.
Holland, C.H. (1981) A geology of Ireland. Scott. Univ. Press, Edinburgh, 335 pp.
Hutton, J.B. (1972) River Main Drainage Scheme, Report of a public inquiry. HMSO, Belfast.
James, K.W. (1982) Catalogue of Giant Irish Deer in the Ulster Museum. Ir. Nat. J. 20: 372–375.
Jessen, K. (1949) Studies in late Quaternary deposits and the flora history of Ireland. Proc. R. Ir. Acad. 52B: 85–290.
Kilroe, J.R. (1888) Directions of ice flow in the north of Ireland. Quart. J. Geol. Soc. Lond. 44: 827–833.
Kilroe, J.R. (1907) The soil-geology of Ireland. HMSO, Dublin, 300 pp.
Kilroe, J.R. (1913) Outline of geological investigations in north-west Londonderry. Proc. Belf. Nat. Fld. Club (2) 6: 634–663.
McAleese, D.M. and McConaghy, S. (1958) Studies on the Basaltic soils of Northern Ireland iii. Exchangeable cation contents of sand, silt and clay separates. J. Soil Sci. 9: 67–75.
McAleese, D.M. and Mitchell, W.A. (1958a) Studies on the Basaltic soils of Northern Ireland iv. Mineralogical studies of the clay separates. J. Soil Sci. 9: 76–80.
McAleese, D.M. and Mitchell, W.A. (1958b) Studies on the Basaltic soils of Northern Ireland v. Cation-exchange capacities and mineralogy of the silt separates (2–20 μm). J. Soil Sci. 9: 81–88.
McAleese, D.M. and Mitchell, W.A. (1958c) Studies on the Basaltic soils of Northern Ireland vi. Cation-exchange capacities and mineralogy of the fine sand separates (0.02–0.2 mm). J. Soil Sci. 9: 289–297.
McCabe, A.M., Coope, G.R., Gennard, D.E. and Doughty, P.S. (1987) Freshwater organic deposits and stratified sediments between early- and late-Midlandian (Devensian) till sheets at Aghnadarragh, Co. Antrim. J. Quat. Sci. 2: 11–33.
McConaghy, S. and McAleese, D.M. (1957a) Studies on the Basaltic soils of Northern Ireland i. Cation exchange properties. J. Soil Sci. 8: 127–134.
McConaghy, S. and McAleese, D.M. (1957b) Studies on the Basaltic soils of Northern Ireland ii. Contributions from the sand, silt and clay separates to cation-exchange properties. J. Soil Sci. 8: 1 35–140.
McCracken, E. (1971) The Irish woods since Tudor Times. David and Charles, Newton Abbot.

McCutcheon, W.A. (1980) The industrial archaeology of Northern Ireland. HMSO, Belfast, 395 pp.

Mitchell, G.F. (1955) The Mesolithic site at Toome Bay. Ulster J. Archaeology 18: 1–16.

Mitchell, G.F. (1976) The Irish Landscape. Collins, London, 240 pp.

Mitchell, G.F. (1980) The search for Tertiary Ireland. J. Earth Sci. R. Dubl. Soc. 3: 13–24.

Mitchell, G.F. (1981) The Littletonian warm stage – post-10,000 BP. In: C.H. Holland (ed.), A Geology of Ireland, Scottish Acad. Press, Edinburgh, pp. 259–272.

Mitchell, G.F., Penny, L.F., Shotton, F.W. and West, R.G. (1973) A correlation of Quaternary deposits of the British Isles. Geol. Soc. Lond. Special Report No. 4, 99 pp.

Newbould, P.J. (1982) Energy resources. In: J.G. Cruickshank and D.N. Wilcock (eds.), Northern Ireland: environment and natural resources. The Queen's University and the New University of Ulster, pp. 241–263.

Ollier, C.D. (1981) Tectonics and landforms. London, Longmans, 324 pp.

O'Sullivan, P.E., Oldfield, F. and Battarbee, R.W. (1973) Preliminary studies of Lough Neagh sediments I. Stratigraphy, chronology and pollen analysis. In: H.J.B.Birks and R.G. West (eds.) Quaternary Plant Ecology, Blackwells, Oxford, pp. 267–278.

Portlock, J.E. (1843) Geological report on Londonderry and parts of Tyrone and Fermanagh. Dublin, Milliken, 748 pp.

Preston, J. (1981) Tertiary igneous activity. In: C.H. Holland (ed.), A Geology of Ireland. Edinburgh, Scott. Acad. Press, pp. 213–224.

Proudfoot, V.B. (1954) Erosion surfaces in the Mourne Mountains. Ir. Geogr. 3: 26–35.

Rainey, N.C. (1982) An economic evaluation of forest recreation in the UK. Unpublished B.Sc. dissertation, The New University of Ulster.

Scoular, J. (1837) Observations on lignites and silicified woods of Lough Neagh. Dubl. Geol. Soc. J. 1: 231–241.

Shaw, J. and Carter, R.W.G. (1980) Late-Midlandian sedimentation and glaciotectonics of the North Antrim end moraine. Ir. Nat. J. 20: 67–69.

Shoard, M. (1980) The theft of the countryside. Temple-Smith, London, 269 pp.

Small, J. (1931) The fenlands of Lough Neagh. J. Ecol. 19: 383–388.

Smith, A.G. (1961) Cannons Lough, Kilrea, Co. Derry: stratigraphy and pollen analysis. Proc. R. Ir. Acad. 61B: 369–383.

Smith, A.G. (1970) Late- and post-glacial vegetational and climatic history of Ireland: A review. In: N. Stephens and R.E. Glasscock (eds.), Irish Geographical Studies. Queen's University Press, Belfast, pp. 65–88.

Smith, A.G. (1981) Palynology of a Mesolithic-Neolithic site in Co. Antrim, Northern Ireland. Proc. IV int. palynol. conf. Lucknow (1976–77) 3: 248–257.

Smith, J. (1957) A mineralogical study of weathering and soil formation from olivene basalt in Northern Ireland. J. Soil Sci. 8: 225–239.

Smith, R.V. (1977) Domestic and agricultural contributions to the inputs of phosphorus and nitrogen in Lough Neagh. Water Res. 11: 453–459.

Stephens, N., Creighton, R. and Hannon, M.A. (1975) The late-Pleistocene period in North-eastern Ireland: An assessment 1975. Ir. Geogr. 8: 1–23.

Symonds, L. (ed.) (1963). Land-use in Northern Ireland. London, University of London Press.

Synge, F.M. and Stephens, N. (1960) The Quaternary period in Ireland; an assessment. Ir. Geogr. 4: 121–130.

Tomlinson, R. (1981) The erosion of peat in the uplands of Northern Ireland. Ir. Geogr. 14: 51–64.

Watts, W.A. (1963) Fossil seeds from the Lough Neagh clays. Ir. Nat. J. 14: 117–118.

Watts, W.A. (1970) Tertiary and interglacial floras in Ireland. In: N. Stephens and R.E. Glasscock (eds.), Irish Geographical Studies. Belfast, Queen's University Press, pp. 17–33.

White, J.M. (1932) The fens of North Armagh. Proc. R. Ir. Acad. 40B: 233–283.

Wilcock, D.N. (1977a) Water resource management in Northern Ireland. Ir. Geogr. 10: 1–13.

Wilcock, D.N. (1977b) The effects of channel clearance and peat drainage on the water balance of the Glenullin basin, Co. Londonderry. Proc. R. Ir. Acad. 77B: 253–267.

Wilcock, D.N. (1978) Afforestation in Northern Ireland since 1970. Ir. Geogr. 11: 167–171.
Wilcock, D.N. (1979) Post-war land drainage, fertilizer use and environmental impact in Northern Ireland. J. Env. Mgmt. 8: 137–149.
Wilcock, D.N. (1982) Water resources. In: J. Cruickshank and D.N. Wilcock (eds.), Northern Ireland: environmental and natural resources. The Queen's University Belfast and The New University of Ulster, pp. 43–72.
Wilkinson, G.C., Bazley, R.A.B. and Boulter, M.C. (1980). The geology and palynology of the Oligocene Lough Neagh clays, Northern Ireland. J. Geol. Soc. Lond., 137: 65–75.
Wilson, H.E. (1972) Regional geology of Northern Ireland, HMSO, London, 115 pp.
Wright, H.B. (1924) Age and origin of the Lough Neagh clays. Q.J. geol. Soc. Lond. 80: 468–483.

3. The morphology, hydrodynamics and sedimentation processes of Lough Neagh

R.W.G. CARTER

3.1. Introduction

As with Chapter 2, this chapter provides some of the essential physical background to the functional ecology of Lough Neagh. The interaction of the water depth and exposure of the lake to mixing forces clearly has a profound influence on the horizontal and vertical homogeneity of the water, and many of the chapters that follow contain explicit or implicit reference to the shallow, well-mixed waters of Lough Neagh, and the effects this has on such processes as photosynthesis and respiration, oxygen distribution and nutrient cycling. Equally the depth of water column turbulence winnows and sorts sediments, producing rocky shores, nearshore sands and deep water silts and clays, which in turn influence, for example, the distribution of shore vegetation or benthic invertebrates. The palaeolimnological 'story' rests on a coherent sedimentary process, retention times are an important component in nutrient budgets, and lough levels and fluctuations have implications for the establishment of fen vegetation, bird breeding and access.

3.2. Morphometry

Lough Neagh has an area of 383 km^2 and is the largest freshwater lake in the British Isles. However the lake is shallow, with a mean depth of only 8.9 m. About 3% of the lake is below 20 m in depth, with a maximum of around 30 m (Bell, 1971). The bathymetric chart (Fig. 3.1) taken from the Admiralty survey of 1835, emphasises the shallow, flat nature of the lake bed. Only in Toome Bay, to the north-west, do any appreciable submarine slopes (2–3°) exist. The narrow, deepwater defile in Toome Bay is known locally as 'The Trench', and may represent a remnant of an old river channel.

With its surface at 12.5 m O.D. Belfast, the lake contains 3.45×10^9 m^3 of water with a mean retention time a little in excess of a year. The shoreline length is approximately 125 km, and maximum fetches are SW–NE 29.5 km

R. B. Wood and R. V. Smith (eds.), Lough Neagh, 35–57.

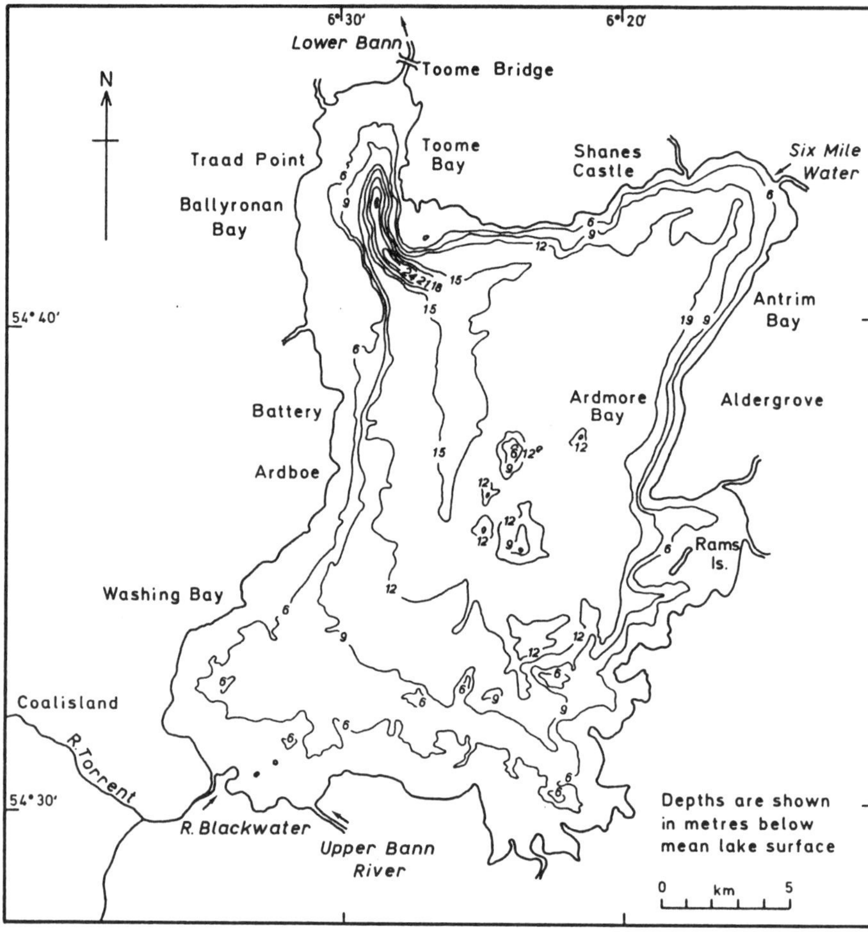

Figure 3.1. Bathymetric chart of Lough Neagh, plus the location of sites mentioned in the text.

and NNW to SSE, 28 km. Table 3.1 includes a number of morphometric and physical characteristics of Lough Neagh and its catchment.

3.3. Water levels and navigation

Tentative comments concerning prehistoric water levels were made in Section 2.2.2. During historical times the lake levels have been subjected to various controls in order to (i) alleviate flooding and (ii) improve navigation.

McCutcheon (1980, p. 71–75) lists the early attempts to improve drainage and navigation through the Lough Neagh system. Most of these involved the removal of natural rock barriers and dredging of the Lower Bann in order

Table 3.1. A checklist[1] of the principal morphometric and physical characteristics of Lough Neagh and its catchment

Item	Value
Latitude	54° 35′ N
Longitude	6° 25′ W
Catchment area	4453 km^2
Area	383 km^2
Maximum dimension	30.5 km SW–NE
Minimum dimension	12.1 km W–E
Height of water surface	+12.5 m O.D. Belfast
Length of shoreline	125 km
Mean depth	8.9 m
Maximum depth	34 m
Volume	3.45×10^9 m^3
Retention time	1.3 yr.
Maximum direct fetch	29.5 km
Maximum effective fetch	17.7 km
Dominant wind direction	SW
Prevailing wind direction	W (30%)
Mean annual wind speed	5 m $second^{-1}$
Average deepwater wave height	0.3 m
Maximum deepwater wave height	1.4 m
Mean wave period	2.5 seconds
Water surface temperature – range	3 to 16° C
Air temperature – range	−2 to 24° C
Mean rainfall – Lough surface	840 mm yr^{-1}
Number of days with more than 1 mm precipitation	155 days yr^{-1}
Number frost free days	184, May to October
Annual heat budget of the lake	13.8×10^3 cal cm^{-2} yr^{-1}

[1] These values often vary from authority to authority. I have included either the most commonly cited value or the most recent (and hopefully most reliable) value.

to reduce the level of the lake. A scheme, first put forward by McMahon in 1826, was undertaken between 1846 and 1858, which resulted in the lake being lowered by 1 m and its variability reduced (Table 3.2). Further lowerings were undertaken in 1942 (The Shepherd Scheme) and 1959 under provisions in the Drainage Act (NI) 1929 and the Lough Neagh and Lower Bann Drainage and Navigation Act (NI) 1955 respectively.

At present lake level is controlled by the operation of sluices at Toome. (For a description of these operations see Bell, 1971, pp. 67–69.) Visually recorded stage gauge data from Toome suggest that current practices maintain a lake level within 0.3 m of 12.55 m O.D. Belfast for at least 99% of the time (Fig. 3.2). Levels appear to rise in winter in response to the higher catchment rainfall, but judicious operation of the sluices can usually effect a rapid drawdown over a period of 2 or 3 weeks. The Lough Neagh Drainage Amendment Act (NI) 1970 contains a provision for further statutory lowering

Table 3.2. Lough Neagh water level fluctuations in m + O.D. Belfast

	Normal	Maximum	Minimum	Extreme fluctuations
Prior to McMahon Scheme 1846–1858	14.8	16.4	not known	c. 3.2
After McMahon Scheme 1860	13.8	14.9	12.8	2.1
After Shepherd Scheme 1942	13.1	13.9	12.3	1.6
After 1952	12.9	13.9	12.3	1.6
After 1959	12.5–12.6	13.1	12.2	0.9
1973–1982*	12.5	13.2	12.2	1.0

Source: Bell (1971) except *.

of the lake; however this could cause serious navigation problems if the level were to fall below 12.2 m O.D. (Bell, 1971).

By the mid-nineteenth century Lough Neagh was the centre of an extensive Ulster canal network (McCutcheon, 1965, 1980). The Lower Bann, Upper Bann, Blackwater and Torrent, near Coalisland, were all utilised for barge traffic, but the network collapsed with the coming of the railways (see Chapter 6). Today only the Lower Bann, the lower reaches of the Upper Bann, Six Mile Water and the lough itself are navigable waterways for small craft.

3.4. Waves, currents and seiches

When wind blows across a water surface it imparts energy and momentum to it, manifest in the generation of waves and currents. The efficiency of conversion of wind energy into wave energy is less than one percent. In turn the lake surface acts to slow down the velocity of the lower atmosphere, creating a frictional drag effect. The frictional coefficient (the ratio of the wind velocity near the surface to the free stream velocity) increases according to the roughness of the surface. Lough Neagh was the site of a number of experimental and empirical investigations into boundary layer meteorology, conducted by Imperial College, University of London, during the 1950s and 1960s (Garrat, 1972; Sheppard *et al.*, 1972). Wind blowing over water forms waves orthogonal to the wind direction, and currents parallel to the direction. The major factors influencing wave and current parameters are the wind duration and the water surface distance (fetch) over which the wind operates. Fetch is usually represented as the effective fetch based on radial dimensions of the water body. This takes account of the lake widths as well as the length. Smith (1979) has calculated the effective fetch of Lough Neagh, which, with distant shores, has an effective fetch not much less than the actual. However

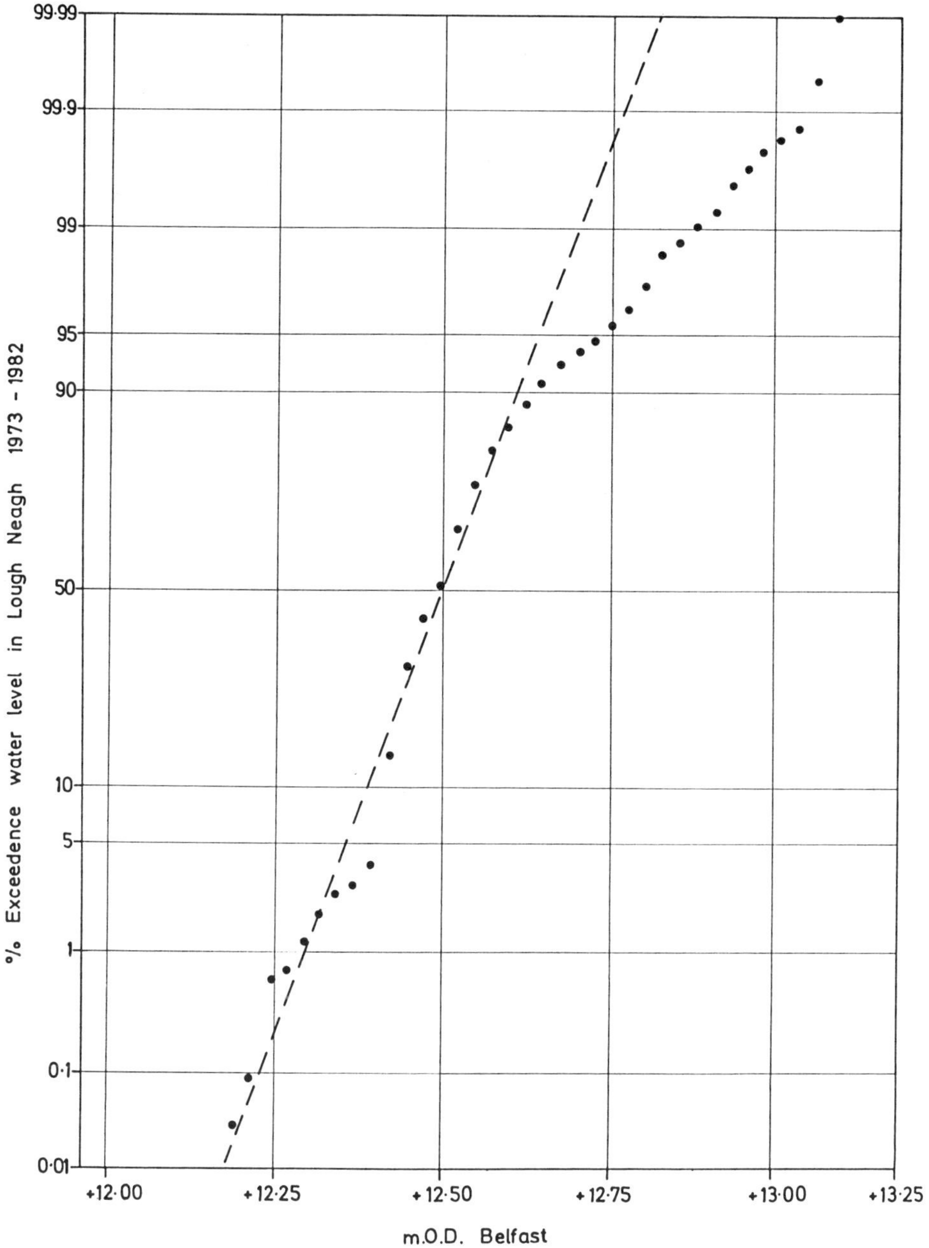

Figure 3.2. Percentage exceedance graph for daily Lough Neagh water levels 1973–1982. The pecked line is the probable Gaussian water level distribution assuming no artificial regulation. Deviations from the Gaussian curve result from management of the Lough level, by opening or closing the sluices at Toomebridge. Of particular interest is the truncation of extreme high water levels (>12.6 m O.D.). Redrawn from Carter (1988).

in a shallow lake like Lough Neagh, wave and current dimensions are limited by depth. Once the wave length (λ) is less than twice the depth (h), energy is dissipated at the sediment/water discontinuity. Eventually, as the λ/h ratio increases this mode of energy loss is sufficient to cause wave decay. Wave height (H) is also limited, as the H/λ ratio becomes unstable at values approaching 0.14 and free spilling occurs (white caps). Figures 3.3a and b show the potential for wave generation in Lough Neagh at various limiting depths.

As waves approach the shore so they are modified by the decreasing water depths. Figures 3.3c and d indicate the changes in H, λ and bottom shear stress (τ) that occur along a wave ray approaching the shore in Ballyronan Bay, in the northeast quadrant of the lake. In deep water (in this case 3.5 m) H and λ are conserved, but in shallower depths, H is reduced from 0.25 to 0.18 m, λ from 9.5 m to 3 m while τ rises exponentially from 0.1 $m^2\,sec^{-2}$ to 0.55 $m^2\,sec^{-2}$. At the point where $H/\lambda = 0.14$, or $H/h = 0.78$, the wave will become unstable and break, and energy is rapidly dissipated as turbulent heat. The comparatively flat, shoaling slopes (often less than 0.5°) around the lake induce multiple breaking, i.e., a wave will break and reform several times before reaching the shore (Carter and Balsillie, 1983). Although energy is dispersed by bottom friction and breaking, momentum is conserved. This leads to the setting-up of a slight, lakeward-facing water surface slope, and the inevitable generation of nearshore sedimentary bedforms produce a very complex pattern of low velocity littoral currents.

On an enclosed water body, particularly one as shallow as Lough Neagh, wave and current spectra are narrow and closely related to prevailing wind conditions. These relationships were investigated by Darbyshire (1956) who determined empirically that

$$H_{max} = 0.0032\,U^2\,m$$

and

$$T_s = 0.67\,U^{0.5}\,sec$$

where H_{max} is the maximum wave height, T_s is the significant wave period (note that $\lambda = 1.56\,T^2\,m$) and U is 1.5 times the free stream wind velocity in m sec^{-1}, measured at Aldergrove 5 km east of Lough Neagh. Darbyshire's relationships are graphed in Figures 3.4a and b. The wind rose (Fig. 3.5) indicates that winds from all directions may occur, but are dominated by those from the southern, southwestern and western sectors. This means that, over time, the greatest wave and current activity will be experienced in the north and north-east of the lake.

Wind-generated currents in Lough Neagh have only recently been the subject of investigations (Soliman, 1986) and even preliminary results are not yet fully analysed. In deep water lakes, Smith (1975) has shown that as surface currents move downwind they generate upwind currents at depth. In this way the water body achieves a degree of equilibrium. In shallow lakes

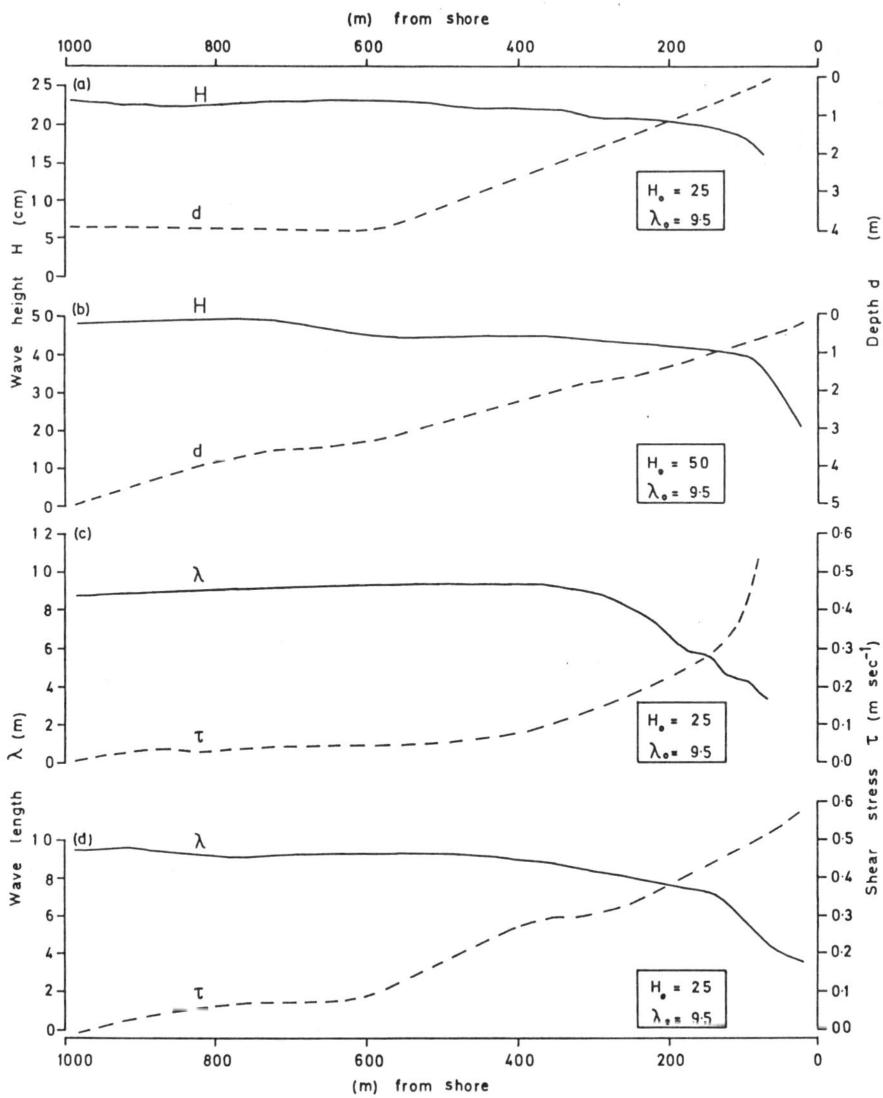

Figure 3.3. Graphs (a) and (b) show the decline in wave height (H) as waves approach the Lough Neagh shoreline. The relatively flat inshore slopes (shown by depths (d)) allow energy to be gradually abstracted from deepwater waves without peaking. In (c) and (d) wave length (λ) decreases rapidly within 200–300 m of the shore, while bottom shear stress (τ) rises to a maximum at the wave breaking point. These graphs are based on results from wave tracking simulations undertaken for Toome Bay.

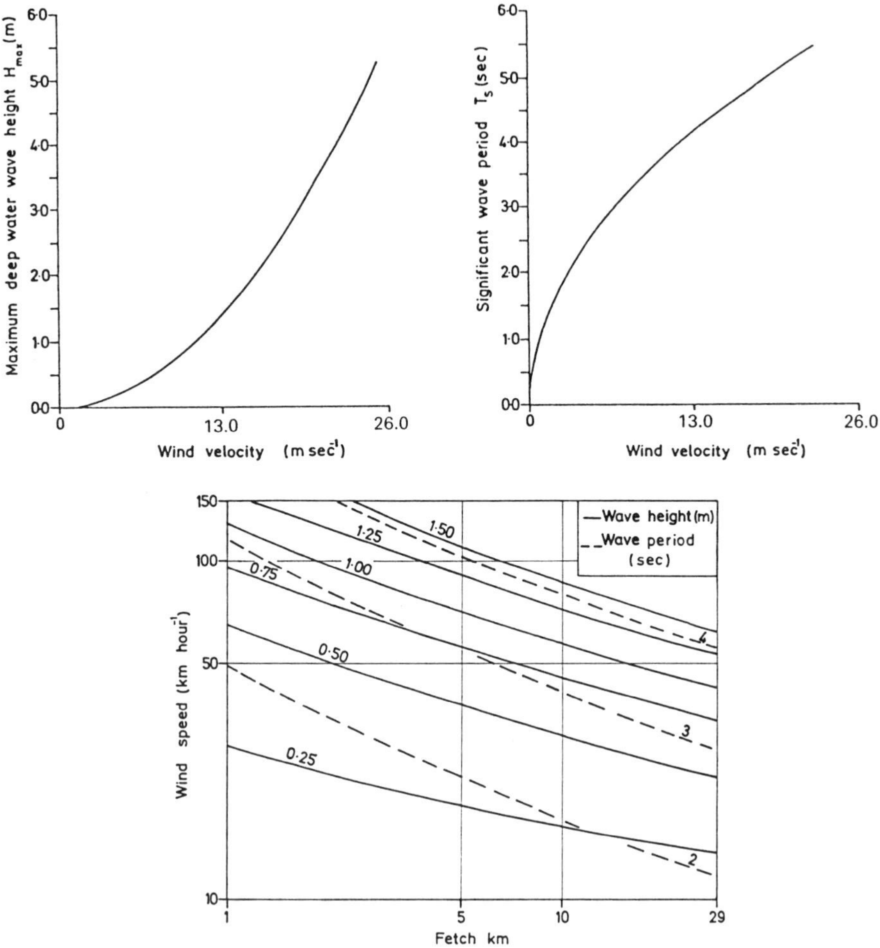

Figure 3.4. (a) Darbyshire's empirical relationships between wind velocity and wave height and period over restricted fetches, based on data collected in Lough Neagh in the late 1940s. (b) Nomograph illustrating the relationship between winds and waves on Lough Neagh.

these deep return currents are replaced by surface currents moving around the lake shore. However, regular patterns are rarely encountered as irregularities in the shoreline or river inflows may cause deflection, the wind is subject to constant changes in strength and direction and secondary circulations may be set up by differential heating and convection. At any point in the lake the three-dimensional pattern of currents is likely to show considerable variation. Direct frictional effects, plus the presence of oscillating waves, cause a deflection of the mean surface vector at depth. Winds may also induce a spiral motion to the surface waters known as the Langmuir circulation (Langmuir; 1938). These act to transfer water slowly downwind and may be detected

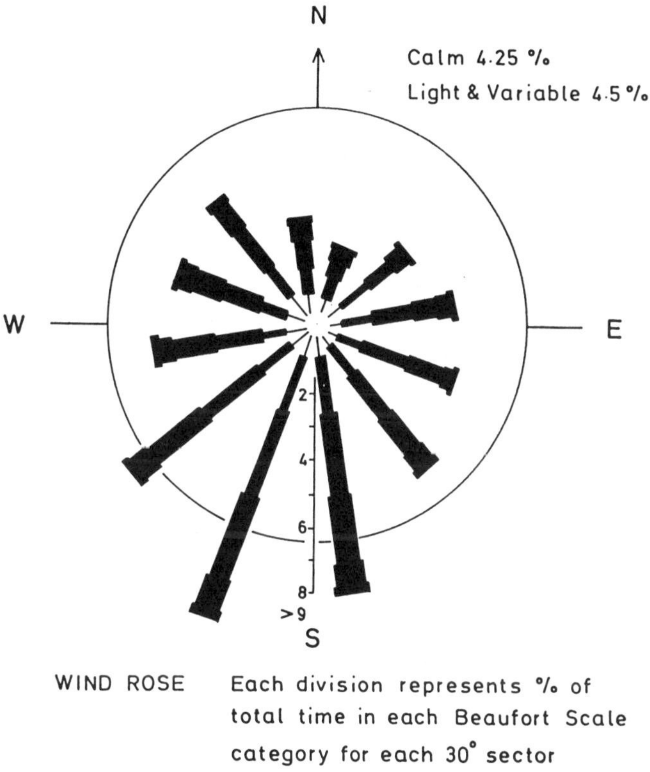

Figure 3.5. Composite wind rose for Lough Neagh from records taken at Aldergrove (4 km east of the Lough) between 1928–1946 and 1949–1960 (from Belfast Meteorological Office, *pers. comm.*).

easily by the presence of parallel foam lines where the Langmuir circulation cells converge. In deep water convergent foam lines in Lough Neagh are 10 to 20 m apart, but on entering the shallower inshore areas the cells progressively divide and may ultimately be only a few centimetres in width.

Important currents may be set up where streams enter the lake. Often, river water has density characteristics different from the lake water, due to variations in temperature, suspended solids or dissolved matter. In Lough Neagh, stream inflows can sometimes be identified some distance offshore due to their lighter colour. Eventually wind-generated water mixing destroys these differences.

Persistent wind from one direction tends to cause water to 'pile-up' at the downwind end of the lake. Darbyshire and Darbyshire (1955) and Knutson (1970) refer to this as 'set-up', although care must be taken not to confuse this with set-up due to wave momentum. Wind set-up on Lough Neagh can be quite dramatic; Knutson (1970, p. 9–10) records instances in which water level at the shore rose 4 cm hr^{-1} following the onset of strong winds.

Set-up (Z) in Lough Neagh is related to wind strength (V) and direction (θ) by the following empirical equations (zero lag) taken from Knutson's work

$$\text{North-South } Z = 0.555 - 0.021\, V^2 \cos\theta - 0.015\, V^2 \sin\theta$$

$$\text{East-West } Z = 0.788 + 0.003\, V^2 \cos\theta - 0.014\, V^2 \sin\theta$$

Cessation of wind stress results in an oscillation of the water surface slope. Seiches in Lough Neagh have been investigated by Darbyshire and Darbyshire (1957) and Parry-Jones (1970). Seiche excitation of the lake surface results in numerous oscillation modes. An example of autographic recorder data from Toome during the passage of a cyclonic depression is shown in Fig. 3.6a. Such records may be resolved into dominant modes; Fig. 3.6b shows the nodal lines for the first four modes determined along NW–SE and NE–SW axes by Parry-Jones (1970). Power spectra indicating oscillation periods are presented in Fig. 3.7 for each of the four autographic recording sites noted on Fig. 3.6b. High frequency modal periods are similar throughout; both Darbyshire and Darbyshire (1957) and Parry-Jones recognise a fundamental period around 90–100 minutes. The investigation by Parry-Jones revealed the presence of very long seiche periods, up to 11 hours in some cases, and seiche amplitudes of up to 30 cm around the shore. Such inshore values are probably slightly accentuated, compared to open-lake amplitudes.

3.5. Annual temperature cycle (by C.E. Gibson)

Because of the shallowness of the water in Lough Neagh the water temperature closely follows that of the overlying air (Fig. 3.8.) with a lag of approximately seven days. In 1969–70, which were fairly typical years, the annual heat budget was 13.8×10^3 cal cm^{-2} yr^{-1} (Gibson and Stewart, 1973). The water temperature normally falls to a minimum of 1° C in February although in mild winters, minima as high as 5° C have been recorded. The summer maximum of an isothermal water column rarely exceeds 18° C. Ice formation is rare and complete or near complete ice cover has only been observed twice in the last forty years (1947 and 1963), although Toome Bay just froze over in 1981–82.

The mean wind strength across the lake surface is approximately 5 m sec^{-1} (Table 3.1), the long term mean measured at Aldergrove Meteorological Station close to the N-E corner of the lake. This wind strength is evidently sufficient to ensure complete mixing of the waters and the lake is virtually isothermal for most of the year. Transient surface heating occurs during calm sunny spells and during June, July and August an incipient thermocline may

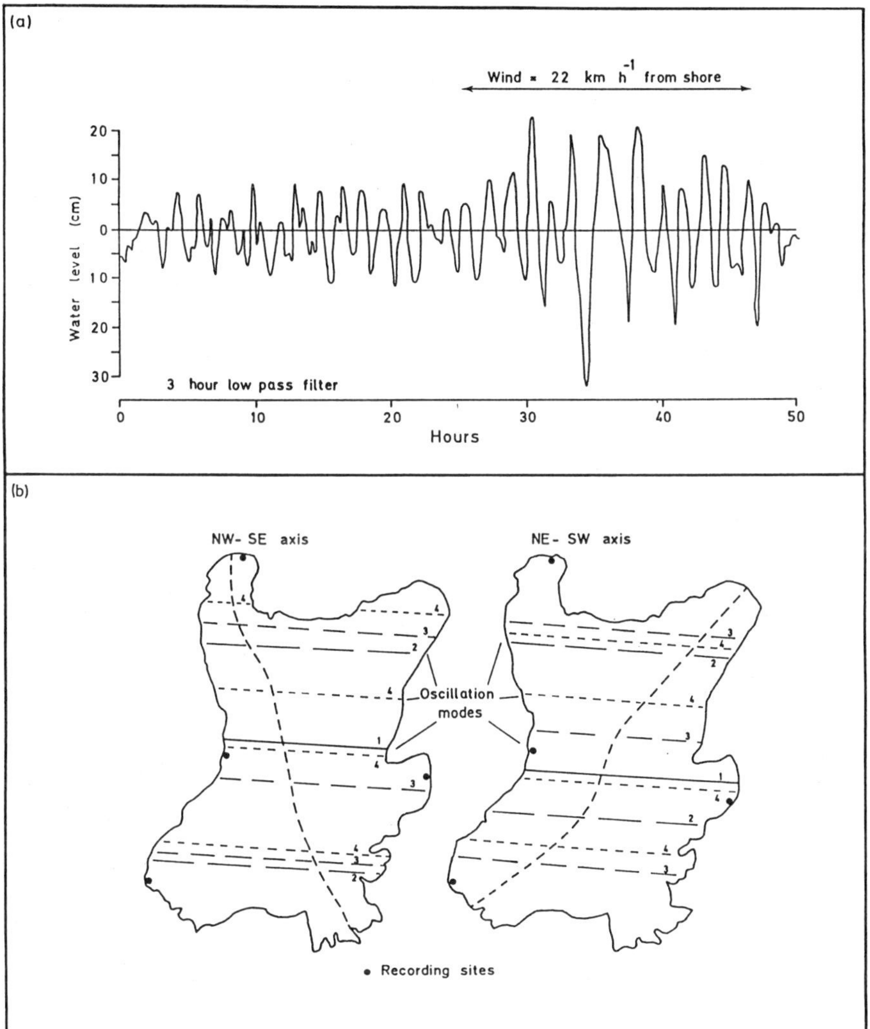

Figure 3.6. (a) Autographic water level recorder trace from Toome Bridge showing seiche oscillations during a southerly storm, 14–16th August 1969. (b) Nodal lines of 1st, 2nd, 3rd and 4th order seiches excited along NW to SW and NE to SW axes. (Both figures after Parry-Jones, 1970).

form. A discernible thermal structure may persist for 3 to 4 weeks and a temperature difference of up to 5° C between the surface and bottom waters may result (Fig. 3.8b). Some ecological implications of incomplete mixing resulting from low wind speeds are discussed in Chapter 9.

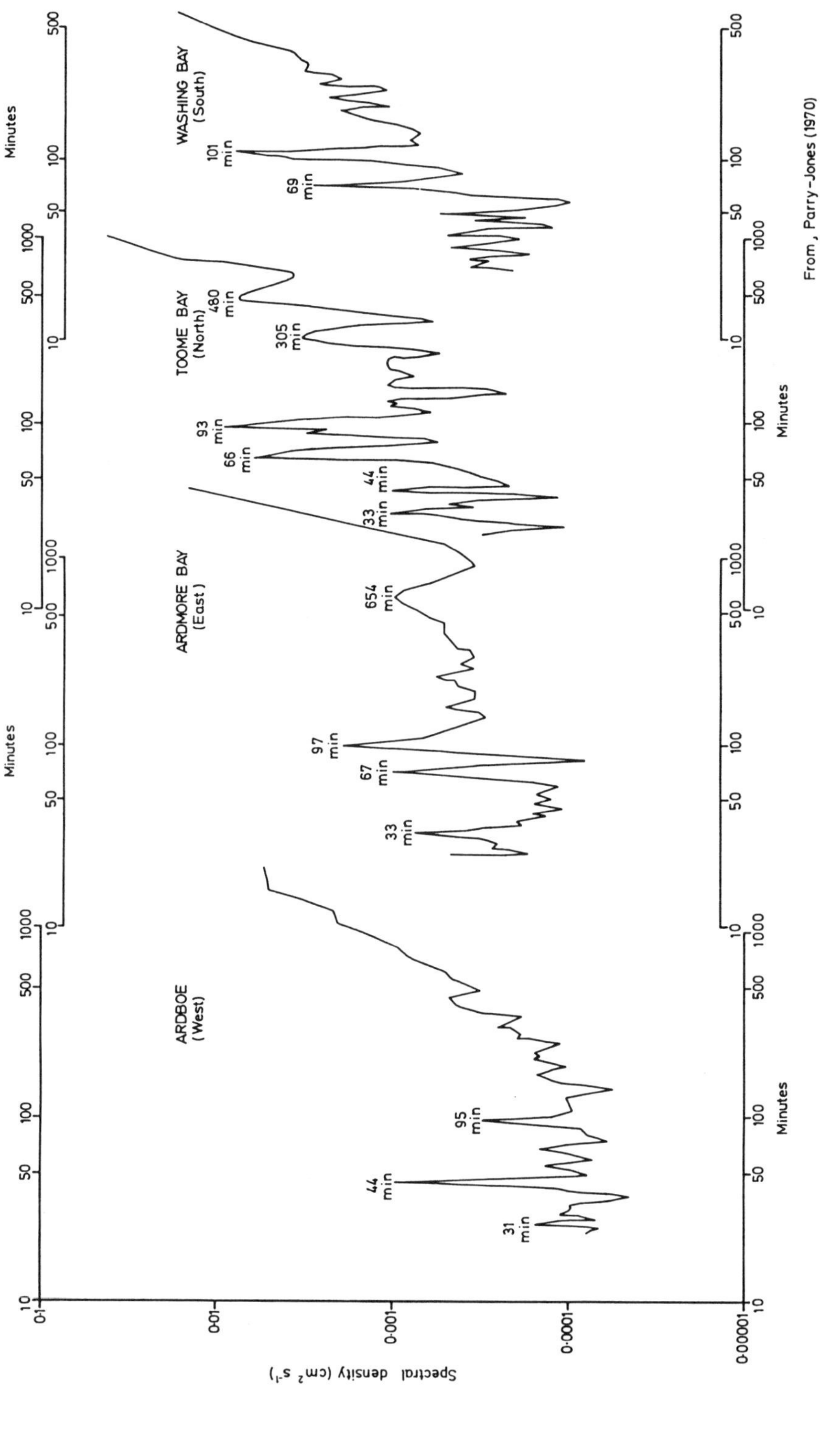

Figure 3.7. Power spectra of water level oscillations from Ardboe (west), Ardmore Bay (east), Toome Bay (north) and Washing Bay (south) based on the work of Parry-Jones (1970). The spectra show clearly the high frequency seiche periods, emerging (to the right) into aperiodic low frequency water level changes.

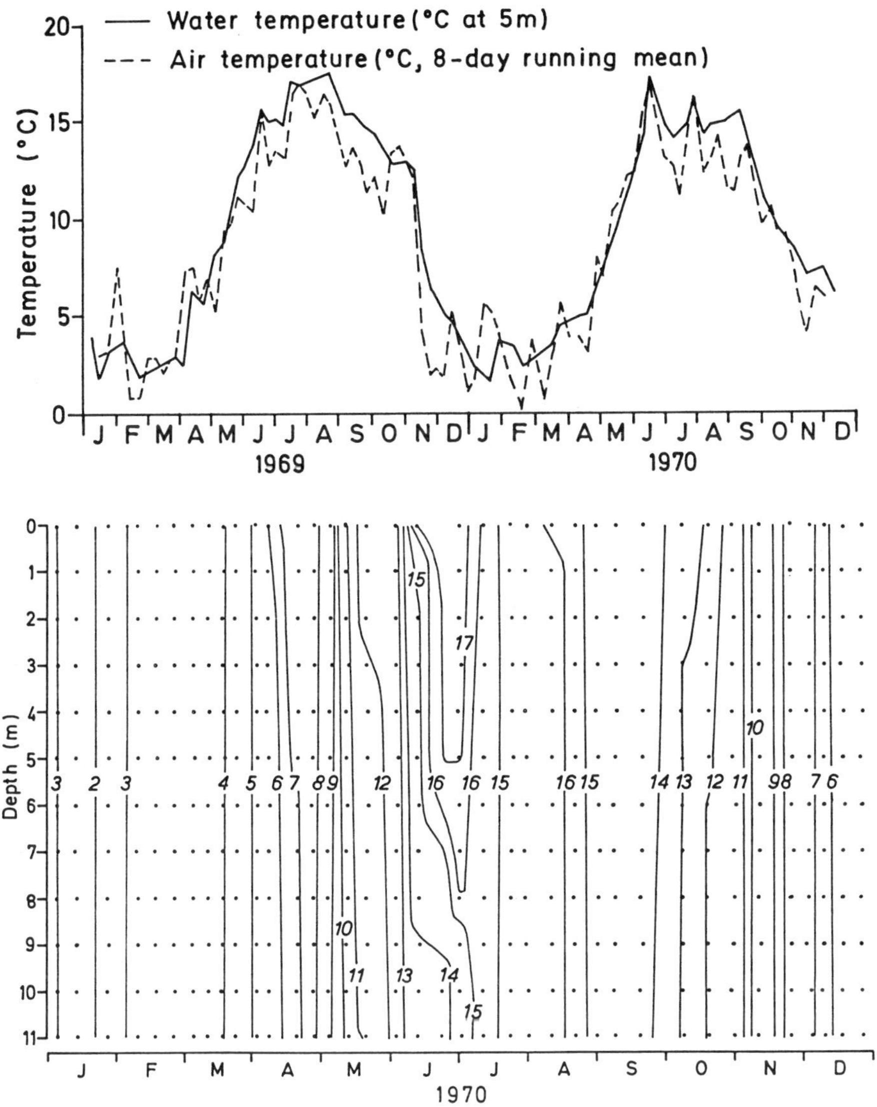

Figure 3.8. (a) Water and air temperatures in Lough Neagh 1969–1974. (b) Depth/Time plot of water temperatures in Lough Neagh in 1970.

3.6. Lake sedimentation

3.6.1. *A simple model of sedimentation*

Very little is known about the mechanics of present day sedimentation in Lough Neagh. Flower (1980) has discussed aspects of biogenic sedimentation arising from problems encountered in the earlier palaeoecological studies (O'Sullivan *et al.*, 1973; Battarbee, 1978a and b).

Flower (see Fig. 2.3, facing p. 13 in Flower; 1980) also presented a grain-size map of the surficial sediments in the lake, upon which Fig. 3.9 is based. This shows that the central part of the lake basin is dominated by clay (mud), and the marginal areas (mainly less than 6 m in depth) comprise sands and mixed sands and clays (silt). It is necessary to relate these data to a simple model of sedimentation in order to examine clearly the processes responsible for such a distribution.

Lough Neagh acts as a sediment trap for material brought down by the catchment streams. Only a very small proportion of this material will pass through the lake. However in such a large lake considerable variations in sediment type and sediment accumulation rates are to be expected. Basically the lake may be split into two major zones, the littoral and the profundal. Within these a series of distinct facies zones may be recognised (Fig. 3.10). The littoral and profundal zones are divided by the wave base line, which marks the position where bottom stress (τ) due to waves just exceeds the entrainment threshold of the mean sediment size. The concept of wave base is a statistical one as both wave and sediment parameters vary enormously. In reality we find a transition zone (4a on Fig. 3.10) marked by the characteristics of both littoral and profundal sedimentation. In Lough Neagh, wave base extends from depths of only a few centimetres during calm periods, to around 7.5 m in extreme storms. The locations of wave base at any one time will differ around the lake due to the influence of fetch. The physical manifestation of wave base is the boundary between mud and sand, clearly seen on Fig. 3.9. The transition zone is marked by mixed grain size populations (see Curve 5 on Fig. 3.11) and interfingering of coarse and fine units. The artificial lowering of the lake level over the last 140 years would have resulted in a lakeward shift in wave base and a general reorganisation in the sedimentation pattern.

Sedimentation in Lough Neagh involves both minerogenic and biogenic material, with the former dominating the littoral zone. Minerogenic particles entering the lake basin begin to settle out as the stream flow disperses radially. In theory the larger, spherical grains should settle first so that sediment should become finer away from the river mouth. However fall velocity depends not only on size, shape and density (relative to the density of the lake water) of particles, but also to the ambient turbulence, which is likely to be considerable where the stream currents interact with wave motions. As a result the near river mouth sediment pattern is often confused. In addition, river mouth accumulations are often unstable and slumping into deeper water is a common occurrence in many lakes, causing further mixing. Dearing and Flower (1982) report a strong positive correlation between rainfall and minerogenic sedimentation in Antrim Bay, which they associated with river bank erosion in the catchment.

Sedimentation of biogenic material occurs throughout Lough Neagh.

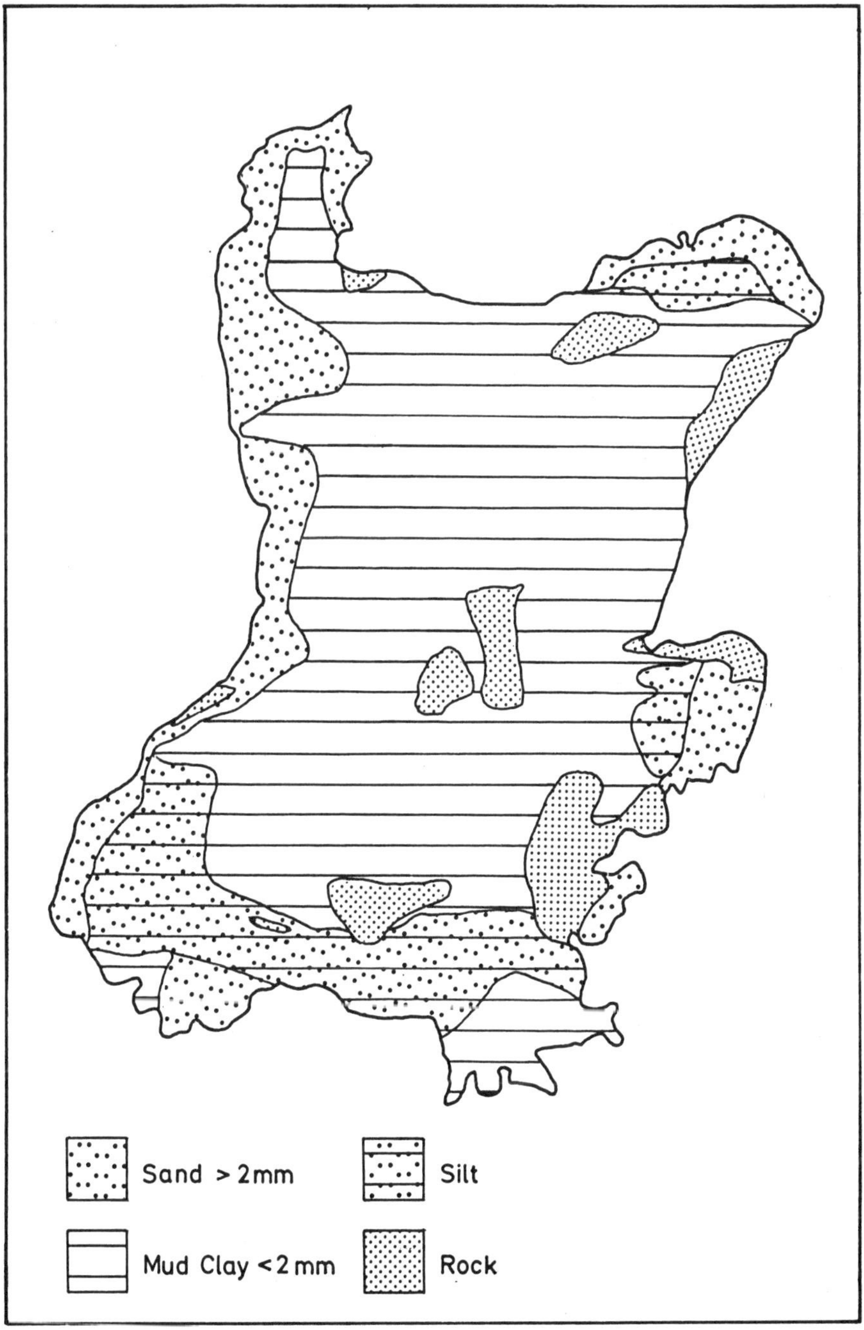

Figure 3.9. The distribution of bottom sediment in Lough Neagh based on grain-size classification.

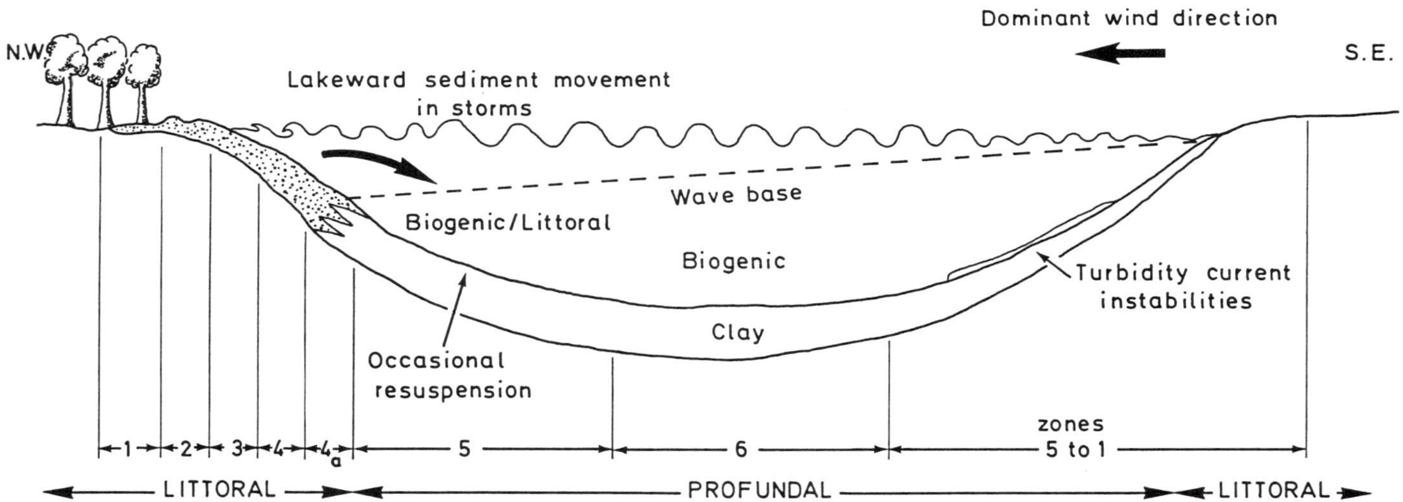

Figure 3.10. Probable sedimentary facies zonation across Lough Neagh, showing six major depositional types. The exact position of zonal boundaries will vary according to lake level and meteorological conditions. In this example the wind is shown as blowing from the SE, along the longest lake axis. Facies codes: 1. Storm deposition; 2. Fairweather shoreline; 3. Shallow littoral; 4. Deep littoral; 4_a. Transitional; 5. Biogenic resuspension; 6. Deepwater biogenic.

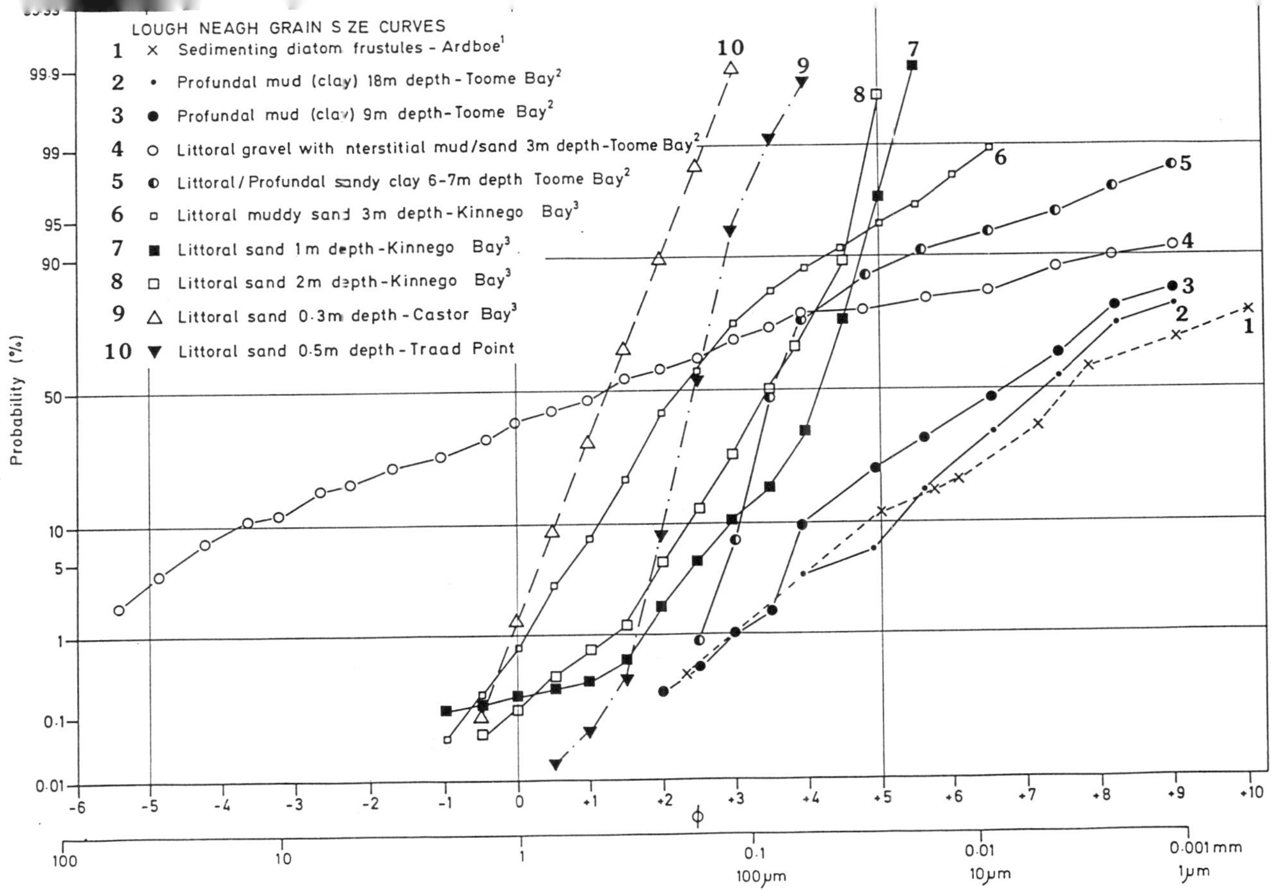

Figure 3.11. Grain size curves from Lough Neagh sediments. The overall 'envelope' of curves covers a range from very coarse, poorly sorted material, to fine clays and silts. The littoral sands appear as the best sorted sediments. Very few of the curves show complex grain size distributions indicative of mixtures. (Note that the phi scale (ϕ) is equivalent to $-\log_2 d$ where d is the grain diameter in millimetres. Thus coarser sediments have negative values; finer sediments, positive values). [1] from Flower (1980), [2] from Old (pers. comm 1970); [3] from Carter and Rippey (1982).

Planktonic diatoms utilise dissolved solids, especially silica, in the lake water to produce frustules. The dissolved silica content in the lake is rapidly depleted each spring (see Chapter 9). Compared with many minerogenic particles, fall velocities of diatoms are low and frustules are held in suspension by even gentle water movements and fairly uniformly spread out across the lake bottom. Net biogenic sediment influx is highest in late spring and summer as the standing crop declines (Flower, 1980).

During storms direct stress on the lake bed above wave base causes entrainment of sediment. Gibson (1979) reports on preliminary observations in Antrim Bay on sediment resuspension during storms. He notes that resuspension occurs throughout the water column (10 m in this case) and that the ratio of inorganic to organic sediment rises slightly, but quickly, as the storm passes. The larger or more buoyant grains will have a tendency to migrate landward particularly as oscillatory wave motions become more elliptical. Smaller minerogenic particles and low density biogenic material are the most likely to be resuspended, and in both cases the time taken to fall back to the bottom will greatly exceed the wave period, the dominant factor in moving particles upwards. Bottom stresses of between 0.2 and 1.0 m sec^{-1} are needed to initiate movement of loose silt and clay particles of normal density (2.65 g cm^{-3}), while most diatom frustules would probably be moved by stresses an order of magnitude less i.e. 0.02 to 0.1 m sec^{-1}. Resuspended material will be diffused throughout the littoral zone, and into the adjacent profundal zone. In the post-storm period this material will settle out, that proportion falling below wave base will be 'lost' to the littoral zone. Gradually a division between coarse and fine sediment will become defined.

Net accumulation of resuspended material will be greatest just lakeward of the wave base line. In time, this will lead to slope instability in the inshore profundal zone (zone 5 on Fig. 3.10) and mass slumpings, 'chuting' or turbidity flows will occur. These processes, which are probably more widespread but less frequent than those occurring off the river mouths, are important in moving material to the deeper parts of the lake basin. From studies elsewhere (Prior and Coleman, 1979) subaqueous slope instabilities are now known to operate on very low angle slopes ($<1°$) due to low, or even negative, pore pressure, low liquefaction thresholds of mud and differential loading. Prior (pers. comm.) considers that Lough Neagh would be an ideal site for such processes. Evidence from sediment cores should be in the form of discontinuities in the accumulation record, perhaps similar to those described by Battarbee (1978b).

3.6.2. *Profundal biogenic sedimentation*

Profundal sediments consist mainly of biogenic muds with mean grain sizes around 4 to 5ϕ (62 to 32 μm). Cumulative frequency curves of grain size distributions (Fig. 3.11) show complex, non-Gaussian forms probably reflecting the heterogeneity of the influx material together with a lack of pre-

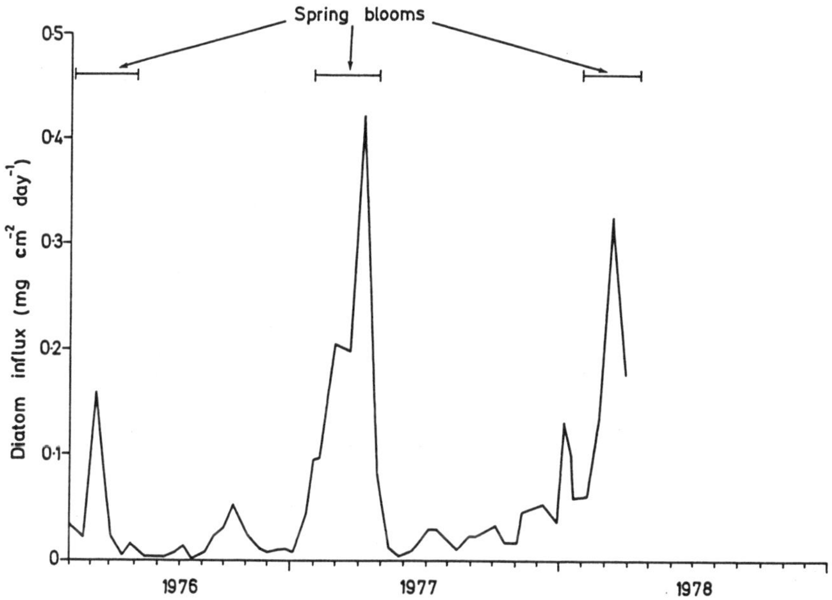

Figure 3.12. Net biogenic sedimentation of diatoms frustules at the Battery site (depth 10.5 m) in Lough Neagh between 1976–1978. The spring peaks are clearly visible, comprising mainly *Melosira italica* sub sp. *subartica* and *Stephanodiscus astraea* (from Flower, 1980).

depositional sorting. From a limited number of X-ray diffractograms, Flower (1980, p. 105) was able to identify illite, chlorite and montmorillonite clay minerals, in the profundal sediments, together with quartz and feldspars. A similar diffraction pattern was obtained from a sample of suspended solids taken from the Ballinderry river.

The biogenic sediment record in Lough Neagh has been the subject of considerable research (Battarbee, 1973, 1978a, 1978b; O'Sullivan *et al.*, 1973; Carter 1977; and Flower, 1980) aimed at reconstructing the recent ecological history of the lake, and to a lesser extent that of the catchment. The lake has become increasingly eutrophic over the last century (Wood and Gibson, 1973; Wood, 1980; Wood and Smith, 1986). These aspects are dealt with in Chapters 7 to 10.

Flower (1980) studied the profundal sedimentation rates using sediment traps at two sites, the Battery (depth 14.5 m) and Antrim Bay (depth 10.5 m). Unfortunately comparative samples could not be taken from other depths. As both sets of traps regularly caught resuspended material we must assume that they either fell within storm wave base for diatom frustules and/or were influenced by profundal fall-out from littoral sediment agitation. Flower's work shows quite clearly the relationships between planktonic diatom productivity in spring and net sedimentation (Fig. 3.12). Non-planktonic diatoms had little influence on sedimentation at Flower's two sites. Effects of storminess are reflected by the amount of resuspended material trapped. Gibson in Chapter 9 argues that a very high percentage of biogenic silica is subsequently redissolved into the water column. Rippey (1977, and in Chapter 9) has determined the environmental factors controlling silica release from the Lough Neagh profundal sediments, and calculates a mean figure of around 15×10^{-3} μg cm^{-2} d^{-1}.

Using ^{210}Pb and ^{137}Cs isotope dating techniques, Battarbee (1978a and b) estimated that average sediment accumulation rates since 1960 have been 9.5 mm yr^{-1}, although he identifies a spectrum of rates from 'fast' to 'slow' varying by a factor of up to three (see Chapter 7). By extrapolation of his sediment trap data, Flower (1980) calculated rates of 8.3 mm yr^{-1} over the same period. Consolidation and compaction processes would act to reduce these figures, perhaps by as much as 40% if complete dewatering took place.

In the absence of wave stress, the profundal sediment suface is unlikely to support extensive development of bedforms, although some ripple-like forms have been reported by divers at depths of 11.0 m in Antrim Bay (Gibson, pers. comm.), although these may become disturbed by mass downslope movements, as mentioned in Section 3.6.1. Such processes might account for the variability in sedimentation rates noted by Battarbee (1978a). Extensive bioturbation, especially by chironomid larvae, which may exceed 10,000 individuals m^{-2} and tubificids (Carter 1976, and Chapter 16) act to integrate minor spatial and temporal changes in sediment deposition over the top 3 cm of mud (see also Rippey and Jewson, 1982). Chironomid faecal pellets are an important constituent of the sediment mass (Flower, 1980).

3.6.3. *Littoral sedimentation and shoreline geomorphology*

Very little is known about the littoral zone of the lake. The sediment comprises mainly well-sorted, angular sand with mean grain sizes between +1.5 and $+2.5\phi$ (0.33 to 0.17 mm) (Fig. 3.11, Curves 7–10). In sheltered localities, particularly near reed beds, sizeable accumulations of organic debris may form 'mats' across the sand surface. In depths of less than 2.5 m, bedform structures abound despite the abundance of chironomid larvae. These range from oscillation ripples to bars. The latter occur widely around the lake margin (Fig. 3.14), in some places, for example Washing Bay, extending over 1 km from the shore. Clayton (1980) has examined the distribution and dynamics of these bars at Traad Point. Here four distinct sets of bedforms parallel the shore, with trough to crest ratios ranging from 1.1 for the deepest (1.5 m) to 1.45 for the shallowest (0.4 m). The bars break and reflect incident waves and reduce wave energy at the shoreline. Clayton (1980, p. 45) and Carter and Balsillie (1983) provide examples showing how wave energy may fall from 1.5×10^6 joules m^{-1} sec^{-1} to 0.3×10^3 joules m^{-1} sec^{-1} over a single bar. The outer bar becomes 'active' during extreme conditions, whereas the inner bars are constantly destroyed and reformed. The innermost bar is also subject to changes brought about by seasonal water level variations.

When the lake level is high and the bar is submerged, the shore-parallel avalanche face is cut by lakeward-moving currents to produce a segmented plan structure. As the water level falls the bar emerges and assumes the characteristics of sub-aerial barrier shoals. These shoals are constantly reorientated as the wind and wave conditions change. Relatively deep water, isolated on the landward side of the shoals is often the site of high organic activity, as seen from the numerous biogenic structures. The relationship of the bar bedform dynamics to the chironomid communities is discussed in Carter and Carter (1983).

The nature of the shoreline itself – vegetation, sandy beach, rocks – has an important bearing on the shallow water and sediment dynamics. For example wave reflection from a rock shore is often over 50% while from a vegetated one, which acts as a baffle, it is virtually nil. Carter (1978) has described the formation of lakeward prograding transverse sand bars at Traad Point which owe their existence to a delicate feedback between wave powered littoral currents, edge waves, shoreline type and bar forms.

Adjustment of the shoreline plan to wave conditions in Lough Neagh may be best described as 'immature'. However, in some places, Antrim Bay, Washing Bay and Ballyronan Bay, the sandy coast has seemingly responded in plan form to long-term wave control. The sweeping shoreline curve of Antrim Bay is particularly reminiscent of sea coast development. Also, small sand spits, indicative of longshore sediment transport, have developed at Traad Point, Shane's Castle and on Rams Island. The shore is low everywhere, with almost no erosion; problems like those experienced in Loch Lomond (Countryside Commission for Scotland, 1979) where the shoreline has retreated dramatically in recent years due to impulsive boat wakes, are mercifully absent in Lough Neagh. The primary reason for this must be the gradual lowering of the lake over the last 140 years which has, by dropping wave base, shifted sediment landward. In some places, for example, between Ballyronan and Traad Point historical emergence and shoreward migration of sand bars has led to the formation of a series of small beach ridges or barriers, allowing some eolian dune formation and the development of an anoxic backswamp environment. Throughout, the shoreline may be considered to be one of accretion.

3.7. References

Battarbee, R.W. (1973) Preliminary studies of the Lough Neagh sediment II: Diatom analysis from the uppermost sediment. In: H.J.B. Birks and R.G. West (eds.), Quaternary Plant Ecology. Blackwells, Oxford, pp. 279–289.

Battarbee, R.W. (1978a) Observations on the recent history of Lough Neagh and its drainage basin. Phil. Trans. R. Soc. Lond. B. 281: 303–345.

Battarbee, R.W. (1978b). Biostratigraphical evidence for variations in the recent pattern of sediment accumulation in Lough Neagh, Northern Ireland. Verh. Int. Ver. Limnol. 20: 624–629.

Bell, P.D. (1971) Lough Neagh Working Group Advisory Report. Vol. 1. Government of Northern Ireland, 93 pp.

Carter, C.E. (1976) A population study of the Chironomidae (Diptera) of Lough Neagh. Oikos 27: 346–354.

Carter, C.E. (1977) The recent history of the chironomid fauna of Lough Neagh from the analysis of remains in sediment cores. Freshwat. Biol. 7: 415–423.

Carter, C.E. and Carter, R.W.G. (1983) Factors affecting the chironomid community of a nearshore sand area. Monogr. Am. Ent. Soc. 34: 47–59.

Carter, C.E. and Rippey, B. (1982) Water quality in four bays in Lough Neagh, N.I. Ir. J. Env. Sci. 2: 51–60.

Carter, R.W.G. (1978) Small-scale transverse bars in Lough Neagh, Northern Ireland. J. Earth Sci. R. Dubl. Soc. I: 205–209.

Carter, R.W.G. (1988) Coastal environments: An introduction to the physical, ecological and cultural systems of coastlines. Academic Press, 617 pp.

Carter, R.W.G. and Balsillie, J.H. (1983) A note on the amount of wave energy transmitted over nearshore sand bars. Earth Surf. Proc. 8: 213–222.

Clayton, J. (1980) A study of parallel longshore sand bars under the influence of waves and fluctuating water levels, Traad Point, Lough Neagh. Unpublished M.Sc. Thesis, The New University of Ulster, 99 pp.

Countryside Commission for Scotland (1979) Shore erosion around Loch Lomond. Perth CCS, 76 pp.

Darbyshire, J. (1956) An investigation into the generation of waves when the fetch of the wind is less than 100 miles. QJR Met. Soc. 82: 461–468.

Darbyshire, J. and Darbyshire, M. (1955) Determination of the wind-stress on the surface of Lough Neagh by the measurement of tilt. QJR Met. Soc. 81: 333–339.

Darbyshire, J. and Darbyshire, M. (1957) Seiches in Lough Neagh. QJR Met. Soc. 83: 93–102.

Dearing, J.A. and Flower, R.J. (1982) The magnetic susceptibility of sedimenting material trapped in Lough Neagh, Northern Ireland, and its erosional significance. Limnol. Oceanogr. 27: 969–975.

Flower, R.J.L. (1980) A study of sediment formation, transport and deposition in Lough Neagh, Northern Ireland, with special reference to diatoms. Unpublished D. Phil. Thesis, The New University of Ulster, Coleraine, 212 pp.

Garratt, J.R. (1972) Studies of turbulence in the surface layers over water (Lough Neagh) Part II. Production and dissipation of velocity and temperature fluctuations. QJR Met. Soc. 99: 680–687.

Gibson, C.E. (1979) Survey Work. In: Freshwater Biological Investigation Unit, Annual Report on Research and Technical Work. Dept. Agriculture, Northern Ireland, pp. 135–136.

Gibson, C.E. and Stewart, D.A. (1973) The annual temperature cycle of Lough Neagh. Limnol. Oceanogr. 18: 791–793.

Knutson, C.E. (1970) Wind set-up on Lough Neagh. Unpublished M.Sc. Thesis, University College of North Wales, Bangor, 59 pp.

Langmuir, I. (1938) Surface motion of water induced by wind. Science 87: 119–123.

McCutcheon, W.A. (1965) Canals of the North of Ireland. David and Charles, Dawlish, 143 pp.

McCutcheon, W.A. (1980) The industrial archaeology of Northern Ireland. HMSO, Belfast, 395 pp.

O'Sullivan, P.E., Oldfield, F. and Battarbee, R.W. (1973) Preliminary studies of Lough Neagh sediments I. Stratigraphy, chronology and pollen analysis. In: Birks, H.J.B. and West, R. G. (eds.) Quaternary Plant Ecology, Blackwells, Oxford, pp. 267–278.

Parry-Jones, W.H. (1970) An investigation into the periods of free oscillation of Lough Neagh and the relationship with wind-stress. Unpublished M.Sc. Thesis, University College of North Wales, Bangor, 61 pp.

Prior, D.P. and Coleman, J.M. (1979) Submarine landslides – geometry and nomenclature. Zeit. für Geomorph. 23: 415–426.

Rippey, B.H. (1977) The behaviour of phosphorus and silicon in undisturbed cores of Lough Neagh sediment. In: Golterman, H.L. (ed.) Interactions between sediments and freshwater. Pudoc, Amsterdam, pp. 348–353.

Rippey, B and Jewson, D.H (1982) The rates of sediment-water exchange of oxygen and sediment bioturbation in Lough Neagh, Northern Ireland. Hydrobiologia 92: 377–382.

Sheppard, P.A., Tribble, D.T. and Garratt, J.R. (1972) Studies of turbulence in the surface layer over water (Lough Neagh) Part 1. Instrumentation, programme, profiles. QJR Met. Soc. 98: 627–641.

Smith, I.R. (1975) Turbulence in lakes and rivers. Freshwater Biological Association Sci. Publ. 29, 78 pp.

Smith, I.R. (1979) Hydraulic conditions in isothermal lakes. Freshwat. Biol., 9: 119–145.

Soliman, M.N. (1986) Wave properties and water movements in Lough Neagh Northern Ireland. Unpublished M.Phil. Thesis, University of Ulster.

Wood, R.B. (1980) Pollution of Lough Neagh: a multidisciplinary approach. Anal. Proc. 17: 369–374.

Wood, R.B. and Gibson, C.E. (1973) Eutrophication and Lough Neagh. Wat. Res. 17: 173–187.

Wood, R.B. and Smith, R.V. (1986) Lough Neagh: 100 years of man's influence on the water quality of a major inland fishery. Proc. 17th Ann. Stud. Course, Institute of Fisheries Managment, Univ. of Ulster, Coleraine, pp. 111–126.

4. The optical properties of Lough Neagh

D.H. JEWSON

4.1. Introduction

There have been many anecdotal references to the colour and turbidity of Lough Neagh since it was reported as turning to blood in AD683 (Annals of Ulster and other references cited by Marshall, 1934). However, quantitiative measurements of underwater light penetration are unknown before 1968. Since this time a range of methods of investigation have been used including Secchi disc (Gibson, 1975), broad-band filters and spectroradiometric data (Jewson, 1977; Jewson and Taylor, 1978).

In the clearest natural inland waters, such as Crater Lake in Oregon, 1% of surface light reaches 75 m and the most penetrating wavelengths are between 400 and 480 nm in the blue part of the spectrum (Smith *et al.*, 1973). Although in recent times Crater Lake has shown signs of deterioration (Larson, 1983) it still contrasts with Lough Neagh which is at the opposite extremes of both turbidity and colour with the 1% light level usually occurring between 2 and 3 m and the most penetrating spectral region varying between 575 and 710 nm at the red end of the visible spectrum. The reasons for this and the implications for the depth limitation of various plant and animal processes are discussed in this chapter. Details of incident solar radiation are given in Table 11.2 and Fig. 11.5 of Chapter 11.

4.2. Light penetration

As light passes down through water it is absorbed and scattered. These two processes not only reduce the quantity but also change the quality and angular distribution of the light environment in Lough Neagh, so that within one metre it has become largely monochromatic and diffuse. The changes in spectral composition with depth are shown in Fig. 4.1.

However, such detailed analyses are only possible during the calmest weather. At other times the amount of wave action on the lake makes measurements very difficult and so most of the work on seasonal variation

R.B. Wood and R.V. Smith (eds.), Lough Neagh, 59–73.

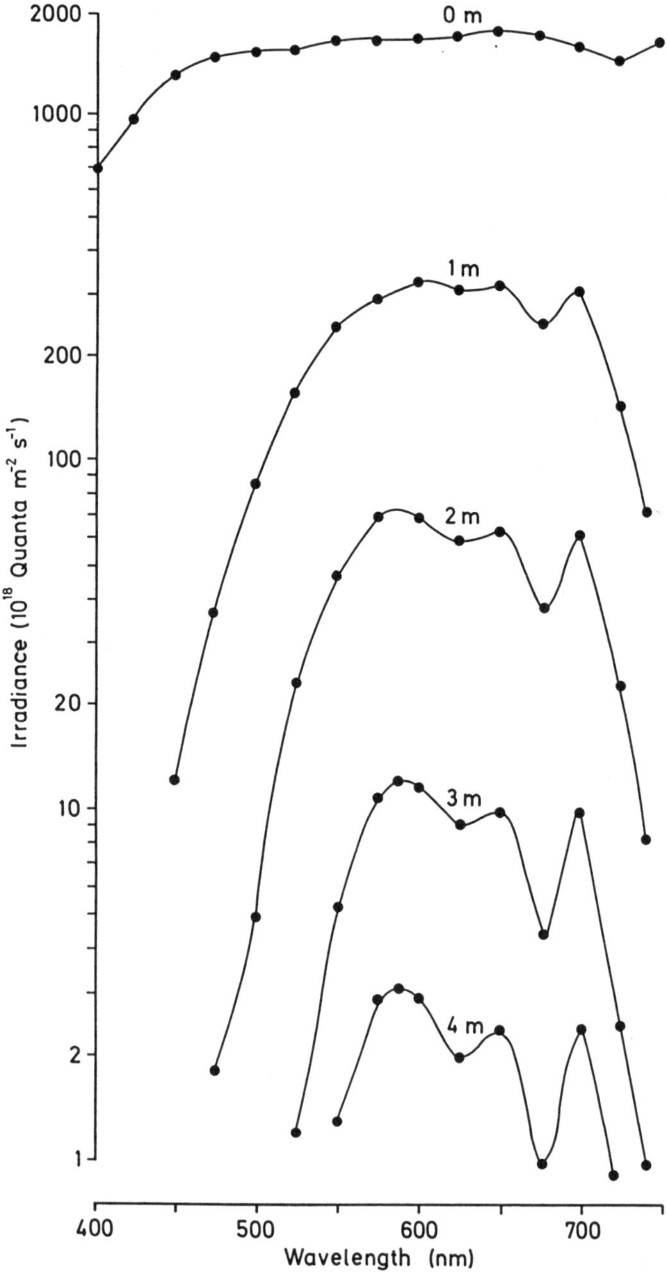

Figure 4.1. Spectral variation of quantum irradiance with depth (24th June 1982), on the open lake of Lough Neagh, plotted on a log scale and measured with a Techtum spectroradiometer (Jewson, unpublished data).

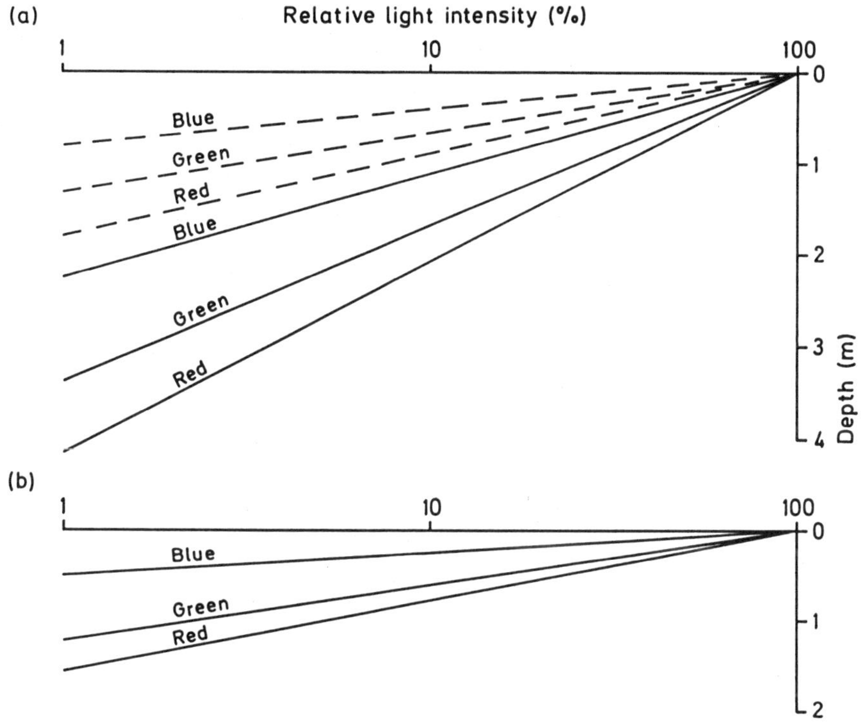

Figure 4.2. Relative light penetration with depth (on semi-log plots) of broad-band filters, blue (BG12), green (VG9), red (RG1). (a) During turbid (13th May 1980 – broken lines) and clear (6th September 1974 – solid lines) conditions measured in the open lake, and (b) Kinnego Bay (10th October 1970) Lough Neagh.

of light quality has been carried out using less sophisticated but more robust coloured glass filters (Jewson, 1977). Two sets of results are shown for downwelling light – representing the clearest and most turbid conditions sampled in the last decade (Fig. 4.2). Three filters are illustrated, blue (BG12, 400–490 nm), green (VG9, 490–585 nm) and red (RG1, 585–700 nm). The red region is the most penetrating in both instances illustrated here but on a few occasions (during crops of *Oscillatoria agardhii*) green has been found to penetrate the furthest. Further details of seasonal changes based on these filters are discussed by Jewson (1977).

Many biological studies require measurements of total 'visible' or photosynthetically available radiation (400–700 nm) and there are now a number of submersible quanta meters available which provide reliable quantitative estimates. However, the selective loss of both short and long wavelengths of light in the surface waters (see Fig. 4.1) means that the decrease of broad spectral bands with depth is not exponential (Fig. 4.3). Because of this it is difficult to formulate a simple mathematical description of the rate of total

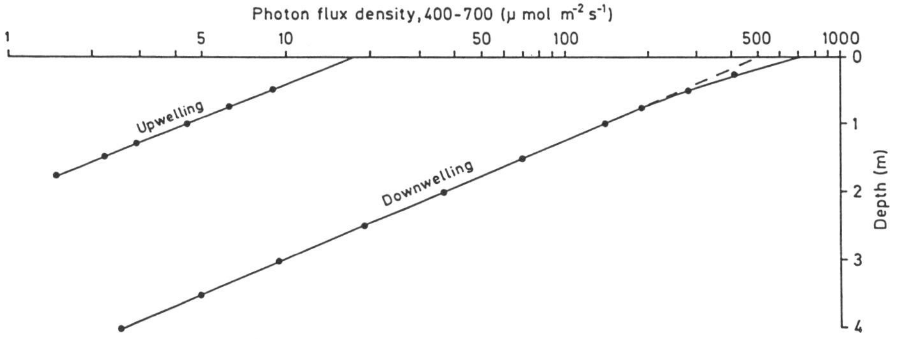

Figure 4.3. Semi-logarithmic plot of the variation in upwelling and downwelling quantum irradiance (400–700 nm) with depth in Lough Neagh (measured with a Li-Cor, Model Li-185).

light (400–700 nm) attenuation that is suitable for all purposes. One approach, which is now becoming widespread, is to force-fit an exponential regression to the whole irradiance-depth profile. This is mathematically incorrect and will introduce large errors if comparisons are made between profiles with varying degrees of completeness. For instance, in Lough Neagh wave height determines how much of the more rapid changes in quality and quantity near the surface is recorded. Therefore, the solution chosen for estimating the rate of attenuation of broad band light has to be chosen carefully. In some cases it is important to take account of the non-linearity, as when estimating absolute light values for photosynthesis (see Jewson, 1976 and Chapter 11). For other applications such as the seasonal comparisons of light penetration, it is permissible (in Lough Neagh) to treat the decrease of quanta (400–700 nm) below 0.75 m as an exponential decay (see Fig. 4.3). It can be measured in all except the windiest weather and correlates well with estimates of the rate of attenuation of the most penetrating spectral region (measured using narrow band filters). As the latter determines the lower limit of many light-related processes it is useful to have a measure of its rate of decrease and that most frequently used by limnologists is the vertical attenuation coefficient. This is derived from the relationship predicting the irradiance for monochromatic light at a given depth

$$I_z = I_o e^{-\epsilon z}$$

where I_z is the irradiance at depth z metres, I_o is the surface irradiance and ϵ is the vertical attenuation coefficient. Typically values of ϵ for red light in Lough Neagh vary between 0.8 for clear water to 3.0 in the most turbid (see

Fig. 4.2). The vertical attenuation coefficient is only an apparent property of the water, which means it can be affected by external factors, such as solar altitude and cloud cover. In reality, changes from these causes are small in turbid waters such as Lough Neagh and can be neglected for many purposes (see Kirk, 1983). This makes it a useful, easily measured parameter for comparative work. The more fundamental (inherent) optical properties of lake water are the absorption and scattering coefficients (see Kirk, 1983). Values of absorption coefficient in Lough Neagh vary from 0.5 to 2.2 m^{-1}, and this variation is mainly due to changes in phytoplankton biomass. The scattering coefficients are typically 3 to 4 times these values but can occasionally go up to 5 to 6 times during crops of the blue-green alga *Oscillatoria agardhii* (Jewson, unpublished data).

One consequence of scattering is that the angular distribution of the light field changes with depth (i.e. it becomes more diffuse). The effect on an individual photon is to increase its pathlength, which also increases its chances of being absorbed. As phytoplanktonic cells can potentially absorb a photon of light from any direction, it is important when comparing lake and laboratory experiments to use sensors which have collecting properties independent of the direction of the light; otherwise large errors may be introduced (Smith and Wilson, 1972; Jewson, 1984).

Some light does escape back towards the surface after multiple scattering. This is called upwelling light and over an annual period varies from 1.8% to 5.9% of the downwelling irradiance in Lough Neagh. Although this may appear low, it is important for many biological processes. For example, as upwelling light has an impoverished spectrum compared to downwelling light (Jewson, 1984), so the zooplankton prey of a fish will contrast differently with the background depending on whether it is seen from the side, above or below.

Although changes in spectral composition and angular distribution are important, it is the rapid decrease in intensity (of approximately one order of magnitude per metre) that characterises the light environment of Lough Neagh. This results in much of the water column being aphotic, that is without light, even though it is a relatively shallow lake (mean depth 8.9 m). The term aphotic is used rather than dark as the various plant and animal processes that involve light have different lower limits (see Fig. 4.4). For instance, only the top quarter of the water column normally has sufficient illumination to permit photosynthesis whereas predation and migration can continue to much greater depths because they use much lower energy levels. So, the following discussions of the optical properties of Lough Neagh are set against a background where light is the one persistent gradient in an otherwise homogeneously mixed water column.

4.3. Euphotic zone

For many primary production studies it is often useful to know the potential depth limit of photosynthesis. The most frequently used method of estimation

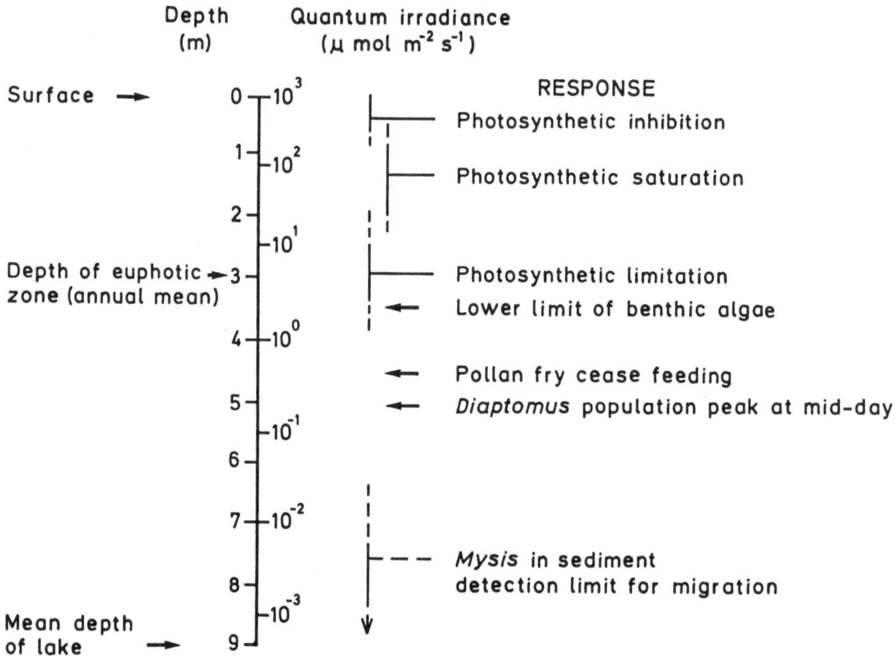

Figure 4.4. The decrease in quantum irradiance with depth and the approximate ranges of light related processes in Lough Neagh for noon on a bright spring day (e.g. see Fig. 9.3, Chapter 9).

is based on the assumption that photosynthesis ceases when the surface light (400–700 nm) is reduced to 1%. In practice, such a general rule can be misleading, particularly in stratified systems where cells can adapt to low light levels. However, in well-mixed lakes like Lough Neagh it can give reliable estimates (see Fig. 11.9, Chapter 11) especially as the contribution to total photosynthesis below the 1% light level is small (Jewson, 1976). Talling (1971) has suggested a general solution to determine the depth of the euphotic zone, $z_{eu} = 3.7/\epsilon_{min}$ where ϵ_{min} is the vertical attenuation coefficient of the most penetrating spectral region. In Lough Neagh the best agreement was found with $z_{eu} = 3.9/\epsilon_{min}$ (Jewson, 1977). The depth of the euphotic zone estimated in this way for the last decade is shown in Fig. 4.5. This is based on monthly averages and also includes estimates of z_{eu} based on Secchi disc measurements (see Gibson, 1975). The range z_{eu} over this period was from 1.3 to 4.6 m. The clearest periods followed the collapse of algal populations in 1974 and 1975 and are referred to in Chapter 10, section 10.5.2.1 in relation to the appearance of *Anabaena flos-aquae*. The most turbid conditions occurred at peak population densities of *O. redekei* during early summer of 1979 and 1980. One consequence of these shallow euphotic

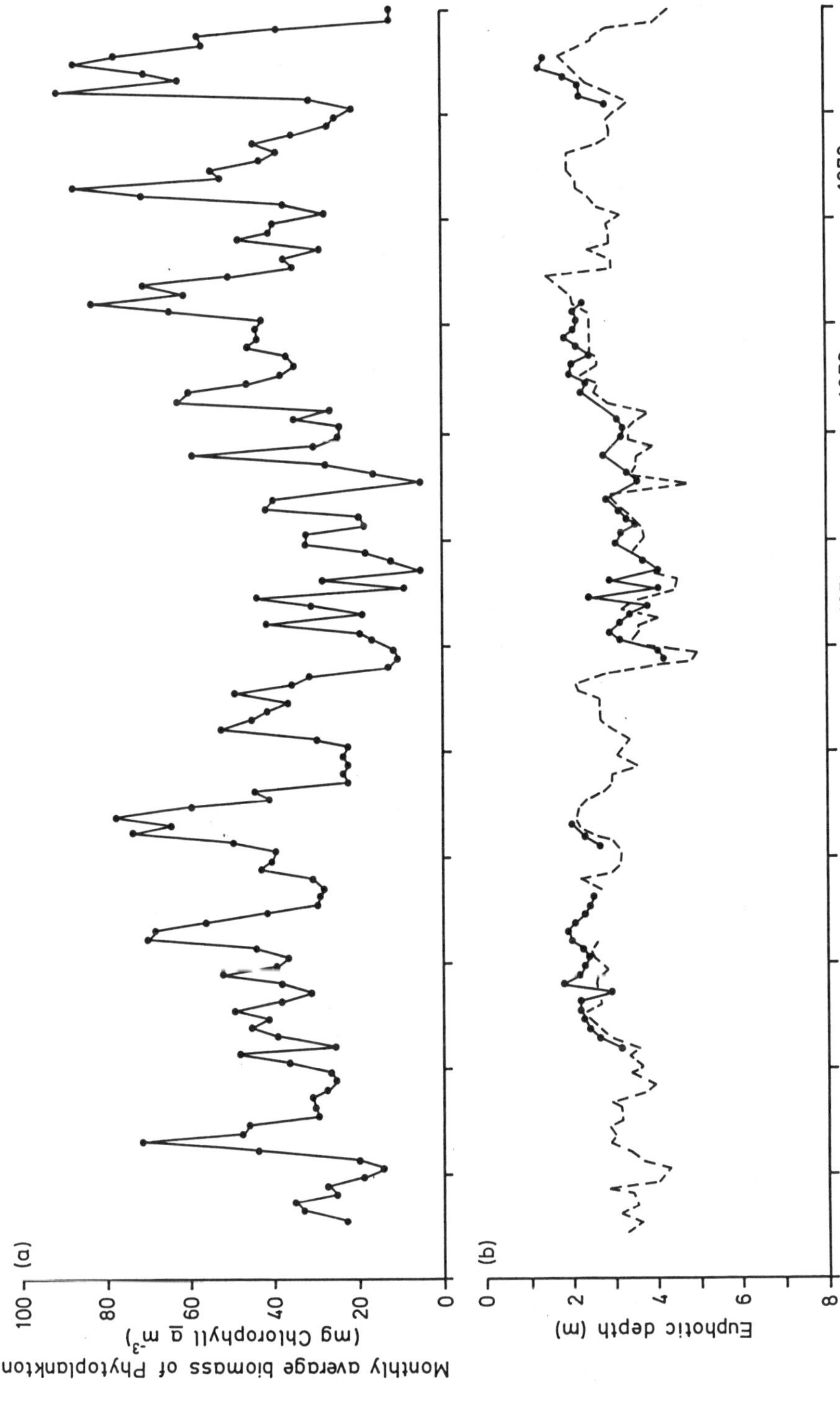

Figure 4.5. Variation of monthly average (a) phytoplankton biomass and (b) euphotic zone (1% light level 1968 to 1979 in the open waters of Lough Neagh. Based on Secchi disc (- - -) (Gibson, unpublished data) and red (RG1) filter (——) (Jewson, unpublished data).

Table 4.1. Examples of the variation in euphotic zone content and light interception during high and low phytoplankton crops in the open lake and Kinnego Bay, Lough Neagh

	Open Lake		Kinnego Bay
	6 Sept. 1974	5 June 1979	10 Oct. 1970
Euphotic zone (m)	4.6	1.32	1.53
Mean depth of water column (m)	8.9	8.9	2.1
Phytoplankton concentration (mg chl *a* m^{-3})	4	136	300
Phytoplankton content of euphotic zone (mg chl *a* m^{-2})	18.6	180	459
Phytoplankton content of water column (mg chl *a* m^{-2})	35.6	1210	798
% of light intercepted	5	72	82
% of population in the euphotic zone	52	15	73
% of population in the aphotic zone	48	85	27

zones is that at any one instant the majority (50–85%) of the population of the open lake is in the dark (i.e. aphotic region)(see Table 4.1 and also Chapter 11).

Higher attenuation is found in inshore areas. For example, Kinnego Bay had values of z_{eu} between 1.0 and 2.1 m in 1970–72 (Jewson, 1976, 1977) and between 0.9 and 2.4 m for 1973–74 (Jones, 1977). The higher ratio of illuminated to non-illuminated portions of the water column (see Table 4.1) means that higher growth rates are obtainable (see Chapter 11).

4.4. The optical environment of animals

Although most of the work on optical properties of Lough Neagh has concerned phytoplankton photosynthesis, there have been several studies on the responses of animals. These are summarised in Fig. 4.4. This is only an approximate guide as the degree of light penetration, season and light quality may all modify the depths at which these processes occur. For instance, Dabrowski (1982a,b) has used laboratory experiments to investigate the effect of light intensity and spectral composition on the feeding activity of juvenile stages of roach, pike, pollan and perch fed on zooplankton. These have been related to Lough Neagh conditions (Dabrowski and Jewson, 1984), where increased sensitivity as the juvenile stages develop may be offset by reduced light penetration due to phytoplankton growth. Time of day and weather conditions also have a marked effect on the volume of water that can be searched during a light period. Over the growing season, the maximum depth of feeding (at midday in sunny weather) for the juvenile stages of pollan varied from 2.6 to 4.1 m, for pike from 2.7 to 4.6 m and for perch from 1.1 to 1.9 m (Dabrowski and Jewson, 1984). Dabrowski (1981) has also discussed the possible effects of high irradiances on the survival of the early stages of pollan.

Other work has concerned the vertical migrations by crustaceans which are often linked to some form of light cue. The diurnal movement of *Diaptomus gracilis* is discussed in the zooplankton chapter (Chapter 14) but it is of interest to note here that the bulk of the population descends to a depth during the day where the irradiance is less than the threshold of feeding of its fish predators (see Fig. 4.4). Extremely low levels of irradiance must also be involved if *Mysis* uses some aspect of the underwater light environment to start its migration to the surface. It lives in and on the surface of the mud below 5 m depth during the day rising to the surface at dusk (see Chapter 14).

4.5. Factors controlling light penetration

There are four main categories of factors attenuating light phytoplankton, Gelbstoff (dissolved and colloidal organic coloured compounds), non-living particulate matter (detritus and inorganic) and water itself. The first three categories show considerable seasonal variation and are discussed below.

4.5.1. *Phytoplankton*

The proportion of the available light intercepted by the phytoplankton varies from 5% at the lowest biomass recorded (4 mg chl *a* m^{-3} on 6 November 1974) to 72% at the highest (136 mg chl *a* m^{-3} on 5 June 1979) (see Table 4.1). The depth of the euphotic zone on these occasions was 4.6 and 1.3 m respectively. So, although there is only a small change (<3 m) in the depth of light penetration there is big change in the proportion of light intercepted by phytoplankton.

A reduction in z_{eu} caused by increases in algal growth is referred to as self-shading and is usually quantified by measuring the incremental increase in the minimum vertical attenuation coefficient per unit increases in biomass (i.e., ϵ_s ln units (mg chl *a*)$^{-1}$ m^{-2}). The value of this increment varies between species. The most frequently recorded limits are between 0.008 and 0.02 ln units (mg chl *a*)$^{-1}$ m^{-2}. The importance of this parameter lies in its use for predicting the maximum possible chlorophyll content of the euphotic zone. In Lough Neagh the lower values of ϵ_s have been found for a small diatom *Stephanodiscus* species (cell volume 750 μ^3). As a result, this species provides some of the highest contents of the euphotic zone recorded (see Jewson, 1977; Jones, 1977; and also Table 4.1) and therefore approaches the theoretical maximum possible (see Talling *et al.*, 1973). This species frequently produces large crops in Kinnego Bay, a shallow almost enclosed area (Jewson, 1976; Jones, 1977), but has only rarely occurred in significant numbers in the open lake during the last decade (see Chapter 10). ϵ_s values

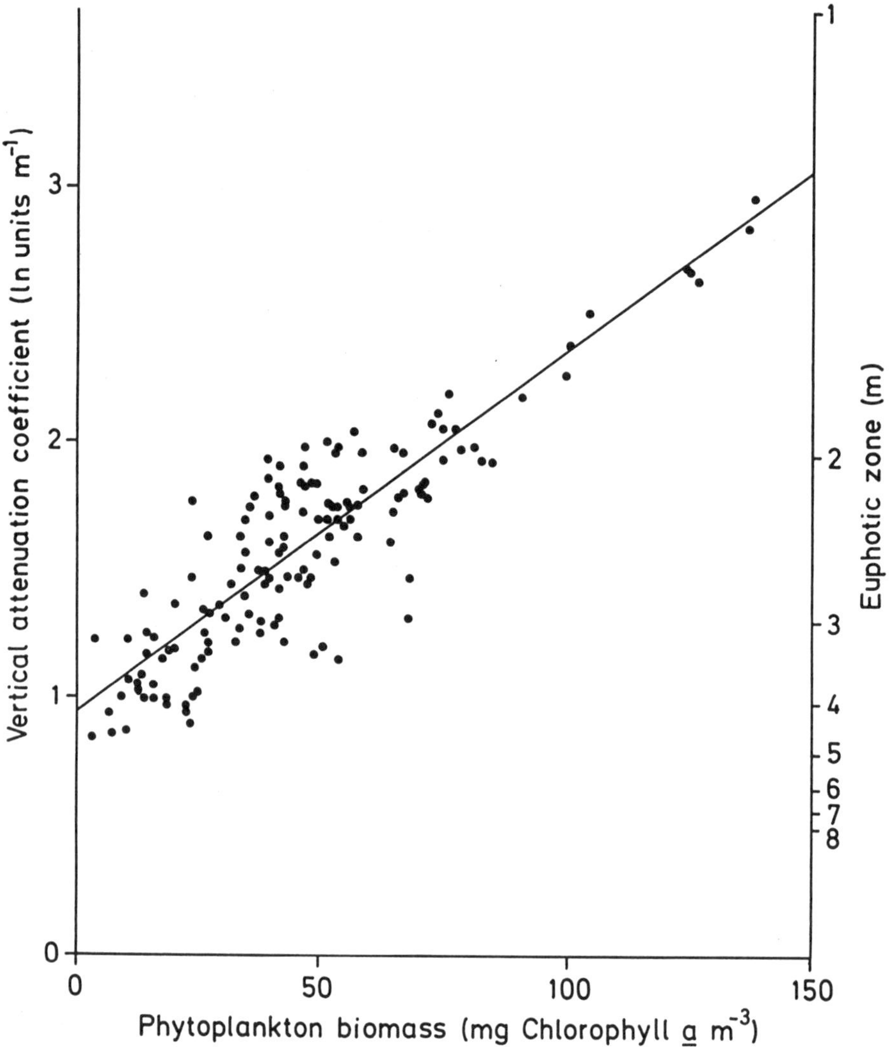

Figure 4.6. Values of the minimum vertical attenuation coefficient (ϵ_{min}) with the corresponding depth of the euphotic zone, measured in Lough Neagh (open water 1970–1979) at various concentrations of phytoplankton biomass. Also shown is the regression of ϵ_{min} on chlorophyll *a* which gives a mean value for ϵ_s of 0.014 ($r = 0.78$) ($y = 0.014x + 0.94$).

for other open lake species show a mean of 0.014 ln units (mg chl a)$^{-1}$ m^{-2} (Fig. 4.6) but this is biased by inclusion of other factors contributing to light attenuation discussed below. The results for individual growth periods suggest 0.011 to 0.014 for *O. redekei* and 0.010 to 0.012 for *O. agardhii* and 0.012–0.014 for *M. italica* (Jewson, 1977, and unpublished data). These are based on increments in the red region. For comparison with other lakes, where measurements are based on either green or blue light, these results should be increased by 1.1 times and 2.5 times respectively.

As phytoplankton grow and become the predominant cause of light attenuation (for details of the periodicity of phytoplankton, see Chapter 10) they also change the spectral composition of the light (Fig. 4.7b,c,d). During the spring growth of diatoms the most penetrating wavelength shifts from 575 to 600 nm (Fig. 4.7b). In dense growths of *S. hantzschii* in Kinnego Bay, ϵ_{min} may even move beyond 700 nm (Fig. 4.7b). The red peak of chlorophyll *a* becomes particularly prominent.

The most penetrating region also shifts from 575 to 700 nm as the blue-green algal crops of *O. redekei* grow during late spring and early summer (Fig. 4.7c). This is because *O. redekei* has two accessory photosynthetic pigments (phycoerythrin and phycocyanin) which absorb in the middle of the spectrum. In some years (1971, 1972), this species shows a change in its pigmentation (Fig. 4.7c), with relatively more phycoerythrin appearing which gives the water a red-brown colour to the human eye. The cause of this change is not yet fully understood but it is more likely to be linked to nutrient stress rather than light adaptation, as on both occasions the change followed the cessation of growth. In late summer (August/September) maximum crops of *O. agardhii* usually occur. The water at this time appears a dirty blue-green hue. This species shows several differences from *O. redekei*; it does not contain phycoerythrin, the red chlorophyll *a* peaks are not as pronounced and there is never the same increase in penetration at 700 nm.

4.5.2. *'Gelbstoff'*

This includes the dissolved and colloidal coloured organic compounds, which are sometimes also called 'Gilvin', or yellow substances (see Kalle, 1966; and Kirk, 1976). Concentrations in Lough Neagh are relatively high (partly reflecting the peaty waters of the catchment) compared to other lakes (see Jewson and Taylor, 1978) and so this contributes, with algal pigments, to the rapid increase in attenuation shown at the shorter wavelengths in all the examples in Fig. 4.7. During the year the concentration varies and summer values may be 40% lower than mid-winter ones (Jewson, 1977).

4.5.3. *Non-living particulate matter*

The major significance of this group is in the changes in light attenuation which result from the resuspension of inorganic and detrital particles during windy weather. As such weather frequently precludes sampling their influence is probably underestimated. Their effect on spectral changes is shown in Fig. 4.7a (Jewson, 1977) where three scans are illustrated for sampling dates before (21 January), during (4 February) and after (18 February) a stormy period at a time when diatom growth was just beginning. The attenu-

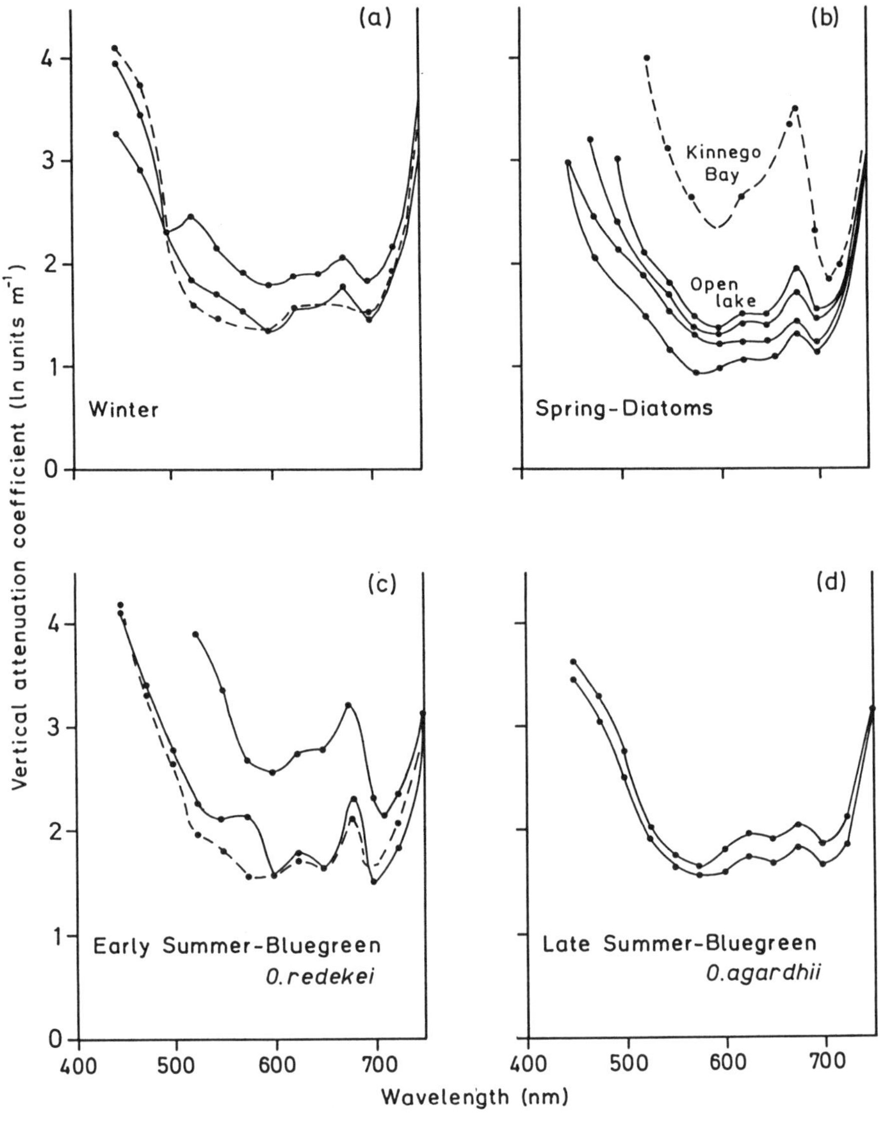

Figure 4.7. Spectral variation (425–750 nm) of the vertical attenuation coefficient measured with an ISCO pectroradiometer in the open lake and Kinnego Bay of Lough Neagh. The most commonly occurring light quality types for 1970–1980 are shown (a) winter storms (b) diatom growth in early spring (c) early summer *Oscillatoria redekei* (d) late summer *Oscillatoria agardhii* (Redrawn from Jewson, 1977; Jewson and Taylor, 1978).

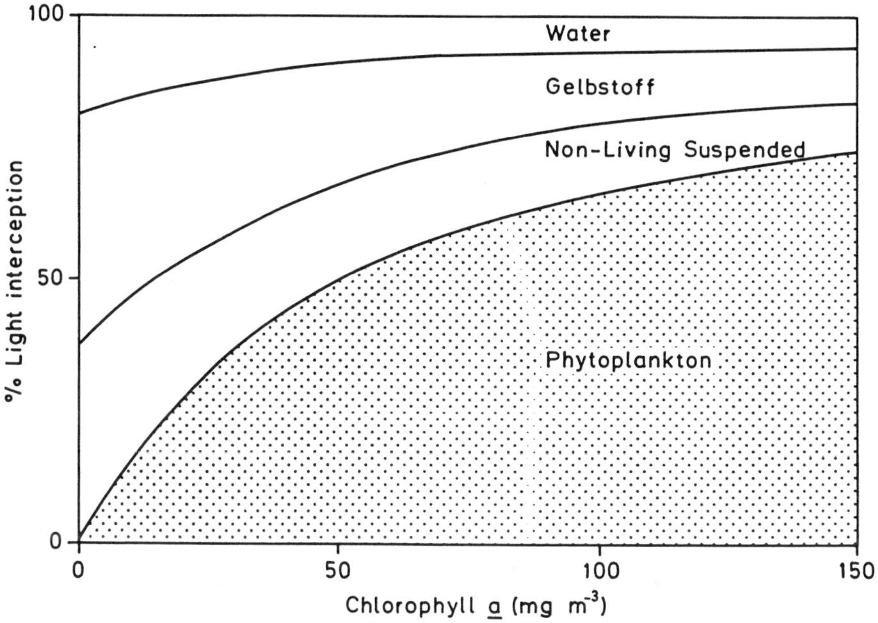

Figure 4.8. Variation in the percentage light interception of the four main components attenuating light in Lough Neagh, at different concentrations of phytoplankton biomass. This assumes light to moderate winds. In stronger winds there is an increase in the contribution of non-living particles which are resuspended from the bottom (see Table 4.2).

ation across the visible spectrum was raised by about 0.5 ln units during the stormy period. This decrease in penetration was sufficient to preclude growth at that time (Jewson, 1976) but with the return of calmer conditions the suspended material dropped out and a net increase in biomass then occurred. This shows up on the scan for 18 February as an increase in attenuation at the wavelength absorbed by the red peak of chlorophyll *a*. Resuspension is one of the major factors resulting in the poor correlation between ϵ and chlorophyll *a* at phytoplankton concentrations below 70 mg chl *a* m^{-3} (see Fig. 4.6).

As the phytoplankton crops increase and calmer summer conditions prevail, the influence of non-living particulate matter on light penetration decreases.

4.6. Interaction of factors controlling light penetration

The relative proportion of light intercepted by each of the major factors contributing to light attenuation in Lough Neagh is shown in Fig. 4.8. This is a generalised picture to illustrate how the phytoplankton come to dominate

Table 4.2. Relative light interception (400–700 nm) at low and high phytoplankton concentration in Lough Neagh during 'normal' (i.e. mixing sufficient to keep the water column isothermal) and very rough conditions.

	At low crops (5 μg chl.a l^{-1})		At high crops (150 μg chl a l^{-1})	
	'normal'	rough	'normal'	rough
Phytoplankton	12	7	87	74
Non-living suspended	27	57	6	17
Gelbstoff	41	24	6	5
Water	19	11	4	3

the light climate at high concentrations. However, this applies only under 'normal' weather conditions, i.e., the moderate winds which prevail for long periods over Lough Neagh. In very rough weather inorganic and detrital materials are resuspended (see Fig. 4.7a) and their relative contribution can increase from about a quarter to over half of the light intercepted (Table 9.2). Such a situation frequently occurs in spring during the period of diatom growth (see the spread in ϵ_{min} at concentrations of chl *a* below 70 mg m^{-3}, Fig. 4.6).

The relative proportion of light absorbed by the algae will also vary according to the species present and the packing effect of its light gathering pigments (see ϵ_s in section 4.5.1 above) but at the highest crops it is in the region of 85%. At such times the proportion of light attenuated by the water is relatively low (less than 5%) and mainly in the red (see Smith and Tyler, 1977) but at lower crops when light penetration is at a maximum the amount intercepted by the water may be nearly 20% (see Table 4.2). In the final group of factors, Gelbstoff, the contribution to light absorption is also low at high algal densities but when phytoplankton populations are very low the water has a characteristic brown hue. Under these conditions, with a z_{eu} of over 4 metres, the Gelbstoff absorbs over 50% of the light in the visible wavelengths. This relative increase in absorption by Gelbstoff at low crop densities acts as a 'buffering effect' and is one of the major reasons why the annual range in the depth of the euphotic zone is so limited.

4.7. References

Dabrowski, K.R. (1981) The spawning and early life history of the pollan (*Coregonus pollan* Thompson) in Lough Neagh, Northern Ireland. Int. Rev. ges Hydrobiol. 66: 299–326.

Dabrowski, K.R. (1982a) The influence of light intensity on feeding of fish larvae and fry. I. *Coregonus pollan* (Thompson) and *Esox lucius*. Zool. Jb. Physiol. 86: 341–351.

Dabrowski, K.R. (1982b) The influence of light intensity on feeding of fish larvae and fry. II *Rutilus rutilus* (L.) and *Perca fluviatilis* (L.) Zool. Jb. Physiol. 86: 353–360.

Dabrowski, K.R. and Jewson, D.H. (1984) The influence of light environment on depth of visual feeding by larvae and fry of *Coregonus pollan* (Thompson) in Lough Neagh. J. Fish Biol. 25: 173–181.

Gibson, C.E. (1975) A field and laboratory study of oxygen uptake by planktonic blue-green algae. J. Ecol. 63: 867–879.

Jewson, D.H. (1976) The interaction of components controlling net phytoplankton photosynthesis in a well-mixed lake (Lough Neagh, Northern Ireland). Freshwat. Biol. 6: 551–576.

Jewson, D.H. (1977) Light penetration in relation to phytoplanktonic content of the euphotic zone of Lough Neagh, N. Ireland. Oikos 27: 74–83.

Jewson, D.H. and Taylor, J.A. (1978) The influence of turbidity on net phytoplankton photosynthesis in some Irish lakes. Freshwater Biology 8: 573–584.

Jewson, D.H. (1984) Comparison of scalar and cosine instruments for measuring photosynthetically available radiation in L. Neagh, N. Ireland. Ver. int. Verein Limnol. 22: 77–81.

Jones, R.I. (1977) Factors controlling phytoplankton production and succession in a highly eutrophic lake (Kinnego Bay, Lough Neagh). II. Phytoplankton production and its chief determinants. J. Ecol. 65: 561–577.

Kalle, K. (1966) The problems of Gelbstoff in the sea. Oceanography and Marine Biology Annual Review 4: 91–104.

Kirk, J.T.O. (1976) Yellow substance (Gelbstoff) and its contribution to the attenuation of photosynthetically active radiation in some inland and coastal South-eastern Australian waters. Aust. J. Mar. Freshwat. Res. 28: 9–21.

Kirk, J.T.O. (1983) Light and photosynthesis in aquatic ecosystems. Cambridge Univ. Press, London.

Larson, D.W. (1983) The Crater Lake study: Detection of possible optical deterioration of a rare, unusually deep caldera lake in Oregon, USA. Verh. Internat. Verein Limnol. 22: 513–517.

Marshall, J.J. (1934) Lough Neagh in legend and in history. Tyrone Printing, Dungannon, 105 pp.

Smith, R.C. and Tyler, J.E. (1977) Transmission of solar radiation into natural waters. Photochem. Photobiol. 117–155.

Smith, R.C., Tyler, J.E. and Goldman, C.R. (1973) Optical properties and colour of Lake Tahoe and Crater Lake. Limnol. Oceanogr. 18: 189–199.

Smith, R.C. and Wilson, W.H (1972) Photon Scalar Irradiance. Appl. Opt. 11: 934–938.

Talling, J.F. (1971) The underwater light climate as a controlling factor in the production ecology of freshwater phytoplankton. Mitt. int. Verein theor. agnew. Limnol. 19: 214–243.

Talling, J.F., Wood, R.B., Prosser, M.V. and Baxter, R.M. (1973) The upper limit of photosynthetic productivity by phytoplankton: Evidence from Ethiopian soda lakes. Freshwat. Biol. 3: 53–76.

5. The major ion chemistry of Lough Neagh

B. RIPPEY

5.1. Introduction

This chapter aims to present and interpret in a general way the major ion concentrations of the six main inflowing rivers of Lough Neagh, of the lake itself and its outflow (Fig. 5.1). The major ion chemistry of the lake and its rivers is perhaps the most obvious link between the terrestrial ecosystems of the catchment and the Lough Neagh ecosystem itself (Likens and Bormann, 1974), so it is appropriate that the major ion chemistry should be presented at this point in the book. The underlying geology of the rivers is shown in Fig. 2.1 of Chapter 2.

Four topics are covered, the first two qualitative and the second two quantitative. Firstly, the variation of major ion composition among the six main inflowing rivers is described. Although the large catchment sizes (301 to 1480 km^2) will tend to lessen the differences in composition between the rivers (Livingstone, 1963), some geological influences on the calcium, magnesium, potassium and silica concentrations are evident. Secondly, the rivers can be classified into broad natural water types which indicate the type of weathering reactions occurring in the catchments. Thirdly, calculation of the major ion fluxes from the six catchments give an estimate of the amount of chemical weathering in the catchments. Finally, the conservative or non-conservative behaviour of the chemical species is examined.

5.2. The data set

The calcium, magnesium, sodium, potassium, sulphate, chloride and conductivity results available were collected as part of a water quality surveillance programme and not to address the four topics outlined in Section 5.1. This means that there are some limitations to the analysis and interpretation of the results, particularly the accuracy of some of the quantitative conclusions that can be drawn.

Table 5.1 gives the number of results for each species and the period of

R.B. Wood and R.V. Smith (eds.), Lough Neagh, 75–90.

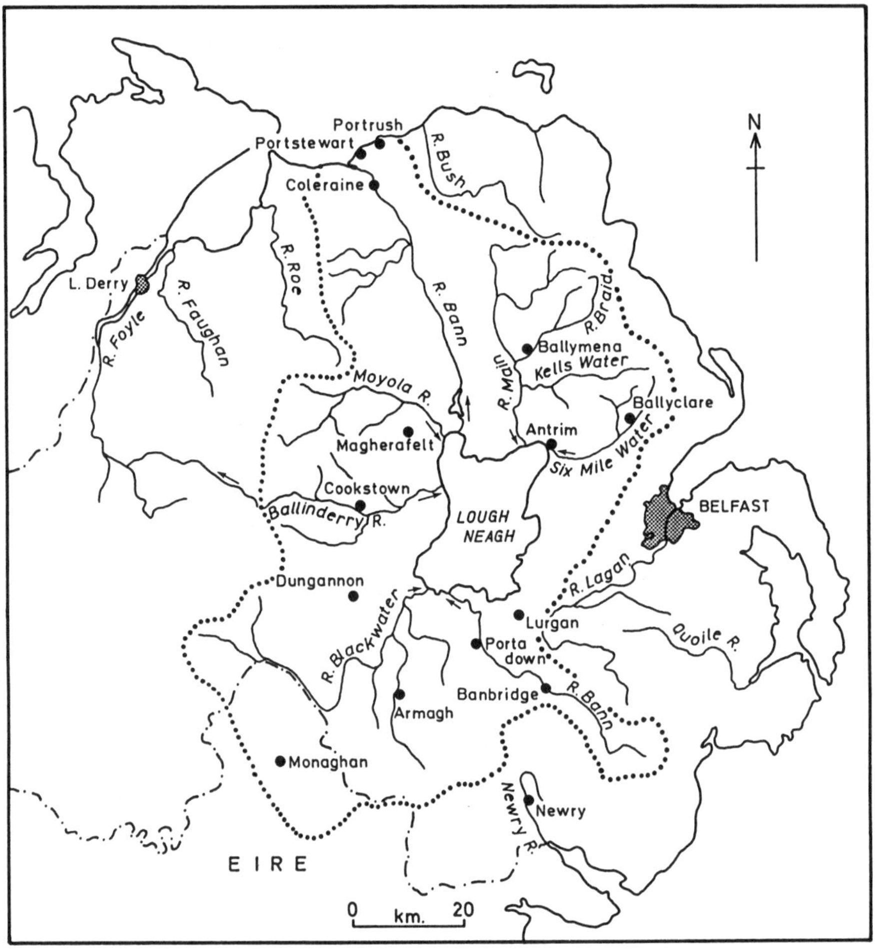

Figure 5.1. The Lough Neagh catchment showing the six major river catchments, the Six Mile Water, Main, Moyola, Ballinderry, Blackwater and Upper Bann. The lake outflow, the Lower Bann, is also shown.

time to which they refer. Results are available for the six main inflowing rivers, the outflow and the lake itself. The cation results were collected quarterly, except in 1975 when roughly fortnightly determinations were made. There were more frequent determinations of sulphate, chloride and conductivity, both in the rivers and in the lake.

Most of the river cation results are for 1975, which was the driest year in the period 1971 to 1979 (Table 5.2). This feature means that calculation of a time average concentration is the only reasonable means of data reduction for the six rivers (Table 5.3). The calculation of a volume-weighted mean,

Table 5.1. The data set for the major ions in the six major inflowing rivers of Lough Neagh, the Six Mile, Main, Moyola, Ballinderry, Blackwater and Upper Bann, the outflow, the Lower Bann, and the lake itself. The number and time period of the results available are given

Species	Information available
Calcium	23 to 28 for 1975, 4 for 1977, 3 for 1978, 4 for 1979 and 3 for 1980 for each of the six inflowing rivers. 4 for 1973, 3 for 1974, 31 for 1975, 35 for 1977, 3 for 1978 and 5 for 1979 for Lough Neagh.
Magnesium	Identical to calcium for the inflowing rivers. 2 for 1973, 3 for 1974, 31 for 1975, 3 for 1977, 3 for 1978 and 5 for 1979 for Lough Neagh.
Sodium	Identical to calcium for the inflowing rivers. 4 for 1973, 3 for 1974, 30 for 1975, 3 for 1977, 3 for 1978 and 5 for 1979 for Lough Neagh.
Potassium	Identical to calcium for the inflowing rivers. 2 for 1973, 3 for 1974, 3 for 1977, 3 for 1978 and 5 for 1979 for Lough Neagh.
Sulphate	175 to 177 for 1968 to 1980 for each of the six inflowing rivers and the outflow. 48 for 1976, 52 for 1977, 45 for 1978 and 7 for 1979 for Lough Neagh.
Chloride	259 to 261 for 1968 to 1980 for each of the six inflowing rivers and the outflow. 22 for 1973, 23 for 1974, 49 for 1975, 49 for 1976, 52 for 1977, 45 for 1978 and 8 for 1979 for Lough Neagh.
Conductivity	347 or 348 for 1968 to 1980 for each of the six inflowing rivers and the outflow. 23 for 1974, 48 for 1975, 48 for 1976, 48 for 1977, 45 for 1978 and 46 for 1979 for Lough Neagh.

Table 5.2. The variation of mean annual flow in the six major inflowing rivers of Lough Neagh and the total inflow from 1971 to 1979, and their means and standard deviations

River	Variation of annual flow, $m^3 \times 10^6$	Mean and standard deviation of annual flow, $m^3 \times 10^6$
Six Mile Water	122–216	175(27)
Main	343–607	490(86)
Upper Bann	224–442	332(82)
Blackwater	475–945	743(158)
Ballinderry	176–362	270(62)
Moyola	165–280	228(40)
Total inflow	1685–3178	2478(413)

for example, is impossible as the stream flow range for which results are available is too restricted by the predominance of the 1975 results in the data set. A time average is nevertheless useful when comparing compositions (Hem, 1970). Inspection of the cation and anion results for Lough Neagh reveals no marked seasonal trends. This is illustrated by the low standard deviations of the means for Lough Neagh, except for sodium in 1973 and 1974, given in Table 5.4 (see also Fig. 5.6).

Table 5.3. The range and arithmetic mean ($\bar{x}$) concentration, standard deviation (s) and number of results (n) for the major ions and conductivity in the six major inflowing rivers of Lough Neagh, the Six Mile Water, Main, Moyola, Ballinderry, Blackwater and Upper Bann, and the outflow, the Lower Bann

		Calcium $mg\,l^{-1}$	Magnesium $mg\,l^{-1}$	Sodium $mg\,l^{-1}$	Potassium $mg\,l^{-1}$	Sulphate $mg\,SO_4\,l^{-1}$	Chloride $mg\,Cl\,l^{-1}$	Conductivity $\mu S\,cm^{-1}$
Six Mile Water	Range	13.7–41.8	5.9–19.4	11.3–46.3	1.0–4.2	16.7–63.0	12.4–77.3	164–745
	$\bar{x}$	31.2	13.2	28.0	2.1	30.5	32.7	390
	s	7.1	3.1	8.7	0.9			
	n	39	38	39	39	177	260	348
Main	Range	9.4–37.5	2.5–18.7	9.5–32.2	0.8–2.2	14.9–41.4	0.8–40.4	125–470
	$\bar{x}$	24.6	11.8	20.6	1.5	23.8	23.8	297
	s	5.6	3.4	5.7	0.3			
	n	39	39	34	32	176	261	347
Moyola	Range	12.9–36.8	3.5–12.6	7.4–23.2	0.7–2.5	12.8–35.1	6.0–32.6	109–385
	$\bar{x}$	25.7	8.2	15.0	1.7	21.1	18.7	249
	s	6.2	2.6	4.5	0.5			
	n	39	39	34	33	177	260	347
Ballinderry	Range	18.1–46.0	3.1–9.6	7.5–25.0	1.2–3.9	13.8–53.4	10.0–36.7	165–560
	$\bar{x}$	33.2	6.9	16.4	2.3	25.7	20.2	311
	s	5.9	1.7	4.4	0.7			
	n	34	34	34	34	177	260	348
Blackwater	Range	15.2–61.2	1.8–11.9	9.2–40.0	1.3–6.2	14.7–48.6	8.9–57.0	200–639
	$\bar{x}$	46.9	6.4	20.7	2.9	30.7	24.9	399
	s	10.3	2.2	8.2	0.7			
	n	34	34	34	32	175	259	347
Upper Bann	Range	15.4–29.5	4.6–9.2	11.1–40.6	1.4–5.6	13.6–41.7	15.0–58.8	140–485
	$\bar{x}$	22.4	6.5	24.7	3.4	28.7	28.9	292
	s	3.6	1.2	7.9	1.0			
	n	33	33	33	33	176	260	347
Lower Bann	Range					15.2–35.5	10.6–32.5	175–331
	$\bar{x}$					25.3	22.7	291
	s							
	n					177	261	348

Table 5.4. The mean annual concentration ($\bar{x}$) in mg l^{-1}, standard deviation (s), and number of observations (n) for the major ions and conductivity ($\mu s\ cm^{-1}$) in Lough Neagh from 1973 to 1979

		1973	1974	1975	1976	1977	1978	1979
Calcium	$\bar{x}$	31.3	34.2	28.3		31.1	31.7	33.0
	s	1.0	2.2	4.7		1.4	1.6	3.4
	n	4	3	31		35	3	5
Magnesium	$\bar{x}$	7.5	9.2	6.9		8.4	8.2	7.9
	s		2.5	1.2		0.4	0.6	0.1
	n	2	3	31		3	3	5
Sodium	$\bar{x}$	16.0	17.3	18.9		12.6	12.5	13.0
	s	7.8	7.6	3.0		1.3	0.2	0.2
	n	4	3	30		3	3	5
Potassium	$\bar{x}$	2.9	2.4			2.3	2.4	2.9
	s		0.5			0.4	0.1	0.2
	n	2	3			3	3	5
Sulphate	$\bar{x}$				23.6	26.1	26.6	25.7
	s				1.4	1.1	1.1	0.9
	n				48	52	45	7
Chloride	$\bar{x}$	20.8	21.4	23.8	24.1	22.3	23.1	22.7
	s	1.2	1.6	1.3	0.9	0.8	0.8	1.1
	n	22	23	49	49	52	45	8
Conductivity	$\bar{x}$		285	295	312	290	293	299
	s		5	14	5	5	9	6
	n		23	48	48	48	45	46

The only method of calculating ion fluxes in the rivers with an arithmetic mean concentration is to use the total flow ($m^3\ yr^{-1}$) $\times$ mean concentration ($g\ m^{-3}$) = flux ($g\ yr^{-1}$) relationship. As most of the cation concentrations decrease with flow (see Section 5.4.1 and Fig. 5.5), this will underestimate the true flux values, but they should be accurate within approximately 15% (Walling, 1978; Stevens and Smith, 1978).

5.3. Qualitative considerations

The mean composition of the six main inflowing rivers to Lough Neagh given in Table 5.3 is examined in Section 5.3.1 for the influence of catchment geology. The type of weathering reactions taking place in the catchments is examined in Section 5.3.2.

5.3.1. *Major ion composition*

Table 5.3 shows that the calcium concentration in the Blackwater is somewhat higher than in the other rivers, as is magnesium in the Six Mile Water and Main. Potassium is somewhat higher in the Upper Bann and Blackwater as is sodium in the Six Mile Water. With the anions, only chloride is somewhat higher in the Six Mile Water. This simple inspection can be supplemented by comparing the river percentage molar compositions with some trends in cation composition found in lakes in Northern Ireland (Rippey and Gibson, 1984). There, the Na/(Na + Ca) weight concentration ratio was used roughly to define the relative amounts of precipitation and rock influence on the composition of freshwater. Fig. 5.2 shows that calcium is relatively high in the Blackwater, magnesium high in the Six Mile Water and Main and, although the Upper Bann's potassium is not high compared to the lakes, it is higher than in the other rivers.

The geological map of the Lough Neagh catchment and its lithology are given in Chapter 2 (Fig. 2.1 and Section 2.2). The large area of limestone in the Blackwater catchment is the calcium-rich parent material producing the high calcium concentrations in the river. Also, the small area of limestone in the Ballinderry catchment appears, similarly, to produce the somewhat higher calcium concentrations in that river (Table 5.3, Fig. 5.2). The basalts of north-east Ireland have magnesium concentrations above the lithosphere average and produce associated natural waters also with high magnesium concentrations (Rippey and Gibson, 1984). These basalts almost totally dominate the Six Mile Water and Main catchments and so produce the high magnesium concentrations noted. The olivine mineral in this rock, responsible for the high magnesium concentrations, together with the plagioclase feldspar and pyroxene, weather relatively rapidly (Curtis, 1976) and are also responsible for the high silica concentrations in the Six Mile Water and Main (see Chapter 9, Section 9.5.1). Shale and granite have higher potassium (and lower calcium) concentrations than the lithosphere average (Bowen, 1979), so the large area of shale and some granite in the Upper Bann catchment, and the area of shale in the Blackwater, appear to produce the higher potassium concentrations in the respective rivers.

The reasons for the higher sodium and chloride concentrations in the Six Mile Water cannot be determined with certainty on the evidence currently available, but some speculations can be made and wastewater, road salt and geological factors may all be involved. Wastewater can in certain circumstances be an important component of the major ion flux in a river (Walling and Webb, 1978), and the Six Mile Water catchment had the highest urban population density and highest wastewater component of the six catchments until 1977, when the Upper Bann exceeded it (Foy *et al.*, 1982). As the sodium and chloride concentrations in the effluent of a small sewage disposal works in Northern Ireland have been shown to be about twice those of the Six Mile Water (Smith, 1976), domestic wastewater may contribute to the higher river concentrations. It can only be partly responsible, for the wastewater flux from the Six Mile catchment, estimated using per capita values in Smith (1976) and the average urban population density during 1974 to 1979 in Foy *et al.* (1982), is around 2 kg ha^{-1} yr^{-1} for both sodium and chloride; and the sodium and chloride fluxes in the Six Mile Water are around

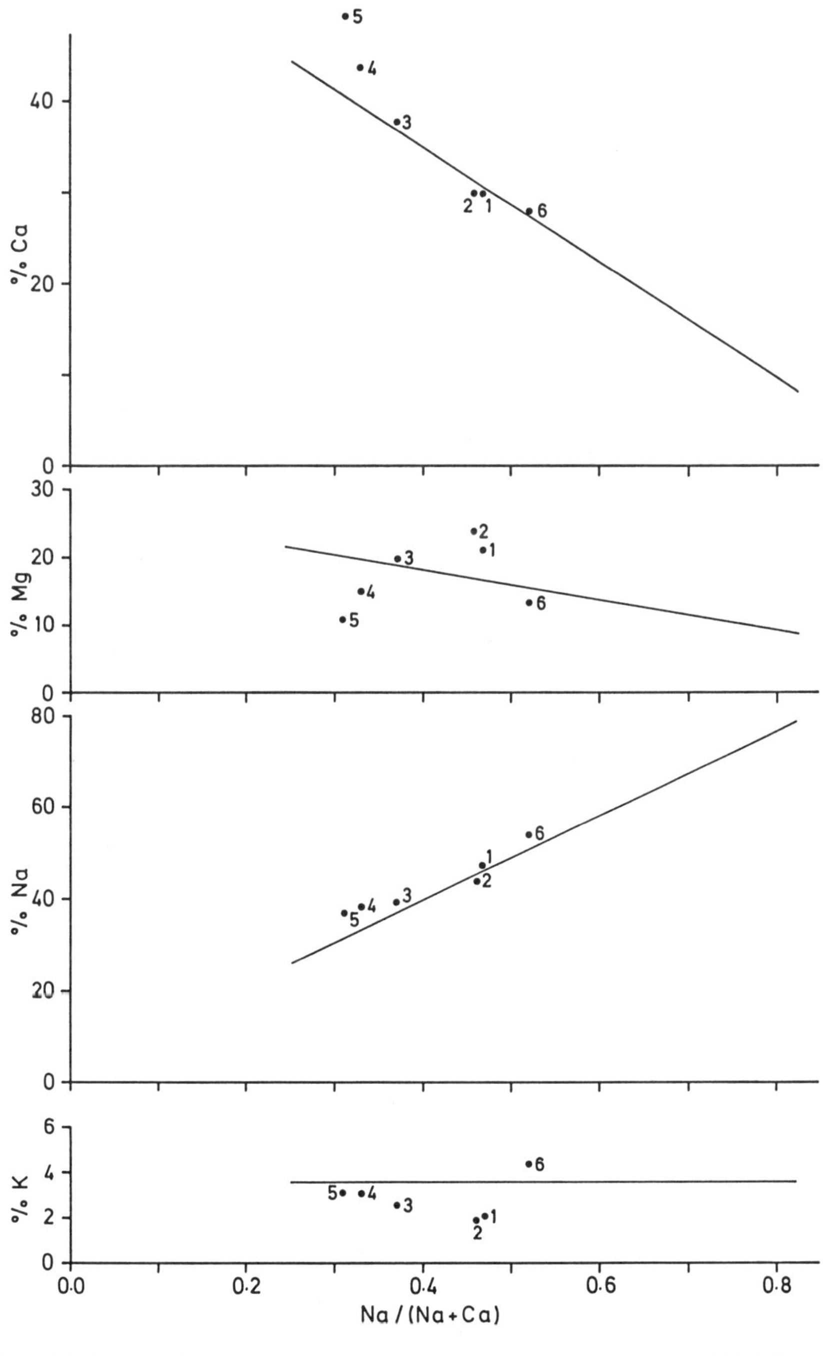

Figure 5.2. The relationship between percentage molar cation composition and Na/(Na + Ca) weight concentration ratio. The lines indicate the relationships found for lakes in Northern Ireland (Rippey and Gibson, 1984) and the Six Mile Water (1), Main (2), Moyola (3), Ballinderry (4), Blackwater (5) and Upper Bann (6) are shown.

Table 5.5. Major ion fluxes in the six major inflowing rivers of Lough Neagh expressed per unit area of catchment (kg ha^{-1} yr^{-1})

	Calcium	Magnesium	Sodium	Potassium	Sulphate	Chloride	Bicarbonate
Six Mile Water	181	77	163	12	177	190	848
Main	170	82	142	10	164	164	836
Moyola	187	60	109	12	154	136	648
Ballinderry	208	43	103	14	161	127	772
Blackwater	235	32	104	15	154	125	768
Upper Bann	112	33	124	17	144	145	562

20 kg ha^{-1} yr^{-1} higher than the Main, a river with a similar geological setting (see Table 5.5). Industrial sources in the towns of Ballyclare and Antrim may also make a contribution. Road salt influences the sodium and chloride concentrations in Lough Neagh (see Section 4.3), so road salt from the fairly dense road network in the Antrim Development Area and the motorway may also be important.

A geological factor may also be present, for within the basalt succession in north-east Ireland there are a number of flows with intermediate compositions, i.e., they have, among other constituents, higher silica, sodium and potassium concentrations than the general basalts which have a basic composition (Lyle, 1980). The available evidence shows that these rocks are found mainly in and around the Six Mile Water catchment and that they have around twice the sodium concentration of the general basalts (see Lyle (1980) and references cited therein). This local variation in the composition may then also contribute to the higher sodium concentrations in the Six Mile Water.

5.3.2. *Classification of the major ion composition*

A graphical procedure based on a trilinear diagram of percentage equivalent composition of cations and anions is used in the geochemical interpretation of water chemistry (Piper, 1944). Dominant water types are differentiated; calcium type, magnesium type, sodium/potassium type or no dominant type among the cations; bicarbonate, chloride, sulphate or no dominant type among the anions. This classification reflects the composition and weathering reactions of the parent materials. Fig. 5.3 shows the trilinear diagram for the six rivers. This was constructed using Table 5.3 and calculating the bicarbonate concentration using the equivalence of anions and cations in freshwater (Livingstone, 1963; Reynolds and Johnston, 1972; Vangenechten, 1980).

Fig. 5.3 shows the rivers are all of the bicarbonate type, which indicates that carbonation is the major weathering reaction over the whole Lough Neagh catchment (Holland, 1978). For the cations, the Six Mile Water, Main, Ballinderry and Moyola are of no dominant type, with the Blackwater a calcium type and the Upper Bann tending to be a sodium/potassium type.

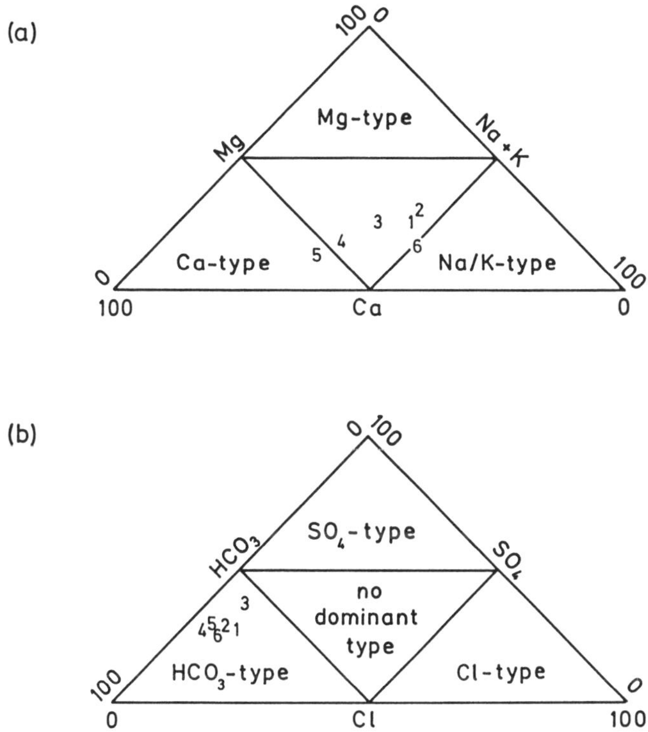

Figure 5.3. The trilinear diagram for (a) cations and (b) anions in the Six Mile Water (1), Main (2), Moyola (3), Ballinderry (4), Blackwater (5) and Upper Bann (6).

A variety of minerals must be present in the catchment parent materials so that sizeable amounts of all the cations contribute to the composition. Additionally, the large amount of limestone in the Blackwater catchment would account for its composition just being a calcium type, and the higher potassium and lower calcium concentrations in the shales and granite of the Upper Bann lead to it being on the fringe of a sodium/potassium type. Carbonation of a variety of silicate minerals and carbonates appears to be the main weathering reaction in the Lough Neagh catchment.

5.4. Quantitative considerations

5.4.1. *General characteristics*

The general limitations of the use of the data set to estimate stream flux were given in 5.2. Here the general quantitative characteristics of the cation results and stream flow are presented.

Over all the six rivers, using all the data points, the cation concentrations are generally uncorrelated except for calcium and magnesium (correlation coefficients beteen 0.56 and 0.91, n = 33 to 39). Fig. 5.4 shows the behaviour of calcium and magnesium. Cation concentration is inversely related to mean daily flow for calcium (coefficients of determination 0.18 to 0.48 for the six rivers, n = 33 to 39), magnesium (0.09 to 0.37) and sodium (0.12 to 0.30), while potassium remains generally constant. Fig. 5.5 illustrates the contrasting behaviour of calcium and potassium. The use of log–log plots does not change the explained variance. Although the use of mean daily flow rather than instantaneous flow at the time of sampling obscures the concentration-flow relationship somewhat (Walling, 1978), the insenitivity of potassium concentration to flow, at least over the flow range experienced, is clear (except in the Six Mile Water at very low flows) and has been found elsewhere (Lewin *et al.*, 1974; Foster, 1980). This is generally thought to be due to the strong interaction of the potassium ion with clay minerals, especially as an interlayer cation in illite (Loughnan, 1969; Grim, 1968).

5.4.2. *Solute loads*

The major ion fluxes in the six main rivers, calculated from mean annual flow (Table 5.2) and mean concentration (Table 5.3), are given in Table 5.5 and they are probably overestimates (see Section 5.2). The bicarbonate was calculated using the equivalence of cations and anions in freshwater (Livingstone, 1963; Reynolds and Johnston, 1972; Vangenechten, 1980). The differences between the catchments are much less with fluxes than with concentration. These fluxes must be corrected for the atmospheric contribution to get true chemical denudation rates. The true atmospheric contribution, wet and dry deposition, for a site in the River Main catchment (Jordan, 1983) and wet deposition for various sites in Northern Ireland (Smith, 1976) are available to estimate the atmospheric component of the river fluxes. Table 5.6 gives these values along with others found elsewhere. The atmospheric component of the total flux is small for calcium and magnesium (perhaps 10%), but significant for sodium and contributing perhaps over half of the potassium. For the anions, precipitation is a significant component of the sulphate and chloride fluxes, which is true generally (Holland, 1978).

Tabie 5.5 does indicate some features of the chemical weathering in the river catchments. The magnesium output from the Six Mile Water and Main catchments is higher than the others, and this is due to the high magnesium concentration of the basalt parent meterial in the catchments (see Section 5.3.1). The higher potassium and lower calcium outputs from the Upper Bann are similarly due to the constitution of the parent materials, in this case the higher potassium and lower calcium concentrations of the shale and granite present (Section 5.3.1). There is a fair uniformity of anion fluxes from the catchments.

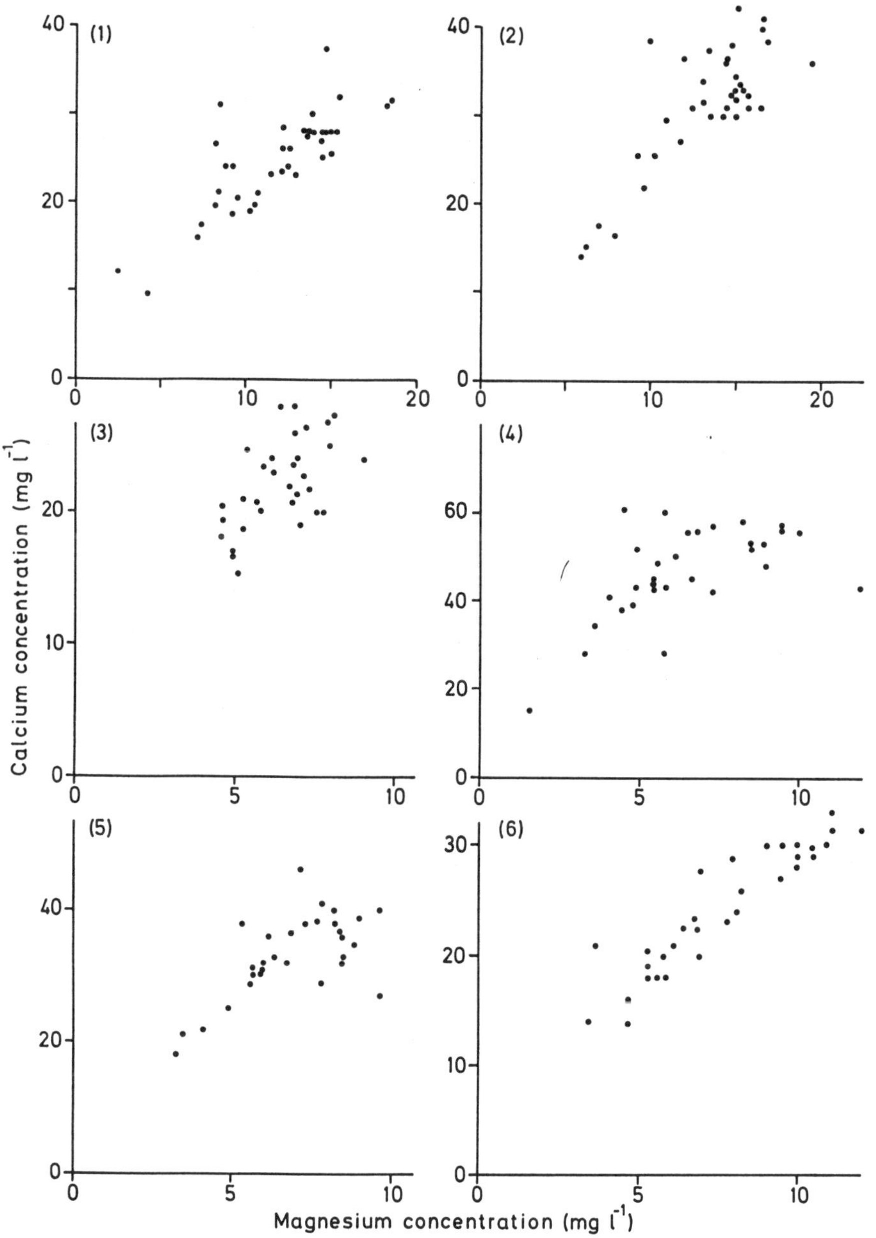

Figure 5.4. The relationship between calcium and magnesium concentrations ($mg\,l^{-1}$) in the Main (1), Six Mile Water (2), Upper Bann (3), Blackwater (4), Ballinderry (5) and Moyola (6).

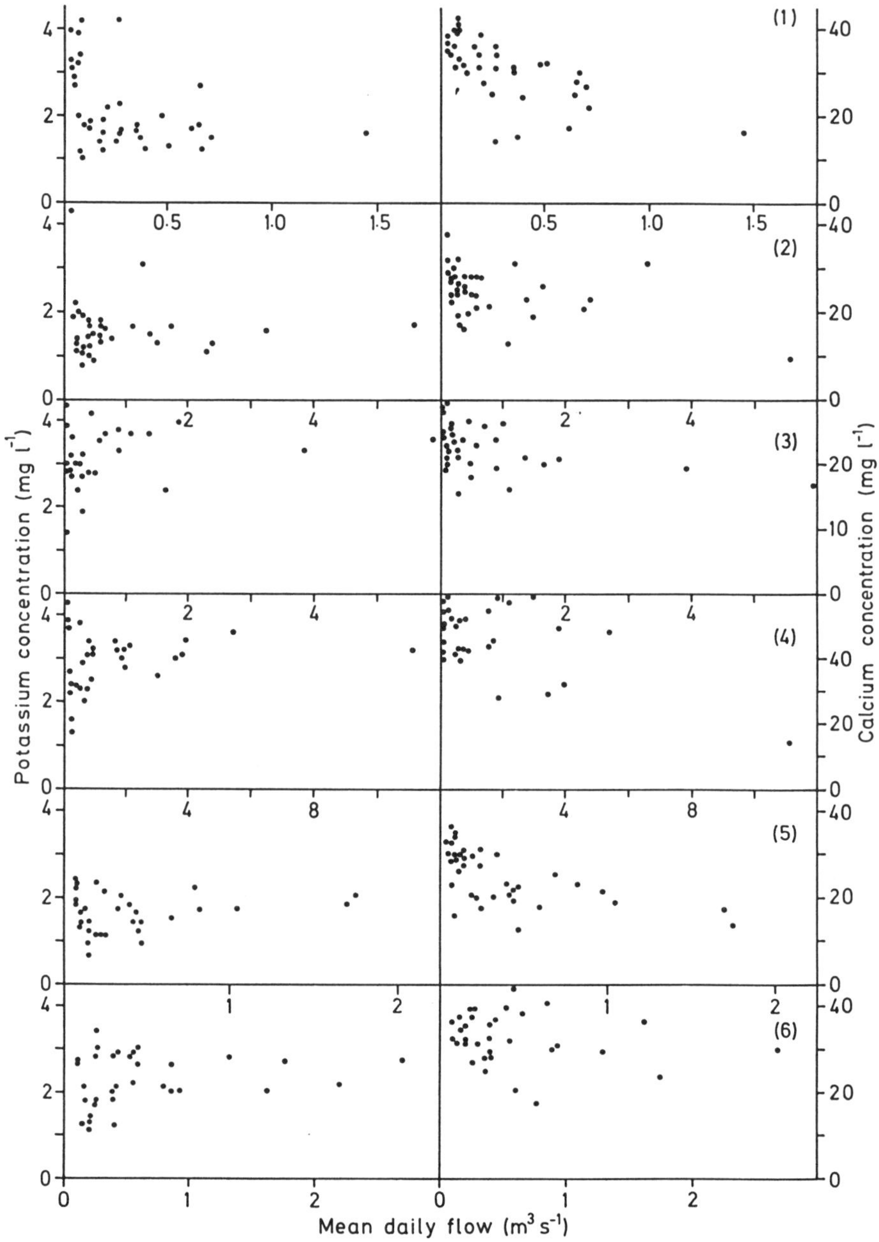

Figure 5.5. The relationship between potassium and calcium concentration (mg l^{-1}) and mean daily flow ($m^3 s^{-1}$) in the Six Mile Water (1), Main (2), Upper Bann (3), Blackwater (4), Moyola (5) and Ballinderry (6).

Table 5.6. Estimates of the atmospheric deposition rates (kg ha^{-1} yr^{-1}) of major ions in Northern Ireland and elsewhere. The rainfall is the annual amount in mm

	A	B	C	D	E	F
Calcium	17	8	9	25	8 to 14	2 to 15
Magnesium	6	3		4	3 to 5	2 to 11
Sodium	37	27	25	27	14 to 29	4 to 63
Potassium	8	1	3	2	5 to 7	1 to 13
Sulphate		50				1 to 21
Chloride		47		46	29 to 53	
Rainfall	1181	1070	2070	1642	551 to 1059	570 to 1640

A, Various sites in Northern Ireland (Smith, 1976).
B, Fourmile-Burn, Northern Ireland (Jordan, 1983).
C, Moor House, England (Crisp, 1966).
D, Maesnant, Wales (Cryer, 1976).
E, Yendacott, England (Foster, 1980).
F, Temperate parts of the world (Likens *et al.*, 1977).

Table 5.7. Comparison of estimated and range of measured mean annual major ion concentrations (mg l^{-1}) and conductivity (μS cm^{-1}) in Lough Neagh during 1973 to 1979

Species	Estimated	Measured
Calcium	33.5	28.3 to 34.2
Magnesium	8.4	6.9 to 9.2
Sodium	21.2	12.5 to 18.9
Potassium	2.4	2.3 to 2.9
Sulphate	28.0	23.6 to 26.6
Chloride	25.5	20.8 to 24.1
Conductivity	334	285 to 312

5.4.3. *Conservative behaviour*

The annual flux of ions in the inflowing rivers and from the atmosphere allows the mean lake concentration to be calculated. As the composition of the lake outflow, the Lower Bann, is similar to the lake itself, (compare Table 5.3 and Table 5.4), the lake concentration can be estimated by a volume weighted mean of the six inflowing rivers (Table 5.2 and Table 5.3) and total atmospheric deposition on the lake surface (Jordan (1983) and mean annual rainfall of 1028 mm yr^{-1}). This procedure neglects evaporation from the lake surface. The results are given together with the range of mean lake concentration from Table 5.4 in Table 5.7.

Table 5.7 shows that the estimated mean concentrations and conductivity either fall within the lake concentrations range or fall just above it. The stream fluxes are probably overestimates (see Section 5.2), so the tendency to overestimate lake concentration is expected, while neglecting evaporation produces a slight underestimate of the true value. The overestimation ranges from 6% for chloride to 22% for magnesium in 1975, the year for which

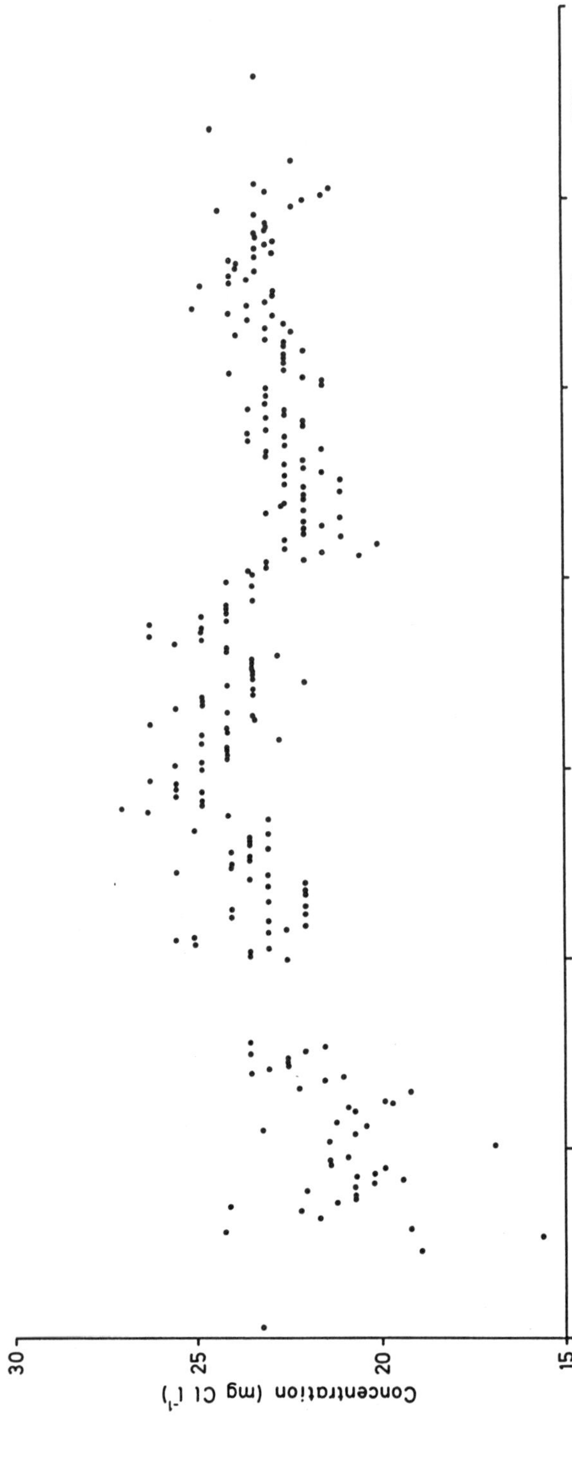

Figure 5.6. The variation of chloride concentration (mg l^{-1}) with time in Lough Neagh (0–10 m mid-lake composite sample) from 1973 to 1979.

most cation results are available. The major ions in Lough Neagh show no large departures from conservative behavior, so it appears there are no strong physical, chemical or biological processes acting on these ions in the lake. This result also indicates that no large ground water influence is present.

The variation of annual cation and anion concentration and conductivity in Lough Neagh from 1973 to 1979 is random, except for sodium, chloride and conductivity (Table 5.4). These increase from 1973 to 1976 and then drop; this is shown for chloride in Fig. 5.6. The drop in concentration takes place mainly during the winter of 1976/7. Gibson and Stewart (1972) found that conductivity and chloride concentration increased in Lough Neagh from 1962 to 1969 and that this was mainly due to the application of road salt in the catchment area. The relationship between the increase of conductivity and chloride was 6.3 $\mu S\ cm^{-1}$ per mg $Cl\ l^{-1}$.

Whether changes in the amount of road salt usage could be responsible for the changes in sodium, chloride and conductivity can be examined by comparing the maximum chloride and conductivity values in late 1976 (11 October) with the minimum values in early 1977 (7 March) (Fig. 5.6). Chloride fell from 26.2 to 20.0 mg $Cl\ l^{-1}$ and conductivity from 315 to 283 $\mu S\ cm^{-1}$. This gives a ratio of 5.3 $\mu S\ cm^{-1}$ per mg $Cl\ l^{-1}$, which is reasonably consistent with that found by Gibson and Stewart (1972). The change in chloride implies a sodium concentration decrease of 4 mg $Na\ l^{-1}$, assuming both derive from road salt, and this is not inconsistent with the drop in mean annual concentration from 1975 to 1977 of 6 mg $Na\ l^{-1}$ (Table 5.4). Changes in road salt usage could then be responsible for the annual variations in lake sodium, chloride and conductivity values.

Acknowledgments

I am very grateful to the staff of the Freshwater Biological Investigation Unit who collected and made available these major ion results, in particular Jim Stevens and Crawford Jordan. Crawford Jordan also suggested the use ot the trilinear diagram.

5.5. References

Bowen, H.J.M. (1979) Environmental chemistry of the elements. London, Academic Press, 333 pp.

Crisp, D.T. (1966) Input and output of minerals for an area of Pennine moorland: the importance of precipitation, drainage, peat erosion and animals. J. App. Ecol. 3: 327–348.

Cryer, R. (1976) The significance and variation of atmospheric nutrient inputs in a small catchment system. J. Hydrol. 29: 121–137.

Curtis, C.D. (1976) Stabillty of minerals in surface weathering reactions: a general thermochemical approach. Earth Surface Processes 1: 63–70.

Foy, R.H., Smith, R.V. and Stevens, J. (1982) Identification of factors affecting nitrogen and phosphorus loading to Lough Neagh. J. Environ. Manag. 15: 109–129.
Foster, I.D.L. (1980) Chemical yields in runoff, and denudation in a small arable catchment, east Devon, England. J. Hydrol. 47: 349–368.
Grim, R.E. (1968) Clay Mineralogy. McGraw-Hill, New York, 596 pp.
Gibson, C.E. and Stewart, D.A. (1972) Changes in the water chemistry of Lough Neagh over a 10-year period. Limnol. Oceanogr. 17: 633–635.
Hem, J.D. (1970) Study and interpretation of chemical characteristics of natural waters. U.S. Geol. Survey Water Supply Paper No. 1473.
Holland, H.D. (1978) The chemistry of the atmosphere and oceans. New York: Wiley-Interscience, 351 pp.
Jordan, C. (1983) The precipitation chemistry of a rural site in Co. Antrim. Record of Agricultural Research (Department of Agriculture for Northern Ireland), 31, pp. 89–98.
Lewin, J., Cryer, R. and Harrison, D.I. (1974) Sources for sediments and solutes in mid-Wales. In: K.J. Gregory and D.E. Walling (eds.), Fluvial processes in instrumented watersheds. London, Institute of British Geographers, pp. 73–85.
Likens, G.E. Bormann, F.H. (1974) Linkages between terrestrial and aquatic ecosystems. Bioscience 24: 447–456.
Likens, G.E., Bormann, F.H., Pierce, R.S., Eaton, J.S. and Johnston, N.M. (1977) Biogeochemistry of a forested ecosystem. New York, Springer-Verlag, 146 pp.
Livingstone, D. (1963) Chemical composition of rivers and lakes. U.S. Geol. Survey Prof. Paper 440 G.
Loughnan, F.C. (1969) Chemical weathering of the silicate minerals. New York, Elsevier; 154 pp.
Lyle, P. (1980) A petrological and geochemical study of the Tertiary basaltic rocks of northeast Ireland. J. Earth Sci. R. Dubl. Soc. 2: 137–152.
Piper, A.M. (1944) A graphic procedure in the geochemical interpretation of water analyses. Trans. Am. Geophys. Union 25: 914–923.
Reynolds, Jr., R.C. and Johnson, N.M. (1972) Chemical weathering in the temperate glacial environment of the Northern Cascade Mountains. Geochim. Cosmochim. Acta 36: 537–554.
Rippey, B. and Gibson, C.E. (1984) The variation of calcium, magnesium, sodium and potassium concentration, pH and conductivity in lakes in Northern Ireland. Arch. Hydrobiol. 101: 345–360.
Smith, R.V. (1976) Nutrient budget of the River Main, Co. Antrim. Tech. Bull. Minist. Agric. Fish Fd. 32: 315–339.
Stephens, R.J. and Smith, R.V. (1978) A comparison of discrete and intensive sampling for measuring the loads of nitrogen and phosphorus in the River Main, County Antrim. Water Res. 12: 823–830.
Vangenechten, J.H.D. (1980) Interrelations between pH and other physico-chemical factors in surface waters of the Campine of Antwerp (Belgium), with special reference to acid moorland pools. Arch. Hydrobiol. 90: 265–283.
Walling, D.E (1978) Reliability considerations in the evaluation and analysis of river loads. Z. Geomorph. N. F. 29: 29–42.
Walling, D.E. and Webb, B.W. (1978) Mapping solute loadings in an area of Devon, England. Earth Surface Processes 3: 85–99.

6. Human settlement and economy of the Lough Neagh basin

P.C. WOODMAN and N.C. MITCHEL

6.1. Introduction

Lough Neagh's drainage basin, nearly 4500 km^2 in extent, takes in over 40 per cent of Northern Ireland's total area; within it live approximately 300,000 people. Given the size and complexity of this basin and the nature of this volume, a detailed analysis of its settlement history would be inappropriate: rather is attention concentrated on the basin's core, Lough Neagh and its immediate shorelands together with Lough Beg and the Lower River Bann. This is an area of outstanding archaeological importance which contains, at Mount Sandel on the Lower Bann, the oldest settlement site so far recorded in Ireland. Excavations at Mount Sandel and at other prehistoric sites indicate an extensive reliance of early man on fish, in particular salmon and eels, which are still the basis of fishing activities (see Chapters 1 and 18). The present eel fishery supports a large and remarkably homogeneous community of fisherfolk, 250 families strong, whose presence suggests links with a distant past. Only in relatively recent times have the Lough's fishermen had to share its waters with other users for to an increasing extent the lough is now recognized to be an asset of great value for domestic, industrial and recreational purposes.

Given this background, in particular the importance of fishing from the earliest days of settlement, some 9000 years ago, the main task of this chapter is to explain the human occupance of the area through this time span in terms of changing relationships between the physical environment of lough, river and shorelands and man's use of the opportunities offered by them which will be seen to vary in time according to his available technology. To understand the present, we must lean heavily on the past which, for a clearer understanding of the processes at work, is here divided into prehistoric and historic periods. Throughout, specific sites mentioned in the text can be located in Figs 6.1 and 6.2.

R. B. Wood and R. V. Smith (eds.), Lough Neagh, 91–111.

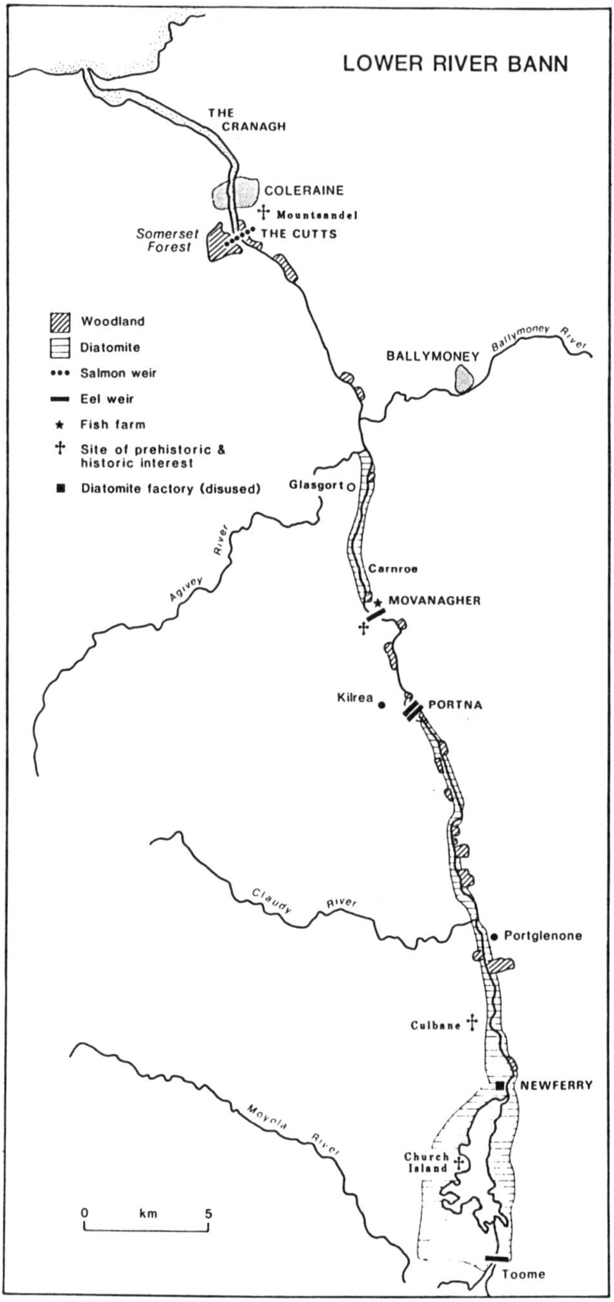

Figure 6.1. The River Lower Bann showing sites of historical settlement and economic importance.

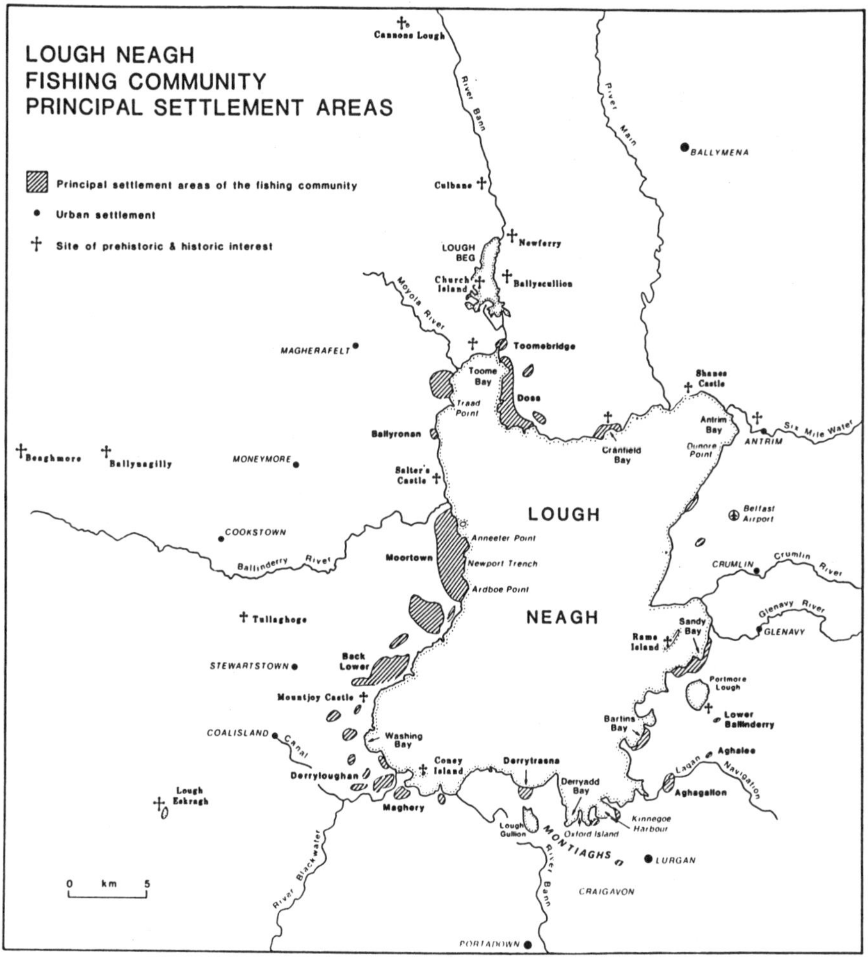

Figure 6.2. Lough Neagh and the principal settlement areas of fishing communities.

6.2. Pre-history

There is no unequivocal evidence that Ireland was occupied by man during the Palaeolithic period. The northern end of the island in particular was heavily glaciated on several occasions with the result that there is little trace of any interglacial deposits and little chance that any traces of human activity would survive. The last period of heavy glaciation – the Midlandian (see Chapter 2) resulted in the presence of a massive Ice Dome centred on Lough Neagh. This spread out over Co. Down as well as forcing its way into the Lower Bann Valley. The northern end of the Bann was in turn blocked by a tongue of ice extending from Scotland (Creighton, 1974). Therefore, perhaps by 20,000 B.P. most of the Lough Neagh Drainage basin was under ice and so the question of dating the arrival of man in this area is one of establishing how soon he arrived after the disappearance of the ice.

We know that by 12,000 B.P. this area had long been free of ice and, while stands of small trees and shrubs gradually made their way north after 12,000 B.P., northern areas around Lough Neagh would have remained at best an open grassland with small stands of Juniper. This grassland was populated by certain mammals, in particular large numbers of Giant Irish Deer which have been found throughout Ireland at this date with some fragments dredged up from the bottom of Lough Neagh (Savage, 1966).

The presence of large mammal populations in Ireland in Late Glacial times suggest that man could also have been here as conditions would have been equally suitable for him. Unfortunately there has been little research into the Late Glacial environment of Lough Neagh. Battarbee's work (Chapter 7) was confined to later periods and the pollen samples taken by Jessen (1949) on the southern end of the Lough extend back only to 9,000 B.P. It has been surmised by Jessen and by Mitchell (1955) that the level of Lough Neagh was lower than 47 ft. OD (Old Irish OD) during the late glacial (see also Chapter 2). The higher levels later would of course remove any chance of finding early shore-line campsites. However, as Britain was still being recolonized during the Late Glacial it is unlikely that there would have been extensive human settlement in this area.

The earliest traces of human settlement so far found in Ireland occur at Mount Sandel (Woodman, 1981, 1985). Occupation of this site had commenced by 9,000 B.P. in the Mesolithic period when Ireland was covered by a so-called 'Boreal' forest, the dominant tree species of which were birch, pine, hazel (and to a lesser extent willow). Trees such as oak and elm were not present in any quantity. From what little evidence is available from pollen cores taken round the Lough it seems to have been a significantly drier period than today so that small areas of open water off the main rivers i.e. those created by ice wastage e.g. at Ballynagilly (Pilcher and Smith, 1979), Cannons Lough (Smith, 1961), or Mount Sandel Kettle hole (Hamilton *et al.*, 1985) were either significantly smaller than might be expected or in the latter case did not exist at that time. Again it can be assumed that the Lough itself was still lower than 47 ft OD and that the River Bann would have been smaller and perhaps in a deeper channel. By this period extensive sand banks had been created in the northern end of the lough and in the bottom of the Bann Valley flood plain.

Excavations from 1973–78 at the Mesolithic site of Mount Sandel have provided important information on the way of life in the Irish Mesolithic and at the same time illustrate the importance of the Lough Neagh Basin and in particular the Lower Bann. The excavation revealed that a small group of hunter-gatherers lived in the Mount Sandel area. Their dwellings have not survived but careful recording of the post holes, pits and hearths

Plate 6.1. Mount Sandel showing motte and site of excavations, 1974–78.

have provided sufficient information to reconstruct in part their campsite. The major problems in interpretation were that the site was re-occupied on numerous occasions thus creating a very complicated pattern while, due to soil acidity and the lack of a constant anaerobic environment, most of the refuse and artefacts have rotted away so that even the animal bones surived only if they were carbonised.

Each occupation may have begun in mid-summer and lasted through to the end of the winter. The central part of the settlement would have been a hut about 6 metres across (Plate 6.1). This was a dome-shaped structure made from small saplings which may have been covered with skins or turves with, at its centre, a bowl-shaped hearth. Outside the hut stone implements were roughed out, food was stored, animals gutted and hides prepared. Traces of these activities were found round the campsite at Mount Sandel which is one of the few Mesolithic sites in N.W. Europe to provide clear evidence as to how these hunter-gatherers organised their camps.

How these people used the River Bann is even more important. As Ireland was an isolated island from perhaps the very beginning of the Holocene, it has a very restricted animal population – in particular there were relatively few native mammals and freshwater fish. This restricts the range of food which hunter-gatherers could procure. The large mammals usually played an essential role as the major source of food for early Boreal hunters. Two of the largest mammals, elk and wild cattle, were absent from Ireland in contrast with the early Mesolithic of Europe where these species along with red deer played a major role. Red deer for some reason does not seem to have been important in the Mesolithic economy of Ireland and, while boar was hunted extensively, boreal forests were not ideally suited to pigs and so their numbers would be relatively low.

With the restricted potential of the mammals, different economic strategies had to be developed. Carbonized remains of plant foods were found at Mount Sandel consisting mostly of hazel nut shells with a few water-lily seeds. These could be collected in the autumn and late summer respectively from the Mount Sandel area and probably were not only consumed at that time but may have been stored as an extra winter resource.

Fish on the other hand were a major source of food. Work by Battarbee *et al.* (1985) suggests that Mount Sandel was occupied at a time of low sea, river and lough levels. The site is now on the upper limit of the estuary, but then it overlooked a large area for catching fish. The site itself has produced a range of salmonid bones as well as eels, a few sea bass, an occasional flounder and one *Coregonus* bone. While it is impossible to distinguish salmon from trout with total certainty it seems probable that the most common fish found on this site is salmon, which could be caught when migrating from the sea. Both salmon and eels have the advantage of running in large numbers at times of year and could be captured with traps, nets or spears, all of which are known from the European Mesolithic. The main crisis for these hunter-gatherers would come in late winter when fish were less plentiful.

Mount Sandel illustrates the importance of the River Bann as a source of food at the beginning of the Mesolithic and other later Mesolithic sites have also been found along the river (Fig. 6.1). During the following 3,000 years the area changed dramatically. The open Boreal forest was replaced by one in which elm and oak were the dominant trees: after 5000 B.C. alder was present in large numbers because the water table had risen. This was partially a climatic phenomenon with greater rainfall but in Lough Neagh and the Lower Bann the rising sea level must have been a contributory factor.

By 4000 B.C. (Carter, 1982) the sea was 8 metres higher than today so that the estuary would have pushed up the Bann Valley with possible back-up effects helping to raise the water levels. At places along the Bann, diatomite flats began to form under water; winter flooding was probably sufficient. At Toome, Mitchell (1955) has noted open water organic muds by 5,700 ± B.C. while at Newferry diatomite first appears at 5000 B.C. at altitude up to 60′ OD. Therefore in spite of suggestions to the contrary (O'Sullivan *et al.*, 1973) the period from roughly 5,500 to perhaps 2,500 B.C. was one of higher water levels which led to areas around the northern end of Lough Neagh and Lough Beg down the Bann becoming filled with diatomite. Deep river channels were gradually converted into marshy diatomite flats. Similarly the water table change may have helped bog development e.g. at Ballyscullion while many of the low-lying areas on both the northern and southern edges of Lough Neagh were filled with peat. Again many little inter-drumlin lakes were closed over and the process which was to lead to the development of large areas of raised bog began.

In spite of these apparent deteriorations in the environment, Mesolithic occupation of the Lower Bann continued. In particular, areas near fords and modern fishweirs have produced immense quantities of Mesolithic material. The most famous site was at Culbane, where tens of thousands of artefacts were uncovered in commerical diatomite cuttings. The range of artefacts had changed from the relatively light microlithic forms to a large heavy-bladed industry. Unfortunately aside from the excavations in the Newferry area this material had the double disadvantage of being dug up without any contextual information surviving and of being in exceptionally acid deposits where few faunal remains survive – e.g. only a few eel and salmonid bones survive from Newferry.

This area has another characteristic in that flint does not occur naturally in the bottom of the flood plain – particularly in the Newferry and Culbane area, yet it must be one of the richest areas for Mesolithic stone artefacts in the British Isles. As there is no evidence of extensive flintworking, these artefacts must have been brought in as blades which could then be converted into a series of tool types. It must have been the continued runs of fish through the Lower Bann which attracted settlement to this area.

Evidence of Mesolithic occupation of the Lough Neagh shorelands is rather scanty, with the possible exception of the area around Toome which in the Mesolithic would have been in the centre of an extended bay of Lough Neagh which included Lough Beg. We know from Mitchell's excavations and from the collection of presumed 7th millenium B.C. material from the shoreline of the Lough at the Maddens site near Toome (Woodman 1978) that there may have been initial occupation on an island made up of the sand banks which Mitchell presumed to be of late glacial origin.[1] Similarly at Newferry, initial settlement was in the centre of the valley on a sand ridge. As the water level rose in the lough this ridge was gradually turned into a marshy flat which was only visited occasionally and occupation seems to have moved to the higher drier lower edges of the drumlins and eskers. Thus the diatomite flats have produced relatively little archaeological material.

The rarity of sites round the shores of the rest of the lough is difficult to explain for it might be expected that residential fish populations in the lough could be extensively exploited during the winter months. Indeed sites may be a misleading term as Mesolithic artefacts are usually found washed out along the present shoreline and not from specific sites. Varying lough levels

[1]This island, documented by Woodman (1978), Fig. 78, lay at a relatively narrow part of the bay and may have accumulated against a drumlin. The River Bann could have flowed past on both sides into what is now called Lough Beg. Toome Bar, which lay across the entrance to the present river, may have been an extension to this island (before water levels were controlled it was possible to walk across this sand ridge). This must have been a prime settlement location at a time when water levels were lower than today. The bar was removed in the 1930s.

throughout the Mesolithic make the situation complex. Initially the lough level was lower than today; rising water levels followed so that only the material from Mitchell's excavations at Toome can be described as being in place. Thus most of the Mesolithic material from Lough Neagh was collected out of context on the modern shoreline during the summer months when lough levels were lower. While some material has been found on the south-western edge of the lough, notably Coney Island, the extensive development of bog around the mouth of the upper Bann has precluded any serious search for early traces of settlement in that area. A detailed listing of sites in the Lower Bann and Lough Neagh area is given by Woodman (1978).

It might therefore be assumed that the paucity of sites is a product of an unstable environment and a lack of suitable collection sites and collectors.[2] Examination of fishing strategies used in other large areas of open water e.g. N. America suggest that waters such as Lough Neagh would only have been exploited effectively when fish were spawning and had moved inshore in large numbers. In the case of Lough Neagh this might mean only Pollan and Char which come inshore to spawn (Chapter 18), as the other species at present in Lough Neagh have been more recent introductions. Perch and Bream are not native to Ireland. Perch, in spite of traditional stories associated with it in Early Christian times and its appearance in a medieval context at Clough Castle (Waterman, 1954), appears to have been introduced into Lough Neagh in the 18th Century (Dubourdieu, 1812) (see Chapter 18, Part A). If only one or two residential fish species existed then alternative environments with a wider choice of food may have had a greater attraction during the Mesolithic so Lough Neagh itself, although potentially very rich in fish, may not have been extensively exploited.

In the upper reaches of the drainage system, later Mesolithic material has been found at Lough Eskragh near Dungannon but little else has been recovered from the more southerly rivers. There is, however, some indirect evidence that these areas were exploited – though not necessarily in the same fashion as the Lower Bann.

For example, numerous baked mudstone axes have been found at Newferry (Woodman, 1977). These are not made from a local rock but come instead from an as yet unknown source or sources within the Palaeozoic peneplain which lies to the south of Lough Neagh. Relatively few traces of settlement have been found throughout this area so it is possible that this rock was being quarried by people who had their home base further north on the Lower Bann. Similarly festooned and carbonated cherts have been found on sites on the Lower Bann and this raw material had to be obtained from areas such as the Antrim coast.

The hunter-gatherer life style seems to have been replaced by a farming economy during the 4th millenium B.C. The mechanism for change is open

[2] Many areas on the east and west shore are rather stoney and small stone artefacts are rarely washed up in this context. Hence few antiquarians collected from the lough shore.

to debate, i.e., invasion or transferral of ideas, but the new way of life brought with it a distinctive range of equipment which gives its name to another period i.e. the Neolithic. The material equipment included pottery and a new range of stone artefacts. The economic changes were the introduction of two large domesticated animals, cattle and sheep, and a range of domesticated plants, in particular cereals. For the first time large burial monuments were erected in Ireland.

The introduction of farming probably brought about a rather more sendentary lifestyle with wooden farm houses being erected. If villages were built then the forest clearance round them would have had a noticeable impact on the environment. The rather denser deciduous forests of the Later Mesolithic would have still been present and small areas would have had to be cleared to plant crops. In fact the palaeo record for the beginning of the Neolithic which is preserved in bogs and lakes shows a drop in tree pollen, an occurrence of cereal pollen[3] and an increase in erosional input.

It might be expected that traces of the activities of these farming communities would be recorded in Lough Neagh but, while the Lower Bann and to a lesser extent Lough Neagh were of crucial importance to hunter-gatherer communities, with the change from hunting/gathering to farming, the Lough assumed lesser importance during the later part of Prehistory. The Atlantic period saw a rise in Lough level and the development of massive peat bogs, in particular along the southern and northwestern edges of the lough. These lowlying water-logged areas must have been of little attraction to the initial Neolithic colonists. Early Neolithic settlements are found on the lighter glacial gravels which lie round the edge of the Lough Neagh Basin. The best and earliest known Neolithic farming settlement in the north of Ireland is found in the eastern foothills of the Sperrins at Ballynagilly near Cookstown (ApSimon, 1976). Traces of early farming are, of course, not confined to the material culture of that period as cereal pollen has appeared at the same date in several pollen cores from this area, e.g. Ballynagilly (ApSimon, 1976), Beaghmore (Pilcher, 1969), Fallahogey (Smith, 1958) and Cannons Lough (Smith, 1961). In lakes near Dungannon (Edwards, 1980) and at Lough Fea (Oldfield, pers. comm.) chemical changes and changes in magnetic susceptibility were found which are thought to indicate land clearances by early farmers.

While the earlier Pleistocene history of the Lough had an impact in attracting early Neolithic settlement to central Ulster no real trace of early agriculture has been found in deposits from the Lough. Battarbee (Chapter 7) has shown that there is an effective mixing of deposits in the lough so that cores from this large area of water tend to produce an average effect for the lough as a whole. Prehistoric forest clearances were so tiny, perhaps

[3]The gathering in of elm branches as fodder, may help to explain this very noticeable decline in elm pollen at the beginning of the Neolithic. The presence of cereal pollen suggests that this decline cannot simply be a product of disease or climatic change.

only a few acres at a time, that they would not influence cores taken from the lough itself.

The arrival of agriculture then marks a shift away from the lough and the river systems as the major source of food although there are occasional traces of shoreline settlement in later periods, in particular in the Later Neolithic which suggest that the Lough and drainage system may have had temporary importance. Significant scatters of Later Neolithic material have been found at Toome, including at least ten flint javelin heads and two decorated, complete Neolithic bowls. These bowls were found during the lowering of Lough Neagh in the mid-1950s. They were embedded in shoreline peat deposits and may represent occupation on a shoreline which was at least as low as that of today. They are not an isolated occurrence as another such bowl was found at Newferry near the levels of the River Bann; a number of others have been recovered from other lowlying contexts throughout Ireland. The presence of a partially drowned dolmen on the west shore of the lough also suggests that the lough level may have been lower during a part of the Neolithic.

Explanations for the occurrence of these lowlying sites in the Later Neolithic vary from a survival of Mesolithic settlement to a failure of the agricultural practices of the Early Neolithic. One can only note the undoubted increase in shoreline settlement which cannot be explained by simple population pressure pushing Neolithic farmers into poorer soils.

There is reason to believe that before the end of the Neolithic period copper objects were in use in Ireland. Certainly by the time of the introduction of Beaker pottery, perhaps around 2000 B.C., metal was in established use. This period is usually referred to as the Bronze Age and from the 2nd millenium BC to ~500 B.C. a simple tentative technology, using quantities of copper in one-piece moulds developed into a complex one using not only alloys but complex multi-piece moulds, clay moulds and sheet metal. In the Dowris phase of the Irish Later Bronze Age there is a wealth of items such as swords, shields and cauldrons as well as axes, hammers and saws. The richness of Irish Bronze Age society which was based at least in part on the exploitation of the native copper ores, is also shown by the range of gold objects belonging to this period.

With regard to the Lough Neagh basin, the pollen records both from the lough and the various sites referred to earlier show that, although clearances were made, there was extensive regeneration of forest and in fact it is not until the historic period that a sufficiently large area was opened to make an impact on the tree pollen percentage in the Lough Neagh sediments (Battarbee 1978).

During the earlier part of the Bronze Age, traces of settlement anywhere in Ireland are relatively rare and it is assumed that the same type of economy as found in the Neolithic continued. However, after 1000 B.C. there are a number of substantial settlement sites with large circular dwellings up to 10 metres across. Addyman (1965) found evidence of early Bronze Age settlement on Coney Island near the southern shore of Lough Neagh. The resources of the lough, notably its fish, must have played an important role in maintaining this settlement on such a tiny island. Another type of lake shore settlement has been found on Lough Eskragh close to the Blackwater. Here the footings of numerous dwellings were uncovered when the lough level was lowered. This site appears to be mostly Later Bronze Age in date (Williams, 1978). A number of other shoreline sites of this period have been found throughout Ireland. They may imply the necessary use of more marginal environments due to population growth.[4]

Sometime in the middle of the last millenium B.C. iron implements came into general use, at first as a few imports but eventually iron technology spread throughout most of Ireland, introduced most likely not by a wave of new settlers but by small groups of craftsmen. It used to be assumed that because of the very distinctive curvilinear geometric art style (known as La Tène after a site in Switzerland) that Ireland was colonized at this time by a massive invasion of so-called Celts. This idea is no longer favoured partially because it is now clear that Celtic material culture, physical type and language do not necessarily coincide.

While numerous objects with La Tène art or derived art styles are known in Ireland, little is known of the habitations and way of life of the peoples of the Early Iron Age, perhaps because archaeologists find it difficult to identify the types of sites these people would have lived on. They did not have a material culture which conveniently produced large quantities of waste material which would have survived to be identified immediately as belonging to this period. We can only note that hill-forts, originating in the Later Bronze Age, may have continued in use and that while pastoralism still existed agriculture was beginning to play a more important role with the appearances of large numbers of rotary cereal grinding querns.

Unlike many parts of Europe where the Romans provide an historically accurate cut off point for the end of prehistory, Irish Prehistory is assumed to continue through to the introduction of Christianity which may have coincided with the beginnings of a period of social upheaval and the collapse of large tribal federations. The society which emerges in the Early Christian (or Early Medieval) period is more fully documented than that which preceded it. It was probably in this period that substantial forest clearances were first made for the pollen records, from both Lough Neagh and the various bogs in its neighbourhood, show that although cereal pollen was evident the landscape remained mostly covered in forest. During this period Ireland continued to be a tribal society centering on a series of Tuaths which changed in size, importance and even location throughout time. The most visible and

[4]There is no evidence of defended artficial islands or crannogs being built at such an early date. Occasional claims have been made for crannogs at Toome but the presence of tree trunks in the large local areas of woody boreal peat deposits, which were exposed when the lough's level was lowered, may be the source of this mistaken identification.

prolific settlement type of this period is the ringfort or rath, of which there are some 30,000 surviving examples in Ireland. These settlements were usually enclosures defended by a high bank and ditch which served as the farmsteads of wealthy farmers. Large multi-vallate examples possibly belonged to the nobility. The excavation of raths and crannogs has given a much clearer picture of Irish society at this time. The raths show that iron ploughs were being used in Ireland while cattle and pigs played an important role. Material culture was still relatively poor though the crannogs, which may have been higher status sites, have produced a richer range of material. We also know from the range of art objects such as the Tara brooch and the Armagh chalice that the early monasteries could produce rich and complex metalwork, manuscripts and stone work. Such art work presumably was not confined to them, but until the arrival of the Vikings in the 9th century, the monasteries were probably the major centres of population and trade.

This cursory examination of the prehistoric record suggests that the Lower Bann valley and at least part of the Lough Neagh shorelands with its vital fords maintained settlements for much, if not all of, the span of Irish prehistory, i.e. from 7000 B.C. until the 5th century A.D. The extensive dredgings of the lower Bann during the mid-19th century and in the 1930s have produced a great wealth of artefacts including a wide range of Stone Age Material and a selection of Bronze Age and occasional Iron Age implements. However, the presence of these implements in the river deposits raises the following questions: (1) have the river and adjacent bogs simply preserved a larger percentage of lost objects than other environments; (2) is there any evidence of ritual deposition and (3) could the River Bann be part of a major trade route? In other words has the dredging of the River Bann created a false picture of its importance as a focus of settlement?

It is possible that some of the artefacts found their way into the river as part of religious ceremonies. In particular a large number of spearheads have been recovered, a number of which are of rather exotic design (Glover, in prep.). Ehrenberg (1977) has noted that these items are unlikely to be accidental losses as spears which retain their shafts will float if they are accidentally dropped and can easily be recovered. On the other hand objects such as the Late Bronze Age cauldron from Portglenone have survived by being buried in peaty deposits. With regard to question (3) it is unlikely that the Lower Bann was a major trading route. Warner (1974) has noted that most major Irish rivers have at least one obstacle to navigation; the Lower Bann was blocked at several points by fast moving shallow water, for example, at Portna, Movanagher, and in particular at the Cutts. There is also the fact that this river does not link any known major centres of production or consumption in the Bronze Age. On balance the evidence still favours continuity of settlement despite varying fortunes.[5]

[5]There is little doubt that the Lower Bann was more important in the Later Bronze Age than in the Earlier period. Axes, by far the most common implement of the Early Bronze Age,

6.3. The historic period

Evidence of settlement and population is much clearer from early Christian times on when oral traditions, documentary evidence and kinship links can be used. Older links to pre-history may exist although recent research shows that the majority of the present 1500 fisher folk have their origins in two Ulster-wide events; settlement by the Scots and English in the 17th century and the 19th century famine (Donnelly, 1981, 1986).

Although the relations between the pre-Iron Age and the later occupants of the raths are uncertain, there is little doubt that in the Early Christian period the fishing community emerges as very much part of Gaelic-Irish society, the shorelands of both lough and river passing from the control of one tribe to another with some frequency. It is also clear that the introduction of Christianity brought monks to the shores of the lough. The evidence of a self-supporting monastic life at Nendrum on Strangford Lough suggests that Lough Neagh's fish might also have been an attraction for these newcomers. According to tradition, St. Colman, who is credited with founding a monastery at Ardboe in the 6th Century, fished for eels both for food and lamp oil. Ardboe is still a key site for fishing, its present-day fishermen attributing their undoubted skills to this monk. Ardboe's church ruins and magnificent 10th or 11th century high cross are still a place of pilgrimage, as is St. Olcan's holy well at Cranfield. Portmore and Magheranagow on the eastern shore, Maghery on the southern, Ram's Island, with its stump of a round tower and Church Island in Lough Beg are other important religious sites where what Evans calls the 'Elder Faiths' merged with Christianity (Evans, 1981). This religious presence may well have encouraged clustering of fisherfolk or perhaps the monks, like missionaries have done in many parts of the world in more recent times, took up residence near existing settlements. Many of these old religious sites are associated with the oldest fishing traditions and settlements.

The Vikings used Lough Neagh as a base for their raids throughout Ulster in the 9th Century A.D. and again in 943 when it is recorded that they were defeated by the O'Neills who were by then in the process of extending their suzerainty over central Ulster. About 1000 A.D. the seat of O'Neill power was transferred from Aileach in the Inishowen peninsula of Donegal to Tulach Og (Tullaghoge, Fig. 6.2). The chief surnames of this western shore area date from this period – Devlin, O'Hagan and Donnelly, see diverse branches of the O'Neills.

The first evidence for extensive forest clearance on the lough's shorelands comes from the pollen record taken from lough sediments and dated at 800 B.P. On the basis of C_{14} dating, Battarbee suggests that this first series

are usually found in mid-Antrim well away from the river. The only exception is a group from Coleraine (the Loughan Island area). By way of contrast the river has produced a large number of swords dating from the Later Bronze Age while few have been found in adjacent areas.

of radical changes may be due to the Anglo-Norman conquest of Ulster (Battarbee, 1978). However, he stresses that, as their influence was not as strong here as in other parts of Ireland, these changes may also be viewed as the natural result of population increase towards the end of the Early Christian period. However, carbon-dating may be too inaccurate for precise use in this context. The Anglo-Norman presence was evident along the eastern shorelands, but the lough and the Lower Bann were very much a frontier between them and the Irish to the west. The motte on Coney Island, the small island near the southwestern shore, was either a Norman outpost or an Irish imitation (Mitchel and Morton, 1971).

Following the Bruce Wars in the 14th Century the Norman Earldom of Ulster collapsed so that in the 14th and 15th centuries the 'great' O'Neills of Tyrone continued to rule over the western and southern shores while the Clandeboye O'Neills established themselves at Shanes Castle. This supremacy was challenged in Elizabethan times when Sir Henry Sidney, the Lord Deputy, attacked Shane O'Neill and for a time gave Lough Neagh a new name – Lough Sidney. In 1597 a force from Carrickfergus captured Antrim which became one of the bases for Sir Arthur Chichester's naval task force with which he attacked the western shore. This exercise in 'sea power' combined with Mountjoy's campaign across the river Blackwater led to the destruction of Tullaghoge and the surrender of Hugh O'Neill in 1603. Mountjoy Castle (Fig. 6.2) was built to consolidate this conquest.

Soon after, Ulster was settled by Scots and English and some Irish were displaced to the hitherto unwanted shorelands of Lough Neagh, such as the Montiaghs along the southern shore and Moortown on the western shore. The great oak forests of Glenconkeyne and Killultagh were cleared and plantations of settlers established. The Salter's Company built a fortified house and bawn at what became Salterstown in Glenconkeyne, and the Mercer's built a comparable bawn at Movanagher. The Company of Drapers had been allocated land near Cookstown and as a result families such as the Corrs moved to Moortown (Ardboe), while Conlans, Croziers and Campbells were relocated at Glenconkeyne (Hutchinson, 1951).

That fishing community remained firmly Gaelic-Irish in character throughout these centuries of instability is shown by the great preponderance of Irish surnames among its members, as much as 90% in Moortown. Only along the northern and eastern shores does the Plantation influence assert itself with between 40 and 50% of surnames betraying Protestant Scots-English origins. However, many family names long associated with fishing have now disappeared completely in the last 130 years, no less than 54 in Moortown, for example, Dillon, Toland, Scullion, Conway and Cassidy (Donnelly, 1981, 1986). Emigration, lack of marriage partners and sons, and alternative employment provide explanations.

Although the fishing community maintained its distinctive character after the Plantation, it lost its communal rights to fish the lough. In the early 17th century the eel fishery of the Bann and Lough Neagh was vested in the

Crown, ownership eventually passing to the Donegall family and their relatives, the Shaftesburys. Not until 1971 was control regained by the fishermen. The intervening 365 years saw many ownership wrangles especially when the fishery began to acquire commercial importance in the latter part of the 19th century, as, with improved communications, markets opened up in Great Britain. Competition for this increasingly valuable resource produced conflict between the fishermen who considered it their traditional right to fish on the lough without restriction and the lessees who had legal control. Court cases were frequent. In 1878 and again in 1910–11 appeals went to the House of Lords and in 1962 the legality of the lessees' rights was tested in the Northern Ireland High Court. At the latter time Toome Eel Fishery Co. Ltd. was owned by a Dutch consortium and judgement was in their favour but although they had the right to impose conservation rules and to market all catch, in practice strict control was impossible. The fishermen had by now formed themselves into the Lough Neagh Fishermen's Association, a registered trade union, which in 1966 was able to buy a share of the Company. Five years later its bid for the whole of the Company's shares was successful. The wheel had turned full circle. The eel fishery was once again in the hands of the fishermen this time to be administered as a co-operative, the Lough Neagh Fishermen's Co-operative Society Ltd. The Society's control also extended to the Lower Bann's important weir fisheries at Toome, Portna and Movanagher.

Historically the Lower Bann is more famous for its wealth of salmon than for its eels although the latter now earn more money. Indeed salmon were an important part of the incentives offered the London Companies to undertake the Plantation of Co. Derry in the early 17th century. The fishery first became a source of revenue in the reign of Henry VIII when it was leased to Otwell Corbett, a merchant of Chester, at the high annual rent of £60 Irish in 1519 (Quinn, 1933–34). In 1521 Corbett was able to claim a rebate from the Crown for damage done by the Irish. In 1542 a garrison was put into Coleraine to guard the fishery. Paper titles meant little in practice at this time, and the fishery remained in effect under the control of the O'Neills, O'Hanlons, McQuillans and O'Donnells who imposed conditions on other users. But the Plantation gave ownership to the London Companies to be administered on their behalf by the Honourable the Irish Society. In 1985 the salmon fishery came under the management of Bann Systems Ltd., after being leased for many years to the Bann Fisheries Ltd.

The Plantation gave the salmon fishery commercial importance on a scale hitherto impossible. Salmon were exported from Coleraine to Bristol and Chester and from there overseas. The fishery commanded what then were enormous rents. By the early 19th century packing in ice rather than salting the catch was general practice and many more fish were exported. The introduction of steam transport in 1861 again increased the value of the fishery which was let in that year by public tender for £5,625 per annum. But the latter part of the century saw diminishing catches interpreted as due to a number of causes including the growth of the linen industry which brought with it pollution from flax dams and bleaching mills, severe poaching on tributaries as the value of salmon increased, and also major drainage and navigation works which were carried out between 1847 and 1858 as part of the McMahon scheme (see Chapter 3).

This scheme proposed a Lower Bann Navigation and the alleviation of the constant flooding problem affecting Lough Neagh's shores. It involved lowering the level of the lough by almost 2 metres and the removal of shoals and sandbars in both the Upper and Lower Bann. The low corridors radiating from the lough to the sea and westwards to the Erne basin had already been exploited for canal building, the development of the Tyrone coalfields provided the stimulus for the first canals in the 18th Century. The Newry Canal – the first major canal in the British Isles – was opened in 1742 to handle cheap coal brought across Lough Neagh by barge. Belfast's growth encouraged the building of the Lagan Navigation which reached Lisburn in 1763 and the lough beyond Aghalee 30 years later. The Coalisland canal was completed in 1787 to cut out overland haulage of coal from the mines to the lough. The least successful of the navigations was the Ulster canal which was built at a high cost between 1825 and 1842, its 67 km linking the lough with the Upper Erne more in the hope of developing that part of the country than on the basis of transport needs. All the navigations were soon in decline, not least because of the network of railways which by 1880 served the area well (McCutcheon, 1980).

During the navigation period a number of small harbours were built and paddle steamers operated on the lough. The first inland navigation steamship in Ireland, the 'Marchioness of Donegall', was launched on the lough in 1821, and was used to tow barges across the lough to the Blackwater. In the 1830s services for goods and passengers linked the western and southern shores. As an example, Ballyronan began life in 1788 as a Huguenot settlement, exported the products of its spade mill, distillery, brewery and corn mill in small vessels of 50 tons to Kinnegoe near Lurgan: in return came barley, timber, slates and groceries. In the 1840s this service gave way to one operated by paddle steamer between Portadown and Newport Trench and occasionally Ballyronan. This service handled a large passenger traffic during the famine years (mid-1800s) when there was much migration from the poorer districts west of the lough.

Diatomite is relevant to the Lough Neagh story, both for its aquatic origins and its socio-economic importance. Used as an abrasive and in filtration, diatomite was processed and factories were set up in Newferry (1906) and Portglenone (1912), the latter moving to Toome in 1920 (MacAfee, 1965). Peak production was reached during the 1939–45 war when other sources were cut off. Work virtually ceased in the 1960s in face of competition from superior overseas deposits.

The 19th century saw the growth of factory industry in many parts of the Lough Neagh basin, particularly the linen industry which here took root in the River Main valley north of the lough, along the Upper Bann and in towns such as Lurgan and Portadown on its southern margins. McCutcheon (1980) provides evidence of the considerable extent to which water power contributed to the linen industry, especially in the Six Mile Water and the River Main; the impact of pollution from retting has already been mentioned. The villages and small towns along the Lower Bann remained as service centres relatively untouched by industrial growth. The physical characteristics of the river must go far to explain this situation; the few stretches of fast flowing water suitable for power generation were difficult of access and tended to be already occupied by fisheries. There were no such obstacles to factory location along the more accessible R. Main.

The changing industrial geography of the period also had a marked effect on Lough Neagh's shorelands. A growing network of third class roads (made possible by the regulation of lough levels and reduced winter flooding of the shallow shorelands) steadily reduced the old isolation. Easier access and the increased opportunities for alternative employment attracted fishermen away from the eastern and southern shores which were closest to the growing urban manufacturing centres. The western shore areas were least affected. When emigration did become prevalent there in the 1920s and 1930s movement tended to be directed further afield, to Great Britain and the United States. Inevitably change has reduced the old self-subsistence qualities of the entire fishing community and once widespread craft industries linked to fishing, such as the weaving of hemp and linen nets and basket-making, have almost vanished.

6.4. The present

6.4.1. *The Lough Neagh fishing community and the river fisheries*

Today fishermen and their families are mainly concentrated along the western shore where Ardboe parish accounts for 46% of the total while Moortown, within that parish, has 26% on a mere 10 kms of shoreline (Donnelly, 1981, 1986). The northern shore has 16% with the main concentration in Toome Bay. Along the eastern shore fishermen mainly operate out of Sandy Bay and Bartins Bay. Few fishermen are to be found nowadays along the southern shore. This distribution is the result not only of the physical and historical factors already mentioned but also to the community's social organisation which for centuries involved clustering of related families, the sons on marriage choosing a site near the parent homestead. Nearness to coves, the channels where the fishermen berth their boats, was long an important locating factor but today fishermen are more mobile thanks to an improved road network and many choose to live further away from the immediate shore. There are about 50 coves still in use: many had to be extended as the lough was progressively lowered during drainage operations.

Space does not permit more than a brief mention of the methods of fishing on the lough. For eels the two main ones are long lines and draft nets, each of which has a long ancestry and whose relative importance has varied at different times. Long lines may be up to six miles in length, each with as many as 2000 hooks which must be baited, a laborious task not in much favour with the younger fishermen. They are used most often by fishermen along the western and northern shores, especially in Ardboe parish, where there is much 'hard bottom', is unsuitable for draft nets unless used with skill. These nets are best suited to sandy bottom which is most widespread in the south and east of the lough. Fishermen in the Sandy Bay, Derrytrasna and Maghery areas are their main exponents. But both methods are interchangeable and in the course of time have spread over the lough from their source areas.

Sail disappeared from the lough in the early 1940s to be replaced by boats with engines. This had social implications for the community for there was no longer need to stay overnight on an opposite shore, a practice which encouraged cross lough inter-marriage. The old type of clinker built fishing boat, 6–8 m in length and with a single thole pin for each oar in the fashion of the ocean curraghs of the western seaboard, has now given way to larger and more powerful boats made of fibre glass. Altogether there are some 200 boats, crewed by about 500 men, fishing for the Co-operative Society, each limited to a set daily quota of eels.

The lough's brown eel catch is collected by agents working for the Co-operative while the weir fisheries at Toome, Portna and Movanagher provide it with a rich harvest of silver eels (see Chapter 18). Three of these river weirs – two at Portna and one at Movanagher – are old type weirs largely made of timber and wattle, the last of their type to be operated on Irish rivers. Nowadays most of the eels caught are sent by air freight to the Netherlands and Germany: at the height of the season some 8 to 9 tonnes may be dispatched daily. Eel fishing is likely to continue to dominate the lough's fishing activities. Perch and trout are of secondary importance and pollan fishing is precarious with markets poorly organised and catches extremely variable (see Chapter 18). The salmon fishery at the Cutts remains important but commercial salmon netting at the Cranagh on the river estuary has been suspended.

The fishing community has shown itself well able to adapt to changing circumstances. In recent decades it has had to face many new demands on its resource base, the waters of the lough, but its catch fetches high prices and as a result its young men are retained in the traditional way of life. Assuming that the use of the lough can be controlled wisely in the future this unique community should be well able to maintain its distinctiveness and vitality.

6.4.2. *Other resource uses*

While fishing remains the link with prehistory, other uses of the lough's waters and shorelands have become of increasing importance in recent years. The lough is now very much a multi-use resource, its waters not only provide the food supply to support one of Western Europe's major eel fisheries and large populations of over-wintering wildfowl but act as a source of supply for both industrial and domestic purposes – the lough now makes an important contribution to Belfast's water supply – and to an increasing extent are used for recreation. While sand and gravel extraction near Toome, involving barge transport across the lough to Sandy Bay on the eastern shore, has been well established for many decades, recent years have seen the possibility of a new extractive activity, namely the mining of lignite. This could affect the western and eastern shorelands and waters of Sandy Bay and Lennymore Bay. There is concern that such open-cast mining and the operation of one or more shore power stations could add to the lough's pollution problem (Jefferson and Mitchel 1991).

The Lower Bann is also coming under increasing development pressures. Applications for building sites along its most attractive stretches, particularly near Coleraine, are increasing as are demands for recreation. Its 19th century navigation system of locks and deep water diversions remains in excellent condition and thus provides good facilities for cruising which is in sharp contrast with the growth of noisy water sports centres on hitherto quiet and isolated stretches. Picnic sites, landing stages and marinas now mingle with Forestry Division's 30 or so small woods which are scattered along the river banks between Somerset Forest near Coleraine and Portglenone. Of mixed species these woods have now grown to attractive maturity on what were known as the 'Bann dumps', the dredged material from the river bed deposited on the banks during the major drainage operations in the 1930s and 1940s. In comparison with the Bann, recreational use of Lough Neagh's shores is concentrated at a few places – Oxford Island in the south, the Ballyronan marina in the northwest and Antrim.

Both Lough Neagh and tiny Lough Beg at its northern end are now designated as Areas of Special Scientific Interest (ASSIs) and the lough itself now has international designation as a RAMSAR site. (Chapters 19 and 20 deal in more detail with ASSIs, especially in relation to fen vegetation and avifauna). The river corridor has no such protection.

In essence this account of settlement and economy is very much a story of resource utilization from the Mesolithic period to the present day. Changing circumstances, among which advances in technology figure prominently, mark its progress. It is increasingly apparent that both lough and river are in need of protection and careful management. No longer is fishing the only use involved and it is clear that its future can be affected by other developments. Luckily there is now growing awareness of the interdependance of economic activities. It is encouraging to find that an umbrella organisation of Local Councils, interested Government departments and voluntary bodies, the Association of Lough Neagh Users (ALNUS), now exists to coordinate efforts to safeguard the lough.

6.5. References

Addyman, P.V. (1965) Coney Island, Lough Neagh. Prehistoric Settlement, Anglo-Norman Castle and Ellzabethan native fortress. Ulster J. Archaeol. 28: 78–101.

ApSimon, A.M. (1976) Ballynagilly and the beginning and end of the Irish Neolithic. Dissertationes Archaeologicae Gandenses 16: 15–30.

Battarbee, R.W. (1978) Observations on the recent history of Lough Neagh. Phil. Trans. R. Soc. (B) 281, 303–347.

Battarbee, R.W., Scaife, R., Phethean, S.J.C. (1984) Palaeoecological evidence for sea level change in the Bann Estuary in the Early Mesolithic. In: Woodman, P.C. (ed.) Excavations at Mount Sandel. Archaeological Research Monograph No. 2. HMSO Belfast.

Carter, R.W.G. (1982) Sea level changes in Northern Ireland. Proc. Geol. Soc. 93: 7–23.

Creighton, J.R. (1974) A Study of the Later Pleistocene Geomorphology of North Central Ulster (unpublished Ph.D. thesis, The Queen's University, Belfast).

Donnelly, D.J. (1981) A Study of the Lough Neagh Fishing Community (unpublished M.A. Thesis, The Queen's Universlty, Belfast).

Donnelly, D.J. (1986) On Lough Neagh's Shores; a study of the Lough Neagh Fishing Community. Publ. The Donnelly Family, Galbally, Co. Tyrone, 333 pp.

Edwards, K.J. (1980) Vegetational changes associated with man in Co. Tyrone. *In* Irish Association for Quaternary Research. Field Guide No. 3 County Tyrone, N. Ireland (ed. K.J. Edwards), pp. 44–51.

Ehrenberg, M.R. (1977) Bronze Spearheads from Berks, Bucks and Oxon. B.A.R. 34: Oxford.

Dubourdieu, J. (1812) Survey of County Antrim. Dublin.

Evans, E.E. (1981) The personality of Ireland. Blackstaff Press, Belfast, 130 pp.

Hamilton, A.C., Dalzell, L.R., Lenehan, B.P. and MacDonagh, B.J. (1984) Palynological investigations of Ketle-hole sediments at Mount Sandel, Coleraine. In: Woodman P.C. (ed.) Excavations at Mount Sandel. Archaeological Research Monograph No. 2. HMSO (NI) Belfast.

Hutchinson, W.R. (1951) Tyrone Precinct: A History of the Plantation Settlement of Dungannon and Mountoy to Modern Times, Belfast 43 pp.

Jefferson, C.W. and Mitchel N.C. (1991). Northern Ireland Lignite. GeoJournal 24: 405–412.

Jessen, K. (1949) Studies in the Late Quaternary deposits and flora history of Ireland. Proc. R. Irish Acad. 52B: 85–290.

MacAfee W. (1965) Diatomite in the Lower Bann Valley. Unpublished B.A. Thesis, The Queen's University, Belfast.

McCutcheon, W.A. (1965) The Canals of the North of Ireland, London, 143 pp.

McCutcheon, W.A. (1980) The Industrial Archaeology of Northern Ireland. HMSO, Belfast, XLV + 395 pp.

Mitchel, N.C. (1965) The Lower Bann Fisheries. Ulster Folk Life, 11: 1–32.

Mitchel, N.C. and Morton R.G. (1971) Lough Neagh: Ulster's Inland Sea, Craigavon Historical Society Review, Vol. 2, No. 1.

Mitchell, G.F. (1955) The Mesolithic site at Toome Bay, Co. Derry. Ulster J. Archaeol. 18: 1–16.

O'Sullivan, P.E., Oldfield, F. and Battarbee, R.W. (1973) Preliminary studies of Lough Neagh sediments. In: H.J.B. Birks and R.G. West (eds.), Quaternary Plant Ecology, London, 267–278.

Pilcher, J.R. (1969) Archaeology, Palaeoecology and C^{14} dating of the Beaghmore Stone Circle site. Ulster J. Archaeol. 32: 73–92.
Pilcher, J.R. and Smith, A.G. (1979) Palaeoecological investigations at Ballynagilly, a Neolithic and Bronze Age settlement in Co. Tyrone, N. Ireland. Phil Trans. R. Soc. (B) 286: 345–369.
Quinn, D.B. (1933–1934) Anglo-Irish Ulster in the Early Sixteenth Century. Publications and Reports of Belfast Natural History and Philosophical Society, p. 67.
Savage, R.J.G. (1966) Irish Pleistocene Mammals. Irish Nat. J. 15: 117–130.
Smith, A.G. (1958) Pollen analytical investigations of the mire at Fallahogey Townland, Co. Derry. Proc. R. Ir. Acad. 59(B): 329–343.
Smith, A.G. (1961) Cannons Lough, Kilrea, Co. Derry: Stratigraphy and pollen analysis. Proc. R. Ir. Acad. 61(B): 369–383.
Warner, R.B. (1976) Some observations on the context and importation of exotic material in Ireland from the 1st Century B.C. to the 2nd Century A.D. Proc. R. Ir. Acad. 76(C): 267–292.
Waterman, D.M. 1954. Excavations at Clough Castle, Co. Down. Ulster J. Archaeol. 17: 103–163.
Williams, B.B. (1978) Excavations at Lough Eskragh, County Tyrone. Ulster J. Archaeol. 41: 37–49.
Woodman, P.C. (1977) Recent Excavations at Newferry, Co. Antrim Proc. Prehist. Soc. 43: 155–200.
Woodman, P.C. (1978) The Mesolithic in Ireland, B.A.R. **58**: Oxford.
Woodman, P.C. (1981) A Mesolithic camp in Ireland. Scientific American, August.
Woodman, P.C. (1984) Excavations at Mount Sandel. Arch. Res. Monographs, No. 2, HMSO Belfast.

7. The recent sediments of Lough Neagh
Part A. Structure, stratigraphy and geochronology

R.W. BATTARBEE and R.J. FLOWER

7A.1. Introduction

Sediments accumulating in the deeper areas of lake basins contain a record of environmental change which can be used to reconstruct aspects of both lake and lake catchment history. In the case of Lough Neagh, the study of the recent sediments has been primarily concerned with the palaeoecological evidence for nutrient enrichment (cf. Part B of this chapter; Battarbee, 1978a; Carter, 1977), and with the provision of a sedimentological and chronostratigraphical framework for palaeoecological analysis. This Part A, describes the progress that has been made in the development of such a framework. A number of other studies have been carried out on other lakes within the Lough Neagh catchment (e.g. Lough Gall, McCallan, 1975; Lough Fea, Thompson *et al.*, 1975) but those studies are not considered further.

7A.2. The contemporary sediments

7A.2.1. *Distribution and composition*

The accumulation of fine-grained organic sediments suitable for biostratigraphic analysis is restricted to water depths below 5–9 m (Fig. 7A.1) (see also Chapter 3, Section 3.2). These sediments are very dark and are composed of silts and clays, organic matter, and diatom frustules. The silts and clays are primarily derived from riverborne suspended material whilst the organic matter originates partly from catchment sources and partly from material produced within the lake. The diatoms are almost entirely of autochthonous origin.

The supply and deposition of sediment and its major constituents varies seasonally in a manner demonstrated by Flower (1980), who exposed sediment traps in the lake for both long and short term periods from 1976–1978. Fig. 7A.2a shows the variation in seasonal sedimentation rate whilst Fig. 7A.2b shows variations in the composition of material caught in traps. Whilst

R. B. Wood and R. V. Smith (eds.), Lough Neagh, 113–132.

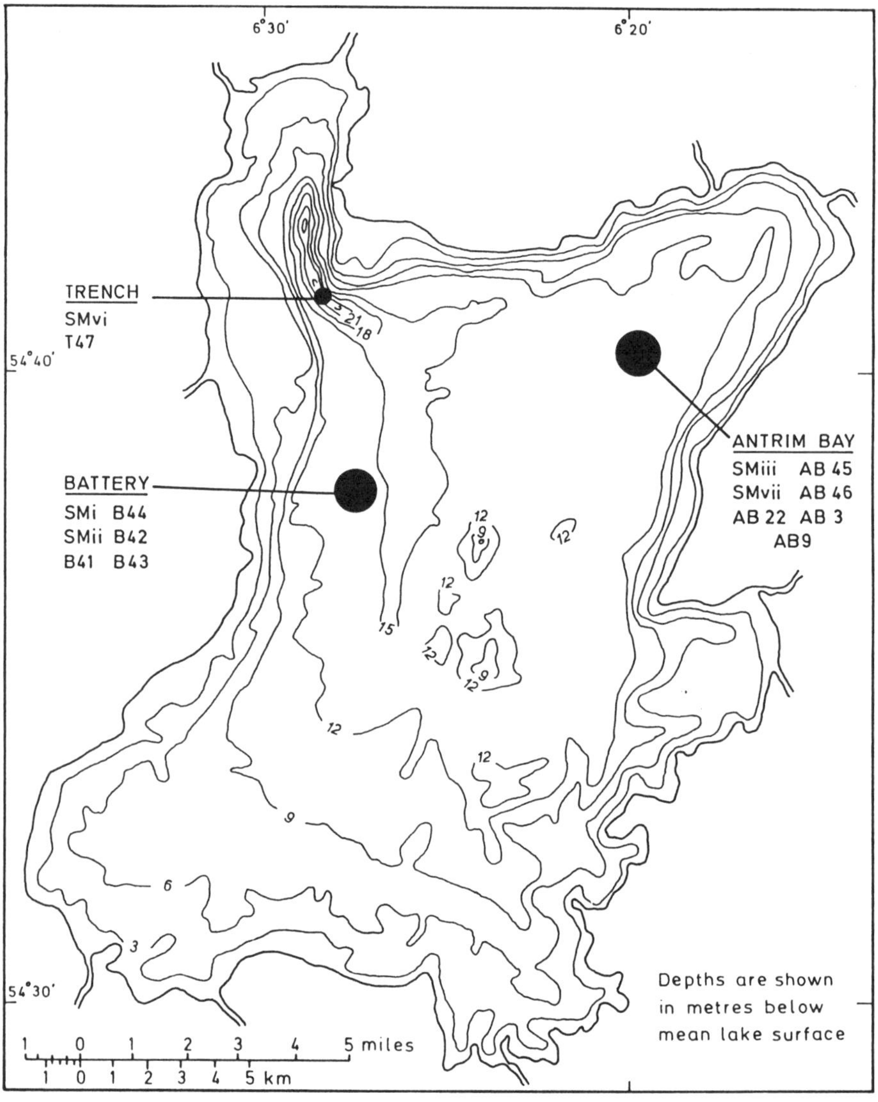

Figure 7A.1. Lough Neagh, bathymetry and location of sediment core sites. Redrawn from Battarbee (1978a).

the values are exaggerated by the inclusion of resuspended material it is clear that peak clastic sedimentation (non-biogenic mineral matter) occurs in the winter months, coincident with the period of high river discharges, and that the main diatom flux (biogenic silica) to the sediment occurs in spring and is associated with the collapse of the planktonic diatom crop in the water column. Summer inputs are much lower but are characterized by a higher

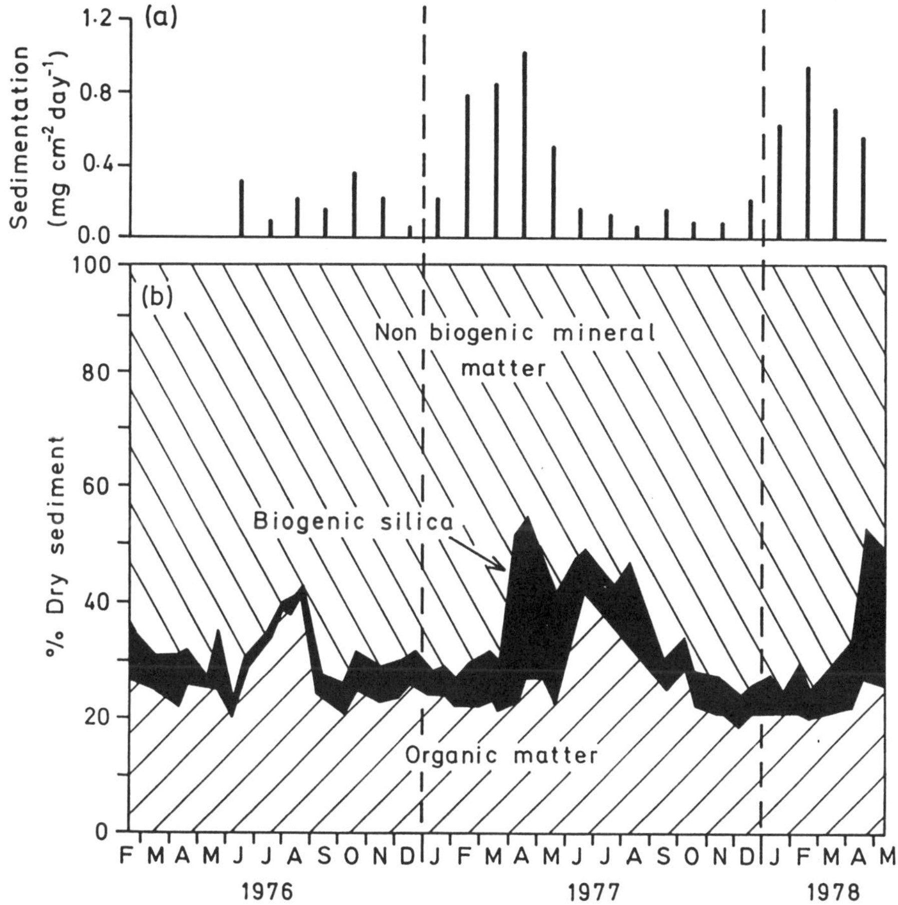

Figure 7A.2. Sediment trap data 1976–1978 showing (a) sedimentation rate changes and (b) seasonal changes in the composition of material caught.

organic content reflecting both the accumulation of larger amounts of detritus from degrading blue-green algal populations as well as the lower input of clastic material.

7A.2.2. *Resuspension and bioturbation*

Despite the seasonal pattern of sedimentation, evidence of seasonality is not preserved in the sediment because of sediment mixing by resuspension and bioturbation. Extremely severe gales can cause sediment to be resuspended throughout the water column (see Chapter 4), and even moderate storms

Table 7A.1. Comparison of sediment caught in traps over a 3 day storm period and a concurrent 23 day period

Accumulation rate	Exposure period	
	3 day	23 day
Periphytic diatoms ($no. cm^{-2} day^{-1}$)	24,100	13,500
Total diatoms ($no. cm^{-2} day^{-1}$)	573,900	256,000
Biogenic silica ($mg cm^{-2} day^{-1}$)	0.24	0.13
Clastic minerals ($mg cm^{-2} day^{-1}$)	0.87	0.38
Periphyton: plankton	0.04	0.05

can resuspend surface sediment to at least 1 m above the sediment surface, the height at which sediment traps were placed (Flower, 1980).

It is difficult to quantify the amount of resuspension that takes place annually but Flower was able to show that over an individual storm period of three days the rate of accumulation in traps was approximately two times that of traps exposed over a concurrent but extended (23 day) period (Table 7A.1). Since the ratio of periphytic to planktonic diatoms for both periods was similar and there was no evidence of increased fallout from the water column Flower concluded that the increase in the rate of accumulation in the sediment trap was almost entirely due to sediment resuspension from the sediment surface in the central part of the lake, and equivalent to a displacement of the top 0.5 mm of sediment. Whilst this is a small amount, continuous resuspension by a succession of storm events, which typically occur during the winter in Northern Ireland, could lead to considerable mixing.

Bioturbation also induces sediment mixing. It is likely to be effective to a greater depth than resuspension and is especially important in summer months. Chironomid larvae and oligochaete worms are the main organisms concerned and both populations reach high densities (cf. Carter 1976 and Chapter 16). Although the work was carried out in a different context, the role of these organisms in breaking down sediment microstratigraphy has been well illustrated by Jewson *et al.* (1981) and Rippey and Jewson (1982). In 1979 the diatom crop collapsed at the beginning of April when silica became limiting. As the population settled out there was a rapid increase in chlorophyll *a* and ATP concentrations in the surface 1 cm of sediment. During the next four weeks both ATP and chlorophyll *a* levels remained high and there was little mixing with lower sediments. However, as the benthic invertebrate population became more active from mid-May onwards, chlorophyll *a* values in the surface 1 cm declined rapidly and by mid-June the top 2 cm of mud were almost completely mixed. Continued mixing of this kind is likely to occur in the summer months, at least until the period of chironomid emergence. The process is repeated each year.

The effects of sediment resuspension and bioturbation on the stratigraphic record are difficult to evaluate. It seems clear that they are responsible for the mixing of at least the upper 2 cm of sediment, thereby destroying any record of seasonal or annual variation. On the other hand the sediment biostratigraphy clearly records pronounced changes consistent from core to core in algal populations that have taken place in the past (cf. Part B of this chapter, Battarbee, 1978a), suggesting that longer-term changes are not destroyed. The effects of resuspension and bioturbation are reduced in the longer term as the very loosely consolidated upper sediment becomes compacted, and as the rapid accumulation of new sediment moves the mixing zone continually upwards. The resolution of the stratigraphic record in Lough Neagh is probably controlled by the variations in the intensity of these interacting processes through time.

7A.3. Stratigraphy

Stratigraphic studies have focussed on the investigation of patterns of sediment accumulation in the lake and on the development of a stratigraphic scheme for dating and core correlation. Litho-, bio-, and magneto-stratigraphic studies of the recent sediment have been carried out. Cores were obtained using a Mackereth pneumatic mini-corer (Mackereth, 1969), which allows approximately 90 cm of the uppermost sediment and the mud-water interface to be sampled without disturbance, and a Mackereth 3 m and 6 m corer (Mackereth, 1958). Coring sites are shown in Fig. 7A.1.

7A.3.1. *Lithostratigraphy*

Except for a marked darkening in colour in the uppermost levels, lithostratigraphic changes are not apparent. The upper metre of sediment is uniform and its overall structure can be described as a brown-black gyttja. Physical analysis, however, shows that small changes in the structure do occur and, in the case of core SMiii from Antrim Bay, these can be seen in Figs. 7A.3 to 7A.5. Fig. 7A.3 shows percentage dry weight and wet density values against depth. The values for dry weight in the uppermost levels reflect the poorly compacted nature of the sediment in immediate contact with the bottom water of the lake. From 3 cm downwards the dry weight values increase slowly to a maximum of 25%, but then decrease again below 50 cm to about 20% at approximately 60 cm. The trend of these values suggests that the water content of the sediment below the top 3 cm is more strongly related to sediment composition, especially the content of organic matter, than to depth below the sediment surface. Fig. 7A.4 shows a linear regression

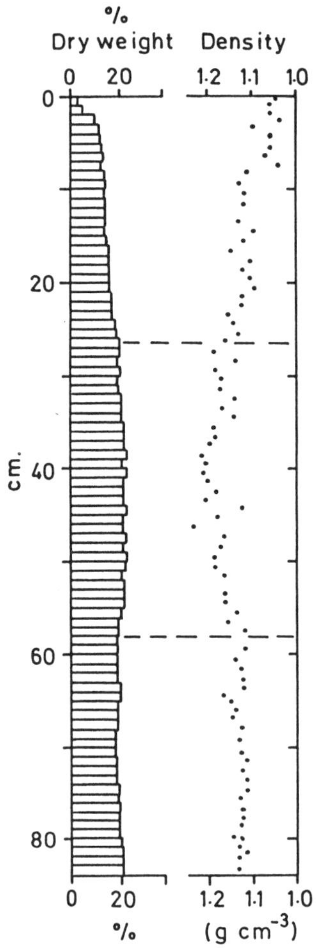

Figure 7A.3. Depth distribution of % sediment dry weight and wet density for core SMiii.

of per cent water content on per cent organic matter to illustrate this relationship.

Fig. 7A.5 shows the proportions of organic matter, diatom silica and mineral matter per unit sediment dry weight against depth. Changes in composition occur at about 55–60 cm and at about 30 cm (cf. Fig. 7A.3). The change in structure at 55–60 cm is indicated by a slight decrease in the proportion of organic matter and (cf. Fig. 7A.3) by an increase in the density of the sediment. The change is probably caused by an increase in the influx of mineral matter to the lake (Battarbee, 1978a; Rippey *et al.*, 1982). The change in composition at 30 cm is due to an increase in the proportion of diatom silica in the sediment (Fig. 7A.5), reaching a maximum of nearly

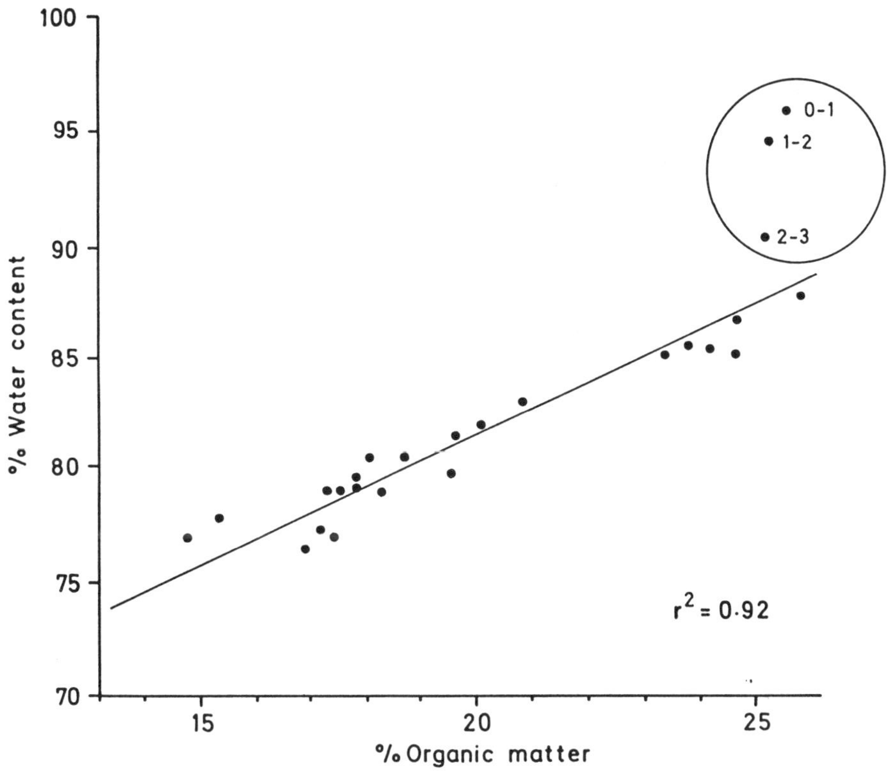

Figure 7A.4. Linear regression of % water content on organic matter (% of dry weight) for core SMiii excluding the 3 uppermost levels (circled).

30% of the dry sediment. The proportion of organic matter is also somewhat higher in this upper zone.

7A.3.2. *Biostratigraphy*

The pollen and diatom biostratigraphy of the recent sediments provides a more detailed means of assessing stratigraphic variation than lithostratigraphy. On the assumption that major changes in the diatom and pollen assemblages occur and have occurred synchronously within the lake these changes can be used to construct a zonation system for core to core correlation (cf. Battarbee, 1978b). Diatom changes have been particularly useful in this respect and the diatom zonation scheme used for Lough Neagh is shown in Table 7A.2.

Diatom biostratigraphies have been drawn up for 14 short cores from the 3 major sampling sites (Fig. 7A.1). If cores from one main site, e.g. Antrim Bay, are compared, it can be seen (Fig. 7A.6) that the sediment thickness

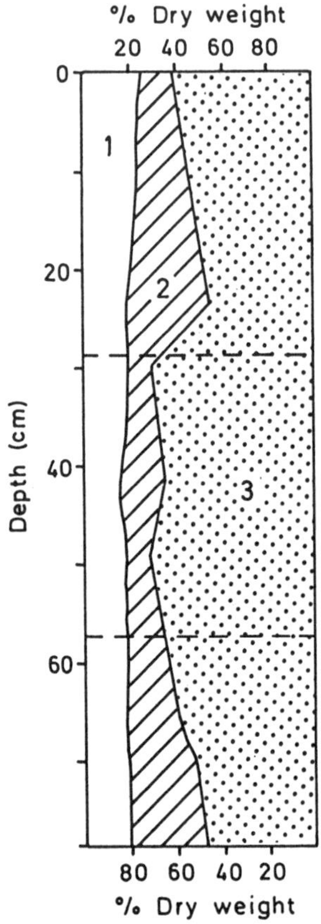

Figure 7A.5. Depth distribution of organic matter (1), diatom silica (2), and mineral matter (3) for core SMiii.

Table 7A.2. Diatom assemblage zones for the recent sediment of Lough Neagh. *Zone Dc is occasionally divided into subzones Dc_1 and Dc_2 where *M. italica* subsp. *subarctica* reaches a minimum value (cf. Chapter 7, Part B, Fig. 7.1)

Zone code	Diagnostic taxa
Da	*Melosira italica* subsp. *subarctica*
	Stephanodiscus astraea
	S. astraea v. *minutula*
Db	*M. italica* subsp. *subarctica*
	S. astraea
	Tabellaria flocculosa
Dc*	*Cyclotella ocellata*
	C. comensis
	T. flocculosa
	Asterionella formosa
Dd	*M. italica* subsp. *subarctica*

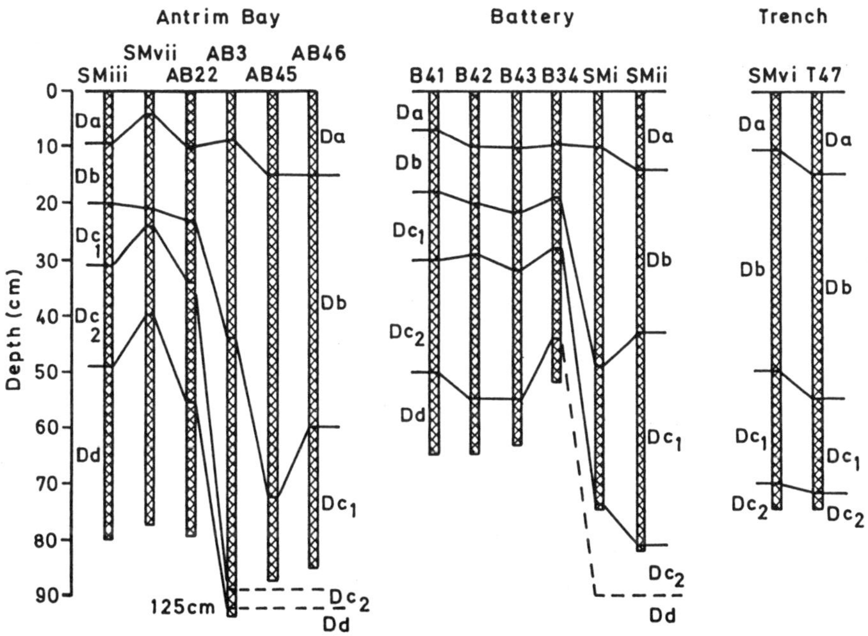

Figure 7A.6. Comparative diatom zonation of 14 cores from the three main sites. Redrawn from Battarbee (1978b).

Table 7A.3. Mean thickness (cm) of diatom biostratigraphic boundaries. Figures in parentheses refer to number of cores. AB = Antrim Bay; B = Battery; T = Trench

Zone	AB	B	T	Mean
Da	9.6 (5)	10.2 (5)	13.0 (2)	10.4 (12)
Db	33.8 (5)	20.2 (5)	39.5 (2)	29.1 (12)
Dc_1	19.0 (4)	17.0 (6)	18.5 (2)	17.8 (12)
Dc_2	22.3 (4)	21.2[2] (4)	–	21.7[3] (8)

[1]Relates to AB3 and 3 slow cores only. [2]Relates to 4 slow cores only. [3]Relates to mean of AB3 and 7 slow cores.

of each zone varies considerably from core to core; zone Db has a thickness of only 12 cm in SMiii, but 62 cm in AB45, despite the proximity of the cores and despite the fact that both cores were taken at the same water depth. A similarly variable pattern of 'fast' and 'slow' accumulation rates can be seen in the cores from the Battery site. (The two cores from the Trench show similar thicknesses but it is not yet known whether this indicates less variability at this deeper water site.) However, if the mean thickness of each zone at the 2 main sites is calculated, similar values are found (Table 7A.3). The mean thickness of Dc_1 at Antrim Bay is 19 cm and at the Battery it is

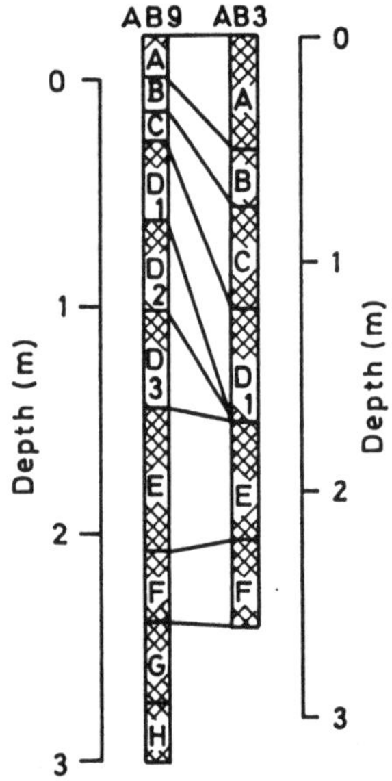

Figure 7A.7. Comparative pollen zonation of the two 3 m cores from Antrium Bay. Redrawn from Battarbee (1978b).

17 cm. These data suggest that the mean accumulation rate of sediment across the floor of the lake is quite uniform. If this is the case then the sediment accumulation rate at individual sites (cores) must alternate in time between relatively fast and relatively slow rates as the sediment builds up. It is not possible to evaluate this inference using diatom stratigraphy since there is little change in the diatom assemblage below the zone boundary Dc_2/Dd. However, the pollen assemblages vary continuously and a comparison of 2 adjacent 3 m cores from Antrim Bay (Fig. 7A.7) shows that alternation in accumulation rates within and between cores does take place. The top metre or so of AB3 has an accumulation rate characteristic of the 'fast' group of minicores (cf. Fig. 7A.6), whilst the top metre of AB9 has a rate similar to that of one of the 'slow' group (cf. SMiii, Fig. 7A.6). Conversely, during pollen assemblage zone D (Fig. 7A.7) AB9 accumulated sediment much more rapidly than AB3, whilst during pollen zones A, B, and C, AB3 accumulated sediment far faster than AB9. Over the total depth of sediment both cores cover an almost identical time span of accumulation.

As yet, no work has been carried out to determine how this alternating pattern of sediment accumulation is produced, but it is probable that resuspended muds are transported laterally by bottom currents, or in response to seiches, and redeposited in shallow mounds and hollows which represent sites of high and low accumulation respectively. The stratigraphic and spatial pattern of sediment accumulation may then be more related to hydrodynamic factors than either to morphometric ones or to spatial variations in the rate

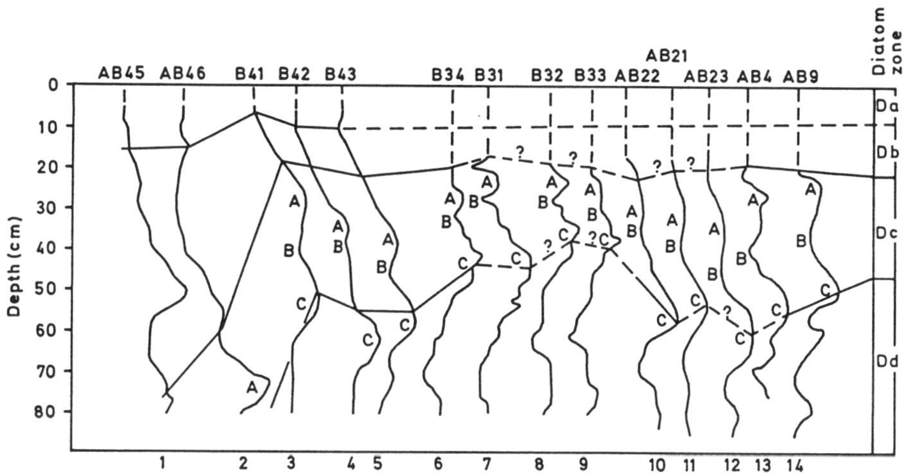

Figure 7A.8. Magnetic susceptibility vs. depth curves for 14 cores from Lough Neagh. No. 1–12 represent data from 1-m cores; No. 13–14 represent the upper portions of 3 m cores. Redrawn from Thompson *et al.* (1975).

of sediment supply. Whatever the mechanism, this kind of variation has a number of implications: occasional hiatuses are likely to occur in cores as sediment is transported from one area to another; estimates of sediment and microfossil accumulation rates need to be based on data from a representative sample of cores rather than from a single core; and replicate cores from closely adjacent sites may not yield identical, or even similar, records of changes in accumulation rate.

7A.3.3. *Magnetostratigraphy*

In order to assess the suitability of downcore variations in magnetic susceptibility as a stratigraphic technique for correlating cores, the susceptibility profiles and diatom biostratigraphies of 14 short cores from both Antrim Bay and the Battery sites were compared (cf. Thompson *et al.*, 1975). Fig. 7A.8 shows that the sequence of susceptibility fluctuations when plotted against sediment depth exhibit variations in the position of swings which might be expected if sediment accumulation rates vary between cores as suggested above. By matching the curves with the position of the diatom biostratigraphic boundaries it can be seen that the swings occur consistently from core to core. In Fig. 7A.8 the downcore swings are marked by the letters A, B, and C respectively, and the position of the diatom assemblage zone boundaries are indicated by heavy lines. Points A and B do not lie close to zone boundaries but they are in good stratigraphic agreement with the zonation system. Point C occurs close to the Dc/Dd boundary and approxi-

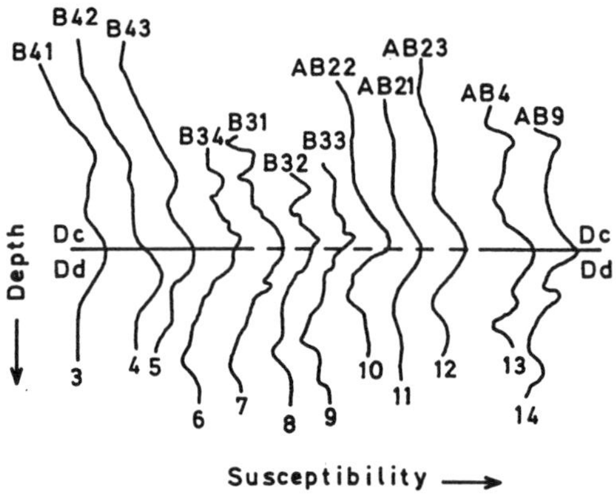

Figure 7A.9. Magnetic susceptibility versus depth curves in relation to the Dc/Dd boundary. Redrawn from Thompson *et al.* (1975).

mately coincides with this boundary in all cases. The stratigraphic consistency of the curves is shown more clearly in Fig. 7A.9 where they are drawn relative to the Dc/Dd boundary. These data suggest that susceptibility fluctuations recorded in cores from widely separated areas of the lake are synchronous. As such they can be used for core correlation and as a means of supplementing stratigraphic information obtained using other techniques. Moreover, since measurements are rapid and non-destructive, comparison of whole core susceptibility traces may be the most useful method of core selection in Lough Neagh where, as noted above, variability in accumulation rate is impossible to assess by lithostratigraphic inspection.

7A.4. Geochronology

The construction of a sediment chronology for Lough Neagh has been based on ^{14}C, ^{210}Pb and ^{137}Cs dating techniques, and has involved a number of different cores. In particular, attempts have been made to date biostratigraphic zone boundaries either directly or by interpolation to enable the construction of a chronostratigraphic system that can be used for any future core throughout the lake.

7A.4.1. ^{14}C *dating*

^{14}C dating was carried out on a number of 10 cm long sections of core AB3, a 3 metre core from Antrim Bay, at the Radiocarbon Dating Laboratory of Queen's University Belfast (Smith *et al.*, 1973). The results are shown in

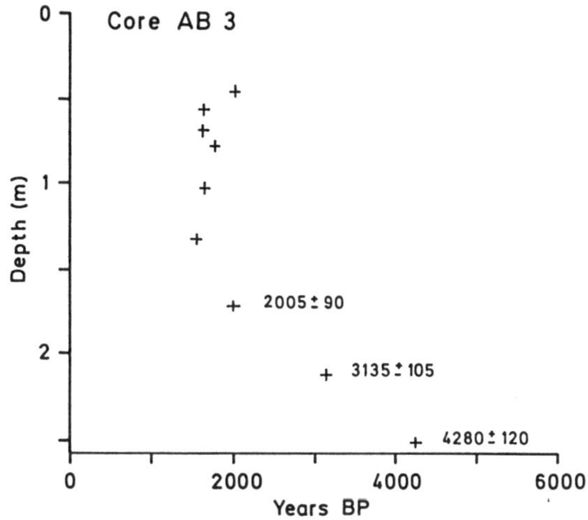

Figure 7A.10. Lough Neagh radiocarbon dates for core AB3, Antrim Bay. Redrawn from O'Sullivan *et al.* (1973).

Fig. 7A.10. The three lowermost dates follow a progressively younger linear trend, but the upper level dates, representing sediments accumulated during the last 2000 years or so, do not give expected values and some are inverted. It seems likely that these are artificially old dates, probably the result of contamination by older organic material inwashed from eroding soils and peats (cf. O'Sullivan *et al.*, 1973). Only the three earliest values have been used in the construction of a chronology.

7A.4.2. ^{210}*Pb dating*

Sediments from two short cores, AB46 and B43, were dated using the ^{210}Pb technique at AERE Harwell (Eakins and Morrison, 1976). Diatom analysis showed that the accumulation rate of AB46 was characteristic of the 'fast' group of cores and that B43 was typical of the 'slow' group (cf. Fig. 7A.6). In the case of AB46, background levels of ^{210}Pb were not reached (Fig. 7A.11a), but unsupported values of ^{210}Pb were obtained after subtraction of the ^{226}Ra concentrations at each level. The dates were calculated from these data using both constant initial concentration and constant rate of supply models (cf. Oldfield *et al.*, 1978; Appleby and Oldfield, 1978). The 14 dated levels for AB46 range from 1906 ± 5 at 68–70 cm to the present (Fig. 7A.11b). In B43 background levels were reached at about 50 cm (Fig. 7A.12a) and the core hence yielded earlier dates from 1844 ± 41 at 40–42 cm to the present (Fig. 7A.12b).

The dates from the two cores are perfectly consistent with the differences in their accumulation rates as indicated by the diatom biostratigraphy. The

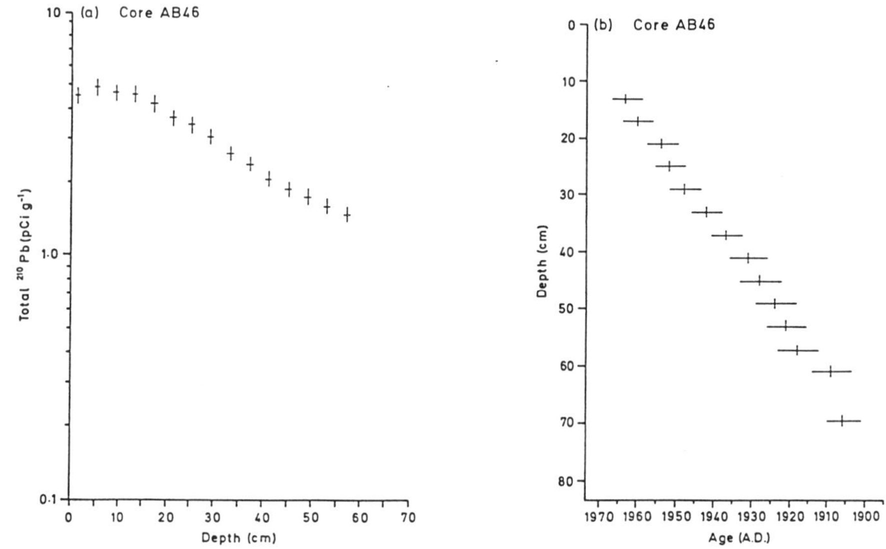

Figure 7A.11. ^{210}Pb dates for core AB46: (a) Total ^{210}Pb vs. depth; (b) Depth (cm) vs. Age (AD).

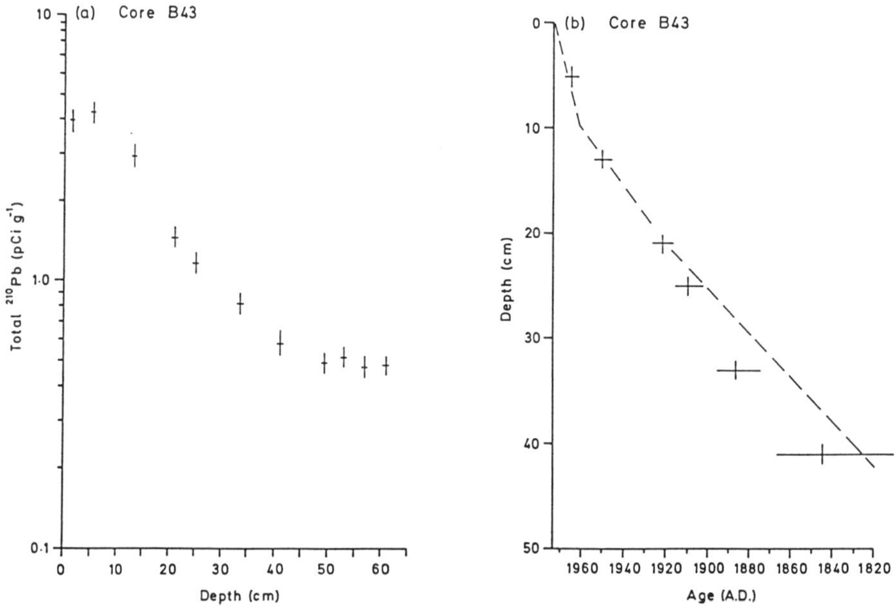

Figure 7A.12. ^{210}Pb dates for core B43: (a) Total ^{210}Pb vs. depth; (b) Depth (cm) vs. Age (AD). Dotted line indicates the somewhat older dates derived from using the crs model.

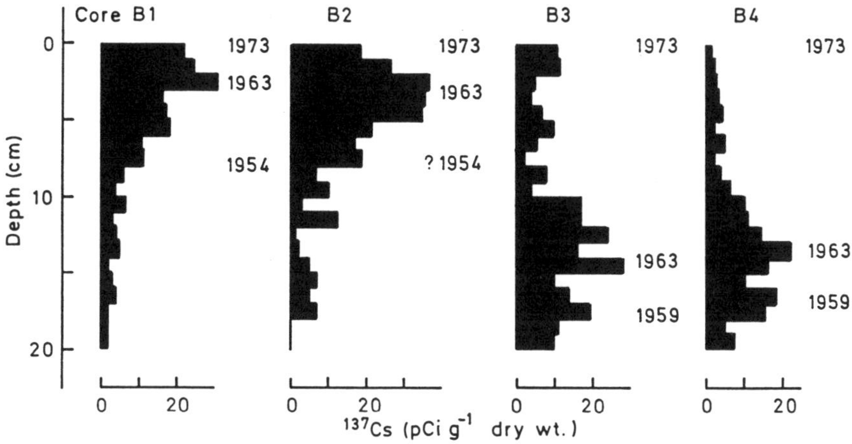

Figure 7A.13. ^{137}Cs distribution in the upper sediment of four cores from Lough Neagh. Redrawn from Battarbee (1978a).

level of the Db/Dc zone boundary occurs at 58 cm in AB46 and at 22 cm in B43. In both cases a date of approximately 1915 is indicated.

7A.4.3. ^{137}*Cs dating*

Further information on recent sediment chronology was obtained from ^{137}Cs analysis of four short cores (Fig. 7A.13). In this case, the differences are shown by the variations in depth of the occurrence of the peak representing maximum ^{137}Cs fallout in 1963. In addition, in B1 and B2 the 1954 increase representing the beginning of fallout can be observed, whilst in B3 and B4 this increase probably occurs below 20 cm. The faster rate of accumulation in B3 and B4, however, allows the subsidiary fallout peak of 1959 to be differentiated from the 1963 peak.

7A.4.4. *Time-depth chronology*

Because of the problems of dating recent sediments using the ^{14}C technique, no dates are available for the time-gap between the youngest reliable ^{14}C date of 2005 B.P. ± 90 and the earliest ^{210}Pb date of 128 ± 21. Until alternative dating methods become available, sediment accumulated during this time period can only be dated by interpolation between the ^{14}C and the ^{210}Pb series (cf. Battarbee, 1978a). This had been done by using diatom and pollen biostratigraphic boundaries to transfer dates from dated cores to the two 3 m cores AB3 and AB9 and then to derive tentative dates for the changes that occur during the undated gap from the resulting time-depth graphs (Fig. 7A.14). The dates for the various biostratigraphic boundaries

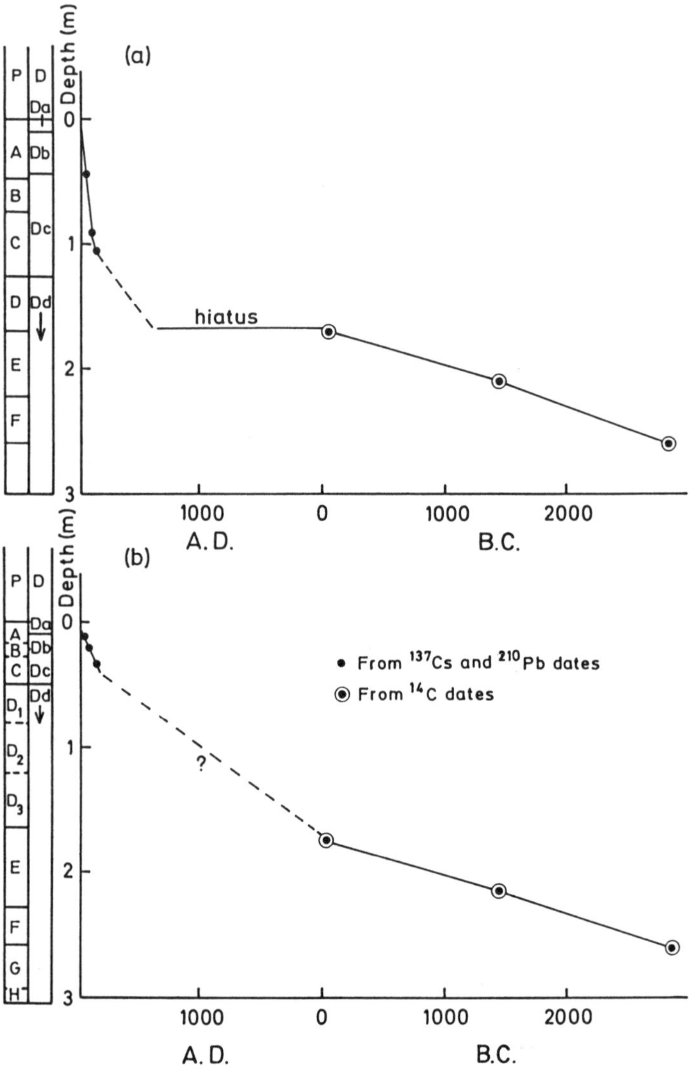

Figure 7A.14. Time-depth curves for (a) AB3 and (b) AB9. The gap between the lowest ^{210}Pb date and the uppermost ^{14}C date is indicated by a dashed line. The two left-hand columns refer to the pollen (based on O'Sullivan *et al.*, 1973) and diatom biostratigraphic zonation schemes respectively. Redrawn from Battarbee (1978a).

can then be assigned (Table 7A.4), and used as a basis for the calculation of accumulation rates.

7A.4.5. *Accumulation rates*

Since accumulation rates vary from core to core, estimates of mean accumulation rate based on data from only a small number of cores can be mislead-

Table 7A.4. Dates assigned to zone boundaries derived from ^{14}C, ^{210}Pb, ^{137}Cs dates, and by extrapolation between them. Dates are subject to at least the counting errors of the original measurements

	Zone boundary		Assigned age
Diatom	Da/Db		A.D. 1960[3]
	Db/Dc_1		A.D. 1915[3]
	Dc_1/Dc_2		A.D. 1880[3]
	Dc_2/Dd		A.D. 1700?
Pollen	–	a/b[1]	A.D. 1890
	–	b/c	A.D. 1800?
	C/D_1[2]	c/d_0	A.D. 1700?
	–	d_0/d_1	A.D. 1620?
	D_1/D_2	–	A.D. 1200?
	D_2/D_3	–	A.D. 500?
	D_3/E	–	50 B.C.
	E/F	–	2100 B.C.

[1] Pollen zonation symbols used by Battarbee (1978)
[2] Pollen zonation symbols used by O'Sullivan *et al.* (1973).
[3] Recent recalculations of ^{210}Pb dates according to the c.r.s. model (Appleby, personal communication) give dates of 1959, 1916, and 1868 for Da/Db, Db/Dc, and Dc_1/Dc_2 respectively, and linear extrapolation of these give 1750 A.D. for Dc_2/Dd.

Table 7A.5. Mean accumulation rates (cm yr^{-1}) T = Total. Numbers in parentheses refer to number of cores analysed

	$\bar{X}(T)$
Diatom zones	
Da	0.95 (15)
Db	0.65 (15)
Dc_1	0.51 (13)
Dc_2	0.12 (9)
Dd	0.08 (2)
Pollen zones	
D_1	0.08 (2)
D_2	0.07 (1)
D_3	0.07 (1)
E	0.03 (1)
F	0.03 (2)
G	0.03 (1)

ing. By using the chronostratigraphic framework described above (see Table 7A.4), an attempt to assess changes in mean accumulation rate has been based on data from all available cores. The mean accumulation rate for each successive pollen and diatom zone can be seen in Table 7A.5. If the mean accumulation rate for each zone is graphed against sediment depth (Fig. 7A.15a) or, more appropriately, time (Fig. 5.15b) it can be seen that substantial increases have taken place over recent centuries and that the mean rate

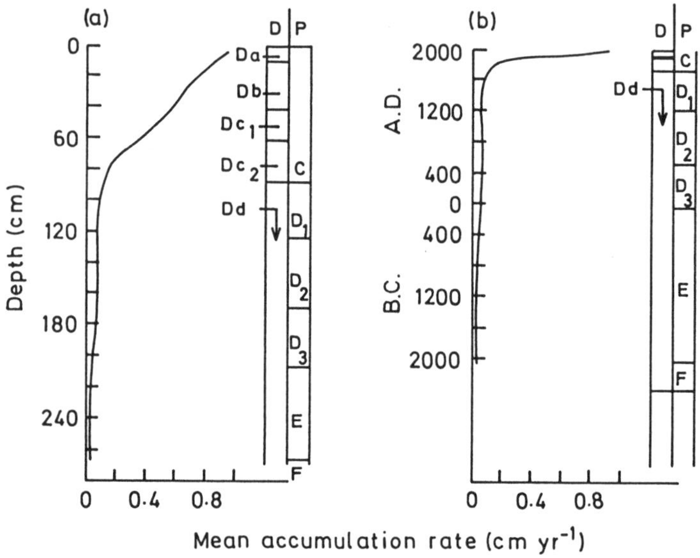

Figure 7A.15. Mean sediment accumulation rate plotted against (a) depth and (b) time, calculated from the data from fifteen cores. Redrawn from Battarbee (1978a).

Table 7A.6. Estimates of contemporary sediment accumulation rate using different methods. The apparently high rates for the stratigraphic marker values may be due to an initial sinking of the sand layer in the uppermost mud

Method	Rate (cm a^{-1})	Sample no.	Dates	Source
$^{210}Pb/^{137}Cs$	0.95	15	1972–73	Battarbee 1978 (a)
Stratigraphic marker	1.90	3	1977–78	Flower 1980
(May 1977)	2.60	3	1978–79	
12 month sediment trap	0.81	2	1976–77	Flower 1980
	1.26	2	1977–78	
	1.11	2	1978–79	

has continued to increase up to the present. The present rate of accumulation estimated in this way is 0.95 cm yr^{-1}, a figure which compares closely with estimates of accumulation rate derived from sediment trapping by Flower (Table 7A.6; Flower, 1980).

7A.5. Conclusion

The surface sediment of Lough Neagh is subject to transportation and mixing processes and rates of sediment accumulation fluctuate spatially by up to a factor of three even for cores located within the same area of the lake and at the same water depth. On the other hand all cores that have been obtained to date contain almost identical biostratigraphic records, and differences in mean sediment accumulation rate between sites are small. As long as a representative number of replicate cores at individual sites are examined, the core record from different parts of the lake is likely to be both conformable and consistent.

The reliability of the chronostratigraphic system is based on the accuracy of the radiometric dating methods used, and on the errors inherent in dating biostratigraphic boundaries by interpolation from time-depth curves. The main weakness at the present time is the lack of a dating system to date sediments which fall between the ^{14}C and ^{210}Pb time-scales, and, until a technique is available for this, the date of 1700 AD assigned to the Dd/Dc diatom boundary remains tentative.

7A.6. References

Appleby, P.G. and Oldfield, F. (1978) The calculation of lead-210 dates assuming a constant rate of supply of unsupported ^{210}Pb to the sediment. Catena 5: 1–8.

Battarbee, R.W. (1978a) Observations on the recent history of Lough Neagh and its drainage basin. Phil. Trans. R. Soc. 281: 303–345.

Battarbee, R.W. (1978b) Biostratigraphical evidence for variations in the recent pattern of sediment accumulation in Lough Neagh, Northern Ireland. Verh. Internat. Verein. Limnol. 20: 624–629.

Carter, C.E. (1976) A population study of the Chironomidae (Diptera) of Lough Neagh. Oikos 27: 346–354.

Carter, C.E. (1977) The recent history of the chironomid fauna of Lough Neagh, from the analysis of remains in sediment cores. Freshwat. Biol. 7: 415–423.

Eakins, J.D. and Morrison, R.T. (1976) A new procedure for the determination of ^{210}Pb in lake and marine sediments. AERE-R 8475. Harwell.

Flower, R.J. (1980) A study of sediment formation, transport and deposition in Lough Neagh, Northern Ireland, with special reference to diatoms. D.Phil. Thesis, The New University of Ulster.

Jewson, D.H., Rippey, B. and Gilmore, W.K. (1981) Losses from sedimentation, parasitism and grazing during the growth, nutrient limitation and dormancy of a diatom crop. Limnol. Oceanogr. 26: 1045–1054.

Mackereth, F.J.H. (1958) A portable core sampler for lake deposits. Limnol. Oceanogr. 3: 181.

Mackereth, F.J.H. (1969) A short core sampler for subaqueous deposits. Limnol. Oceanogr. 14: 145–151.

McCallan, M.E. (1975) A study of the Late-Flandrian sediments of Lough Gall, Co. Armagh, in relation to local agricultural history. D. Phil. Thesis, The New University of Ulster.

Oldfield, F., Appleby, P.G. and Battarbee, R.W. (1978) Alternative ^{210}Pb dating: results from the New Guinea Highlands and Lough Erne. Nature 271: 339–442.

O'Sullivan, P.E., Oldfield, F. and Battarbee, R.W. (1973) Preliminary studies of Lough Neagh sediments. I. Stratigraphy, chronology and pollen analysis. In: H.J.B. Birks and R.G. West (eds.) Quaternary Plant Ecology. Oxford, Blackwell, pp. 267–278.

Pennington, W., Cambray, R.S. and Fisher, E.M.R. (1973) Observations on lake sediments using fallout ^{137}Cs as a tracer. Nature 242: 342–346.

Rippey, B. and D.H. Jewson (1982) The rates of sediment-water exchange of oxygen and sediment bioturbation in Lough Neagh, Northern Ireland. Hydrobiologia 92: 377–382.

Rippey, B., Murphy, R.J. and Kyle, S.W. (1982) Anthropogenically derived changes in the sedimentary flux of Mg, Cr, Ni, Cu, Zn, Hg, Pb, and P in Lough Neagh, Northern Ireland. Environ. Sci. Technol. 16: 23–30.

Smith, A.G., Pearson, G.W. and Pilcher, J.R. (1973) Belfast Radiocarbon Dates V. Radiocarbon 15: 212–228.

Thompson, R., Battarbee, R.W., Oldfield, F. and O'Sullivan, P.E. (1975) Magnetic susceptibility of lake sediments. Limnol. Oceanogr. 20: 687–698.

7. The recent sediments of Lough Neagh
Part B. Diatom and chironomid analysis

R.W. BATTARBEE and C. CARTER

7B.1. Introduction

A comparison of present conditions in Lough Neagh with those described by Dakin and Latarche (1913) and West and West (1902, 1906) suggest that substantial shifts in the flora and fauna of the lake have taken place during the twentieth century. However, because of the unreliability of some of these early data and since there are no detailed studies between 1910 and the present (1968 onwards) it is impossible to use the early records on their own as a means of evaluating either the nature or time-scale of the changes that have taken place. Attempts to trace changes in the recent ecology of Lough Neagh have therefore been mainly based on microfossil analysis of the recent sediments (Battarbee, 1978; Carter, 1977) and on an interpretation of the differences between the early diatom records of Dakin and Latarche (1913) and the diatom record in the sediment (Battarbee, 1976, 1979).

7B.2. Diatoms

7B.2.1. *Modern diatom assemblages*

The diatom assemblage found in the surface sediment is mainly composed of planktonic forms. The abundance of plankton (usually more than 90% of the total) reflects both the high productivity of the diatom phytoplankton (cf. Gibson *et al.*, 1971; Gibson, 1981) and the restriction of periphytic diatoms to shoreline habitats. The periphytic forms found in sediment cores have therefore been transported considerable distances before being incorporated in the sediment and only poorly represent the composition and diversity of periphytic communities. The periphytic spectrum is consequently biased in favour of small epipelic and epipsammic forms such as *Opephora martyi*, *Fragilaria construens* and varieties, *Achnanthes clevei*, *Amphora ovalis* v. *pediculus*. The planktonic component of the assemblage, on the other hand, clearly reflects the floristic composition and pattern of dominance of

R.B. Wood and R.V. Smith (eds.), Lough Neagh, 133–147.

Table 7B.1. Lough Neagh diatom phytoplankton 1968–1982 (see also Table 10.2, p. 207)

Lough Neagh diatom phytoplankton 1968–1982
Melosira italica subsp. *subarctica* O. Mull.
M. ambigua (Grun.) O. Mull.
M. islandica subsp. *helvetica* O. Mull.
Stephanodiscus astraea (Ehr.) Grun.
S. astraea v. *minutula* (Kütz) Grun.
S. dubius (Fricke) Hust.
S. hantzschii Grun.
S. tenuis Hust.
S. invisitatus Hohn and Hellerman
Cyclotella comta Kütz.
C. pseudostelligera Hust.
C. atomus Hust.
C. meneghiniana Kütz.
Asterionella formosa Hassall
Tabellaria flocculosa (Kütz) Knudson

[1]The epithet 'astraea' has been shown to be illegal (Håkansson and Locker, 1981; Round, 1981). This taxon is *not* synonymous with *S. rotula* Kütz. as described by Round (1981) and at this moment has not been assigned a new specific epithet. [2]= *Cyclostephanos dubius* (Fricke) Round (1982). The nomenclature of Battarbee (1978) is used here and in the text.

the diatom phytoplankton as recorded since 1968 by the FBIU and Traad Point Laboratories (Gibson *et al.*, 1971; Gibson, 1981). The flora is dominated by *Melosira italica* subsp. *subarctica* O. Mull. and *Stephanodiscus astraea* (Ehr.) Kütz. Many small *Stephanodiscus* and *Cyclotella* species also commonly occur (Table 7B.1).

7B.2.2. *Past diatom assemblages: evidence from old algal records*

References to the diatom flora of Lough Neagh ranging from the mid-nineteenth century to the present occur in the scientific literature (Table 7B.2). Of these the papers by West and West (1902, 1906) and Dakin and Latarche (1913) are the most detailed. However, because of taxonomic

Table 7B.2. References to diatom samples from Lough Neagh from 1853 to 1960

Collector	Publication date	Sampling date
Foged	1960	1953
Round	1959	1953
Macan and Lund	1954	1951
Dakin and Latarche	1913[1]	1910–11
West and West	1902,[1] 1906	1900–01
O'Meara	1875	?
Smith	1853[2]	?

[1]Some samples extant; [2]Some slides extant.

problems these early descriptions can be difficult to interpret (cf. Battarbee, 1976, 1979, 1981). Fortunately some of the original samples of West and West and Dakin and Latarche are still extant in the British Museum (Natural History), and a number of W. Smith's slides are in the Van Heurck collection in Antwerp. This material is being re-examined in order to make a direct reconstruction of past diatom assemblages, and as a means of evaluating the documentary accounts. So far only the Dakin and Latarche material has been examined in detail.

Dakin and Latarche took plankton samples at approximately three-week intervals between February 1910 and February 1911. The diatoms they reported are listed in Table 7B.3b. If this list is compared to the list of planktonic taxa found in the sediment which accumulated at about the same time (Table 7B.3a) a number of differences can be observed. Some of the differences are due to the use of synonyms. *Melosira crenulata* is a synonym for *Melosira italica* (Ehr.) Kütz, and *Tabellaria fenestrata* var. *asterionelloides* has been renamed *Tabellaria flocculosa* var. *asterionelloides* by Knudson (1952). Korner (1970) regards *Asterionella formosa* and *A. gracillima* as conspecific, and *Synedra revaliensis* Lemm is synonymous with *S. acus* var. *delicatissima* (W. Sm.) Grun, a diatom which is only rarely found in the sediments (Battarbee, 1973b). On the other hand, *Coscinodiscus lacustris*, a

Table 7B.3. Comparison of (a) sediment, (b) plankton as recorded by Dakin and Latarche (1913) and (c) probable plankton after re-examination of Dakin and Latarche samples

a. Sediment 1910 ± 5 (Battarbee, 1978)	b. Plankton 1910–1911 (Dakin and Latarche, 1913)	c. Probable plankton 1910–1911 (Dakin and Latarche, 1913; Battarbee, 1979)
Melosira italica subsp. *subartica* O. Mull.	*M. crenulata* Kütz.[1]	*M. italica* subsp. *subarctica*
M. ambigua (Grun.) O. Mull.	*M. italica* subsp. *subarctica*	*M. ambigua*
M. islandica subsp. *helvetica* O. Mull.	*M. islandica* subsp. *helvetica*	*M. islandica* subsp. *helvetica*
Cyclotella comta (Ehr.) Kütz.	*St. astraea*[2]	*C. comta*
C. comensis Grun.	*Coscinodiscus lacustris* Grun.[3]	*St. astraea*
C. ocellata Pant.	*T. fenestrata asterionelloides* Grun.[4]	*St. astraea* v. *minutula*[7]
Stephanodiscus astraea (Ehr.) Grun.	*A. gracillima* (Hantzsch) Heib.[5]	*St. dubius*
St. astraea v. *minutula* (Kütz.) Grun.	*A. formosa* Hass.	*T. flocculosa* v. *asterionelloides* (Grun.) Knudson
St. dubius Hust.	*F. crotonensis*	*A. formosa*
Tabellaria flocculosa (Grun.) Knudson	*Sy. revaliensis*[6]	*F. crotonensis*
Asterionella formosa Hass.		*Sy. acus* v. *delicatissima* (W. Sm.) Grun.
Fragilaria crotonensis Kitt.		
Synedra rumpens Kütz.		

[1]Synonym for *M. italica*; [2]probably equivalent to *C. comta*; [3]equivalent to *St. astraea*; [4]synonym for *T. flocculosa* v. *asterionelloides*; [5]equivalent to *A. formosa*; [6]synonym for *S acus* v. *delicatissima*; [7]not reported by Dakin & Latarche but present in Nov. 11, 1910 sample.

species reported by Dakin and Latarche and by West and West (1902, 1906), has no taxonomic equivalent in the sediment.

Two of Dakin and Latarche's samples, 11th November, 1910, and 3rd February, 1911, have been traced, and since Dakin and Latarche listed their identification and counts sample by sample (Dakin and Latarche, 1913) direct comparisons with the reported counts can be made. *Coscinodiscus lacustris* is absent from the samples and it is apparent that the taxon concerned refers to *Stephanodiscus astraea* (Battarbee, 1976, 1979), an error perpetuated by Dakin and Latarche (1913) stemming from switched figure captions in the earlier descriptions by West and West (1902, 1906). It is consequently presumed that the *Stephanodiscus astraea* referred to by Dakin and Latarche is equivalent to *Cyclotella comta*. Table 7B.3c lists the probable species list for 1910–11 using the taxonomic concepts of Battarbee (1973a, 1978).

If the Dakin and Latarche data are compared to the sediment record on this basis (cf. Table 7B.3a and c) the agreement is very close. The main exceptions refer to *Synedra acus*, perhaps more common in the plankton than in the sediment because of poor preservation, and *Cyclotella comensis* and *C. ocellata* which are not present in the Dakin and Latarche samples but are present in the earlier samples of West and West (Battarbee, unpub.). In this case the difference may be due to sediment mixing or to a small error in the ^{210}Pb dating (cf. Battarbee, 1979).

A comparison of the 1910–11 diatom plankton (Table 7B.3c) with the present (Table 7B.1) shows that there has been an assemblage change towards *Stephanodiscus* forms. Moreover, *Tabellaria flocculosa* and *Asterionella formosa*, two of the dominant taxa in 1910–11 are hardly ever seen in modern samples (Gibson, pers. comm.).

7B.2.3. *Past diatom assemblages: evidence from sediment cores*

A more continuous record of change in the diatom flora of Lough Neagh can be obtained from microfossil analysis of the sediment. Fig. 7B.1 is a diatom diagram of sediment core SMiii and shows how the percentage frequency of the planktonic taxa varies over the upper 80 cm of sediment. The diagram has been divided into 4 biostratigraphic zones, Da, Db, Dc, and Dd, and the spectrum representing approximately 1910–11, the Dakin and Latarche sample year, occurs just below the Db/Dc boundary.

The earliest zone, Dd, represents a period of extreme stability. The flora is dominated by *Melosira italica* subsp. *subarctica*, with *Cyclotella comta*, *C. comensis*, *C. ocellata*, *Asterionella formosa*, *Tabellaria flocculosa* and *Stephanodiscus* sp. *B* present throughout at low but consistent levels. The upper boundary of the zone has been dated by interpolation (see Part A of this chapter) to about 1700 A.D. The date of the base of the zone is not known but on the evidence from longer cores, the assemblage is almost exactly the same downwards for a period representing at least 4000 years (Battarbee, 1978). Despite the fact that considerable population expansion,

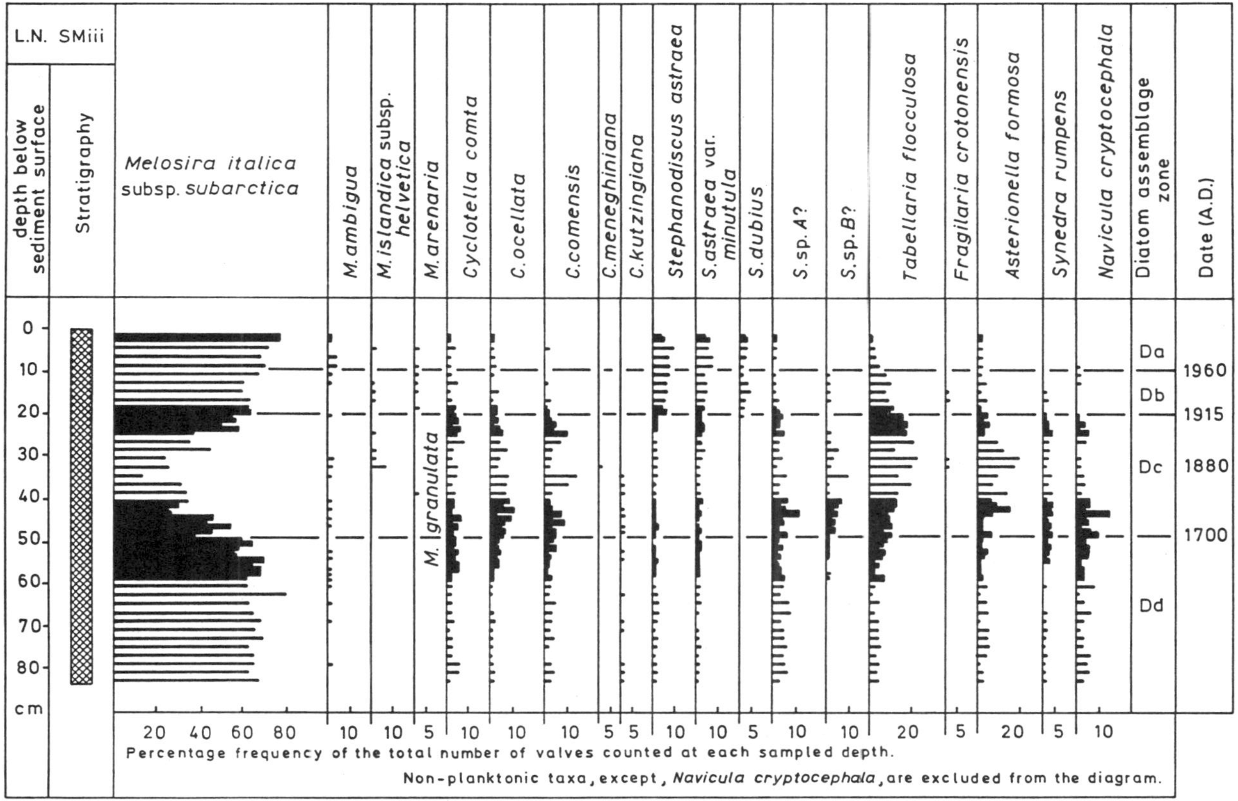

Figure 7B.1. Diatom diagram for core LN SMiii showing the position and approximate dates of zone boundaries referred to in the text. Redrawn from Battarbee (1978).

forest clearance, and agricultural activity was occurring during this time, it seems that little or no change in the ecological character of the lake was caused. The diatom flora is characteristic of an alkaline mesotrophic lake.

A major change in the diatom assemblage is registered in the diagram at a depth of 50 cm (Dc/Dd zone boundary). It is characterized by the rapid decline of *Melosira italica* subsp. *subarctica*. It is likely that this change, coming after such a long period of stability, reflects the beginning of human impact on the lake. Documentary historical evidence indicates that unprecedented forest clearance and agricultural expansion took place in the 17th and 18th centuries as Scots and English farmers settled in Ulster (Chapter 6), and the effect of these events can be recognised in the pollen record for core SMiii (cf. Battarbee, 1973a). Slightly before the Dd/Dc boundary, oak pollen values decline rapidly, and a decline in the proportion of *Isoetes*, an aquatic plant which does not tolerate silting, suggests there was an increase in soil erosion. A decrease in loss on ignition values (Part A of this chapter) and sharp increases in cation concentrations at the same time (Rippey *et al.*, 1982) point in the same direction. These conditions persist during the early part of zone Dc and it is tempting to hypothesise that the increase in *Asterionella formosa* and the continued decline of *Melosira italica* subsp. *subarctica* to a minimum at about 35 cm reflects continued soil disturbance as arable cultivation developed to a maximum in the early 19th century. If this is so then the recovery of *Melosira* values and decline in *Asterionella* between 35 and 25 cm may be related to the restabilisation of soils as agriculture shifted from cereal to grassland cultivation following the mid-nineteenth century famines in Ireland. This interpretation is speculative but it is consistent with the ^{210}Pb date for the *Melosira* minimum of c. 1880 A.D.

The diatom spectrum at the end of the nineteenth century (i.e. at about 25 cm in SMiii) is quite similar to that found during the stable conditions represented by zone Dd, and it may indicate that the lake was beginning to return to its pre-1700 state. However, further changes in the diatom flora took place early in the twentieth century and represent the beginning of a new influence on the ecology of the lake viz. nutrient enrichment from sewage effluent. The diatom changes are indicated in the diagram by the Dc/Db zone boundary which is dated to about 1915 (Fig. 7B.1). There are decreases in the proportion of *C. ocellata* and *C. comensis*, species mainly found in oligotrophic to mesotrophic lakes, and increases in *Stephanodiscus astraea*, *S. astraea* var. *minutula*, and *S. dubius*, taxa associated with eutrophic lakes. *Navicula cryptocephala*, a common benthic diatom also decreases markedly, perhaps because of a decrease in the area of lake bottom receiving light as the growth of larger plankton crops reduce the depth of the euphotic zone (see Chapter 4). Historical evidence shows that many of the small towns in Ulster began to discharge sewage directly into streams and rivers in the catchment during the late 19th and 20th centuries and the gradual extension of water closet sanitation in the early decades of the 20th century probably led to a gradual but continued increase in the rate of nutrient enrichment (see Chapters 8 and 9). A more rapid increase in this progressive eutrophication process is indicated by the increase in *S. astraea* var. *minutula* and decline of *Tabellaria flocculosa* which occurs at about 10 cm in Fig. 7B.1 (Db/Da) and is dated to about 1960. Small *Stephanodiscus* forms are especially characteristic of highly eutrophic systems (Battarbee, 1978; Bradbury, 1975) and the near elimination of *Tabellaria flocculosa* from the lake may be due to the inability of this species to compete for available silica in the spring. In more recent years other small *Stephanodiscus* species, *S. hantzschii*, *S. invisitatus*, and *S. tenuis*, and two small *Cyclotella* species characteristic of nutrient rich lakes, *C. atomus* and *C. pseudostelligera*, have begun to occur. These are not represented in the diagram (Fig. 7B.1) because core Smiii was taken in 1969.

This post-war, and especially post-1960, eutrophication may be due to a combination of factors: the more rapid installation of main sewage services since the war, the construction of large numbers of fully serviced local authority housing estates in the 1950s and 1960s, the development of septic tank sanitation for rural populations, and the introduction and spread of phosphate-rich synthetic washing powders from the mid-1950s. This interpretation is consistent with the sources of nutrients described by Smith in Chapter 6.

7B.2.4. *Changes in diatom accumulation rate*

Further information on the history of nutrient enrichment can be gained by considering the past productivity of planktonic diatom communities. If it is assumed that diatom production in a lake increases with nutrient enrichment, at least until silica becomes a seasonally limiting nutrient, then changes in the accumulation rate of planktonic diatoms in the sediment should be closely related to past changes in production (Battarbee, 1978). However, the accumulation rate of diatoms in the sediment also depends on changes in the magnitude of outflow and preservation losses and on post-depositional movement of sediment within the lake basin. For Lough Neagh, outflow losses have been calculated to about 6% in 1969 and 8% in 1970 (Battarbee, 1978), and although silica is released from the sediment annually (Chapter 9), preservation appears good. The origin of the released silica is uncertain, but were it to be derived from diatoms the quantities concerned might be accounted for by the dissolution of girdle bands and valve rims, the etching of valve surfaces and the widening and coalescence of pores. This is possible without either the entire valve being lost or appearing poorly preserved in the light microscope. The pitted but otherwise well preserved appearance of micro-fossil material examined with the SEM gives some support to this view (Battarbee, 1978). The variability of sediment distribution in Lough Neagh has already been noted (Chapter 7A, Battarbee, 1978). Whilst the mean accumulation rate of sediment appears to be approximately constant throughout the basin, the substantial local variations in accumulation necessitate the

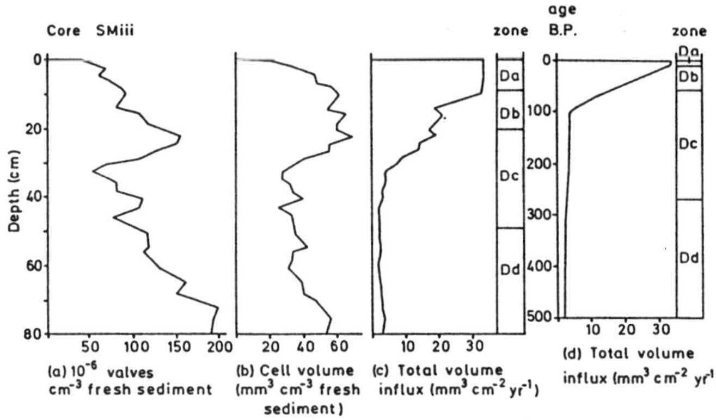

Figure 7B.2. (a) Diatom concentration for core SMiii, converted to (b) cell volume per cm^3 fresh sediment (main planktonic taxa only); (c) data of (b) corrected for accumulation rate and expressed in terms of annual accumulation against depth; (d) annual diatom accumulation plotted against time. Redrawn from Battarbee (1978).

analysis of a range of cores in order to determine mean diatom accumulation rates through time. So far data are available for only one core, SMiii, a core with a relatively slow accumulation rate (Part A of this chapter, Table 7A.5). An interpretation of the results is necessarily tentative.

Diatom accumulation rates are calculated using data for diatom concentration, species cell volume, and sediment accumulation rate. In order to determine the diatom concentration of sample levels for core SMiii the evaporation tray method was used (Battarbee, 1973b). Single diameter traverses of eight replicate coverslips were made and the concentrations of valves per cc fresh sediment were calculated (Fig. 7B.2a). Duplicate counts were carried out at 10 cm intervals and in all cases the mean of the duplicate value fell within the 95% confidence interval of the first sample (Battarbee, 1973b). These data were then corrected for differences in the cell size of the diatoms contributing to the count (Table 7B.4) enabling the values of Fig. 7B.2a to be expressed in terms of diatom volume (mm^3) per cm^3 fresh sediment (Fig. 7B.2b). The effect of this correction is clear, as the large diatom *Stephanodiscus astraea* becomes more important in the upper 30 cm.

If the values represented in Fig. 7B.2b are multiplied by the sediment accumulation rate calculated for each level (see Part A of this chapter), the data can be expressed in terms of diatom volume accumulation per cm^2 per year. Figure 7B.2 shows these values plotted both against depth (c) and against time (d). Similar expressions can also be calculated for individual taxa (Battarbee, 1978).

Despite the uncertainties over preservation and the representativity of only one core these data show increases in the total biovolume of diatoms accumulating annually in the sediment which might be expected on the basis of the floristic changes that have taken place. Data from cores with differing

Table 7B.4. Diatom cell volumes of selected taxa from Lough Neagh

Taxon	μm^3
Melosira italica subsp. *subarctica*	494
M. ambigua	280
M. islandica subsp. *helvetica*	3,825
Cyclotella comta	1,700
C. comensis	100
C. ocellata	534
Stephanodiscus astraea	11,536
S. astraea var. *minutula*	695
S. dubius	511
Tabellaria flocculosa	2,137
Asterionella formosa	408

accumulation rates are needed but it is probable that the spatial variations in diatom concentration and sediment accumulation rate are not large enough to mask the general trends of the curves shown here. Accordingly the large increase in diatom accumulation recorded in SMiii over the last 100 years is likely to represent a real increase in planktonic diatom production during that period.

7B.3. Chironomids

The use of chironomid head capsules in the interpretation of lake histories has been reviewed by Frey (1964), Hofmann (1971a) and Stahl (1969) for long term studies, and has been done by Warwick (1975) and Wiederholm (1979) in recent studies comparable to those on Lough Neagh. There are difficulties with the amounts of material recovered, and identification of the head capsules, but the technique has proved useful (e.g. Hutchinson, 1970; Warwick, 1975). This is particularly so in studies like those on Lough Neagh where there have been complementary palaeolimnological studies on the same cores (O'Sullivan *et al.*, 1973; Battarbee, 1978; Battarbee, Part 7B.2 of this chapter) and also a study of the present day fauna (Carter, 1976; Carter, 1978; Carter and Murphy, Chapter 6).

Partial disintegration of head capsules in the sediment means that generally only the mentum is available to aid identification which is therefore usually possible only to genus level. The keys of Bryce (1960), Bryce and Hobart (1972), Chernovskii (1949) and Hofmann (1971b) were used and, where more than one instar was present, only fourth instar capsules were counted to avoid duplication. Results were obtained from four cores: Smiii, Smv; B43; and AB46. The first three cores had a similar 'slow' sediment accumulation rate and the last a 'fast' rate, approximately twice that of the others (see Chapter 7A for details of 'fast' and 'slow' accumulation rates).

7B.3.1. *Past chironomid communities*

The genera present at each sampled level in the four cores are shown in Fig. 7B.3. In all cases *Chironomus* and Tanytarsini were the most abundant members of the fauna at all depths, followed by *Glyptotendipes*, *Microtendipes* and *Procladius*. The material available from Smiii was rather sparse but *Chironomus* was present throughout the core, Tanytarsini from 27 cm upwards and *Stempellina* did not occur above 30 cm. More material was available from the other cores. In Smv *Chironomus* was abundant above 27 cm but much less so below, while Tanytarsini were present throughout the core, comprising the main group below 27 cm, less than 25% for 27–17 cm and then 40–46% of the fauna, between 17 cm and sediment surface. Orthocladiinae were more abundant below 20 cm than above and *Stempellina* reached a maximum at 29 cm and was not found above 21 cm. AB46 ('fast' accumulation rate) was very similar, with *Chironomus* and Tanytarsini the major components, 25–50% of the fauna each throughout the core. Orthocladiinae were more abundant below 40 cm and *Stempellina* was not present above 38 cm except for a single occurrence at 10–12 cm. The decrease in Orthocladiinae and *Stempellina* in these two cores corresponds to the Db/Dc boundary of Battarbee (see Part A of this chapter) where he found a decrease in taxa associated with oligotrophic-mesotrophic lakes and an increase in ones characteristic of eutrophic lakes. This is a similar change, Orthocladiinae being associated with oligotrophic lakes and *Stempellina* favouring detritus-poor sediment (Thienemann, 1954). Core B43 shows the same pattern as the other cores for *Chironomus* and Tanytarsini, but Orthocladiinae and *Stempellina* persisted further up the core. This seems unlikely to be due to a littoral influence because prominent littoral species like *Stictochironomus* were not found.

Changes in chironomid taxa are not as obvious in these cores as those shown by the diatoms, but the composition of the community has perceptibly changed. Below 20 cm (40 cm in Ab 46), *Chironomus* was generally less abundant than Tanytarsini, Orthocladiinae and *Stempellina* were consistently present and a variety of other Chironomidae were present, although *Glyptotendipes* was less abundant than higher up the core. This type of community would be characteristic of a mesotrophic lake in the typology systems of Thienemann (1913) and Brundin (1958). As discussed in Section 7B.2 the lough catchment at that time was subject to the effects of forest clearance and increasing agriculture, and the chironomid community that existed prior to that is not known. The Dc/Db boundary (1915) approximates to the beginning of nutrient enrichment at which Orthocladiinae and *Stempellina* became much less abundant, Chironomus increased and Tanytarsini decreased to some extent, and *Glyptotendipes* also became more abundant. This is a community characteristic of a eutrophic lake (Brundin, 1958). Battarbee could distinguish a further change (Db/Da boundary) in the diatom community consistent with an increasing rate of nutrient enrichment but this

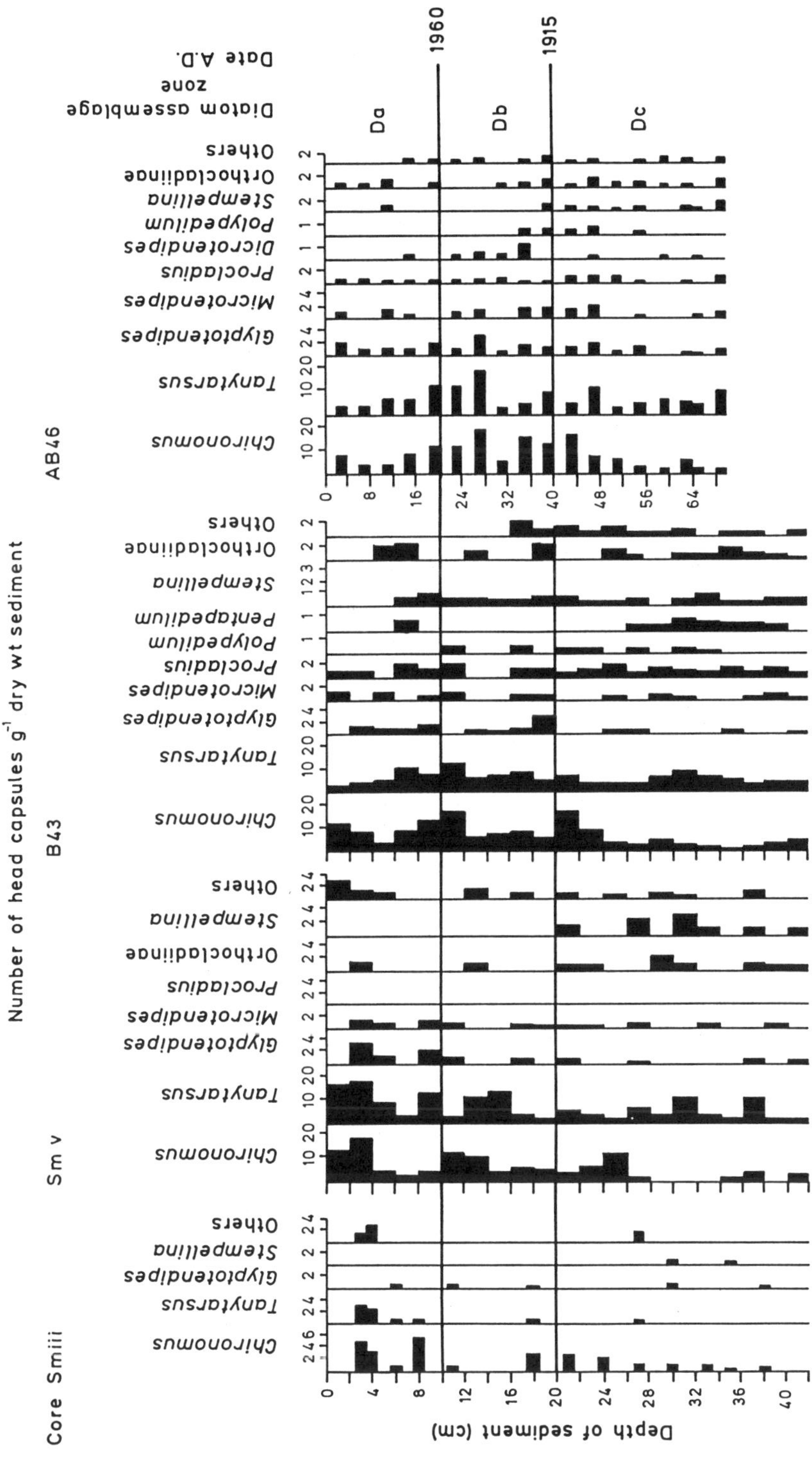

Figure 7B.3. Species compositioı of chironomid fauna at sampled depths in the sediment. Redrawn from Carter (1977).

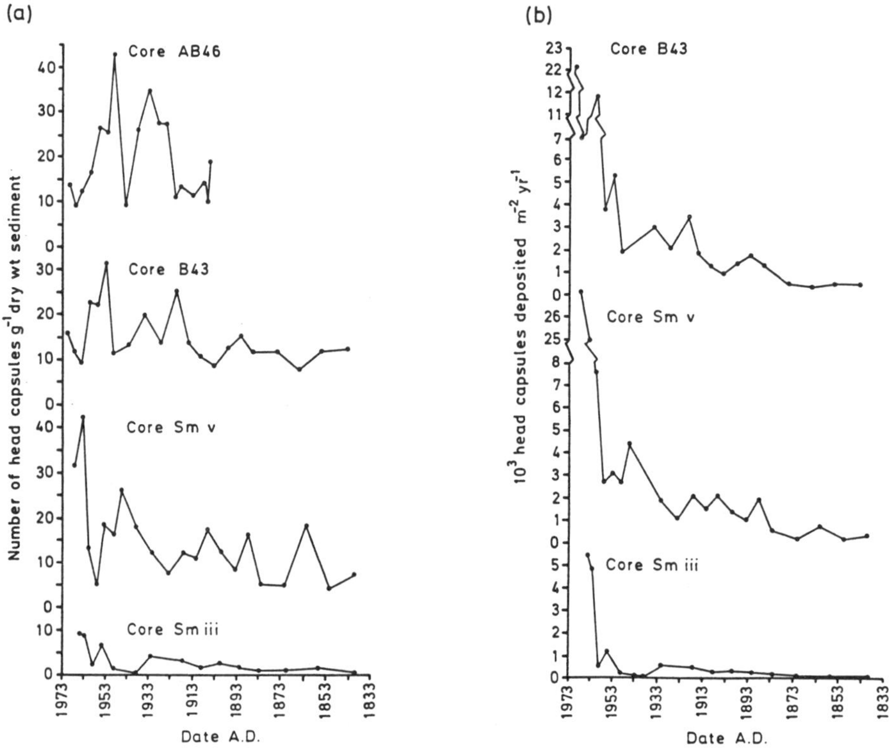

Figure 7B.4. (a) Total number of fourth instar head capsules. (b) Number of fourth instar head capsules accumulated in the sediment. Redrawn from Carter (1977).

is not reflected in the qualitative composition of the chironomid community. The species composition of the community at the surface of the cores is a good approximation to the living community (Carter 1976; Carter 1978, also Chapter 16). Tanytarsini seem to be overrepresented in the sediment record compared to their recent percentage abundance but this could be due to the fact that, when they are numerous, the commonest species appears to have three generations a year, compared to the one a year of *Chironomus* (Chapter 16) and so the Tanytarsini numbers may be proportionally exaggerated.

7B.3.2. *Changes in chironomid accumulation rate*

If numbers of head capsules in the sediment at each level are plotted in relation to time (Fig. 7B.4a), there are wide fluctuations in numbers, but certain common trends can be distinguished. Numbers in Smiii are low, probably because of the small sample size, but the other cores generally had 5–45 head capsules g^{-1} dry wt sediment. Numbers fluctuated around a fairly steady level until c. 1910–1920 when a slow increase began. This continued, steepening towards the present, until near the surface of the cores, where there was a fall in the number of head capsules found. Accumulation rate of head capsules can be calculated if the sediment accumulation rate is known, in the same way as it was for diatoms (see Section 7B.2.4). The results are shown in Fig. 7B.4b. Data are available for Smiii and these have also been applied to Smv and B43, the cores with similar sediment accumulation rates. This reveals that the trend of increasing number of head capsules up the core is very pronounced. Smiii shows a slight increase in accumulation rate from c. 1880 and a rapid one from c. 1950, and the other cores both show increases post-1880 with a very rapid increase post-1945. The first increase slightly predates Battarbee's Dc/Db boundary but the later acceleration of accumulation rate corresponds to the Db/Da boundary when a change in diatom taxa was observed. The chironomid fauna here has responded quantitatively rather than qualitatively. The drop in numbers seen at the surface of the cores in Fig. 7B.4a is revealed as due to the low sediment density at the surface rather than an actual decrease in deposition. For comparison with contemporary fauna, the peak fourth instar chironomid population in Lough Neagh in 1970 was 10,000 m^{-2} (Carter, 1976).

7B.4. Conclusion

Although diatoms and chironomids represent different trophic positions in the structure of the Lough Neagh ecosystem, the sedimentary record of both groups indicates a progressive enrichment of the Lough from the end of the nineteenth century to the present day. This is seen both in changes in species and changes in abundance. The beginning of the enrichment process seems to have begun in the late nineteenth century when the sediment records an increase in the percentage of *Chironomus* and increases in the abundance of both diatoms and chironomids. This was followed in the early part of the twentieth century by a pronounced change in the structure of the diatom and chironomid communities. *Stephanodiscus* sp. replaced *Cyclotella* sp. in the diatom record and *Stempellina* disappeared from the chironomid record. The latest change occurred in post-war years, probably about 1960 when *Tabellaria flocculosa* declined strongly and large increases in the accumulation rate of both chironomids and diatoms took place.

Other palaeolimnological analyses of the recent sediments would be valuable. A recent preliminary study of zooplankton remains has shown the disappearance of both *Bosmina longirostris* (O.F. Muller) and *B. coregoni* Baird which were the summer dominants in 1910–11 (Dakin and Latarche, 1913) but not found today (Chapter 14), confirming from another trophic level the recent and rapid changes in populations in Lough Neagh. Palaeopigment analyses might allow the history of the blue–green algal expansion in the lough to be traced.

The data already available, however, clearly indicate that the present highly enriched state of the Lough is the result of recent cultural eutrophication. In terms of lake management (see Smith, Chapter 8), a reduction in nutrient levels to those pertaining prior to 1960 is likely to cause a significant decrease in productivity, although a return to a mesotrophic system requiring a nutrient reduction to the nineteenth century level, could well be much more difficult.

7B.5. References

Battarbee, R.W. (1973a) A pollen and diatom study of the Late-Flandrian sediments of Lough Neagh, Northern Ireland. D.Phil. Thesis, The New University of Ulster.

Battarbee, R.W. (1973b) A new method for estimating absolute microfossil numbers with special reference to diatoms. Limnol. Oceanogr. 18: 647–653.

Battarbee, R.W. (1976) *Coscinodiscus lacustris* in Lough Neagh – A case of mistaken identity? Br. Phycol. J. 11: 305–307.

Battarbee, R.W. (1978) Observations on the recent history of Lough Neagh and its drainage basin. Phil, Trans. R. Soc. 281: 303–345.

Battarbee, R.W. (1979) Early algological records: help or hindrance to palaeolimnology? Nova Hedwigia Beiheft 64: 379–394.

Battarbee, R.W. (1981) Changes in the diatom microflora of a eutrophic lake since 1900 from a comparison of old algal samples and the sedimentary record. Holarctic Ecology 4: 73–81.

Bradbury, J.P. (1975) Diatom stratigraphy and human settlement in Minnesota. Geol. Soc. Am. Special Paper 171, 74 pp.

Brundin, L. (1958) The bottom faunistical lake type system and its application to the southern hemisphere. Moreover a theory of glacial erosion as a factor of productivity in lakes and oceans. Verh. internat. ver. Limnol. 13: 288–297.

Bryce, D. (1960) Studies on the larvae of the British Chironomidae (Diptera), with keys to the Chironomidae and Tanypodinae. Trans. Soc. Brit. Ent. 14: 19–62.

Bryce, D. and Hobart A. (1972) The biology and identification of the larvae of the Chironomidae (Diptera). Ent. Gaz. 23: 175–217.

Carter, C.E. (1976) A population study of the Chironomidae (Diptera) of Lough Neagh. Oikos 27: 346–354.

Carter, C.E. (1977) The recent history of the chironomid fauna of Lough Neagh, from the analysis of remains in sediment cores. Freshwater Biol. 7: 415–4233.

Carter, C.E. (1978) The fauna of the muddy sediments of Lough Neagh, with particular reference to eutrophication. Freshwat. Biol. 8: 547–559.

Chernovskii, A.A. (1949) Identificaton of larvae of the midge family Tendipedidae Izdat. Akad nauk SSSR Transl. E. Lees (1961) (K.E. Marshall (ed.). National Lending Library for Science and Technology, Boston Spa, Yorkshire.

Dakin, W.J. and Latarche M. (1913) The plankton of Lough Neagh. Proc. R. Ir. Acad. B. 30: 20–96.

Foged, N. (1960) Notes on diatoms. II Cymbellonitzschia diluviana in Denmark, Northern Ireland and Iceland. Bot. Tidsk. 55: 289–295.

Frey, D.G. (1964) Remains of animals in Quaternary lake and bog sediments and their interpretation. Arch. Hydrobiol. Beih. Ergeb. Limnol. 2: 1–114.

Gibson, C.E. (1981) Silica budgets and the ecology of planktonic diatoms in an unstratified lake (Lough Neagh, N. Ireland). Int. Revue ges. Hydrobiol 66: 641–664.

Gibson, C.E., Wood, R.B., Dickson, E.L. and Jewson, D.H. (1971) The succession of phytoplankton in Lough Neagh 1968–1970. Mitt. int. Verein. theor. angew. Limnol. 19: 146–160.

Håkansson, H. and Locker, S. (1981) *Stephanodiscus* Ehrenberg 1846, a revision of the species described by Ehrenberg. Nova Hedwigia 35: 117–150.

Hofmann, W. (1971a) Die postglaziale Entwicklung der Chironomiden-und Chaoborus Fauna (Dipt.) des Schohsees. Arch. Hydrobiol. Suppl. 40: 1–74.
Hofmann, W. (1971b). Zur Taxonomie und Palokologie subfossiler Chironomiden (Dipt.) in Seesedimenten. Arch. Hydrobiol. Beih Ergeb. Limnol. 6: 1–50.
Hutchinson, G.E. (1970) Ianula: An account of the history and development of the Lago di Monterosi, Latium, Italy. Trans. Am. Phil. Soc. 60: 1–178.
Knudson, B. (1952) The diatom genus *Tabellaria* 1. Taxonomy and Morphology. Ann. Bot. 16: 421–440.
Korner, H. (1970) Morphologie und Taxonomie der Diatameengattung *Asterionella*. Nova Hedwigia 20: 559–724.
Macan, T.T. and Lund, J.W.G. (1954) Records from some Irish Lakes. Proc. R. Irish Acad. 56B: 135–157.
O'Meara, E. (1875–77) Report of the Irish Diatomaceae. Proc. R. Ir. Acad. 2: 235–425.
O'Sullivan, P.E., Oldfield, F. and Battarbee, R.W. (1973) Preliminary studies of Lough Neagh sediments 1. Stratigraphy, chronology and pollen analysis. In: H.J.B. Birks and R.G. West (eds.) Quaternary Plant Ecology, Blackwell, pp. 267–278.
Rippey, B., Murphy, R.J. and Kyle, S.W. (1982) Anthropogenically derived changes in the sedimentary flux of Mg, Cr, Ni, Cu, Zn, Hg, Pb and P in Lough Neagh, Northern Ireland. Environ. Sci. Technol. 16: 23–30.
Round, F.E. (1959) A comparative survey of the epipelic diatom flora of some Irish loughs. Proc. R. Ir. Acad. B60: 193–215.
Round, F.E. (1981) The diatom genus *Stephanodiscus:* An electron-microscopic view of the classical species. Arch. Protistenk 124: 455–470.
Round, F.E. (1982) *Cyclostephanos* – a new genus within the Sceletonemaceae. Arch. Protistenk. 125: 323–329.
Smith, W. (1853–1856) A synopsis of the British Diatomaceae; with remarks on their structure, function and distribution. John von Voorst, 2 vols.
Stahl, J.B. (1969) The uses of chironomids and other midges in interpreting lake histories. Mitt. Int. Verein. Theor. Angew. Limnol. 17: 111–125
Thienemann, A. (1913) Der Zusammenhang zwischen dem Sauerstoffgehalt des Tiefen Wassers und der zusammensetzung der Tiefen fauna unserer Seen. Int. Rev. Hydrobiol. 6: 243–249.
Thienemann, A. (1954) Chironomus. Leben, Verbreitung und wirtschaftliche Bedeutung der Chironomiden. Die Binnengewasser 20: 1–834.
Warwick, W.F. (1975) The impact of man on the Bay of Quinte, Lake Ontario, as shown by the sub-fossil chironomid succession (Chironomidae, Diptera), Verh. int. Ver. Limnol. 19: 3134–3141.
West, W. and West, G.S. (1902) A contribution to the freshwater algae of the North of Ireland. Trans. R. Irish Acad. B32: 1–100.
West, W. and West, G.S. (1906). A comparative study of the plankton of some Irish lakes. Trans. R. Irish Acad. B33: 77–116.
Wiederholm, T. (1979) Chironomid remains in recent sediments of Lake Washington. Northwest Science 53: 251–256.

8. Phosphorus and nitrogen loadings to Lough Neagh and their management

R.V. SMITH

8.1. Introduction

The earliest limnologists tended to regard lake metabolism as self contained processes. This viewpoint was expressed by Forbes (1887): 'The lake forms a little world within itself—a microcosm within itself—a microcosm within which all elemental forces are at work'. It is now abundantly clear that, at the very minimum, it is the lake and its catchment which function as a single system and the enrichment of a lake with nutrients arises because of an intensification of human activity in the lake catchment area. Natural terrestrial ecosystems have been able to establish nearly closed circulation patterns to economise on nutrients which are often in limited supply. However, human activity can induce fluxes of carbon, nitrogen and phosphorus that are comparable or exceed in magnitude the natural nutrient loading rates from catchment systems to lakes.

To assess the importance of human activity on the nutrient loadings to lakes, Stumm and Baccini (1978) calculated a capita loading parameter for a range of important lakes (Table 8.1). The number of inhabitants per unit lake volume is the product of the population density in the catchment area of the lake and the surrounding factor (drainage area/lake area) and the reciprocal of the mean depth of the lake. It can be seen that Lough Neagh has a human loading factor of 105 inhabitants per 10^6 m^3, lake volume. In contrast, Lake Baikal and Lake Superior, which are deep lakes situated in remote areas with low population densities have human factor loadings of 0.6 and 0.05, respectively.

Lough Neagh first gave particular cause for concern in the summer of 1967 when a large crop of the blue-green alga *Anabaena flos-aquae* formed a surface water bloom. The algae were blown downwind into Antrim Bay and their accumulation interfered with three of Lough Neagh's important uses, water abstraction, fisheries and recreation. Antrim Bay is the site of the water intake for Belfast and is also an important recreational area for Antrim. The algae caused the death of eels held in cages, perhaps by algal

R.B. Wood and R.V. Smith (eds.), Lough Neagh, 149–169.

Table 8.1. Comparison of loading parameters of other lakes with Lough Neagh. Data for lakes other than Lough Neagh and Lough Erne (Gibson *et al.*, 1980) are abstracted from Stumm and Baccini (1978)

Lake	Country	(1) Surrounding factor*	(2) Mean depth (m)	(3) Inhabitants (km^{-2})	Inhabitants per 10^6 m^3 lake volume $\frac{(1) \times (3)}{(2)}$
Greifensee	Switzerland	15	19	441	348
Plattensee	Hungary	10	3	60	200
Lough Neagh	*Northern Ireland*	*11.6*	*9.0*	*80*	*103*
Lough Erne	Northern Ireland	38.4	11.9	26	83
Lake Washington	USA	15	18	50	42
Lake Constance	Switzerland/Germany/ Austria	19	90	114	24
Lago di Lugano	Switzerland/Italy	11	130	264	22
Lake Biwa	Japan	4.5	41	150	16
Lago Maggiore	Switzerland/Italy	26	175	92	13
Lake Winnipeg	Canada	35	13	3	8
Lake Geneva	Switzerland/France	13	154	94	7.9
Lago di Como	Italy	22	185	60	7.1
Lake Titicaca	South America	14	100	40	5.6
Lake Victoria	Africa	3	40	70	5.1
Lake Baikal	USSR	17	730	5	0.6
Lake Superior	USA/Canada	1.5	145	5	0.05

* Drainage area/lake area.

toxins, but more probably by the decaying mass of algae removing the oxygen from the water and so suffocating the fish (see Chapters 1 and 9).

This spectacular growth of algae may be interpreted as a response to nutrient fluxes of nitrogen and phosphorus to the lake which both in their relative proportions (N:P ratio) and their magnitude were abnormal. The high human loading factor to Lough Neagh which contributes to the nutrient flux can be reduced only by appropriate environmental protection measures (Smith, 1977). The purpose of the present chapter is to assess which nutrients in Lough Neagh are critical for the growth of algae, to calculate loadings of these nutrients to the lake, to estimate the main sources of these nutrients in the catchment and assess their possible management.

8.2. The study area

Fig. 8.1 shows the Lough Neagh catchment area which covers 4,453.4 km^2 and has an average rainfall of 1028 mm (1941–70 average). Six inflow rivers: Main, Six Mile Water, Upper Bann, Blackwater, Ballinderry and Moyola drain 88% of the total catchment. The geology of the catchment has been described by Wilson (1972) and is summarised in Chapter 2. The soil material is mostly basaltic glacial till and the principal soil associations are acid brown earth and gleys on the lowland and climatic peat above 200 m. Over the study period (1968–80 for concentration data and 1971–1979 for loading data) the major land use was grassland and rough grazing which together covered 85–87% of the area with only an additional 7–9% under cultivation. This is very similar to the data for Northern Ireland as a whole (Fig. 8.2). Farming mainly consists of relatively small mixed livestock farms with milk, beef and pig production being most important. In 1971 the total human population was estimated at 333,600 of which 200,600 were connected to sewage disposal works. By 1979 the total population had increased to 358,500 of which 229,600 were connected to sewage disposal works. The increase in sewered population was greater than the increase in total population, reflecting a general shift of population to towns. The growth in urban population has been concentrated mainly in two towns: Antrim (+11,300) and Craigavon (+7,100). Craigavon Sewage Disposal Works discharges directly to Lough Neagh, while that of Antrim was diverted from the Six Mile Water directly to Lough Neagh in 1976. As a result the urban population whose sewage was discharged within the six river catchments increased by only 3.3% during the study period compared to an increase of 14.4% for the Lough Neagh catchment as a whole.

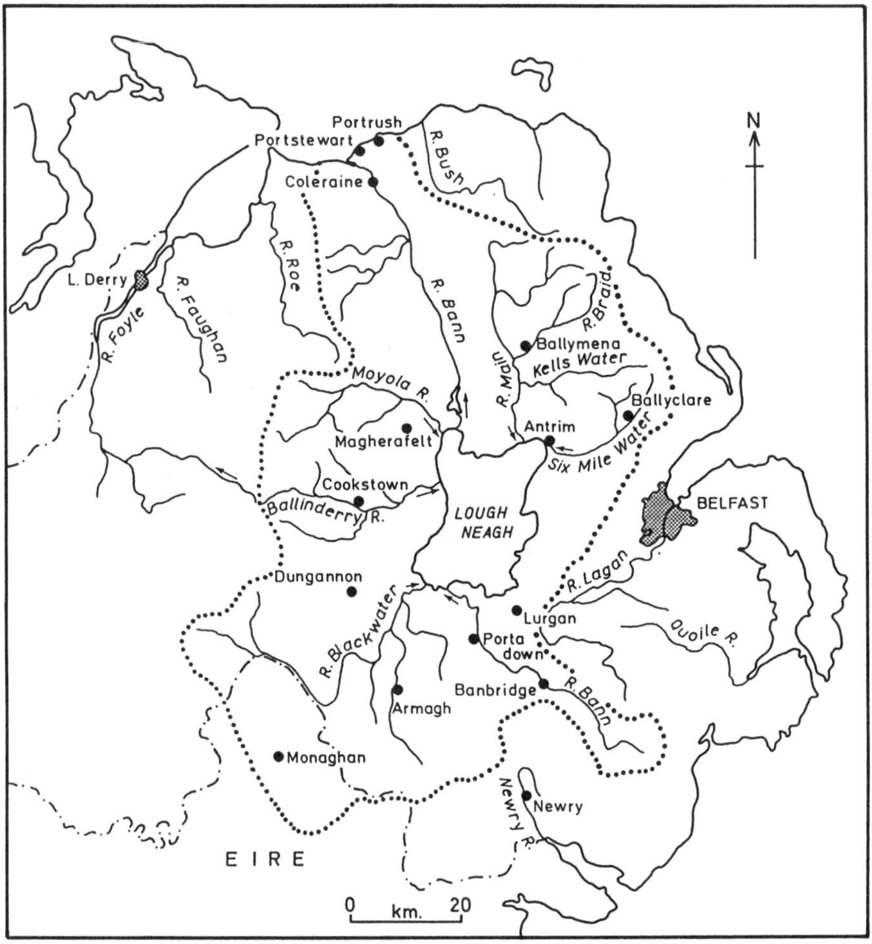

Figure 8.1. The Lough Neagh catchment area.

8.3. Chemical analyses

In the period 1971–74, 0.45 μm membrane filtered samples were stored deep-frozen prior to analysis. Soluble reactive phosphorus (SRP) was determined by the Technicon Auto Analyser 1 method Hendriksen (1965). Kjeldahl N was determined by sulphuric acid digestion followed by steam distillation (American Public Health Association 1971). Nitrate was determined by a Technicon Auto Analyser I method where nitrate was reduced to nitrite by copper sulphate and hydrazine followed by diazotization of the nitrite with N-(-naphthylethylenediamine (Chapman *et al.*, 1967). Soluble organic P (SOP) particulate P (PP) determinations were not made in 1971–1974. In the remaining period 1975–1979, filtered samples were not frozen except for

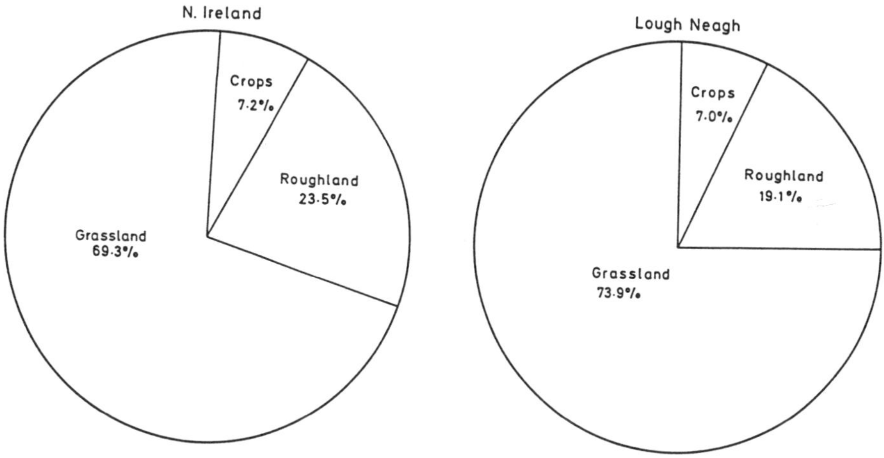

Figure 8.2. Land use in the Lough Neagh catchment area and Northern Ireland during 1979.

nitrate which was analysed as before. SRP was determined manually on the day of sample collection by the Murphy and Riley (1962) method using the 0.45 μm membrane filtered samples. Total P and total soluble P (TSP) were determined by the acid–persulphate autoclave digestion technique of Eisenreich *et al.* (1975). SOP was determined as the difference between TSP and SRP and the difference between TP and TSP was called PP (see Fig. 8.3). Kjeldahl N was determined by a semi-automated ammonia probe method (Stevens, 1975).

8.4. Limiting factors for the growth of algae

Detailed investigations into the limiting factors for the growth of algae in the Lough have been carried out to confirm that P is the critical nutrient in the Lough. Comparison of atomic ratios of the major nutrient elements in the River Main and nutrient elements in algae indicates that P, N and Fe are the elements most likely to be limiting (Table 8.2); other elements are available in excess in the input water.

A second approach used to identify limiting nutrients is to examine correlations between algal crops and nutrients in the water column of Lough Neagh. Observations made in 1977 clearly showed that SRP was depleted more rapidly by the algal crop than nitrate and in this case P appeared to be the critical limiting nutrient (see Chapter 10). However, a more detailed scrutiny of soluble P and algal crop data in other years suggests that the inter-relationships between the two is very complex and cannot be used definitely to indicate nutrient limitation. Analysis of planktonic suspended material for N and P has shown that for most of the time the N:P ratio

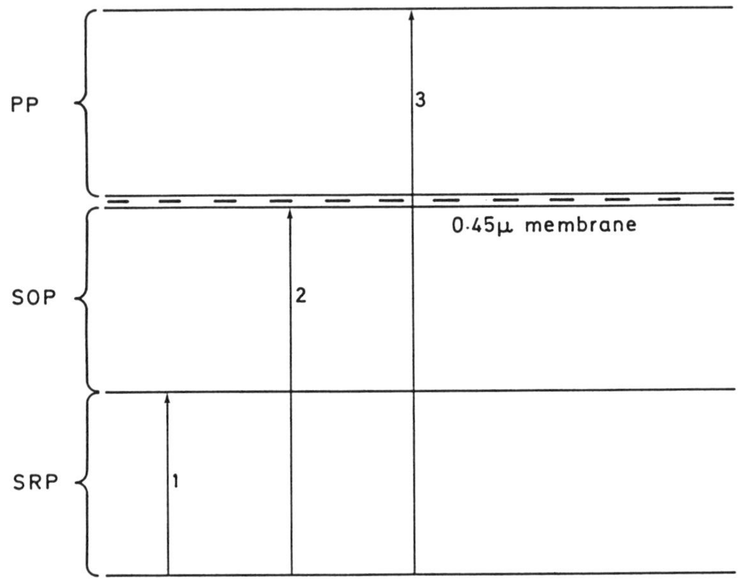

Figure 8.3. Fractionation of P by analytical procedures. The analyses carried out are (1) SRP, (2) total soluble P and (3) total P. PP and SOP are calculated by difference.

Table 8.2. Comparison of atomic ratios of nutrient elements in algae (Healey, 1973) and River Main water (1968–1980 mean concentrations)

Element	Relative atoms		*Algae*
	R. Main	Algae	R. Main Ratio
Mg	332.2	2.1	0.006
Na	298.4	2.4	0.008
Ca	278.8	2.0	0.007
Nitrate N	61.5	35.3	0.573
K	16.4	4.0	0.243
SRP	1.59	3.2	2.012
Fe (Total soluble)	1.00	1.00	1.00

approximates 15:1 (by atoms) which is close to the commonly found values for algae but during some parts of the summer months the ratio rises to 70:1 thus showing good evidence of P limitation. Another indicator of P limitation is reputed to be the production of increased concentrations of the enzyme alkaline phosphatase by P-limited algae. Alkaline phosphatase activity was measured weekly in Lough Neagh during 1974 and 1975 but the activity was found to be associated with algal crops rather than limiting P concentrations. As significant enzyme activity was also found during periods of the year when SRP concentrations were high and algal numbers were low, it is possible

that part of this activity was from river inputs (Stevens and Parr, 1977) and throws considerable doubt on the value of this approach to identifying P as a growth limiting nutrient.

In another approach to this problem algae were used in monthly bioassays of Lough Neagh water to assess nutrient limitation (Parr and Smith, 1976). It was found that all bioassays showed dependence on P for a significant growth response and as the nitrate N to SRP ratio of the inflowing water was 17:1 in the study year (1975) compared to 57:1 in the years 1975–1979, the study emphasised the importance of P as a growth limiting element for Lough Neagh algae.

8.5. Phosphorus availability

Phosphorus, the key element controlling the growth of algae in Lough Neagh, exists in fresh waters in many different physical and chemical forms. The arithmetic mean concentrations from approximately weekly sampling of the 6 major rivers during 1975–1979 were 170, 38 and 56 $\mu g\, l^{-1}$ of SRP, SOP and PP, respectively. From nutrient budget studies, SRP, SOP and PP constituted 54, 16 and 30% of the total P inputs (Table 8.5). There is some debate as to whether SRP is synonymous with orthophosphate because of the acidic conditions of the SRP test. Experimental evidence that SRP was greater than orthophosphate by about 20% was obtained by developing an enzymatic assay for orthophosphate (Stevens, 1979). It is generally accepted, however, that algae can use P measured as SRP for growth. The chemical nature and availability for algal growth of SOP and PP are in the main unknown. As nutrient control programmes are based primarily on lowering the SRP inputs from point sources, the fraction of SOP and PP that becomes available for algal growth is therefore extremely important in considering control measures for diffuse sources of P. The chemical nature of SOP and PP have been investigated and algal availability by bioassay studies have been performed (Stevens and Steward, 1982a and b)

The major difficulty in assessing the availability of P to algae was in relating laboratory results to the river or lake situation. Bioassay studies indicated potentially available P (over the period of the test, which is usually about 2 weeks) and not available P in the natural water body. Factors which eventually affect the availability of PP when discharged to a lake are the temperature, SRP concentration in the lake, the rate of settling or resuspension of the PP, and the density and 'size' of the PP particles. The mean residence time of PP in the photic zone is considered to be the most important factor in any assessment of bioavailability of PP in the actual environmental situation. Drawing on the work of Stevens and Stewart (1982a and b) an estimate of available P in river waters discharging to lakes may be taken as the sum of SRP, 10% of SOP and 20% of PP. The estimate of SOP availability is more likely to be a minimum value while PP availability in

Table 8.3. Relative contribution of effluent from sewage disposal works to the observed nutrient output from the River Main

Nutrient	Sewage effluent loading (kg ha^{-1} yr^{-1})	River loading (kg ha^{-1} yr^{-1})	Contribution of sewage effluent (%)
Soluble reactive P	0.40–0.53	0.65	62.0–82.0
Total P	0.53	1.08	49.0
Nitrate N	1.22	13.68	9.0
Total N	1.66	21.02	8.0
Cl	1.50	135.2	1.0
Na	1.00	95.7	1.0
K	0.45	11.7	4.0
Ca	0.67	132.1	0.5
Mg	0.25	66.5	0.4
Soluble Fe	0.00	0.34	0.0
Soluble Mn	0.00	0.13	0.0

the lake may be negligible because, from budget studies, settling without resuspension appears to occur there (see Chapter 9).

As high alkaline phosphatase activity occurred during the algal bioassays, further availability of SOP due to the degradation of pesters in L. Neagh is unlikely. Degradation by UV light is more probable. Since SRP constitutes the largest component (54%) of the total P input to the Lough and as PP is unavailable, the 10% of the SOP that is available amounts to only an additional 2% of the total P input possibly being available to algae.

8.6. Nutrient budget of the River Main

Given the status of SRP as the critical nutrient in Lough Neagh the next important consideration is what proportion of the SRP and other nutrient inputs to Lough Neagh originate from point sources such as sewage treatment works. Smith (1976) made an intensive budget study of one river catchment, the Main, which drains to Lough Neagh. It has a total size of 709 km^2 and an average rainfall of 1,181 mm. The observed outputs of SRP and total P from the River Main during 1974 were in good agreement with the predicted outputs based on land drainage measurements and intensive surveys of sewage treatment works. The budget indicated that 62% of the SRP and 44% of the total P outputs were derived from sewage disposal works effluent. If as a result of degradation in the river all the P from sewage effluent was in the form of SRP, the relative contribution from this source would be increased to 82% (See Table 8.3). It was observed that less than 10% of the loading of nitrate in the river was derived from sewage effluent. For nutrients other than N and P, sewage effluent made an insignificant contribution to the river

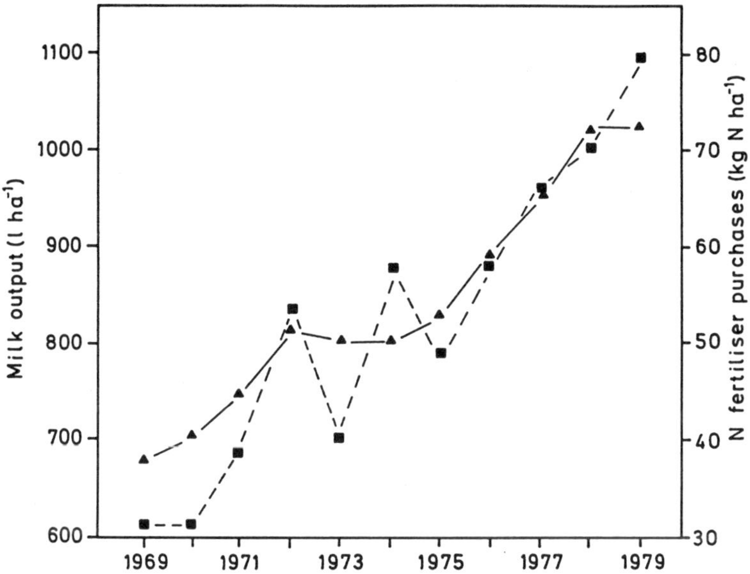

Figure 8.4. Nitrogen fertiliser purchases (■----■) and milk output (▲——▲) in Northern Ireland during the period 1969–1979. Redrawn from Foy *et al.* (1982).

output. Therefore P is the only critical nutrient which is amenable to a reduction in river loading by nutrient removal at sewage treatment works.

8.7. Calculation of nutrient loadings to Lough Neagh

8.7.1. *Sampling*

During the study period (1971–1979) samples were taken every 7 or 15 days between 10.00 and 14.00 h from the 6 major rivers entering Lough Neagh. The sampling stations were at convenient locations as close to Lough Neagh as possible while avoiding backwash from the Lough. Flow data were collected from 7 flow gauging stations and mean daily flows were computed from observations at 8 h intervals. Conversion factors to predict flows entering Lough Neagh are shown in Table 8.4.

8.7.2. *Nutrient load estimation*

In order to predict the annual nutrient loadings, the loads for days on which flow data only were available were estimated using the regression equation (1) which was considered by Smith and Stewart (1977) to be the most useful for predictive purposes.

Table 8.4. Conversion factors to compute total flows entering Lough Neagh from flows at the gauging stations

	(a) Area of basin (km^2)	(b) Rainfall of (a) ($mm\ yr^{-1}$)	(c) Area above gauging location (km^2)	(d) Rainfall of (c) ($mm\ yr^{-1}$)	Conversion factor $\frac{(a) \times (b)}{(c) \times (d)}$
Six Mile Water	301.4	1070	277.2	1080	1.076
Upper Bann	660.6	1023	335.5	1106	1.821
Main	709.2	1181	646.7	1238	1.046
Moyola	312.6	1261	306.4	1261	1.020
Blackwater	1480.3	964	951.3	990	1.515
Balllinderry	430.1	1096	419.4	1110	1.012

$$\text{Load} = a + b \log Q \tag{1}$$

where, Q = mean daily flow.

Regression equations were derived annually for individual rivers from days on which flow and chemical concentration data were available and, by using daily flow observations, daily loads were predicted. Annual loadings were estimated by addition. This method was used for the SRP, SOP, PP and Kjeldahl N loads but was not considered suitable for the nitrate loading as the strong seasonal trend in river nitrate concentrations (Smith *et al.*, 1982) reduced the predictive value of equation (1). The nitrate N loadings were estimated using equation (2).

$$M\ \text{Load} = MQ \times NO_3\text{–N conc} \tag{2}$$

where, M Load = monthly nitrate N load,
MQ = monthly flow,
NO_3–N conc = mean monthly nitrate N concentration.

A comparison of the nitrate N loading predictions using equation (2) and those obtained from intensive daily flow and concentration data from the River Main showed good agreement (Stevens & Smith, 1978). The SOP and PP loadings were available only from 1975–1979 but the remaining inputs have been estimated for the period 1971–1979 (see Table 8.5). Two of the inputs, Kjeldahl N and SOP, showed little change over the period of measurement, but the remaining inputs all showed mimimum values in 1975 which coincided with the low flows of that year. The PP input appears to be strongly correlated with flow but the SRP remained, apart from 1975, quite constant at approximately 210 ± 20 tonnes yr^{-1}. The nitrate N loading showed an upward trend, as would be expected from the increasing concentrations of the period (Smith *et al.*, 1982) with the mean input for the period 1976–1979 of 6,344 tonnes yr^{-1} showing an increase of 53% compared to the 1971–1974 period mean of 4,529 tonnes yr^{-1}.

Table 8.5. N and P loadings from the major rivers to Lough Neagh

	1971	1972	1973	1974	1975	1976	1977	1978	1979
						(tonnes yr^{1})			
Nitrate N	2747	5398	3939	4529	2634	5120	6562	7997	5722
Kjeldah N	1704	2984	2728	1549	1479	2819	2693	2678	3144
SRP	195.0	195.9	220.2	214.6	126.8	187.4	197.4	246.7	247.1
SOP					54.9	65.9	55.3	46.3	68.2
PP					42.2	114.1	120.9	150.4	143.1
Total P	299.7	423.7	413.6	444.2	223.9	367.5	373.6	443.4	458.4
Flow	1781	2200	1989	2473	1510	2223	2407	2703	2824

Table 8.6. Catchment land use variables entered in regression analysis

Variable	Description
1. URBAN POPULATION	Population connected to sewage treatment works. (nos ha^{-1})
2. RURAL POPULATION	Unsewered population (nos ha^{-1})
3. ABATTOIR OUTPUT	Combined output of abattoirs and meat plants in catchment. (kg dead carcase weigh ha^{-1})
4. CREAMERY OUTPUT	Combined output of dairies and milk processing factories. (1 ha^{-1})
5. GRASS DCE*	Combined animal population of cattle and sheep (DCE ha^{-1})
6. INDOOR DCE*	Combined pig and poultry population. (DCE ha^{-1})
7. FARM MILK	Milk output from farms (1 ha^{-1})
8. CROPS	% of catchment under cultivation.
9. GRASS	% of catchment under grass.
10. ROUGHLAND	% of catchment assessed as rough grazing, forest and other land.
11. FLOW	Annual catchment runoff. (m)

* DCE = Dairy Cow Equivalent.

8.8. Multivariate analysis of nutrient loadings

8.8.1. *Soluble reactive phosphorus*

A statistical analysis of the data was undertaken using a stepwise multiple regression analysis to determine which land use and population density variables (Table 8.6) were significantly correlated with chemical area loss rates (Foy *et al.*, 1982). The SRP regression equation (3) is given below.

$$\text{SRP Y} = 0.451 + 0.573\ \text{URBAN POPULATION} + 0.000043\ \text{CREAMERY OUTPUT} - 0.127\ (\text{FLOW})^{-1} \quad (3)$$

$$R^2 = 0.693$$

Urban population density was the most highly correlated variable reflecting the high P content of domestic sewage. The regression coefficient of 0.573 was equivalent to a daily per capita input of 1.6 g P $person^{-1}\ d^{-1}$, a figure in line with direct measurements of domestic sewage in the catchment area (Alexander & Stevens, 1976). $(\text{FLOW})^{-1}$ was also strongly correlated with SRP loss rates suggesting increased leaching from the soil under high flow conditions. However, differences in land use or stocking density were not found to be significant, possibly because of the relatively uniform nature of the six catchments, which are predominantly grassland systems although there are differences in intensity of management which have affected the nitrate-N loss rates. The third significant variable was the output of dairy and milk processing factories (Creameries), which, like the urban population, were point sources. The magnitude of this source however was small as the average output of these plants is 1.23 litres 10^3 ha^{-1} giving an SRP loss rate of 0.054 kg ha^{-1}.

8.8.2. *Nitrate*

The nitrate regression equation (4) showed that although flow was a significant factor, the remaining variables were land use types.

$$\text{Nitrate N } Y = 8.01 + 0.0173 \text{ FARM MILK} + 13.61 \text{ LOG FLOW} - 13.60 \text{ ROUGHLAND} \quad (4)$$
$$R^2 = 0.801.$$

Although nitrate loss rates have increased in all rivers over the period 1971–1979 in line with a period of increasing use of N fertiliser in Northern Ireland, figures for fertiliser usage for individual river catchments are not available. Milk production appears to reflect the increasing use of fertilisers as it is the only sector of agricultural output to show a consistent increase over the study period and the R^2 value of 0.567 suggests that it is associated with over 50% of the variation of the nitrate loss rate. The negative regression coefficient of ROUGHLAND suggested a lower loss rate from unproductive hill land. However, the significance of this variable was low as measured by the F-ratio and the reliability of such values must be in doubt. The high levels of significance of the other two variables in the regression equation suggests that those correlations are in fact real.

8.8.3. *Soluble organic phosphorus and Kjeldahl nitrogen*

Of the three remaining chemicals two, SOP and Kjeldahl N produced very low R^2 values with only one dependent variable showing even a low level of significance. The regression equations (5 and 6) are given below.

$$\text{SOP loss rate} = 0.871 \text{ CROPS} + 0.089 \quad (5)$$
$$R^2 = 0.188$$

$$\text{Kjeldahl N loss rate} = 2.809 \text{ LOGFLOW} + 7.932 \quad (6)$$
$$R^2 = 0.133$$

The Kjeldahl N regression equation had a higher level of significance because there were more observations than for SOP. The relationship between LOGFLOW and Kjeldahl N can be interpreted as due to increased loss rates from the soil under high flow conditions. The relationship between SOP and the proportion of arable land in a catchment is not understood and because of the low level of significance may not in fact be real.

8.8.4. *Particulate phosphorus*

The remaining equation (7) for PP is given below and shows the highest R^2 value found.

$$\begin{aligned} \text{PP Y} &= 0.117 - 0.191\ \text{FLOW}^{-1} + 0.937\ \text{GRASS} \\ &\quad + 0.000554\ \text{ABATTOIR OUTPUT} \\ R^2 &= 0.897 \end{aligned} \tag{7}$$

Flow again with a highly significant factor but the reason for the inclusion of GRASS was not clear and may have been due to the fact that the two rivers with the highest proportion of grassland, U. Bann and Blackwater, were also the most slow flowing of the six rivers and may have a high solids (= particulate) load due to algal growth in the summer months.

8.9. Upward trends in nitrate concentrations

The mean concentration of nitrate in the major rivers has increased from 1.41 mg N l^{-1} in 1969–1970 to 2.41 mg N l^{-1} in 1978–1979 (Smith *et al.*, 1982). The maximum observed concentrations of nitrate have increased from 2.9 mg N l^{-1} in 1969–1970 to 7.5 mg N l^{-1} in 1979–1980. However, all values observed were well below the WHO (1971) permissible nitrate concentration of 11.3 mg N l^{-1}.

During the study period, fertiliser usage has markedly increased in Northern Ireland from 42 kg N ha^{-1} in 1969 to 106 kg N ha^{-1} in 1979. The relationship between nitrate concentration and fertiliser usage have been assessed by regression analysis:

$$\begin{aligned} &y = 0.803 + 0.16x \\ &\text{where, } y = \text{annual nitrate concentration in mg N l}^{-1} \\ &x = \text{the rate of fertiliser usage in kg N ha}^{-1} \\ &\text{and } \tau = 0.835\ (0.001 < P < 0.01,\ 9\ \text{d.f.}). \end{aligned} \tag{8}$$

As it is standard agricultural practice in Northern Ireland to apply fertiliser to the land in the spring period, the changing annual pattern of nitrate concentration was assessed for the periods 1969–1974 and 1975–1979. If there were direct losses of fertiliser to land drainage waters, one might hypothesise that the spring concentration would show a relatively larger increase in the latter half of the study period compared with concentrations

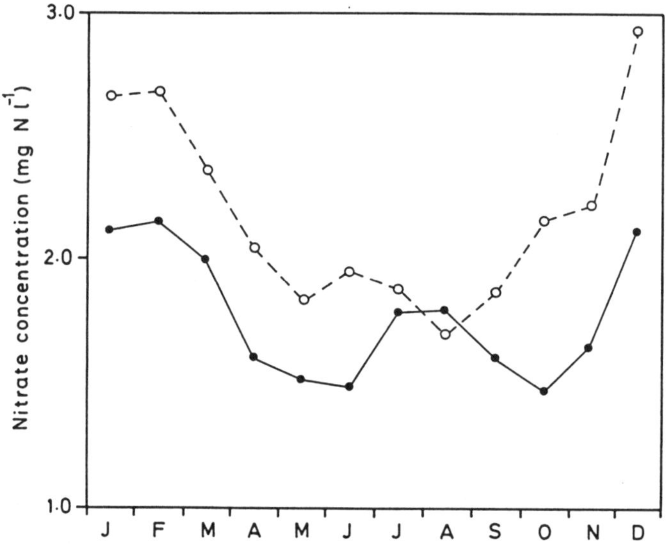

Figure 8.5. Mean montly nitrate concentrations in the rivers discharging into Lough Neagh for the periods 1969–1974 (●——●) and 1975–1979 (○----○). Redrawn from Smith *et el.* (1982).

at other times of the year. However, Figure 8.5 clearly shows that the increase in spring concentration for 1975–1979 was no greater than for other periods in the year. The seasonal change in concentration was similar for both periods. It appears that there was a lower concentration of nitrate in the summer months which may be due to evapotranspiration exceeding precipitation leading to a reduction in leaching and also greater uptake at this time by plant growth. In contrast, the winter period was characterised by poor uptake by plants and a consequential leaching of residual soil nitrate when soil field capacity was exceeded. One must conclude therefore that the increase in nitrate concentration observed in the rivers studied was a consequence of the increase in the intensity of grassland management during the study period and the increase in fertiliser usage, although not directly contributing to the loss of nitrate from the land, nevertheless was probably a major factor in raising the nitrogen status of the soil.

8.10. Management strategies

Since the SRP input is the principal component of the P input to Lough Neagh available for algal growth, it is important to quantify the input sources. Using 1979 URBAN POPULATION and CREAMERY OUTPUT variables for the whole catchment and the 1971–1979 mean flow value, a SRP budget

Table 8.7. Predicted SRP budget to Lough Neagh calculated from equation (3)

Source	Tonnes P yr^{-1}	% Contribution
Sewage treatment works*	137	54
Creamery output	16	6
Background**	103	40
Total	256	100

* Sewage treatment works output includes 5.6 tonnes of the creamery output discharged direct to sewage disposal works.
** Calculated using mean (1971–1979) run-off of 0.576 m.

has been constructed (Table 8.7) which shows that 54% of the SRP originates from STWs, 6% from creamery discharges and the remainder as a background contribution, the main component of which is probably land drainage. The only practical way of reducing the SRP input is to reduce the loading from STWs but removal at all works is difficult as there are 196 in the catchment. However, eighteen STWs serving populations >2,000 persons account for 77% of the urban population and, if 90% P removal was achieved at these STWs, the SRP input would be reduced by 42%

An alternative strategy to P removal at STWs is to impose a restriction on the P content of detergents. Legislation to this effect has been passed in the USA, notably in certain states in the Great Lakes catchment area, which has led to removal efficiencies in the region of 60%. In the Lough Neagh catchment a P detergent ban, apart from being difficult to enforce is a small area, would not be particularly effective as the detergent P content of the United Kingdom sewage appears to be lower than that found in the USA. Assuming a low dietary *per capita* contribution of 1.0 g P day^{-1} to the SRP *per capita* of 1.6 g P day^{-1} derived from the regression, a ban would be expected to reduce the *per capita* values by only approximately 40% and the SRP input to Lough Neagh by less than 20%.

Although P is considered to be the limiting nutrient for algal growth in Lough Neagh, N inputs can affect the balance of algal species (see Chapter 9). We have suggested (Smith *et al.*, 1982) that lowering N inputs may be harmful as this control strategy would lower the N/P ratio and hence favour gas-vacuolate, N-fixing species of blue-green algae. P reduction at STWs would have the added advantage of raising the N/P ratio of the Lough Neagh inputs. As the dominant variable controlling the nitrate input, FARM MILK, increased by 45% over the study period compared to an increase of 15% for the dominant SRP variable, URBAN POPULATION, there has been a steady increase in the N/P ratio of the Lough Neagh input (Fig. 8.6). This increase may be an important causal factor in the decline over the study period of the N-Fixing blue-green algae in the Lough (Fig. 8.7).

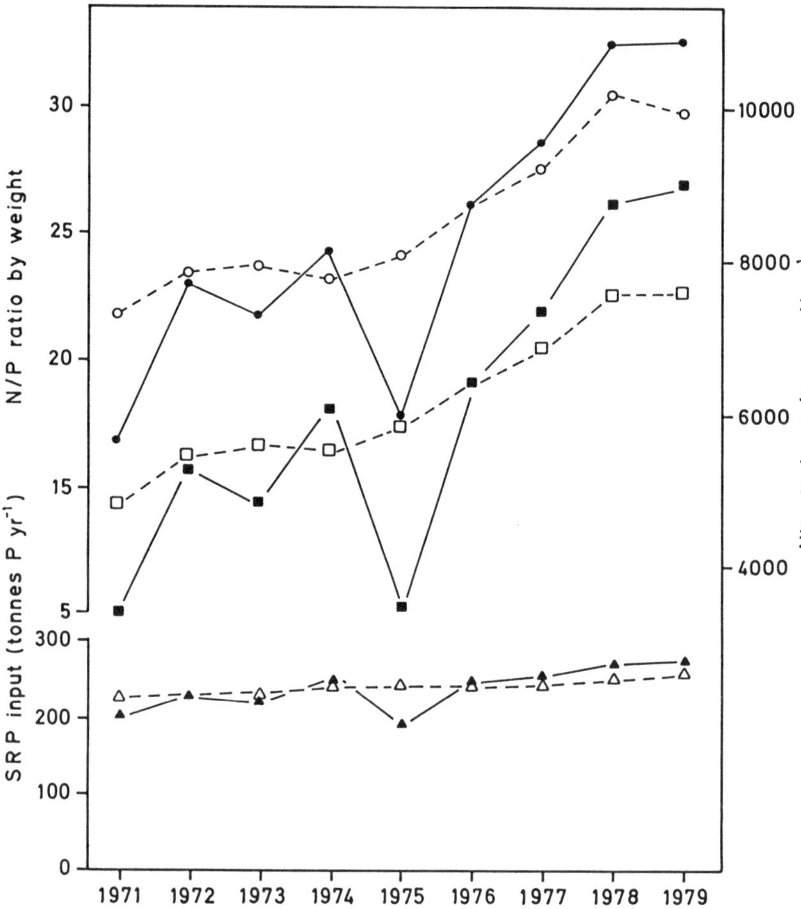

Figure 8.6. Predicted SRP (▲—▲), nitrate (■—■) and N/P ratio (●—●) of inputs to Lough Neagh during the period 1971–1979. Open symbols calculated using mean runoff of O.576 m. Redrawn from Foy *et al.* (1982).

8.11. Postscript: the phosphorus reduction programme

It has been established that the growth of algae in Lough Neagh is limited by the availability of P. Furthermore results from bioassays have shown that the only significant available form of P is SRP and from budget studies PP from the rivers is mostly sedimented in the Lough and does not take any further part in the algal growth cycle,

The major source of SRP in the rivers and effluents discharging into Lough Neagh is that from the major sewage treatment works (STWs). Although attempts are made in a number of ways to reduce phosphorus loading, e.g., by requiring for farm slurry adequate storage and land for disposal,

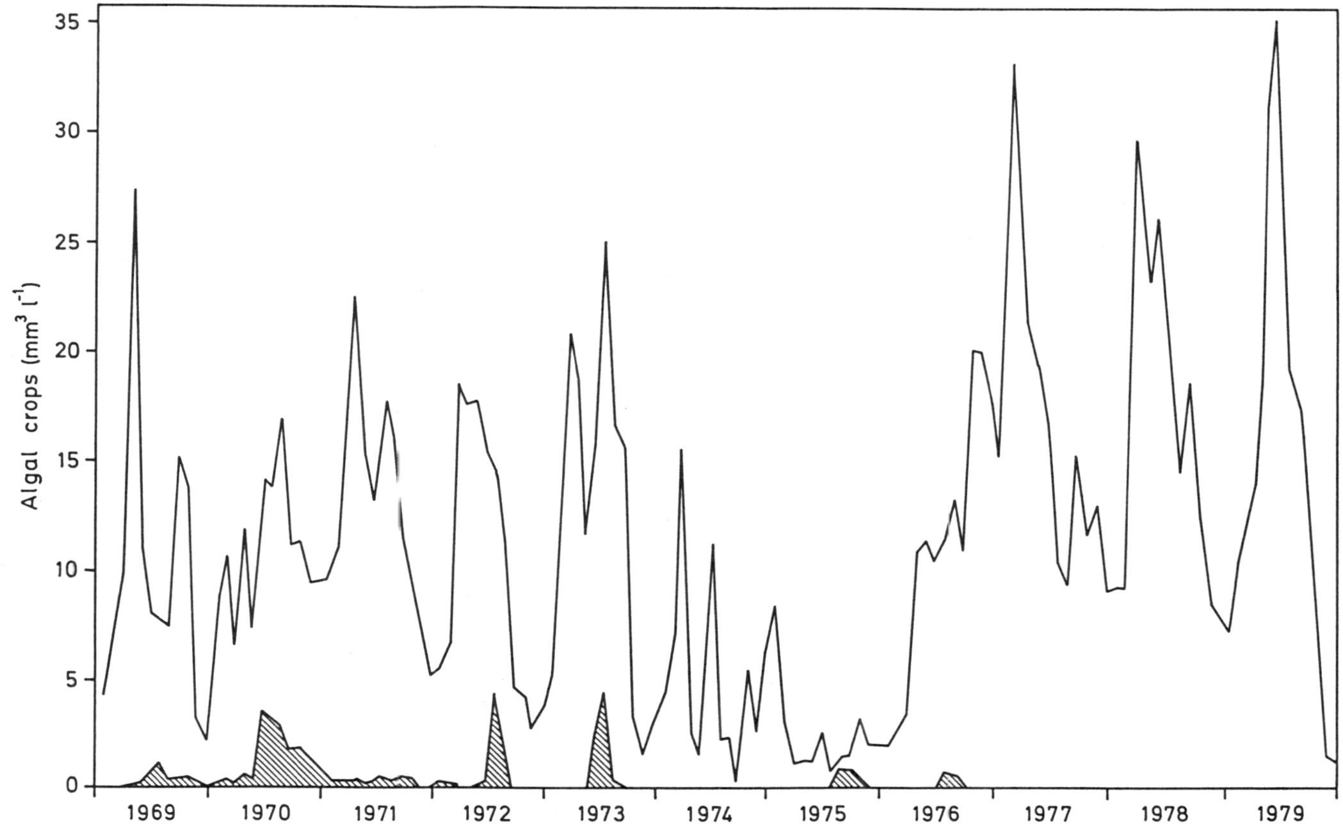

Figure 8.7. Phytoplankton volume changes in Lough Neagh for the period 1969–1979. Nitrogen-fixing species are shaded. Redrawn from Smith *et al.* (1982).

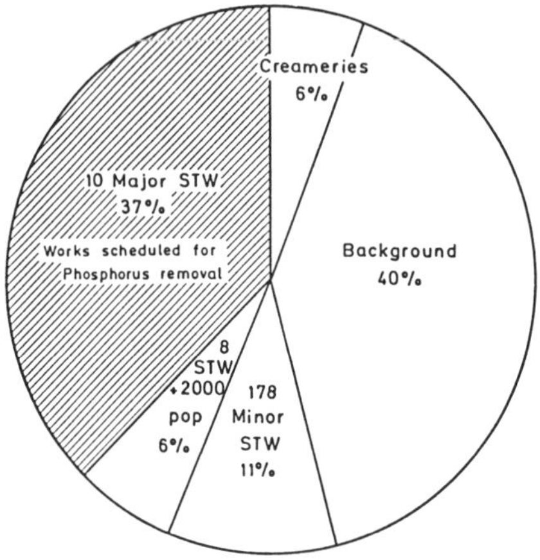

Figure 8.8. Predicted SRP budget to Lough Neagh.

and by advising septic tank soak away, where suitable, to land rather than directly into water courses, the most effective strategy for reducing the SRP load is the introduction of P-reduction treatment at the STWs. Drawing on these findings the Department of Environment (NI) has started a programme to install P-reduction at all of the ten major works which together contribute about 37% of the SRP load to the Lough (see Fig. 8.8). Most of the SRP in the influent sewage is precipitated out at the primary settlement stage by dosing the incoming sewage with DIFCO (an alumino-ferric liquor available locally and developed by the Industrial Science Division of the Department of Commerce) in amounts such that the effluent finally discharged to the rivers contains no more than 1 mg l^{-1} total P.

In 1981 the first major sewage treatment works in the Lough Neagh catchment to benefit from P-reduction was that at Ballymena whose effluent discharges into the River Main. The FBIU has undertaken an intensive monitoring programme to compare SRP concentrations in the River Main with the corresponding 1971–1980 values. During the treatment period it was observed that SRP values in the River Main in 1981 were often more than 50% lower than the corresponding mean values recorded over the previous years (see Fig. 8.9). In contrast, an identical analysis of the River Moyola, whose catchment area lies adjacent to that of the River Main and which, therefore, can be expected to experience similar SRP contributions from sources other than effluent from sewage treatment works showed no statistically significant difference between the 1971 to 1980 and 1981 SRP concentrations throughout most of the year (Table 8.8). This comparison suggests that the significant decrease in SRP concentrations observed in the River Main during the spring and autumn of 1981 should be attributed to the P-reduction treatment. With the extension of the P-reduction programme

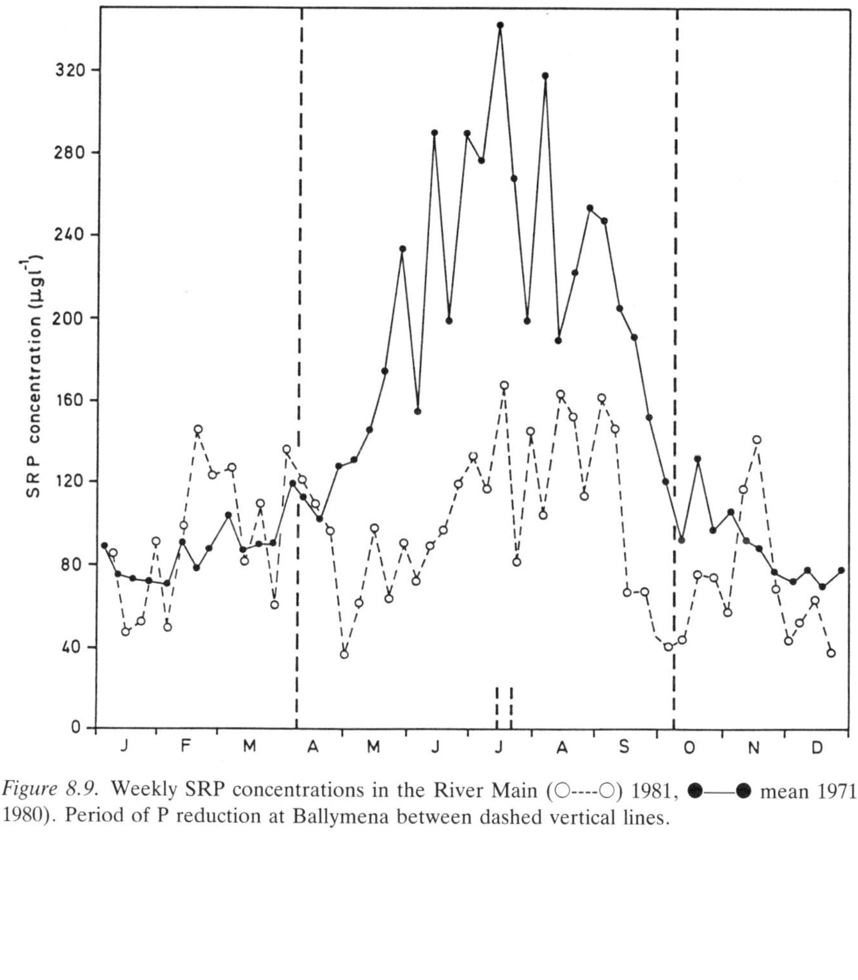

Figure 8.9. Weekly SRP concentrations in the River Main (○----○) 1981, ●—● mean 1971–1980). Period of P reduction at Ballymena between dashed vertical lines.

Table 8.8. Significance levels of the arithmetic differences in concentration of SRP between corresponding paired weeks in the periods 1971–1980 and 1981 for the Rivers Main and Moyola.

Season	No of Observation	River Main t-value	River Main Significance level	River Moyola t-value	River Moyola Significance level
Spring (Feb–May)	17	−1.56	NS	1.01	NS
Summer (June–Sept)	17	−4.27	***	−1.36	NS
Autumn (Aug–Nov)	17	−4.89	***	1.47	NS
Winter (Dec–Mar)	18	−0.32	NS	2.18	*

NS not significant at $P < 0.05$ level
* $0.01 < P < 0.05$
** $0.001 < P < 0.01$
*** $P < 0.0001$.

to all the major STWs in the Lough during 1982 a significant reduction in the nutrient enrichment of Lough Neagh and the benefits that that will bring now seem assured (See also Chapter 21 Epilogue).

8.12. References

Alexander, G.C. and Stevens, R.J. (1976) Per capita phosphorus loading from domestic sewage. Water Res. 10: 757–764.

American Public Health Association: American Water Works Association and Water Pollution Control Federation (1971) Standard Methods for the Examination of Water and Waste Water. 13th edition. Washington.

Chapmann, B., Cooke, G.H. and Whitehead, R. (1967) Automated analysis: the determination of ammonical nitrous and nitric nitrogen in river waters, sewage effluents and trade effluents. J. Inst. Wat. Pollut. Control. 77: 478–491.

Eisenreich, S.J., Bannerman, R.T. and Armstrong, D.E. (1975) A simplified phosphorus analysis technique. Environ. Letters 9: 43–53.

Forbes, S.A. (1887) The lake as a microcosm. Bull. Scient. Assoc. Peoria, 1887: 77–87. (Reprinted: Illinois Nat. Hist. Surv. 15: 537–550, 1925).

Foy, R.H., Smith, R.V., Stevens, R.J. and Stewart, D.A. (1982) Identification of factors affecting nitrogen and phosphorus loadings to Lough Neagh. J. Environ. Management 15: 109–129.

Gibson, C.E., Foy, R.H. and Fitzsimons, A.G. (1980) A limnological reconnaissance of the Lough Erne System, Ireland. Int. Revue ges. Hydrobiol. 65: 49–84.

Healey, F.P. (1973) Inorganic nutrient uptake and deficiency in algae. CRC Crit. Rev. Microbiol. 3(1): 69–113.

Hendriksen, A. (1965) An automated method for determining low-level concentrations of phosphates in fresh and saline waters. Analyst 90: 29–34.

Murphy, J. and Riley, J.P. (1962) A modified single solution method for the determination of phosphate in natural waters. Analytica Chim. Acta 27: 31–36.

Parr, M.P. and Smith, R.V. (1976) The identification of phosphorus as a growth limiting nutrient in Lough Neagh using biossays. Water Res. 10: 1151–1154.

Smith, R.V. (1976) Nutrient budget of the River Main, Co. Antrim. Tech. Bull. Minist. Agric. Fish Fd., London. No. 32: 315–339.

Smith, R.V. (1977) Domestic and agricultural contributions to the inputs of phosphorus and nitrogen to Lough Neagh. Water Res. 11: 453–459.

Smith, R.V., Stevens, R.J., Foy, R.H. and Gibson, C.E. (1982) Upward trend in nitrate concentrations in rivers discharging into Lough Neagh for the period 1969–1979. Water Res. 16: 183–188.

Smith, R.V. and Stewart, D.A. (1977) Statistical models of river loadings of nitrogen and phosphorus in the Lough Neagh system. Water Res. 11: 631–636.

Stevens, R.J. (1975) Semi-automated ammonia probe determination of Kjeldahl nitrogen in freshwaters. Water Res. 10: 171–175.

Stevens, R.J. (1979) Evaluation of an enzymatic method for orthophosphate determination in freshwaters. Water Res. 13: 763–770.

Stevens, R.J. and Parr, M.P. (1977) The significance of alkaline phosphatase activity in Lough Neagh. Freshwater Biol. 7: 351–355.

Stevens, R.J. and Smith, R.V. (1978) A comparison of discrete and intensive sampling for measuring the loads of nitrogen and phosphorus in the River Main, Co. Antrim. Water Res. 12: 823–830.

Stevens, R.J. and Stewart, B.M. (1982a) Concentration, fractionation and characterisation of

soluble organic phosphorus in river water entering Lough Neagh. Water Res. 16: 1507–1519.
Stevens, R.J. and Stewart, B.M. (1982b) Some components of particulate phosphorus in river water entering Lough Neagh. Water Res. 16: 1591–1596.
Stumm, W. and Baccini, P. (1978) Man-made chemical perturbation of lakes. In A. Lerman (ed.) Lakes Chemistry, Geology, Physics. Springer-Verlag, New York. pp. 91–126.
Wilson, H.E. (1972) Regional Geology of Northern Ireland. H.M.S.O., Belfast.
World Health Organisation, (1971). International Standards for Drinking water. WHO, Geneva.

9. Nutrient cycles in Lough Neagh

C.E. GIBSON and D.A. STEWART

9.1. Introduction

The homogeneity of the water mass and its relatively slow turnover time (average 1.2 years) makes Lough Neagh an ideal subject for budget studies. The water balance is also simple because the lake is maintained at a constant level and comparison of the river inflow and outflow water volumes shows that there are no measurable ground water inputs (Table 9.1).

The aim of this chapter is to describe the annual cycles of phosphorus, nitrogen and silica in Lough Neagh and to investigate the factors that determine the course of the cycle. A final section gives an account of the changes in the water quality of the lake that have been observed since 1969.

9.2. Methods

Methods of analysis of nitrogen and phosphorus compounds and their fractionation are described earlier (Chapter 8).

Silicate was determined by a Technicon Autoanalyser II adaptation (industrial methods No. 105-71 W/B) of the method of Mullin and Riley (1955) giving an estimated standard deviation of 0.2 mg SiO_2 l^{-1} in the 0–10 mg SiO_2 l^{-1} range.

Budget calculations were carried out assuming a lake volume of 3.45×10^9 m^3 and a surface area of 383 km^2. The precision of the volume estimate is unknown since it depends upon planimetric integration of the British Admiralty Chart Surveyed in 1835, corrected to present lake level. There is unfortunately no independent means of checking the value obtained. Substance budgets were calculated on the basis of computed river inputs (see Chapter 8) corrected for drainage components. The correction involved increasing the input by 14% to allow for ungauged portions of the catchment and adding in direct sewage discharges to the lake, which varied between years. Outflow was calculated from weekly measurements of Lower Bann substance concentrations and daily water flow measurements from Movan-

R.B. Wood and R.V. Smith (eds.), Lough Neagh, 171–201.

Table 9.1. Annual water inflows and outflows in Lough Neagh for the calendar years 1971–1979. Values are 10^6 m^3 and assume a direct drainage inflow of 14%

Year	Inflow	Outflow	Input/Output
1971	2017	1985	1.02
1972	2619	2471	1.06
1973	2292	2207	1.04
1974	2721	2834	0.96
1975	1804	1861	0.97
1976	2554	2444	1.05
1977	2769	2775	1.00
1978	3039	2725	1.12
1979	3261	3547	0.92
Mean	2564	2539	1.01

agher Weir corrected for inflow between Toome outflow sluices and Movanagher. Storage in the lake was calculated from three 0–10 m composite samples from Antrim Bay, Central Station and South-East Station. Budget calculations were carried out on a computer by daily interpolation between data points.

9.3. Phosphorus

9.3.1. *The annual phosphorus cycle in Lough Neagh*

The annual cycle of phosphorus in the lake is illustrated by data from 1979 (Fig. 9.1). A fuller account has recently appeared (Gibson *et al.*, 1988).

9.3.1.1. *Total phosphorus*

The TP cycle falls into four phases. At the beginning of the year, the concentration of TP changes little and input and output are almost in balance. From the middle of March onwards, the TP concentration decreases as phosphorus is deposited onto the sediment surface by the settlement of particulate matter, predominantly planktonic diatoms. Some settlement continues during the next phase but it is masked by the much greater release of phosphorus from the sediment into the water column. The deposition and release phases occur at a time of low river flows and export down the outflow is small. In 1979, net release of phosphorus had ceased by the end of August and the concentration in the lake stayed relatively constant until the end of the year. In that year the concentration at the end of the year was slightly higher than at the beginning but this was not always the case in other years.

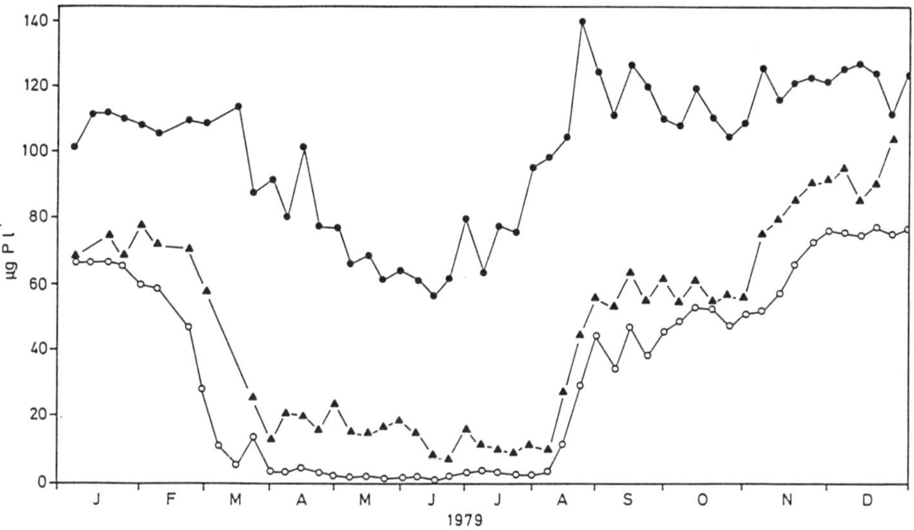

Figure 9.1. The annual of phosphorus in Lough Neagh ●—● TP; ▲—▲ TSP; ○—○ SRP. SOP and PP values may be obtained by difference.

9.3.1.2. *Soluble reactive phosphorus*

SRP concentrations are more variable than TP concentrations because uptake by the phytoplankton diminishes the SRP fraction without necessarily affecting the TP concentrations. At the beginning of the year, SRP concentrations begin to decline when the rate of uptake by the phytoplankton is greater than the rate of supply from the catchment inflows. In 1979 SRP reached the limit of detection at the end of March and in other years this happened both earlier and later depending on the timing of the spring phytoplankton outburst. SRP concentrations remained undetectable until the beginning of July and there was a period when TP was increasing but SRP remained undetectable because SRP was being released from the sediment but immediately absorbed by the phytoplankton. By the middle of July, the supply of SRP from sediment release exceeded the demand from the phytoplankton and the concentration in the lake began to increase rapidly. As with TP, the rise in concentration continued until the end of August, then reached a plateau. In November and December, the concentration again began to increase due to greater catchment inputs.

9.3.1.3. *Soluble organic phoshorus*

There is evidence from the phytoplankton studies (Chapter 10) that some of the SRP taken up by the phytoplankton is excreted as SOP, but the SOP data are somewhat unsatisfactory because they are rather variable and are calculated by difference. There is some evidence that SOP increases when SRP is being removed during the spring outburst. SOP is not a chemically

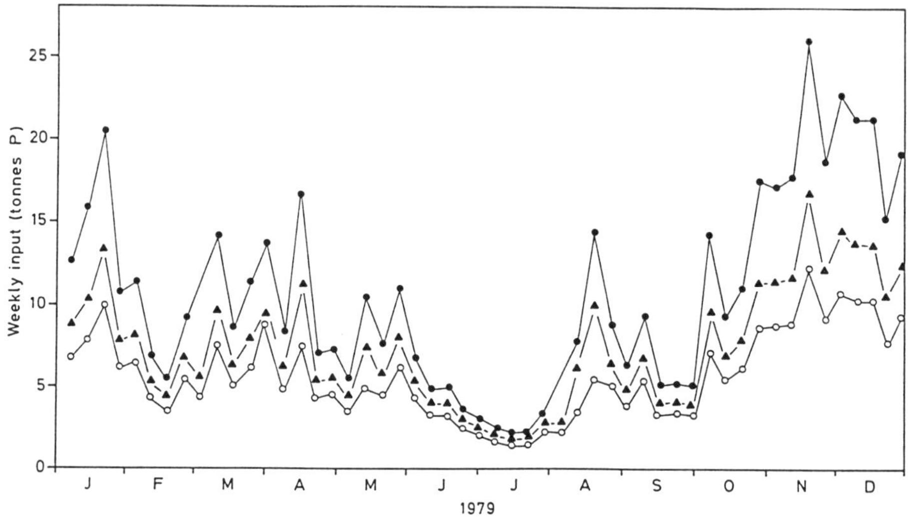

Figure 9.2. Annual cycle of river inputs of phosphorus; symbols as Fig. 9.1.

defined fraction and consists of a number of different types of compounds, some washed in from the catchment and some generated within the lake. Since there has been little investigation of SOP in Lough Neagh, it is not proposed to discuss it further. SOP fractions in river water are discussed in Chapter 8.

9.3.2. *Phosphorus input from the catchment*

Because water flow plays such an important part in determining load, inputs to the lake show a strong seasonality (Fig. 9.2), especially with TP, where there is a first order relationship between load and flow. The least variable fraction is SRP which is predominantly from point sources. During the spring outburst of phytoplankton, the average SRP input is 0.8 tonnes P d^{-1} which, if phytoplankton require 50:1 C:P, would satisfy a photosynthetic carbon fixation rate of only 100 mg m^{-2} d^{-1}, much less than the meaured value (see Chapter 10). PP inputs are more erratic than SRP and floods bring in a good deal of suspended material. For example, during November 1979, a weekly input of 26.1 tonnes TP included 9.4 tonnes PP (Fig. 9.2). PP inputs in the rivers show more variability than PP concentrations in the lake and it is of interest to examine the fate of the PP brought in by the rivers.

9.3.3. *The fate of inwashed particulate phosphorus*

In order to follow the fate of inwashed (allochthonous) PP it is necessary to distinguish between it and PP generated within the lake (autochthonous) by the growth of phytoplankton. This is not possible from the analytical data

Table 9.2. Fate of inflowing particulate phosphorus calculated as stated in 9.3.3, from mid-October to mid-March

	1977–1978	1978–1979
Particulate P input (tonnes)	76.9	90.3
Inorganic PP export (tonnes)	19.3	24.7
Sedimentation	57.6	65.6
Sedimentation %	75.0	73.0

but one solution is to estimate the phytoplankton PP from independent estimates of biomass. Regression analysis of data from weekly analysis of centrifugable seston and chlorophyll *a* suggested that although there was significant seasonal variation in phytoplankton phosphorus contents (see Chapter 10), on average 1 mg chlorophyll *a* was associated with 1 mg PP. This relationship was used to derive an estimate of the phytoplankton contribution to PP and the allochthonous PP was deduced by difference. After a storm, this procedure may result in a significant overestimate since PP resuspended from the sediment will be included in the allochthonous fraction.

Budgets for allochthonous PP over the period October–March were calculated for 1977–1978 and 1978–1979 (Table 9.2). PP concentrations were the same at the beginning and end of both periods so that changes in storage may be neglected. Very little of the inflowing PP left the lake and as it did not accumulate in the water column, it must have settled rapidly to the bottom.

9.3.4. *Sediment release*

Fig. 9.1 demonstrates the existence of sediment release in the annual phosphorus cycle. The mechanism of phosphorus release will be discussed later (Section 9.6) and the aim of this section is to quantify phosphorus release and to assess its significance to the annual budget. The quantity of phosphorus released from the sediment can be calculated from the increase in concentration in the water column corrected for inflow and outflow. Expressed as an increase in concentration, the result does not depend on any assumptions about the lake volume, but a water volume estimate is essential for the calculation of absolute quantities. The volume assumed is unlikely to be seriously in error but the value assumed for sediment area in order to calculate areal release is more questionable. It is likely that phosphorus is released only from areas with organic sediment below a certain depth of water so that, as the area of the whole lake bed (383 km^2) has been used, the rates obtained are certain to be underestimates. Periods were chosen for the calculation of release rates which showed a steady increase in concentration and these varied both between years and between SRP and TP in the same year. The longest possible period was chosen and this ranged from 39

Table 9.3. TP and SRP release rates from Lough Neagh sediments

	tonnes d^{-1}		$\mu g\, l^{-1}\, d^{-1}$		$mg\, m^{-2}\, d^{-1}$	
	TP	SRP	TP	SRP	TP	SRP
1974	5.2	12.2	1.5	3.5	13.3	30.7
1975	3.7	9.1	1.1	2.6	9.8	22.9
1976	7.1	10.1	2.1	2.8	18.7	25.4
1977	2.7	8.6	0.8	2.4	7.1	21.6
1978	1.7	4.0	0.5	1.1	4.5	10.1
1979	3.4	3.5	1.0	1.0	8.9	8.8
Mean	4.0	7.9	1.2	2.4	10.4	19.9

to 81 days for TP, 18–44 days for SRP. Release did not proceed smoothly and, over short time intervals, more rapid release rates could be calculated. In the data presented, there was no direct connection between the length of data run and the size of the release estimate.

There are obvious differences between release rates in different years (Table 9.3) but the most consistent fact is that release of SRP is more rapid than the apparent release of TP. This paradox implies either that unreactive phosphorus is transformed to SRP in the water column or that there is settlement of TP from the water column. Inspction of the data suggests that the latter is the main cause. There are two phases to the sediment release (Fig. 9.1). In the first phase, phytoplankton uptake transforms released phosphorus into the particulate fraction. In the second phase, the SRP demand of the phytoplankton is satisfied and SRP appears in the water column. It is probable that the settlement of particles is promoted by calm weather which would also enhance the sediment release (see Section 9.6).

9.3.5. *The significance of sediment release to the annual budget*

Phosphorus is released from the sediment in quantities comparable to the river inputs and the rate of release is greater than the maximum rate of river input. In 1979, the maximum rate of input from the rivers was 3.7 tonnes d^{-1} for TP and 1.74 tonnes d^{-1} for SRP which compare with the lowest calculated sediment release rates, occurring over a more extended period than the high river inputs. The release of phosphorus has an important effect on the phosphorus budget and is equivalent to a large summer loading.

It is possible to calculate the concentration of phosphorus that would be attained in the lake if there were no sediment release. Starting at a time when the SRP was undetectable in the lake, SRP loads were used to calculate a theoretical rise in SRP concentrations and outflow was calculated on the basis of outflow volume and SRP concentration. Uptake or release by phytoplankton was ignored. Fig. 9.3 shows the large discrepancy between concentration derived from river inputs and the observed concentration in the

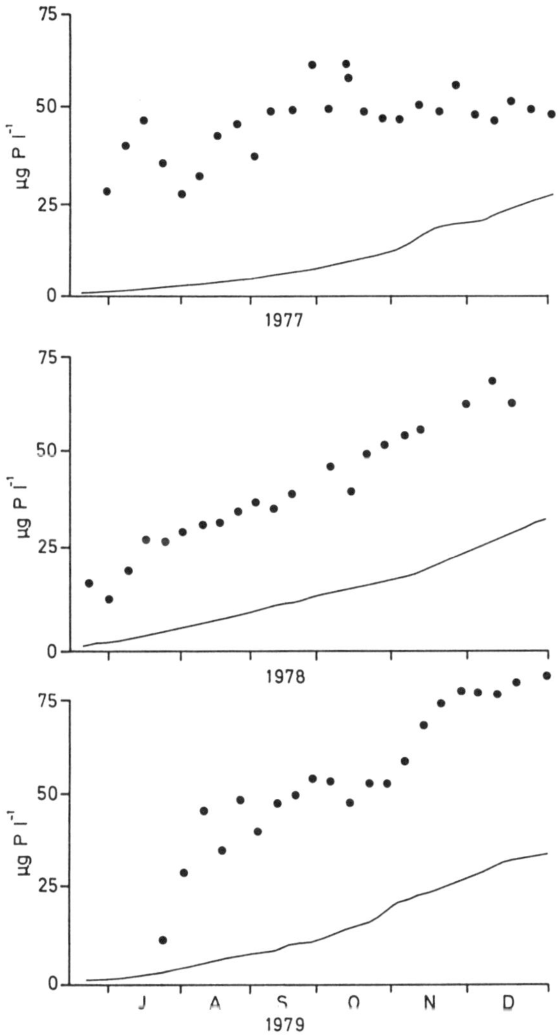

Figure 9.3. Observed SRP concentrations (●) and concentration calculated from river inputs (—) in second half of 1977–1979.

lake for the years 1977–1979 which was 53%, 53% and 43% in 1977–1979 consecutively.

Sediment release may well have been significantly less in the past than it is today and this would have had important consequences for the concentration of phosphorus available for phytoplankton growth in the spring. If no sediment release had occurred, the spring SRP concentration would only have been half the present value, even if the catchment input had been the same.

Table 9.4. TP budgets for Lough Neagh in the calendar years 1975–79. Sedimentation is calculated as excess of input over outflow corrected for change in storage. Budget quantities are tonnes (10^3 kg) P yr^{-1}. Loading and sediment increment are g P m^{-2} yr^{-1}

	1975	1976	1977	1978	1979	Mean
Input	269	428	448	503	557	441
Outflow	175	280	300	258	370	276
Change in storage	+101	+137	−18	−38	+25	–
Sedimentation	−9	11	166	283	162	165
Sedimentation % input	0	2.5	37	56	29	37
Loading	0.70	1.12	1.17	1.31	1.45	1.15
Sediment increment	0	0.03	0.43	0.74	0.42	0.43

9.3.6. *Annual phosphorus budget*

All the components of the annual phosphorus cycle interact to produce the overall phosphorus budget (Table 9.4). Most of the phosphorus entering the lake does so in the river waters or in direct discharges of sewage effluent to the lake but a small quantity also enters in precipitation falling directly to the lake surface. Whilst the precipitation volume may be calculated reasonably accurately, the calculation of the TP carried by the rainfall is more difficult. The best data available are those of Jordan (1987) who measured TP in precipitation in the Four Mile Burn catchment, a tributary of the Six Mile Water, during 1980–1982. Based on his analysis and the calculated precipitation volume, the likely precipitation input of TP is 7 tonnes yr^{-1}. This has been added as a constant quantity to the inputs in Table 9.4. The budget was calculated from input (rivers + precipitation) − outflow (Lower Bann flow × concentration), calculated daily. The input budget utilised the load flow model (see Chapter 8), the outflow was calculated as the summed products of outflow water volume and substance concentration. The difference between these two, corrected for changes in the amount of phosphorus held in the lake water (storage) gave an estimate of the quantity of phosphorus retained on the bottom sediments.

A close relationship exists between TP input and river flow. A regression calculated from the data in Tables 9.1 and 9.4, together with a precipitation phosphorus input value at zero flow gave the highly significant regression equation:

$$\text{Load} = 0.17 \times \text{flow} - 7.9 \ (r^2 = 0.9936,\ n = 6)$$

where the units are as in Tables 9.1 and 9.4. A crude estimate of TP input under modern catchment conditions could therefore be made from total river flow. This would be of little help in calculating the spring phosphorus concentrations however, since the sedimentation varied considerably between years. The range of values in Table 9.4 is between 0 and 57%, the mean value for the period 1975–1979 was 37%.

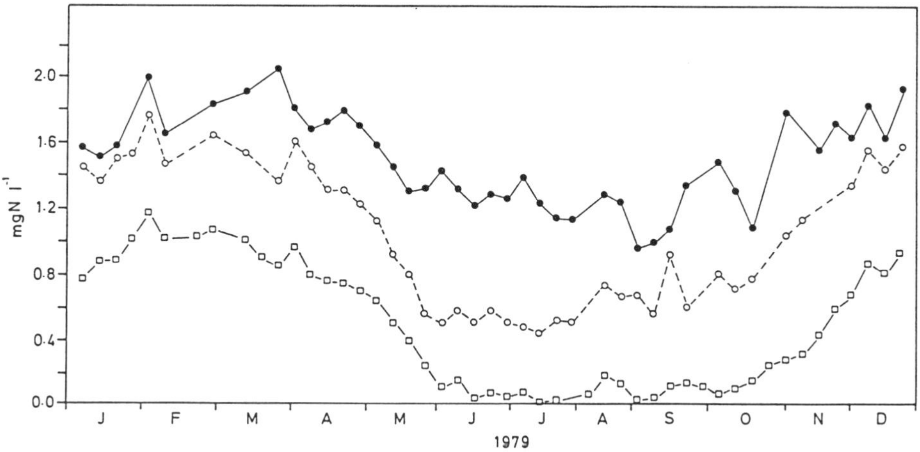

Figure 9.4. Annual cycle of nitrogen in Lough Neagh ●—● total N; ○—○ total soluble N; □—□ nitrate N. Other fractions can be obtained by difference.

9.4. Nitrogen

9.4.1. *The annual nitrogen cycle in Lough Neagh*

The annual cycle of combined nitrogen in the lake is shown in Fig. 9.4 using data for 1979. The nitrogen cycle may be divided into three phases – the input phase in winter and early spring; the uptake phase in spring and early summer and the stagnation phase when lake concentrations are low and river inputs are small.

9.4.1.1. *Nitrate nitrogen*

In the first part of the year, the nitrate-nitrogen concentration usually increases slightly due to contributions from the rivers, then begins to decline as phytoplankton uptake exceeds the rate of supply. Unlike SRP, nitrate-nitrogen is not taken up by the phytoplankton in excess of immediate requirements. In 1979, the major decline in SRP concentration ended by the beginning of April whereas the nitrate-nitrogen concentration continued to decline until the end of July. In further contrast to SRP, apart from one small concentration increase in August, nitrate-nitrogen was at less than 10% of the maximum concentration until significant replenishment from the inflows was resumed in October.

9.4.1.2. *Organic nitrogen*

The organic nitrogen fractions vary in concentration between 0.6 and 1.2 mg l^{-1}. Particulate nitrogen varies most and is composed primarily of phytoplankton. The phytoplankton contains between 6% and 12% N (as

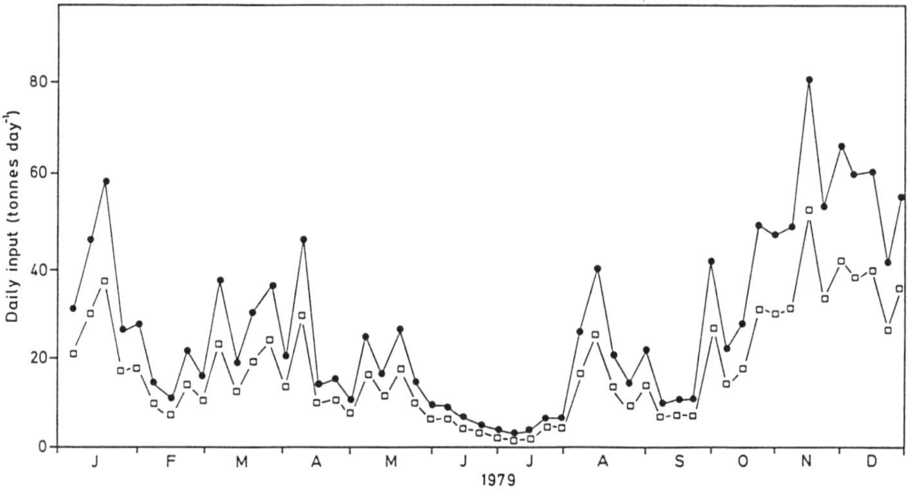

Figure 9.5. Annual cycle of river nitrogen inputs to Lough Neagh. Symbols as Fig. 9.4.

percent ash-free dry weight) and blue-green algae have a higher content than diatoms. In 1979 (Fig. 9.4), the greatest concentration of particulate nitrogen was found during the blue-green maxima in the summer. The total nitrogen cycle shows the same features as the nitrate-nitrogen and lacks the rapid summer increase typical of the phosphorus cycle.

The soluble organic nitrogen is composed primarily of combined nitrogen compounds; ammonia is present in only trace amounts for most of the year. Regular ammonia analysis in 1978 showed that concentrations were consistently below 50 $\mu g\, N\, l^{-1}$ and often below the limit of detection. Low ammonia concentrations could arise from rapid phytoplankton uptake and are not in themselves evidence of a lack of release from the sediments, but there is no evidence of a rise in total nitrogen which would point to such an event.

9.4.2. *Nitrogen input from the catchment*

The seasonal variation in river input is shown for 1979 in Fig. 9.5. There is a high degree of flow dependence and nitrate-nitrogen dominates the nitrogen input. After a period of dry weather, high river flows in August caused a sharp increase in nitrogen inputs and this was accompanied by a modest increase in the concentration of nitrate in the lake.

9.4.3. *Sediment release*

Fig. 9.4 shows that over the period of time when phosphorus and silica are released from the sediment, there is no rise in the total nitrogen concentra-

Table 9.5. Mean concentrations of inflows to Lough Neagh (total load ÷ flow) in mg l^{-1}

	1975	1976	1977	1978	1979
SRP	0.088	0.088	0.088	0.094	0.092
NO_3-N	1.5	2.0	2.4	2.6	2.0
SiO_2	8.2	8.4	8.5	8.6	8.4

tion. Budget calculations for the periods of nitrogen increase in the lake (January, August, October to December) suggest that the computed input of nitrate cannot entirely explain the observed rises in lake concentration, but this is not true in other years and it is safest to conclude that the discrepancy arises from uncertainties in the budget calculation rather than sediment release. If any combined nitrogen is released from the sediment, it must be a very much smaller proportion of the river load than in the case of either phosphorus or silica. The loss of nitrogen from the sediment via the gas phase has not been studied directly, but it can be estimated from a comparison of the nitrogen and phosphorus budgets. From Tables 9.6 and 9.4, it can be seen that between 1975 and 1979, 4307 tonnes N and 165 tonnes P were 'retained' in the lake. If this were all in the sediment, the sediment N:P ratio would be 4307:165 or 26.1:1. Analysis of 53 sediment samples taken in Antrim Bay over a 13 week period and sampling from sediment more than 3 cm below the sediment:water interface yielded a consistent N:P ratio of 6.6:1. Therefore, much of the nitrogen apparently retained in the sediment cannot be found there and is presumably lost by denitrification. Based on this calculation, there is a net loss of 75% of the sediment nitrogen input, equivalent to 3212 tonnes or approximately 34% of the input from the catchment. In the period 1983 to 1985 the comparable net loss of N was ~44%, see Section 9.10 and Fig. 9.17.

Because little combined nitrogen is released from the mud, the spring concentration of nitrogen is smaller in relation to the input than with either phosphorus or silica. The mean annual inflow concentration (load divided by flow) would be attained in the lake if there were no uptake or sedimentation. Table 9.5 shows the calculated mean concentration of SRP, nitrate and silicate in different years. The calculated values for SRP and silica only slightly exceed the observed maximum concentrations in the lake whereas the predicted nitrate concentrations are at least twice those observed in the lake (Fig. 9.4).

9.4.4. *The annual nitrogen budget*

Annual nitrogen budgets, calculated in the same way as for TP budgets, are shown in Table 9.6. Inputs rose steadily between 1975 and 1978, due for the most part to increased nitrate loadings in the rivers. An attempt was made to calculate direct inputs from precipitation and planktonic nitrogen fixation

Table 9.6. Total N budgets for Lough Neagh. 'Difference' is the loss by sedimentation and denitrification

	1975	1976	1977	1978	1979	Mean
Input	5208	9488	10600	12109	10458	9572
Outflow	2695	4294	6536	5042	5652	4843
Change in storage	+174	+1950	−1163	−44	+ 833	+422
Difference	2339	3244	5226	7181	3973	4307
Difference %	45	34	49	59	38	45
Loading $g\,N\,m^{-2}\,yr^{-1}$	13.5	24.5	24.5	31.5	27	25

Table 9.7. N:P ratios by weight of input and outflow. Values in italics are atomic ratios

	1975	1976	1977	1978	1979	Mean
Input	18	21	23	23	18	21
	41	*48*	*51*	*52*	*41*	*46*
Outflow	15	15	22	19	15	17
	34	*34*	*48*	*43*	*34*	*38*

but no satisfactory data yet exist to calculate the precipitation input. In this budget a value of 500 tonnes yr^{-1} has been assumed, based upon the Four Mile Burn data (Jordan, 1987) as for TP. The precipitation TN represents approximately 5% of the annual input, similar to the precipitation input of phosphorus. Riddolls (1974, 1985) estimated the planktonic nitrogen fixation input in 1973 to be less than 5 tonnes yr^{-1}. Although her calculation depended on several assumptions, the value is nugatory and has not been included in the present budget. Crops of nitrogen-fixing blue-green algae have declined since Riddoll's observations and at the present time, the nitrogen fixation input is probably nearly zero.

Changes in storage are not as significant in the nitrogen budget as in the TP budget and the difference between input and outflow quantities is more consistent. The average percentage of inflow nitrogen that does not leave via the outflow is 45%. This may be an underestimate since there is an unexplained anomaly in the outflow quantities – Lower Bann nitrate concentrations are sometimes significantly higher than those at the mid-lake station. Part of the reason for this may be found in short circuiting of the inflow from the River Moyola which enters relatively near the outflow, but this is unlikely to be the whole explanation. No other sources of nitrate are known in the outflow area.

9.4.5. *N:P ratios in the lake budget*

The mean ratio of the inputs of total N and total P is 46:1 N:P by atoms (Table 9.7). Analyss of the seston in the lake (Fig. 9.6) show that under nutrient-sufficient conditions, the N:P ratio lies between 15:1 and 25:1, so

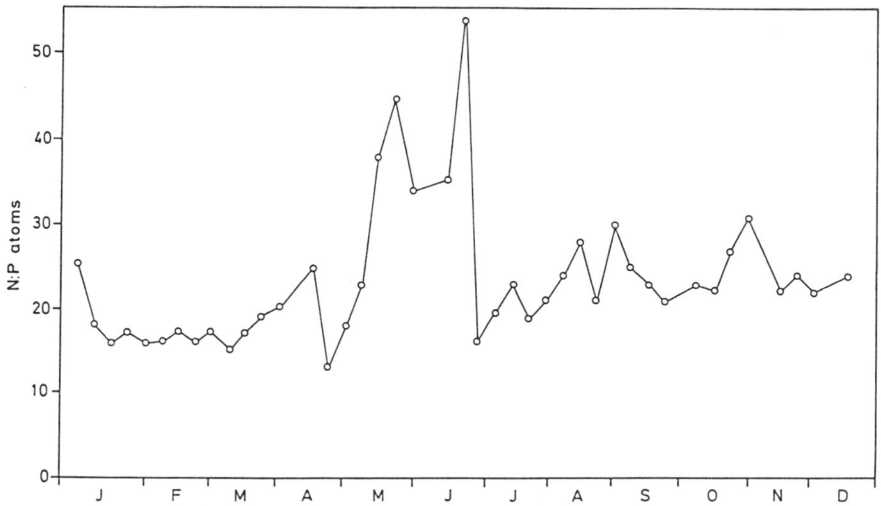

Figure 9.6. N:P ratio of seston in Lough Neagh 1977.

that the catchment supply is deficient in phosphorus for the requirements of algal growth (and more so if most of the particulate – P is not available for algal growth (see Chapter 8). The mean N:P ratio of the outflow is significantly less than that of the inflow because of the smaller percentage of the nitrogen than the phosphorus input is exported down the outflow.

9.5. Silica

Few determinations of total silica have been made in Lough Neagh and the following account is restricted for the most part to soluble silica. Soluble silica is of great significance in the nutrient cycles of Lough Neagh because it limits the spring outburst of diatoms (see Chapter 10). Diatom growth also utilises nitrate and SRP so that the silica cycle has a direct bearing on the nitrogen and phosphorus cycles. Since silica was not considered in the river budgets in Chapter 8, a brief account of the sources of the silica input to the lake will be given here. Further details may be found in Gibson (1981) and in Dickson (1975).

9.5.1. *Silica loads in the rivers*

Silica loads may be modelled by a simple zero order equation because the concentration of silica in any one river is relatively constant. Although a detailed study of the River Main (Gibson, 1981) showed that silica concentration declined with flow, the increase in flow was very much greater than the decrease in concentration and flow was the overwhelming determinant of

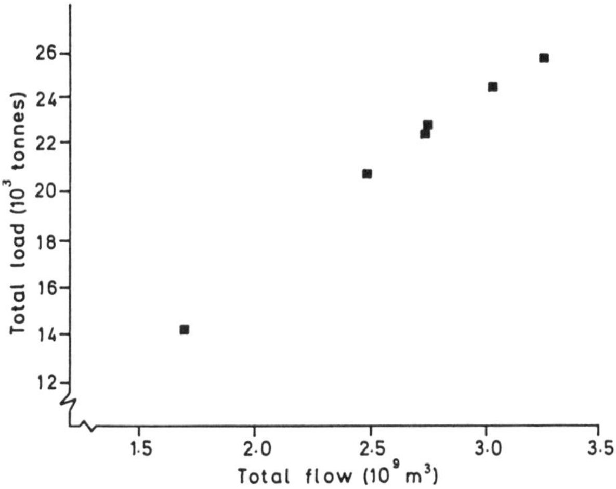

Figure 9.7. Relationship between river flow and annual dissolved silica load 1975–1980.

Table 9.8. Dissolved silica and water yields in the inflowing rivers

	Mean dissolved SiO_2 mg l^{-1}	Areal loss SiO_2 km^{-2} yr^{-1}	Water yield m yr^{-1}
Six Mile Water	15.8	9.6	0.61
Main	13.4	9.5	0.71
Moyola	7.2	5.5	0.76
Ballinderry	7.9	5.3	0.67
Blackwater	6.8	3.6	0.53
Upper Bann	8.5	4.5	0.53

load. For this reason, there is good agreement between the overall flow into the lake and the annual silica load (Fig. 9.7) and inputs are considerably reduced in drought years such as 1975.

The concentration of silica in the rivers reflects the geological characteristics of the catchment and the River Main and the Six Mile Water which drain the basaltic Antrim Plateau contain significantly higher silica concentrations than the other rivers. Table 9.8 gives the mean silica concentrations and the areal loss rates in the six major rivers. Silica concentrations are similar in the Moyola, Ballinderry, Blackwater and Upper Bann but the low water yields of the two latter rivers result in lower areal loss rates.

9.5.2. *The annual silica cycle in Lough Neagh*

Eleven years of weekly data for the dissolved silica concentration in the lake are presented in Chapter 10 Fig. 10.3 which shows that the maximum

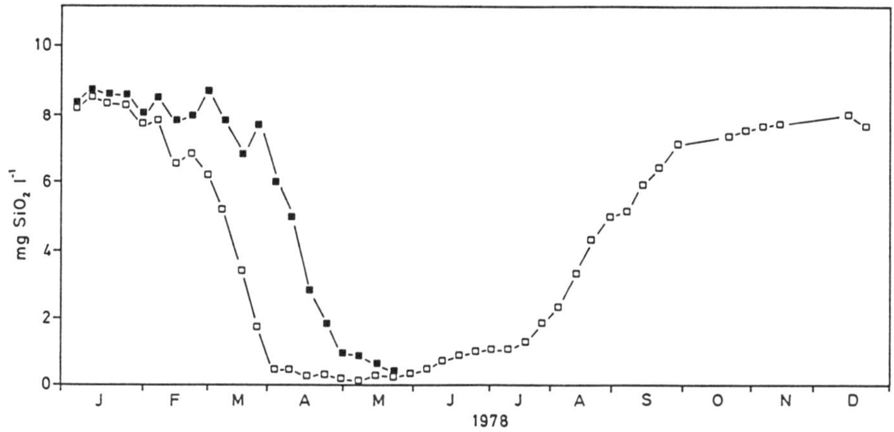

Figure 9.8. Dissolved (□—□) and dissolved + diatom (■—■) silica in Lough Neagh 1978.

concentration varied considerably in different years. More detailed data for 1979 are given in Fig. 9.8 which shows the measured dissolved silica concentration and the computed particulate diatom silica (see Chapter 10). The sum of the two variables gives an estimate of the 'total' silica. 'Total' silica remained relatively constant until soluble silica had been depleted to less than 0.5 mg SiO_2 l^{-1}, then diatoms began to settle out rapidly and the 'total' silica dropped to 0.5 mg SiO_2 l^{-1}. Diatom standing crops were insignificant after this and the dissolved silica concentration increased very slowly as a result of river inputs until silica was released from the sediments at the end of July. The concentration rose steadily until the end of September, then levelled out between 7 and 8 mg l^{-1}. The annual silica cycle is therefore like the phosphorus cycle in that there is a rapid alternation between high and low concentrations, caused by the exchange between the mud and the water column through the agency of the planktonic diatoms. Because there are usually no significant planktonic diatom crops during the summer, sediment release is not masked by uptake in most years, but the timing of the soluble silica release resembles SRP rather than TP.

In most years, the cycle can be divided into four phases: the equilibrium phase from September to February, uptake and deposition in the spring, stagnation in the early summer and sediment release from June to September or October.

9.5.3. *The timing of silica inputs from the catchment*

As flow is the main determinant of load, the input of silica is very seasonal. Floods can bring in large quantities of silica and over 10% of the annual load may enter in one week. In peak flow conditions, more than 1000 t have

Table 9.9. Sediment release of silica in Lough Neagh calculated by regression

Year	Starting date	Finishing date	Calculated release rate	
			$t\,d^{-1}$	$g\,m^{-2}\,d^{-1}$
1975	10/7	17/9	100	0.26
1976	20/5	27/10	127	0.33
1977	15/7	11/11	137	0.36
1978	12/7	20/10	309	0.81
1979	31/7	9/10	222	0.58
Mean			179	0.47

been calculated to enter in a single day but such loads are not sustained for long periods. On the other hand, input over the summer months is very low and the average input is less than 8 $t\,d^{-1}$ so that during this period, sediment release is the dominant source of silica.

9.5.4. *Sediment release*

Silica is released from the sediment in a very similar pattern to phosphorus (Section 9.3.4) and sediment release is correspondingly important in the annual budget. The rate of release observed is a minimum figure not allowing for uptake by diatoms, but apart from 1975, the timing of diatom growth made this a very minor error. Release rates were calculated by fitting a linear regression to a period of smooth increase in silica concentration and correcting for input and output. This procedure gives a more representative value for a sustained rate of release than calculating by difference. Comparison of the regression and difference methods shows that both gave very similar answers but that regression analysis resulted in slightly greater values in most cases. The apparent release rates obtained are given in Table 9.9 together with the period over which the calculation was based. Release seems to continue well into the autumn, as late as 11 November in 1977, and may well go on longer but the river flows become great at this time and the calculation of release is less certain. The rates calculated for 1978–1979 are considerably higher than in the previous years. The reason for this is not known, but there are several possible contributory factors. Firstly, there may be a connection between the rate of release and the amount of silica deposited in the spring, as measured by the maximum concentration of silica observed in January (Fig. 9.9), although the relationship is not perfect. Secondly, the activity of the benthic organisms may have been greater in the latter years because the unsettled windy weather of those summers prevented the formation of significant thermal gradients and no serious oxygen depletion took place at the sediment-water interface. Low oxygen conditions present a serious check to the chironomid populations and Carter (1975) observed that the stagnation of the water column in 1975 was sufficiently prolonged to

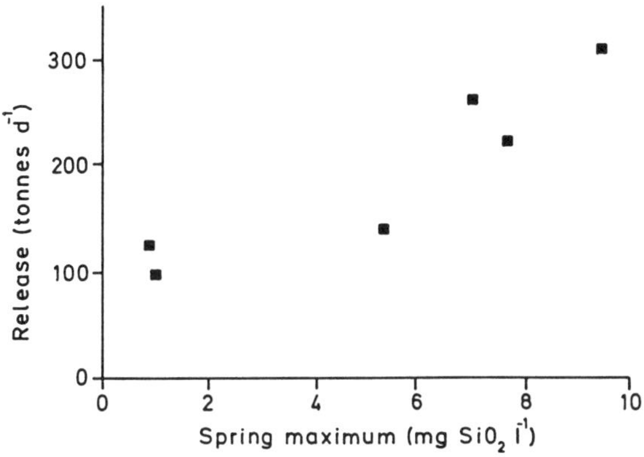

Figure 9.9. Relationship between rate of silica release from sediment and maximum spring concentration of dissolved silica 1974–1979.

delay the chironomid hatch the following year. Thirdly, it is conceivable that the rate of release was more similar than appears from the calculation, but uptake by diatoms reduced the net release. Certainly this is true in 1975 and although planktonic diatom crops were not significant in the summers of 1974, 1976 and 1977, benthic diatom uptake is a possible cause of loss. The light climate was more favourable for benthic growth in those years as the mean chlorophyll *a* concentrations show. Over the period of sediment release, mean chlorophyll *a* values were 12, 35, 34, 43 and 74 for the years 1975–1979 in order (see also Chapter 12).

9.5.5. *The significance of sediment release*

The significance of sediment release is as great in the silica cycle as in the phosphorus cycle and without the release of large quantities of silica during the summer, the silica concentration would be considerably lower the following spring. An effect similar to lack of release was produced by silica uptake in the late summer of 1975. Fig. 9.10 shows the observed lake silica concentration and the concentration predicted on the basis of river inputs in 1975. Release took place in a manner similar to other years, but diatom uptake reduced the concentration on two occasions, first in August, then later in September when diatom growth removed all the soluble silica. The concentration attained in the spring of the following year therefore depended solely on the river input of silica over the winter and as a result, the diatom crop failed in 1976. More details of this event are given in Chapter 10.

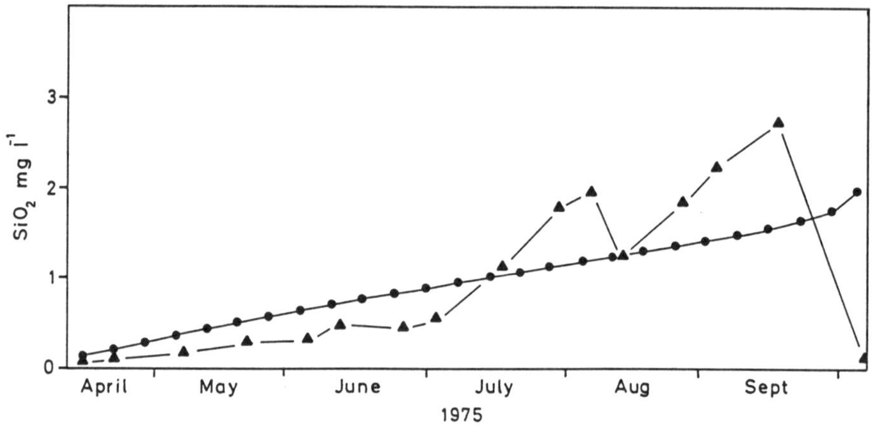

Figure 9.10. Observed concentration of dissolved silica (▲—▲) and concentration calculated from river input (●—●). Redrawn from Gibson (1981).

Table 9.10. Annual silica budgets in Lough Neagh (tonnes yr^{-1})

	1974	1975	1976	1977	1978	1979
Input	22812	14119	20826	22561	24490	25801
Outflow soluble	10189	877	7927	11413	17386	20059
diatoms	3725	1802	530	4987	2201	2347
Change in storage	−12170	−11859	+16024	+13410	−5704	+3173
Sedimentation	21068	23299	−3655	−7249	10607	222
Diatom sedimentation %	85		82	72	93	91

9.5.6. *The annual silica budget*

Annual silica budgets are as variable as annual P budgets (Table 9.10). The budgets have been constructed on soluble silica data but in order to estimate the quantity of silica retained in the sediment, an allowance has been made for the export of diatom silica. The amount retained in the sediment varies widely in different years and as in the phosphorus budgets, the storage term plays a major role in determining the sedimentation for individual years. Calculating a budget over 1974–1979 suggests an overall sedimentation of 33%, very similar to the value obtained for total phosphorus. Calculated over the same five year period (1975 to 1979) as the TP budget gives a value of only 20%. Table 9.10 also gives the percentage of the spring diatom crop that is retained in the sedimentation phase and it is obvious that the export of diatom frustules is relatively small. However, the massive release of silica from the sediment suggests that relatively little of the deposited diatom silica remains permanently in the sediment although that which does remain is sufficient to provide the excellent palaeo-record given in Chapter 7.B. The variability in sedimentation between years and the change in the calculated mean value by including 1974 suggests that the final mean figure may not accurately reflect the situation over a long period of time.

9.6. The sediment release process (by B. Rippey)

The rate of release of material from sediment depends upon the type of release mechanism involved. Berner (1978) identifies three types of mechanism which may limit the rate of release: detachment of the ion or molecule from the solid phase, transport away from the solid phase or a combination of both these. The detachment mechanism from the solid phase is specific to each substance in the sediment, but transport through the upper sediment layers is brought about by processes common to all substances, and depends upon the type of sediment water interface. Three general types of sediment-water interface have been recognised (Lerman, 1979). The first, simple, type occurs under a homogeneous water column when the sediment is undisturbed and transport through the pore waters is by molecular diffusion. The second type resembles the first, but in the boundary layer above the sediment transport is by molecular diffusion and not by turbulence as in the rest of the water column. The third type has an undisturbed sediment layer where molecular diffusion governs transport overlain by an upper sediment layer where particle mixing occurs and transport is by turbulent or advective means.

9.6.1. *Disturbance of Lough Neagh sediments*

Three pieces of evidence indicate that the sediment water interface in Lough Neagh is of the disturbed type. In laboratory experiments with intact cores of Lough Neagh sediments, Rippey (1983) found that over one month the concentration of silica in the supernatant water rose linearly with time regardless of temperature and even though the volume of the supernatant water decreased during the experiments. This was because silica concentrations in the pore water increased approximately as the square of distance from the sediment surface due to disturbance of the superficial layers by the activity of benthic animals and by turbulence in the boundary layers.

A comparison of silica release rates measured in the laboratory with those observed in the field reveals a discrepancy which also implies a disturbed sediment-water interface. Rippey (1983) found that in the laboratory, the release rates ranged from 36 to 120 mg SiO_2 m^{-2} d^{-1} in the temperature range found in Lough Neagh. Estimates derived from field data (Table 9.9) gave rates between 260 and 810 mg SiO_2 m^{-2} d^{-1}. Assuming that the dissolution characteristics of the silica source are unchanged when taken into the laboratory, the discrepancy is probably caused by higher transport rates of silica in the pore waters of sediments in the lake. Transport may be by turbulence or advection at the disturbed sediment-water interface. There is also direct evidence of disturbance in the superficial sediment of Lough Neagh. Rippey and Jewson (1982) followed the settlement of a spring diatom crop to the sediment and, by determining how long it took for the chlorophyll *a* to be distributed evenly through the mixed layer of the sediment, they calculated that the biodiffusion coefficient for particles in the upper 2 cm of sediment was $6 \times 10^{-8}\,cm^{-1}\,s^{-1}$. Particle mixing was thought to be due mainly to larvae of *Chironomus anthracinus*, which were also thought to be responsible for the increased transport of oxygen into the sediment indicated by the depth profiles of redox potential from May onwards. The redox profiles showed that in spite of an 8-fold increase in the rate of oxygen uptake in the sediment, the aerobic layer deepened from 10 mm to 30 mm. The irrigational activities of the midge larvae and tubificids transport oxygen into the upper sediment layer and material out of the sediment from the upper pore waters. Both the particle and pore water components of the upper sediment in Lough Neagh are disturbed.

9.6.2. *The silica release process*

The spring diatom crop settles to the sediment surface when soluble silica is exhausted in the water column of Lough Neagh (Gibson, 1981; Jewson *et al.*, 1981; Chapter 10). These diatoms constitute the most important source of the silica which is released from the sediments. Over 90% of the diatom crop is sedimented (Jewson *et al.*, 1981; see also Table 9.10) and dissolution of the biogenic silica initiates the sediment silica release process. This conclusion is supported by laboratory and field results. The relationship between pH and sediment silica release rate measured in the laboratory indicates that biogenic silica is the active component, not quartz, primary silicates, clays or a desorption reaction (Rippey, 1983). This is corroborated by a temporary increase in the rate of silica release after the input of biogenic silica to the sediment (Rippey, 1983).

9.6.3. *The phosphorus release processes*

In contrast to silica, the rate of phosphorus release is limited by the detachment process in Lough Neagh. Stevens and Gibson (1977) have argued that the main phosphorus release episode is triggered by the reduction of the hydrous ferric oxides in the sediment and the subsequent release of the associated phosphate molecules. The rapid change from aerobic conditions in the upper 30 mm of sediment to anoxic conditions in all but the upper 5 mm or so in late July 1979 (Fig. 9.11) can be seen to coincide with the major phosphorus release episode in that year (Fig. 9.1). A comment on the tendency towards oxygen-depletion follows below.

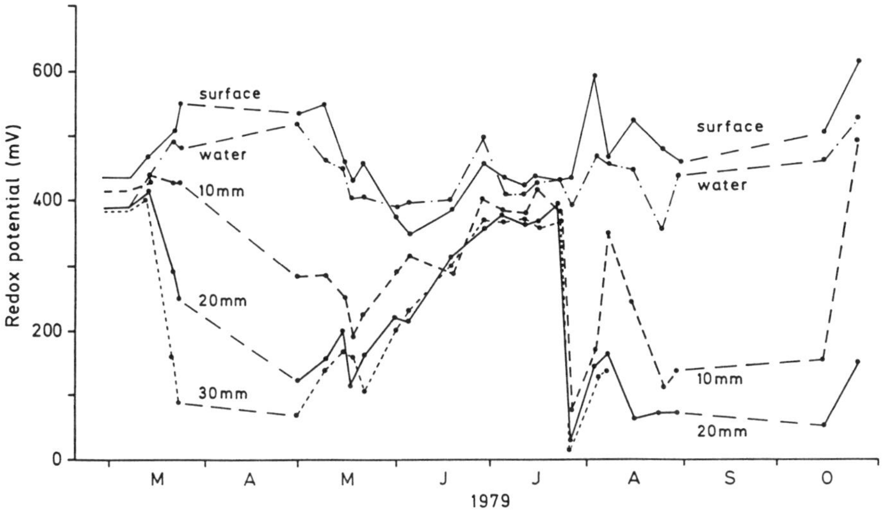

Figure 9.11. Redox potential at different depths of the bottom sediment from March to October 1979.

9.7. Oxygen consumption by the sediments

For most of the year, the water column is virtually isothermal and this results in a very uniform distribution of oxygen with depth (Fig. 9.12), although the overall oxygen balance may still be below air saturation (see Chapter 1 and Fig. 1.2) for protracted periods. During calm spells, a temperature gradient establishes and the resulting stagnation of the water column reveals a gradients of oxygen production near the surface and consumption by the bottom sediment (Fig. 9.12, May and July). Such episodes occur more or less briefly every year and they have been studied in greater detail. A continuous oxygen monitor was positioned in the middle of Antrim Bay with the oxygen electrode stationed 1 m above the mud. The electrode was anchored to a block set in the mud and kept aloft by a subsurface buoy. The record of oxygen saturation (Fig. 9.13), showed a close relationship between the degree of mixing caused by the wind over the lake surface and the degree of oxygenation in the bottom water (Fig. 9.13). It seems that in Antrim Bay, if the wind speed drops below ~2 m sec^{-1}, a thermal gradient can establish and oxygen depletion begins near the mud. A more detailed plot of an oxygen depletion episode in July 1977 (Fig. 9.14) shows that oxygen can be entirely consumed from the bottom 1 m layer of water within 5 days. The course of the depletion is not completely smooth and even on the plot of hourly values (Fig. 9.14), noticeable ripples occur on the downward trend due to transient minor increases in oxygen concentration. An even more complicated pattern is apparent in the continuous trace. The most likely cause of these phenom-

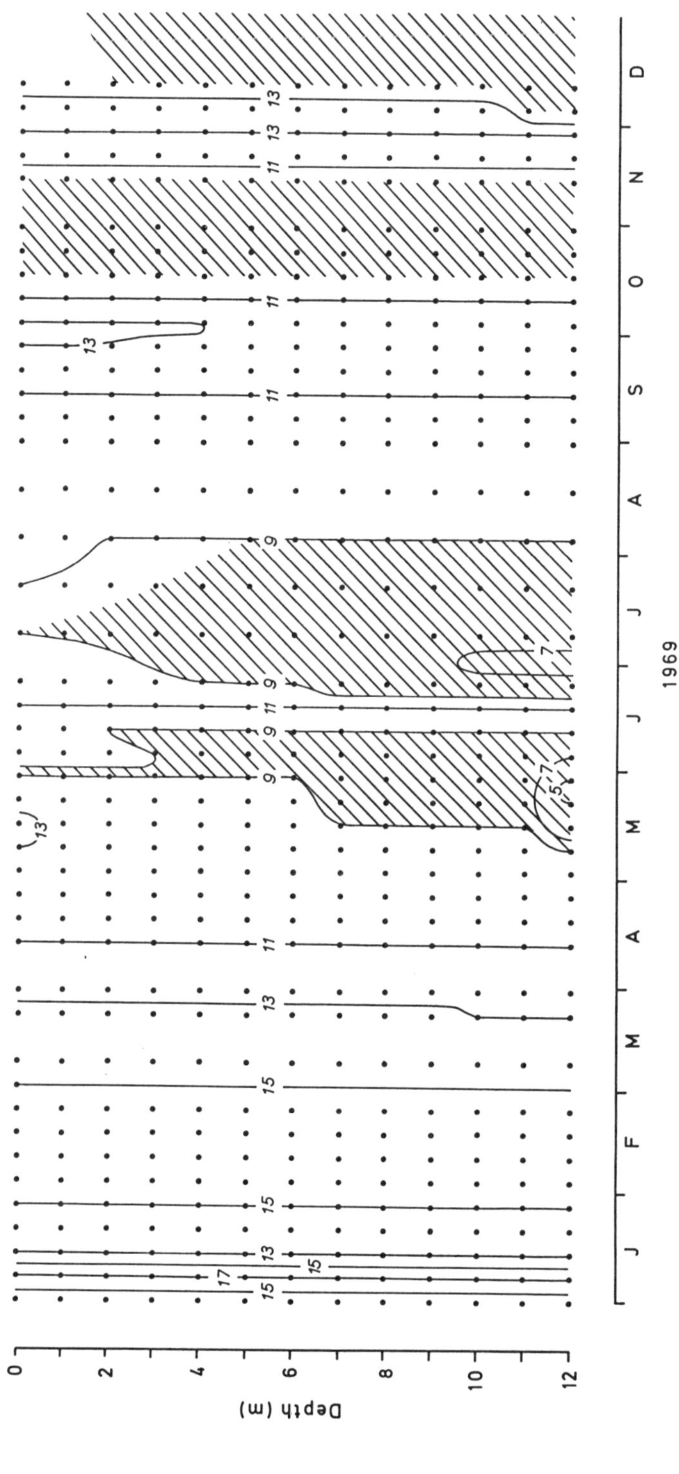

Figure 9.12. Time Depth plot of dissolved oxygen concentrations in Lough Neagh central station. Hatched area shows the occurrence of values below air saturation.

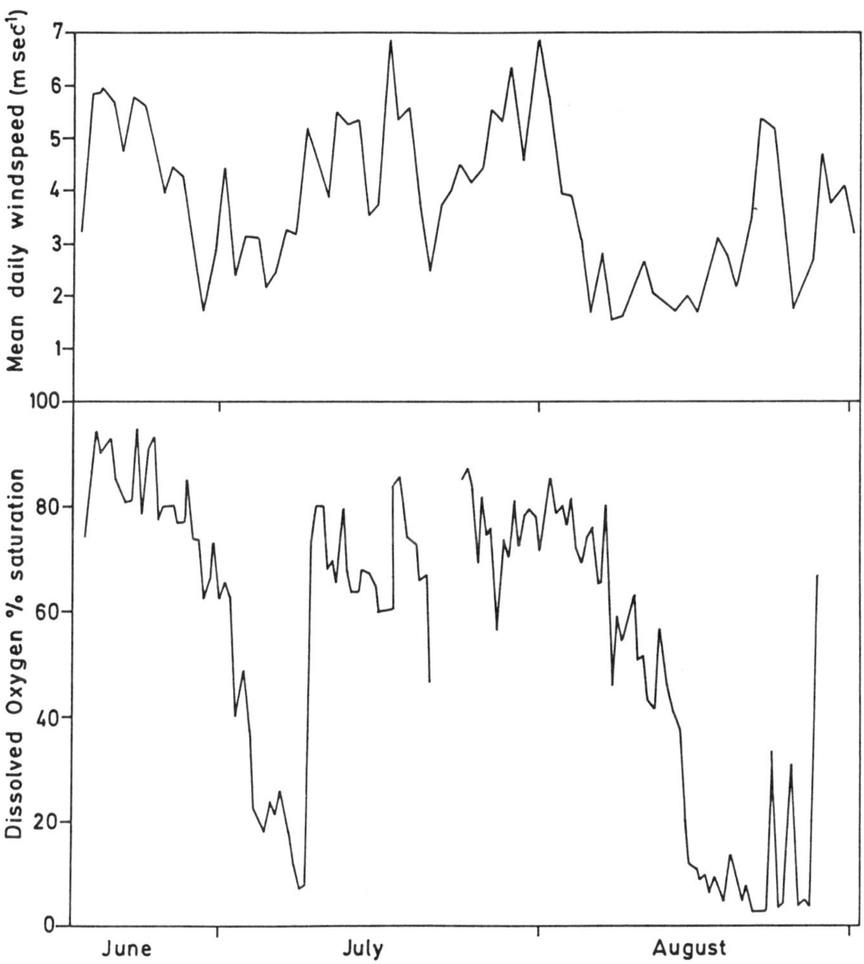

Figure 9.13. Continuous oxygen monitor and wind strength data Summer 1976. The oxygen electrode was situated 1 m above the mud in 12 m of water in Antrim Bay.

ena is the ingress of water containing more oxygen and such a possibility cautions against the calculation of an oxygen budget based only on vertical gradients of oxygen. A clearer understanding of the movement of the bottom water layer is called for.

9.8. Long-term changes in water quality

Since systematic data collection began in 1969, several significant changes have occurred in the quality of Lough Neagh water. Because of the interven-

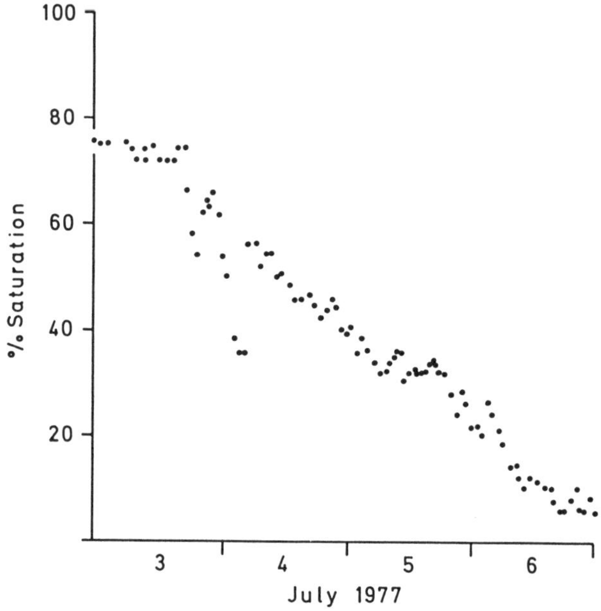

Figure 9.14. Depletion of oxygen in bottom water during a calm spell in July 1977. Hourly oxygen values plotted from a continuous oxygen record from an electrode situated 1 m above the mud in 12 m of water in Antrim Bay.

tion of low river flows and unusually settled summer weather in the years 1974–1976, comparisons which include these years are rather confused. It is therefore better to compare the beginning and end of the study period and in the discussion below, data will be presented for the periods 1969–1972 and 1977–1980. The comparisons are based on mean values of all the data available for each period, calculated as weekly means and presented in graphical form as 3 weekly running means.

9.8.1. *Nitrate-nitrogen*

One of the most striking changes has been in the concentration of nitrate-nitrogen (Fig. 9.15). Perhaps as a result of the greatly increased input from the catchment, the spring concentration of nitrate was higher in the second period than in the first. This resulted in a more prolonged decline with the minimum values being reached later. The low values were then maintained for longer and concentrations only began to rise again in October. The significance of the difference between the periods is confirmed statistically (Table 9.11). Although the increased spring concentration may be explained by the increased river inputs, the difference in the summer values requires

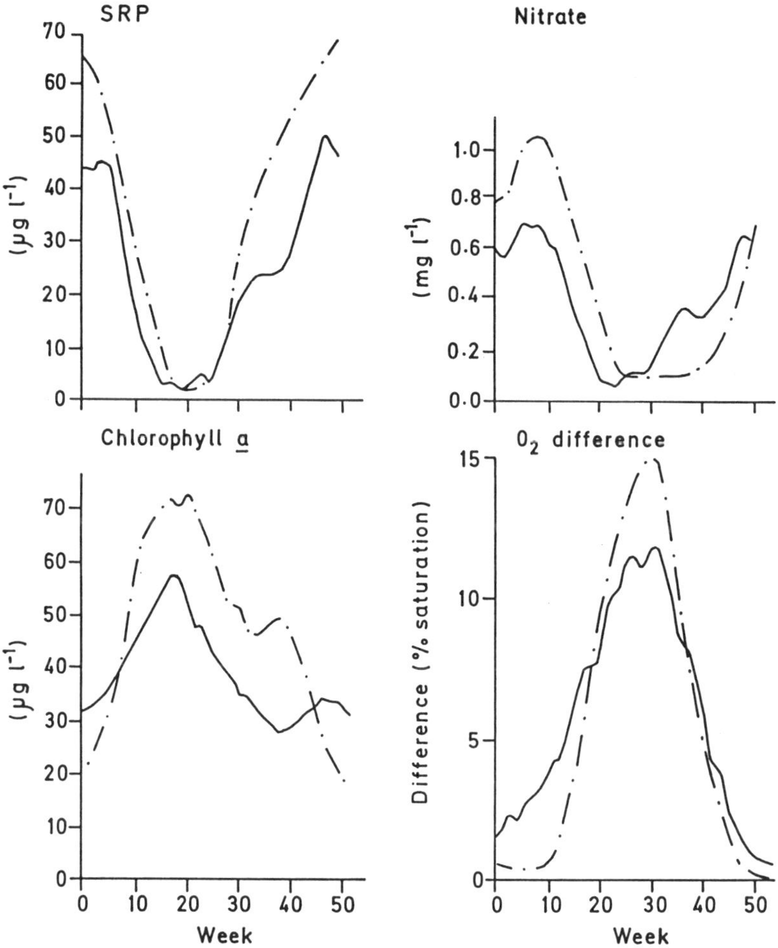

Figure 9.15. Comparison of 1969–1972 (——) and 1977–1980 (–·–) smoothed by 3 week running means. O_2 difference is the difference in oxygen saturation between 0 m and 10 m depth at the central station.

a further explanation. It is clear from the chlorophyll *a* data (Fig. 9.15) that summer algal crops have increased and it seems likely that there has been increased uptake of nitrate by the phytoplankton which at this time of year is dominated by *Oscillatoria agardhii*.

9.8.2. *Chlorophyll* a

Chlorophyll *a* concentrations have increased at most times of the year (Fig. 9.15). The most striking increase has occurred in the spring crop when mean

Table 9.11. Comparison of period means for some variables in Lough Neagh. T tests are significant as follows; $T = 2.13$ $p = 0.05$; $T = 2.95$ $p = 0.01$; $T = 4.07$ $p = 0.001$

Variable	Period 1	Period 2	Spring	Summer	Autumn	Winter
SRP	69–72	77–80	2.18	2.28	7.08	3.66
NO_3	69–72	77–80	10.87	−0.84	−6.10	2.17
SiO_2	69–73	76–80	−2.53	3.66	4.42	−1.97
Chl*a*	69–72	77–80	−2.22	6.55	3.40	−5.03
Secchi	69–72	77–80	−2.42	−4.36	−1.66	0.75
S Solids 105°	69–73	76–80	2.76	12.79	9.46	−0.97
Phytovolume	69–72	77–80	7.42	9.17	8.11	3.66

values have almost doubled. This has been caused partly by increases in diatom crops and partly by the maxima of diatoms and blue-green algal crops coinciding more closely in recent years. The winter crops were somewhat greater in the earlier period, due almost entirely to larger overwintering crops. Confirmation of the increase in algal crops comes from data on phytoplankton volumes and suspended solids (Table 9.11), both of which exhibit very significant increases in a similar manner to chlorophyll *a*.

9.8.3. *Soluble reactive phosphorus*

Like nitrate, SRP shows an increase in spring concentrations (Fig. 9.15) but due to the variability of the data, much of the difference is removed in the smoothing. There is better evidence of a more rapid concentration increase in the summer which may reflect changes in the rate of sediment release.

9.8.4. *Dissolved oxygen gradients*

There was a highly significant increase in the gradient of dissolved oxygen with depth during the summer months (Fig. 9.15). This was not caused by a more frequent incidence of temperature stratification since an analysis of temperature data shows no changes which might explain the oxygen data. This conclusion is supported by wind data which shows that, if anything, wind speeds were slightly higher during the second period than during the first. The most likely cause of the difference is more rapid consumption of oxygen by the sediments during the second period, but there are no direct observations in confirmation.

9.8.5. *Nitrate and soluble reactive phosphate inputs*

Nitrate and SRP loadings calculated from load flow models are shown in Table 9.12, which clearly illustrates the upward trend in nitrate loadings reaching a peak in 1979. There is also a less marked upward trend in SRP loads and a comparison of the average 1969–1972 loadings with the 1977–

Table 9.12. Calculated loads of SRP and nitrate entering Lough Neagh (tonnes yr^{-1}) including areas of direct drainage and sewage works discharging directly to the lake

	1969	1970	1971	1972	1973	1974	1975	1976	1977	1978	1979	1980
SRP	217	240	200	228	221	248	192	242	256	270	276	268
NO_3	3920	5388	3373	5254	4818	6051	3393	6345	7324	8794	9008	8551

1980 loadings show an increase of 88% in nitrate loadings and 21% in SRP loadings. The effect of river flow is clearly evident in both variables, particularly in 1975 which had the lowest loadings of any year in the set.

9.9. Discussion

This chapter began with a list of the qualities of Lough Neagh which make it suitable for the calculation of mass balances, but perhaps the fact that data are available for several consecutive years is the most important property of all. In the budgets for silica and phosphorus there were considerable differences between years and only the calculation of the average value over several years is likely to result in a value which represents the 'true' state of the lake. During the period of data collection, one unusually dry period and several very wet years occurred and the input budgets of all the major nutrients were affected.

A general concept has been outlined in the separate accounts of phosphorus, nitrogen and silica which divides the annual cycle into phases delimited by the direction of mass flow of the nutrients. This direction is primarily determined by solar radiation, which sets the seasons. The nutrient input phase occurs when solar radiation and hence evaporation is low and river flows are high. At this time, the growth of the phytoplankton is slow and there is little uptake of nutrients. When daylength increases in the spring, phytoplankton growth causes increasingly rapid uptake of nutrients and the demand soon exceeds the rate of supply so that the concentrations of SRP and soluble silica drop below the limit of detection. Nitrate-nitrogen appears to be present in excess since concentrations remain detectable for a considerable period after phytoplankton growth has been limited by lack first of silica, then of SRP. In the case of diatoms, nutrient limitation causes rapid settlement of cells to the mud surface in a resting state. A much smaller quantity of blue-green algae settles out over a more protracted period. A period of stagnation then occurs in the lake: low river flows cause low nutrient inputs and nutrient limitation prevents a net increase in the standing crop of phytoplankton so that settlement to the bottom gradually decreases the population. When the newly hatched Chironomid larvae grow to a sufficient size, they graze on the settled algal cells which then become increasingly liable to bacterial attack (see Chapter 17 for information on heterotrophic organisms.) The oxidation of the organic matter in the cells results in reducing conditions in the sediment causing the re-solution of the iron-phosphorus complexes and phosphorus is released into the water column. Whether because of the reducing conditions or simply as a result of the mechanical damage to the frustules and the removal of protective organic layers, soluble silica is also released into the water column. There is evidence (Patrick, 1981) that bacteria play an active role in the re-solution of silica from diatom frustules.

Initially the SRP is taken up by the phytoplankton but the demand is soon satisfied and SRP is then detectable in the water column. Little silica is taken up at this time in most years, perhaps due to the constraints of an unfavourable light climate and rapid settlement on the growth of planktonic diatoms (see Chapter 10) and the silica concentration soon reaches a steady state. Exchange between the water and sediment is the dominant flux during this period and it is only when river flows increase again in October that the input from the catchment dominates the nutrient flow.

The nitrogen cycle is not as clearly understood since the backflow of nitrogen from the sediment to the water has not been studied. The evidence available points to the fact that less nitrogen than phosphorus is released from the sediment and no rapid increase in the concentration of combined nitrogen analogous to that of SRP and soluble silica has been observed. Further evidence comes from the fact that the N:P ratio of the outflow is higher than that of the inflow. Bearing in mind that much of the particulate phosphorus coming in from the rivers settles to the bottom soon after entering the lake, the retention of the soluble phosphorus fractions is small. In contrast to the phosphorus input, relatively little nitrogen enters in particulate form so that comparison of the TN to TP ratios underestimates the difference between the retention of the active fractions of the two elements.

A notable feature of the phosphorus and silica budgets is the instability of the sedimentation coefficient of the input which varied widely between years. The result of this instability is that the system is difficult to model. The attractively simple model of Edmondson and Lehman (1981) cannot cope with the situation in Lough Neagh where there may be big differences in the lake concentrations at one end of the year and the other. Edmondson and Lehman's model calculates the outflow of substance from the initial quantity in the lake and the yearly water outflow. In Lake Washington, for which the model was intended, this assumption was shown to be reasonable because the concentration in the lake only dropped during low flow periods when outflow was quantitatively unimportant. Their model has been tried on Lough Neagh data for TP. The model is of the form

$$Pi + 1 = Ii(1 - f) + Pi(1 - Qi/V)$$

where
Pi = the quantity of TP at the start
Pi + 1 = the quantity of TP one interval later

Ii = the P loading over that period
Qi/v = the water throughput as a fraction of lake volume
f = a coefficient of loss by sedimentation and operates as a fractional loss of input not related to the concentration in the lake.

This model would seem to answer well to the situation in Lough Neagh where the retention of P in the sediments appears to be related to the input of particulate phosphorus. Applied to the years 1977–1979, the model gave sedimentation coefficients of 36, 48 and 33% respectively compared to the arithmetically computed values of 39, 57 and 31%, the greatest error therefore being in 1978. In the years when phosphorus retention was small or negative, the model gave completely erroneous estimates.

It has been shown (Section 9.8) that significant changes have occurred in the lake over the last decade and some of these changes are undoubtedly due to increased nutrient inputs. Part of the reason for the increased phosphorus input is to be found in the somewhat increased loading of sewage because of larger urban populations (see also Chapter 8, Section 8.2), and

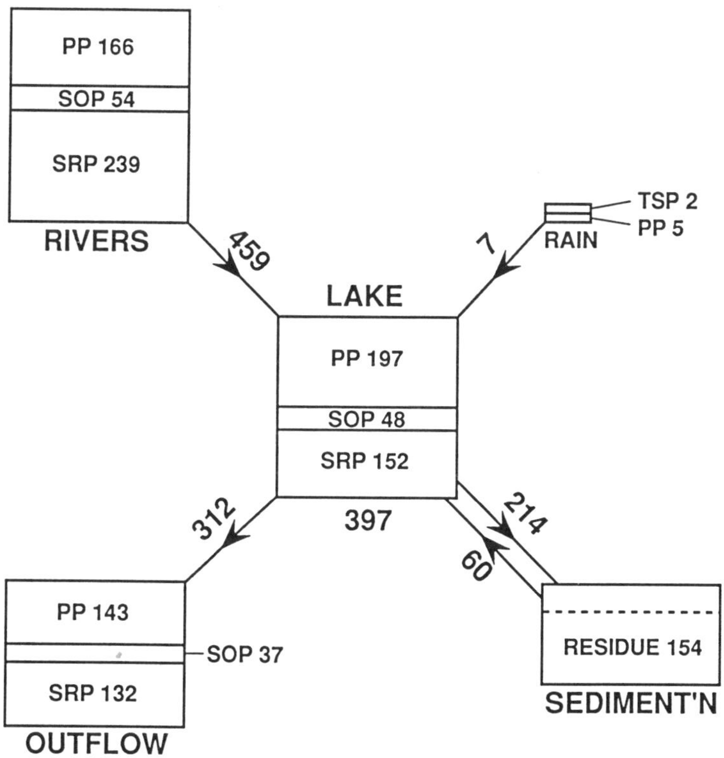

Figure 9.16. Average annual P fluxes (tonnes P) in Lough Neagh 1983–85 inclusive. Totals for each compartment and direction of the flux are shown on the connecting arrows.

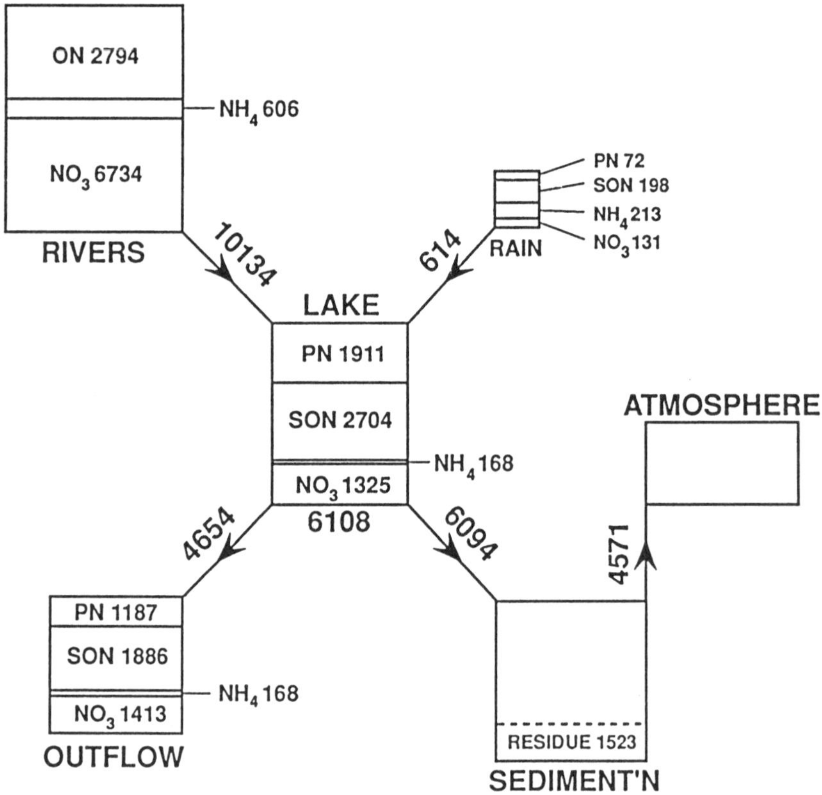

Figure 9.17. Average annual N fluxes (tonnes N) in Lough Neagh 1983–1985 inclusive. Totals for each compartment and direction of the flux are shown on the connecting arrows.

increased losses from land drainage are important in the case of nitrate. In both cases the increased river flow rates have played a significant part in increasing the loadings.

9.10. Summary of nitrogen and phosphorus cycles

Figs. 9.16 and 9.17 summarise the annual nitrogen and phosphorus cycles, using the mean values for the years 1983–1985 inclusive. The size of the boxes is proportional to the average annual fluxes (given numerically by the arrows) to and from the lake except in the case of the lake box where the values are the average lake content (concentration × volume). The sediment box is scaled to the major spring sedimentation period, the residual quantity after solubilisation (P) or denitrification (N) is indicated by the dashed line.

9.11. References

Berner, R.A. (1978) Rate control of mineral dissolution under earth surface conditions. Am. J. Sci. 278: 1235–1252.

Carter, C.E. (1978) Fauna of the muddy sediments of Lough Neagh with particular reference to eutrophication. Freshwat. Biol. 8: 547–559.

Dickson, E.L. (1975) A silica budget for Lough Neagh 1970–1972. Freshwat. Biol. 5: 1–12.

Edmondson, W.T. and Lehman, J.T. (1981) The effect of changes in the nutrient income on the condition of Lake Washington. Limnol. Oceanogr. 26: 1–29.

Gibson, C.E. (1981) Silica budgets and the ecology of planktonic diatoms in an unstratified lake (Lough Neagh, N. Ireland). Int. Revue ges. Hydrobiol. 66: 641–664.

Gibson, C.E., Smith, R.V. and Stewart, D.A. (1988) A long term study of the phosphorus cycle in Lough Neagh, Northern Ireland. Int. Revue ges. Hydrobiol. 73: 249–257.

Jewson, D.H., Rippey, B. and Gilmore, K. (1981) Loss rates from sedimentation, parasitism and grazing during the growth, nutrient limitation and dormancy of a diatom crop. Limnol. Oceanogr. 26: 1045–1054.

Jordan, C. (1987) The precipitation chemistry at rural sites in N. Ireland. Record of Agric. Res. (DANI) 35: 53–66.

Lerman, A. (1979) Geochemical Processes: water and sediment environments. Wiley, New York, 481 pp.

Mullin, J. and Riley, F.P. (1955) The colorimetric determination of silicate with special reference to sea and natural waters. Analyt. Chim. Acta. 12: 162–176.

Patrick, S. (1981) The role of bacteria in the solubilization of diatom frustules. PhD Thesis. University of Edinburgh.

Riddolls, A. (1974) Aspects of nitrogen fixation by blue-green algae in Lough Neagh. D.Phil. Thesis. New University of Ulster.

Riddolls, A. (1985) Aspects of nitrogen fixation in Lough Neagh 1. Acetylene reduction and the frequency of Aphanizomenon flos-aquae heterocysts. Freshwat. Biol. 15: 289–297.

Rippey, B. (1983) A laboratory study of the silica release process from a lake sediment (Lough Neagh, N. Ireland). Arch Hydrobiol. 96: 417–433.

Rippey, B. and Jewson, D.H. (1982) The rates of sediment-water exchange of oxygen and sediment dioturbation in Lough Neagh, Northern Ireland. Hydrobiologia 92: 377–382.

Stevens, R.J. and Gibson, C.E. (1977) Sediment release of phosphorus in Lough Neagh, Northern Ireland. In: Interactions between sediment and freshwaters, Junk/Pudoc, The Hague/Wageningen, pp. 343–347.

10. The phytoplankton populations of Lough Neagh

C.E. GIBSON

10.1. Introduction

Lough Neagh is ideally suited for phytoplankton population studies because the homogeneity of Lough Neagh's waters allows a relatively small number of samples to represent the phytoplankton populations over the whole lake and because the long turnover time of the water means that loss of phytoplankton crops down the outflow is insignificant from week to week. The aim of this chapter is to outline: (1) the changes in the phytoplankton over the 70 years between the beginning of the century and my own observations; (2) the changes over the first 12 years of continuous observations made at the FBIU; and (3) the predictable annual changes associated with the seasonal cycle in the lake at the present day. A wider historical perspective is given in Chapter 7B, and more detailed accounts of phytoplankton physiology and production ecology will be found in Chapters 11 and 13.

10.2. Changes in the phytoplankton

10.2.1. *Previous records*

In the early part of this century, two important sets of observations were made. In 1900 and 1901, the Wests visited the lake and took net samples from a steam launch setting out from Toome. Their first report (West and West, 1902) remarks on the clarity of the water. "Although the water was apparently perfectly clear, a mass of minute plankton life gradually accumulated within the nets; and at intervals of from half an hour to an hour, the nets were drawn in and the water was allowed to filter gradually away"; a later report (West and West, 1906) tabulates the species found in the samples. Ten years later, Dakin and Latarche (1913) sampled Antrim Bay at 3-weekly intervals, and their work seems to have been the first study on plankton periodicity in a lake in the British Isles.

Battarbee (1979, also Chapter 7) has discussed the difficulties of using

R.B. Wood and R.V. Smith (eds.), Lough Neagh, 203–223.

Table 10.1. Species recorded from the plankton of Lough Neagh by both West and West (1906) and Dakin and Latarche (1913). (See also Table 7B.2 and 7B.3, pp. 134–135)

Anabaena flos-aquae (Lyngb.) Breb
Asterionella formosa Hass.
Ceratium hirundinella O. F. Mull.
Chroococcus sp.
Dinobryon divergens Imhof.
Eudorina elegans Ehr.
Fragilaria crotonensis Kitton
Gomphosphaeria sp.
Microcystis sp.
Oscillatoria agardhii Gom.
Pediastrum boryanum (Turp.) Menegh.
P. duplex Meyen
Sphaerocystis schroeteri Chodat
Stephanodiscus astraea (Ehr.) Grun.
Synedra sp.
Tabellaria flocculosa Knud.

historical algal records to complement sediment studies and similar problems arise when comparing historical records with the present day phytoplankton data. One difficulty is in the identification of species, either where species were insufficiently known hitherto or where there is a mis-identification. For example, both West and West (1906) and Dakin and Latarche (1913) record *Coscinodiscus lacustris* Grun. from Lough Neagh but Battarbee (1976) has shown this to be a mis-identification of *Stephanodiscus astraea* Ehr. (Grun). There are also difficulties in recovering quantitative estimates from net tow data. Dakin and Latarche (1913) used a 35 cm diameter net hauled vertically for 40 feet (12.13 m); the net therefore theoretically strained a water volume of 1.17 m^3. According to their Table of results, (pp. 88–89) the maximum catch of *Tabellaria* was 78.81×10^6 cells, which represents a crop of 67 cells ml^{-1} in the lake, which does not seem reasonable and does not agree with the reported settled volumes of plankton which were in the range 4–10 ml per haul.

Close analysis of these early results is therefore unjustified, but the data do give a general picture of the phytoplankton and it is interesting to compare the species recorded at the beginning of the century with those found at the present time. Table 10.1 lists the species that were found both by West and West (1906) and Dakin and Latarche (1913). Certain simplifications and changes of nomenclature have been made; *Coscinodiscus lacustris* has been merged with *Stephanodiscus astraea*, *Tabellaria fenestrata* has been re-named *T. flocculosa* Knud, *Asterionella formosa* Hass and *A. gracillima* Heib. have been merged and *Dinobryon cylindricum* var *divergens* (Imhof) Lemm has been called *D. divergens* Imhof. Even excluding those which they rated as rare or very rare, West and West (1906) noted 45 species, including all those listed by Dakin and Latarche. Desmids seem to have been relatively well

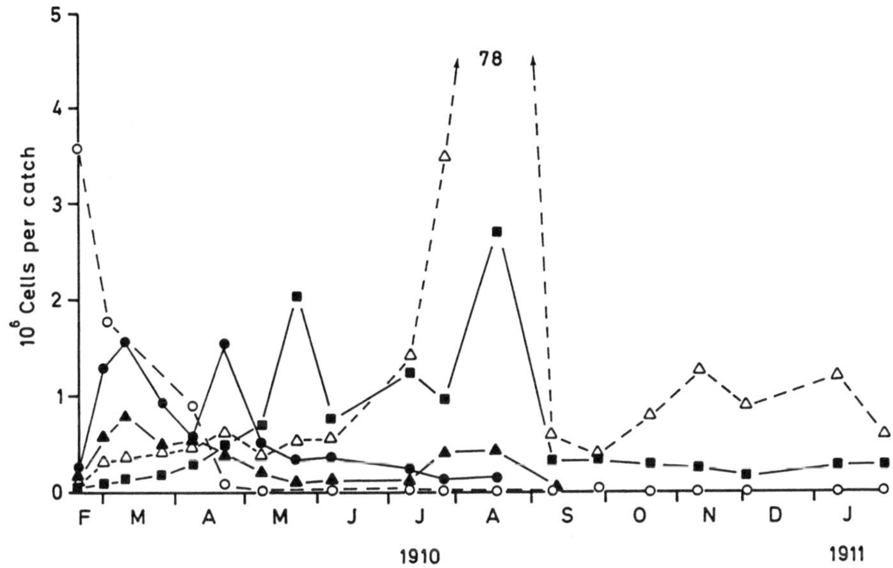

Figure 10.1. Net tow catches of diatoms in 1910–1911, from Dakin and Latarche (1913). ■ = *Asterionella formosa*; ▲ = *Fragilaria crotonensis*; △ = *Tabellaria flocculosa*; ○ = *Melosira spp*. ● Stephanodiscus astraea. Note Melosira plotted at $\frac{1}{5}$ actual values.

represented, and altogether West and West found 3 species of *Closterium*, 7 species of *Cosmarium* and 12 of *Staurastrum*.

Some features of the phytoplankton periodicity at the beginning of the century can be deduced from Dakin and Latarche's results and Fig. 10.1 shows their catches of 5 diatom species. Early in the year *Melosira*, *Stephanodiscus* and *Fragilaria* were abundant; there was then a bloom of *Asterionella* followed in August by *Tabellaria*. Diatom growth extended almost throughout the year unlike the present day, where the main diatom crop occurs in the early spring and only exceptionally do diatoms form a significant summer or autumn maximum (see Section 10.4).

No further observations were published until Lund (Macan and Lund, 1954) reported on a net tow from Toome Bridge which contained *Oscillatoria agardhii*, *O. agardhii* var *isothrix* Skuja, *Melosira italica subarctica* (Ehr.) Kutz. *Cyclotella comta* and *Stephanodiscus astraea*. Further examples from Lough Neagh were examined by Lund in 1952, 1958, 1963 and 1967. (Lund, pers. comm.) *Oscillatoria redekei* van Goor was present in 1963 (78 threads ml^{-1}), but was not seen in earlier samples and *Aphanizomenon* sp. was tentatively identified in 1952.

10.2.2. *Changes between 1969 and 1981*

Several changes took place in the phytoplankton even during this comparatively short period. In 1969–1970 *Staurastrum paradoxum* Meyen was relatively common in the phytoplankton counts and large numbers were taken in the net hauls, but in 1981 it was not seen even in the net hauls. Other desmids have also declined, notably *Closterium aciculare subpronum* W. and G.S. West, populations of which are greatly reduced. On the other hand, there has been an increase in the numbers of small centric diatoms. These are very ill-defined taxonomically and, as a matter of practical technique, it is impossible to distinguish between them in routine counts because their fine structure requires cleaned mounted preparations and high magnification for satisfactory study. They certainly include *Stephanodiscus astraea* var *minutula* (Kütz.) Grun., *S. hantzschii* Grun., *S. dubius* Hust., *S. invisitatus* Hohn and Hellerman, *Cyclotella comensis* Grun. and *Thalassiosira guillardi* Hasle. In the early 1970s *Stephanodiscus hantzschii* was common in Kinnego Bay but rare in the plankton of the open water. In recent years, it and the other small centric diatoms have formed a second peak after the collapse of the main diatom crop in the spring. Perhaps the most striking change in recent times is the relative scarcity of blooms of blue-green algae at the present day.

Unfortunately, their occurrence before 1967 is very poorly documented, but they were said to be a regular occurrence in post-war years and a spectacular bloom of *Anabaena flos aquae* Bréb. (B. and F.) covered much of the lake surface in 1967.

There is no exhaustive species list of the present day Lough Neagh phytoplankton but a partial list is given in Table 10.2. Considerable scope remains for study in this field, particularly of the nannoplankton where there are several relatively common organisms that would repay close attention.

The changes that have taken place over the last 80 years resolve themselves into three main effects:

1. the disappearance of many though not all species of desmid and of species of other groups e.g., *Dinobryon*
2. the contemporary pre-eminence of the spring diatom peak which utilises all the available silica
3. the appearance of almost perpetual crops of *Oscillatoria redekei* and *O. agardhii*.

All these changes are no doubt connected directly or indirectly with the increased quantities of dissolved nutrients entering the lake. More subtle and complex changes also occurred and during the last 12 years some normally rare species have occasionally formed a large crop after a change of conditions in the lake. This was particularly noticeable during 1975 when populations of the main species collapsed and the lake water suddenly became clearer. Crops of coccoid green aslgae and flagellates formed, presumably as a result of the improved light climate. Later that year the filamentous green algae *Planctonema lauterbornii* Schmidle developed a significant crop, although it had never previously been a prominent member of the flora. At the present time it has again become rare.

Table 10.2. Lough Neagh phytoplankton in 1980 (see also Table 7B.1, p. 134)

Actinastrum gracillimum G.M. Smith
Anabaena flos-aquae Bréb.
Ankistrodesmus falcatus (Corda) Ralfs
Aphanizomenon flos-aquae fa *gracile* Lemm.
Asterionella formosa Hass.
Botryococcus braunii Kütz.
Ceratium hirundinella (O.F. Mull) Dujardin
Closteriopsis longissima
Closterium aciculare subpronum W. and G.S. West
C. limneticum Lemm.
Cryptomonas spp.
Cyclotella comensis Grun.
Diatoma elongatum Agardh
Dictyosphaerium pulchellum Wood
Didymocystis
Didymogenes palatina Schmidle
Gymnodinium helveticum achroum Skuja
Lagerheimia quadriseta (Lemm.) G.M. Smith
Melosira granulata (Ehr.) Ralfs
M. italica subarctica (Ehr.) Kütz
Micractinium pusillum Frésenius
Nitzschia
Oocystis lacustria Chodat
Oscillatoria agardhii Gom.
O. limnetica Lemm.
O. redekei van Goor
Pandorina morum (Mull.) Bory
Pediastrum duplex Meyen
Planctonema lauterbornii Schmidle
Quadrigula chodatii (Tanner–Fullman) G.M. Smith
Rhodomonas minuta nannoplanktica Skuja
Scenedesmus dimorphus (Turp) Kütz
S. quadricauda (Turp) Bréb.
Stephanodiscus astraea (Ehr.) Kütz
S. astraea minutula (Kütz) Grun.
S. dubius Hustedt
S. hantzschii Grun.
S. invisitatus Hohn & Hollerman
Tabellaria flocculosa Grun. (Knudson)
Thalassiosira guillardi Hasle

10.3. Ecological observations on minor species

One of the most curious members of the phytoplankton is *Gymnodinium helveticum* fa *achroum* Skuja which occurs in the plankton throughout the year and forms a peak crop in November and December. Crops have increased in recent years and in November 1980, an exceptional crop of 659 individuals ml^{-1} was found. Since this species contains no photosynthetic pigment, growth is presumably either heterotrophic or phagotrophic and

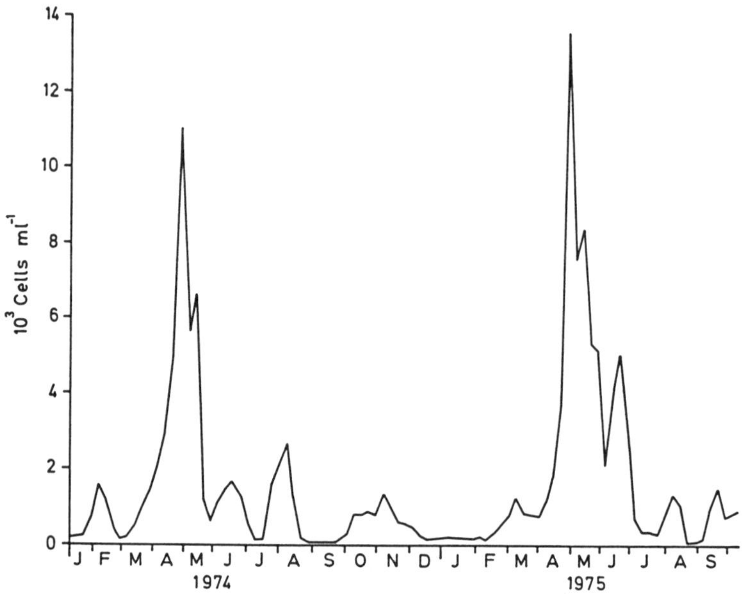

Figure 10.2. Periodicity of *Rhodomonas minuta mannoplanktica* in 1974–1975.

engulfed diatoms are sometimes seen in the cells. The cell volume of *G. helveticum* is probably nearly 70,000 μm^3 so that the maximum crop observed may equate to as much as 46 $mm^3 l^{-1}$. A more typical maximum would be 3 $mm^3 l^{-1}$.

The cryptophytes *Rhodomonas minuta* var *nanoplanktica* Skuja and *Cryptomonas ovata* Ehr. are also characteristic of the Lough Neagh plankton. They usually form a spring crop closely coinicident with *Oscillatoria redekei*. *Rhodomonas minuta* var *nanoplanktica* may occur in large numbers. Fig. 10.2 shows its periodicity in 1974–1975. In 1974, the average rate of increase was 0.116 doublings day^{-1} (28 February–24 April) but at its fastest it approached 0.16 doublings day^{-1} (28 February–20 March). After the main peak was attained, there was a rapid decline in the crop which, expressed as a negative growth rate, was -0.21 doublings day^{-1}. Throughout the remainder of the year, the crop waxed and waned several times, which suggests a rapid rate of growth with heavy losses, presumably from grazing.

Although there has been a decline in some desmid species, others continue to form part of the phytoplankton. Because of the relatively low numbers of these organisms, only rather imprecise estimates are made in the routine counts. Maximum crops amount to approximately 20 cells ml^{-1}, although in June 1981, 130 cells ml^{-1} of *Closterium limneticum* were recorded. A subsequent infection by a chytrid parasite decimated the crop on that occasion.

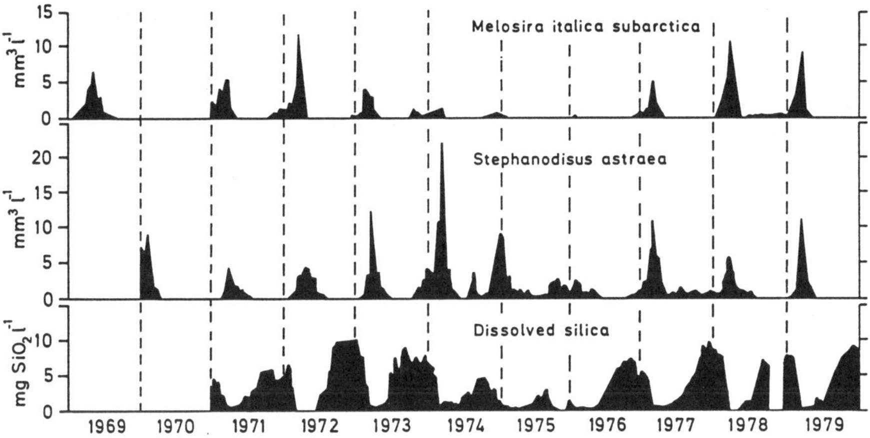

Figure 10.3. Diatoms and dissolved silica in Lough Neagh 1969–1979. Redrawn from Gibson (1981).

10.4. Diatoms

The most abundant diatoms in the plankton are *Melosira italica* subsp. *subarctica* (hereafter called *M. italica*) and *Stephanodiscus astraea* (hereafter *S. astraea*). Weekly counts have been made of both species since 1969 and it is on them that the emphasis of this account will be placed. Other diatoms, notably *Asterionella formosa*, also occur in the plankton, but there is insufficient knowledge of them to give a worthwhile account. Gibson (1981) gives a more detailed account of the inter-relationship between diatoms and silica in Lough Neagh, while Gibson *et al.* (1971) included diatoms in their paper on phytoplankton periodicity published near the beginning of the recent, sustained period of Lough Neagh study.

10.4.1. *Periodicity*

Fig. 10.3 shows the crops of *M. italica* and *S. astraea* together with the dissolved silica in the water column. In 1969, the crop consisted almost entirely of *M. italica* and in 1970 almost entirely of *S. astraea* (Gibson *et al.* (1971), but in succeeding years, a crop of *M. italica* was followed immediately by a crop of *S. astraea*. The relative importance of *M. italica* decreased, until in 1975 and 1976 only trace crops were seen. The timing of the crops also altered in this period. Whereas initially maximum crops were formed in the period February to April, the maxima occurred progressively earlier in subsequent years so that in 1974 and 1975 they were formed in the autumn. In 1978 a very large crop of *M. italica* was again formed in February and March and a similar pattern was seen in 1979. In the mid 1970's, there

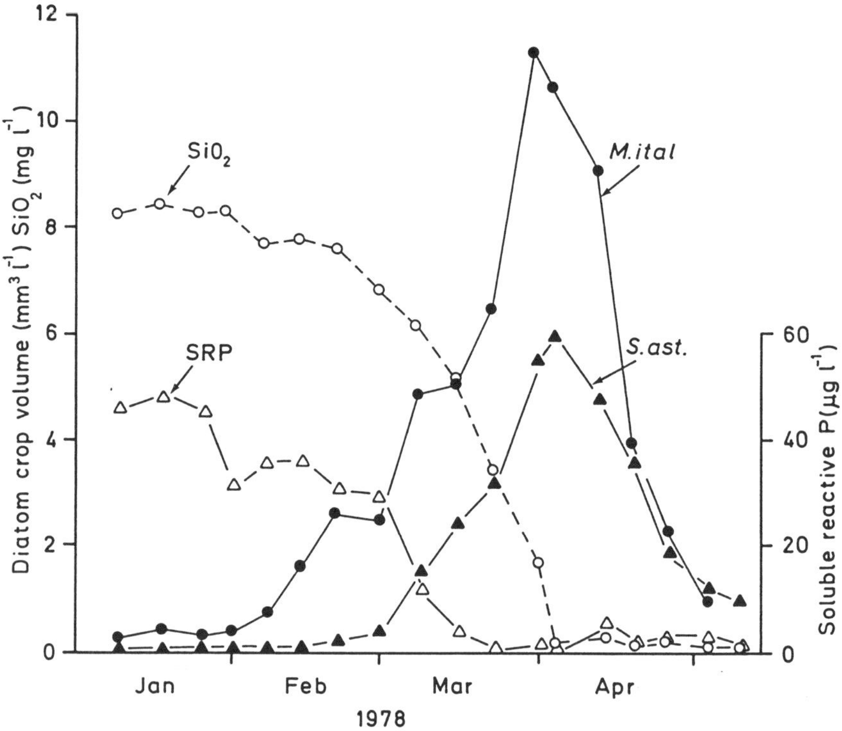

Figure 10.4. ●–● = *M. italica*; ▲–▲ = *S. astraea* crop volumes, ○--○ dissolved SiO_2, △–△ and SRP in Lough Neagh 1978. Redrawn from Gibson (1981).

was a diversification of the diatom flora: *Tabellaria flocculosa* and *Diatoma elongatum* appeared briefly in the late spring, after the main diatom crop had crashed and *Asterionella* formosa also formed a small maximum. Small centric diatoms (see Section 10.2) also began to appear during this period.

10.4.2. *Silica usage and crop limitation*

Fig. 10.3 shows that there is a close relationship between the dissolved silica and diatom crops. This is particularly clear in spring 1976, when the abnormally low dissolved silica concentration in the lake led to an almost complete failure of the spring diatom crop, Fig. 10.4 shows the relationship between diatoms, silica and phosphorus in more detail for spring 1978. When diatom growth started at the beginning of February, silica concentrations began to fall and trace levels were observed by the first week in April. From that time, the crops of *M. italica* began to decline. The crop of *S. astraea* increased slightly for one further week and then it too began to decline.

Although the soluble reactive phosphorus (SRP) was depleted before dissolved silica, there were considerable intracellular reserves of phosphorus (see Section 10.5.2.1) so that silica, not phosphorus, limited the diatom crop.

10.4.3. *Usage of phosphorus in relation to silica*

Because diatoms form the first crop of the year, they are able to take up large quantities of SRP, much of which therefore becomes unavailable for the growth of succeeding crops. Since the size of the diatom crop relates to the amount of dissolved silica, the amount of dissolved silica in the water column at the beginning of the year indirectly regulates the amount of SRP available for the succeeding crops of blue-green algae.

Gibson (1981) calculated that the uptake of 1 mg SiO_2 by *M. italica* and *S. astraea* was accompanied by the disappearance of 14.9 μg SRP from the water column. This did not tally with the appearance of only 9.8 μg particulate phosphorus, and although the settlement of diatoms contributes to the discrepancy, much of the difference can be explained by the appearance of soluble organic phosphorus. By adding in this fraction, the unreactive phosphorus was increased to 13.1 μg, which cannot be shown to differ significantly from the uptake of SRP. The usage of phosphorus by the diatoms is therefore between 9.8 and 14.9 μg P/mg SiO_2 depending on how much of the SOP is available.

10.4.4. *Recruitment of cells from the sediment*

In a stratified lake, the importance of thermal stratification in determining the periodicity of *M. italica* has been clearly shown (Lund, 1954, 1955), but in Lough Neagh there is no stable stratification. At the end of the spring growth period, the majority of the cells of both *S. astraea* and *M. italica* settle out from the water column and enter a resting state similar to that described for *M. italica* in the English Lake District by Lund (1954). Once on the bottom, invertebrates consume many of the settled cells, and few are left to form an inoculum for the next spring (Jewson *et al.*, 1981). The plankton counts show no evidence of abrupt rises in diatom numbers prior to the spring outburst which suggests that there is no significant recruitment of cells from the sediment. This conclusion was supported by intensive sampling of the seston during a winter storm (Fig. 10.5). The seston ash weight increased during the storm reaching a peak value of 17.7 mg l^{-1} at 19.00 hrs then decreased again to approximately 4.5 mg l^{-1} by 04.00 hrs next morning. A very similar pattern was seen in the numbers of dead cells which increased several-fold between 13.00 and 19.00 hrs, then decreased to their original value. There was a small increase in the numbers of live *M. italica* cells, peaking at 19.00 hrs, but no significant change in the small numbers of live cells of *S. astraea*. The validity of these results was confirmed by phytoplank-

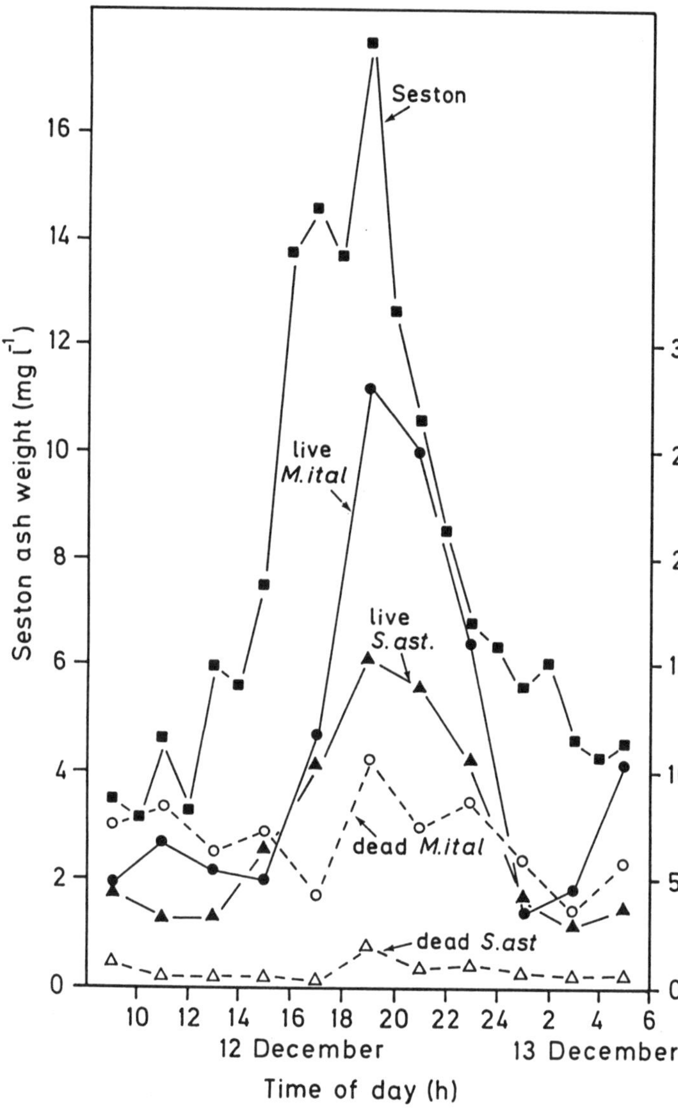

Figure 10.5. Seston resuspension during a storm ■—■ Ash weight; ○ and ● *M. italica*, △ and ▲ *S. astraea*. Filled symbols dead cells, open symbols live cells. Redrawn from Gibson (1981).

ton counts before and after the storm, which showed no significant changes in the cell numbers of either species.

The relationship between 'inoculum size' and spring crop can be examined by plotting the number of cells of *M. italica* and *S. astraea* present in the first week of January against the maximum crop reached that year (Fig.

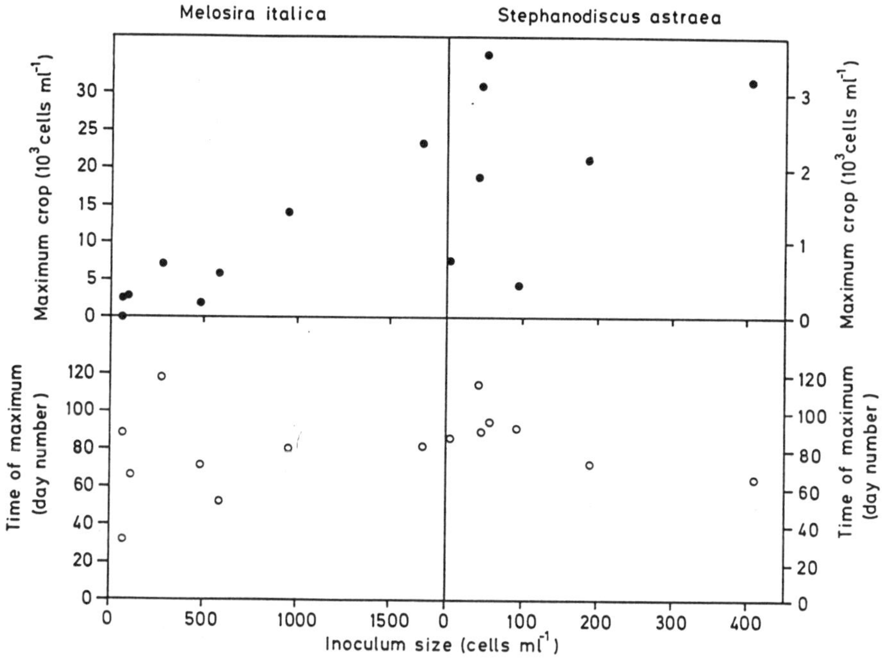

Figure 10.6. Relationship between 'inoculum size' (populations at beginning of January) and timing or size of crop maximum in *M. italica* and *S. astraea*. Redrawn from Gibson (1981).

10.6). There is a clear trend for large crops of *M. italica* to result from large inocula, but this is not so with *S. astraea*. The same data for inoculum size are plotted against the timing of the maximum, expressed as number of days from 1 January (Fig. 10.6). There is no evidence of a relationship between the two variables with *M. italica* but with *S. astraea* smaller inocula seem to form later maxima.

10.4.5. *Infection by parasites*

In Lough Neagh, *M. italica* filaments are frequently attacked by the chytrid parasite *Zygorhizidium melosirae* Canter. Fungal attack was particularly noticeable in spring 1978 (Fig. 10.7) when 53% of the filaments (7% of the cells) were infected. The parasite grew rapidly between 28 February and 7 March, but subsequently, both the percentage of filaments infected and the total number of infected cells decreased. Even at the peak of the epidemic, the majority of infected filaments had only one cell infected; on 28 February and 7 March only 8% of filaments had 2 infected cells and less than 1% had 3 infected cells.

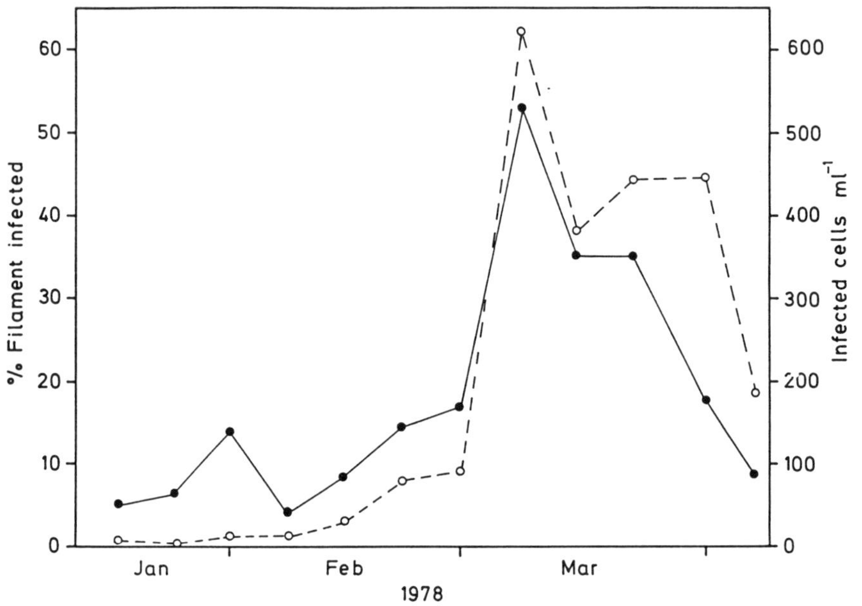

Figure 10.7. Progress of *Zygorhizidium melosirae* infection on *M. italica* crop ●—● % filaments infected; ○—○ infected cells ml^{-1}. Redrawn from Gibson (1981).

10.5. Blue-green algae

Blue-green algae dominate the phytoplankton of Lough Neagh. Although the diatoms form a major peak in the spring, blue-green algae are almost always present in significant amounts and between them *Oscillatoria redekei* and *Oscillatoria agardhii* form an almost perpetual crop. A detailed account of the ecology of the planktonic blue-green algae is given in Gibson and Fitzsimons (1982).

10.5.1. *Periodicity*

Fig. 10.8 shows the weekly estimates of cell volume for the 4 blue-green algae counted in the years 1969–1979. Three main features are apparent from the data. Firstly, in spite of the massive bloom of *Anabaena flos-aquae* which occurred in 1967, the alga has been virtually absent since, being abundant only in 1974 and 1975. *Aphanizomenon flos-aquae* Ralfs formed summer crops in 1969–73 and 1976 but subsequently no significant crops of this or any other heterocystous species appeared. Riddolls (1985a) has reported on nitrogen fixation by *Aphanizomenon flos-aquae* during the earlier period. Secondly, *Oscillatoria redekei* and *Oscillatoria agardhii* both showed a great decline in the middle years of the study. *O. agardhii* almost vanished

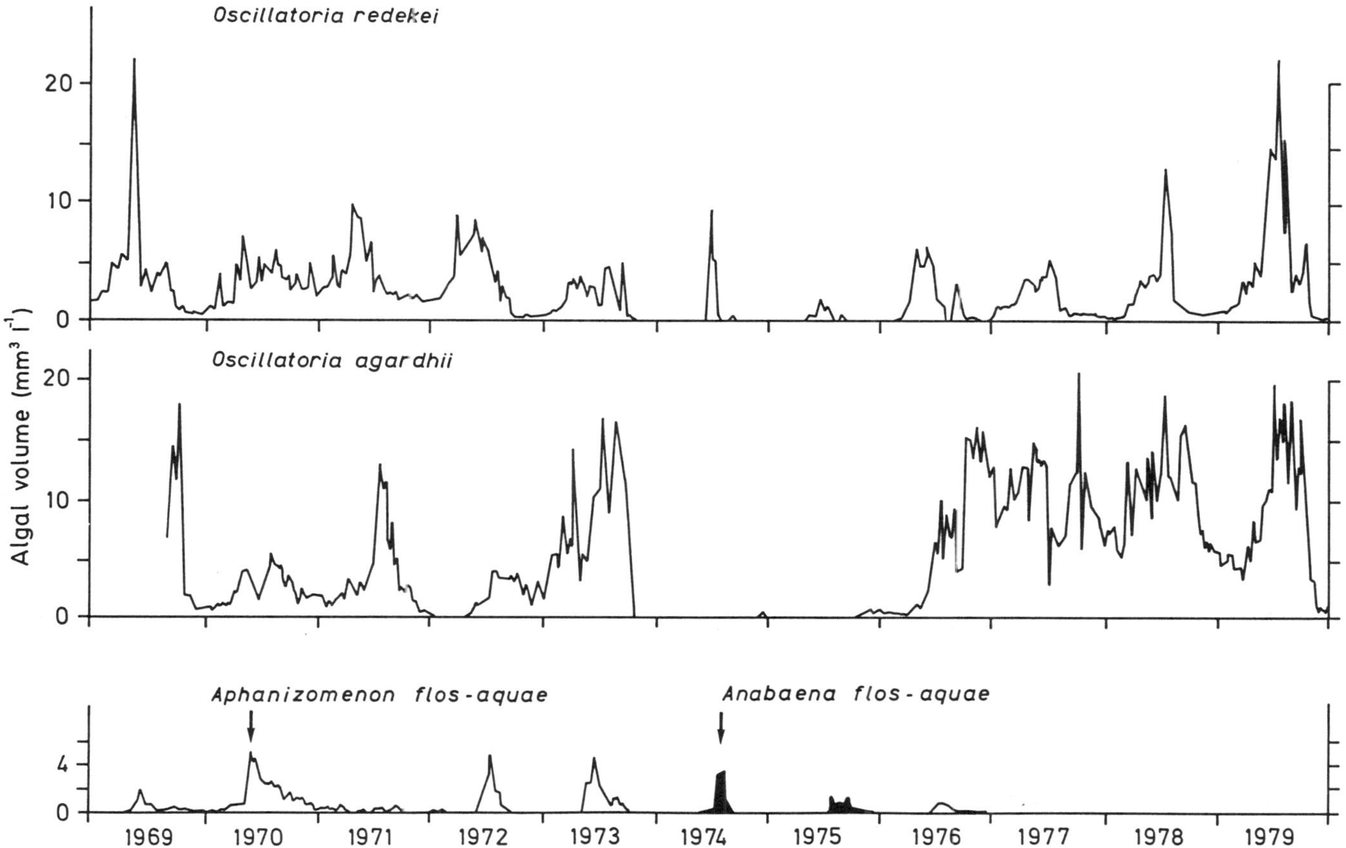

Figure 10.8. Planktonic blue-green algae in Lough Neagh 1967–1979. Hatched areas in lower panels denote *Anabaena flos-aquae*. Redrawn from Gibson and Fitzsimons (1982).

in 1974–1975 but subsequently recovered. The third feature concerns the magnitude of the yearly changes of population size. In the early years, *O. redekei* showed relatively small changes in population size with overwintering populations doubling no more than 5 times to the maximum and in some case doubling only 2.5 times. In 1974, 1975 and 1976 the population doubled 10, 9 and 6 times respectively, in 1977 only 1.5 times and in subsequent years approximately 4 times. *O. agardhii* showed marked periodicity with pronounced peaks and troughs, although the size of the maximum varied from 4.4 $mm^3 l^{-1}$ (1972) to 20.3 $mm^3 l^{-1}$ (1977). In 1976, 1977 and 1978 large crops were maintained throughout the year and the population scarcely decreased below 5 $mm^3 l^{-1}$ at any time.

10.5.2. *Growth rate and inoculum size*

O. redekei is the only blue-green alga for which extensive data on growth rate in Lough Neagh exists. An earlier report (Foy *et al.*, 1976) gave some growth rates from Lough Neagh and more recently Foy (1980) has measured growth rates of a range of species at 15° C in the laboratory. In Lough Neagh, *O. redekei* usually begins to grow in February at an average apparent rate of 0.098 doublings day^{-1} (10.25 days $doubling^{-1}$), with variation in individual years between 0.067 (1979) and 0.126 doublings day^{-1} (1971). Losses at the time of growth are considered to be small. No equivalent data exist for *O. agardhii* since there were very few consistent increases in crop over a sustained period, but a calculation over the early period of observation (Foy *et al.*, 1976) suggested a somewhat lower growth rate than *O. redekei* (0.091 doublings day^{-1} average). Similar calculations for *Aphanizomenon flos-aquae* and *Anabaena flos-aquae* gave higher growth rates of 0.151 and 0.226 doublings day^{-1} respectively.

The time taken to reach the maximum crop of *Oscillatoria redekei* varied from year to year and this was partly determined by the size of the overwintering crop (Fig. 10.9): large inocula obviously result in earlier maxima. A regression of time of maximum on $\log_2$ inoculum size gives a highly significant relationship ($r^2 = 0.9266$, $n = 8$) and predicts that a doubling of inoculum size advances the maximum by 7.5 days.

Because the maxima of *O. agardhii* occur much later in the year than those of *O. redekei*, there are greater uncertainties in the relationship between overwintering inoculum and time of maximum. Similarly with inoculum and size of crop maximum, the only safe conclusion is that very small inocula (1974 and 1975) were followed by crop failures.

10.5.2.1. *Nutrient limitation*

The relationship between inoculum and crop maximum of *O. redekei* is due in part to the relatively constant size of the crop maximum. From January onwards, the rate of supply of nutrients from the inflows is less than the demand from the phytoplankton and concentrations of dissolved nutrients

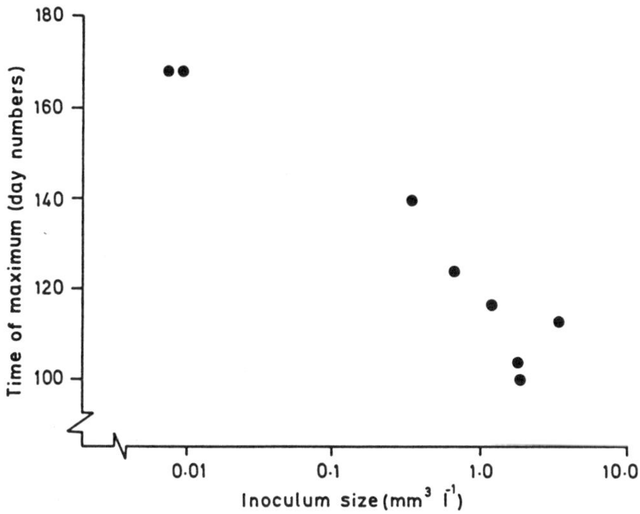

Figure 10.9. Relationship between 'inoculum size' (crop at beginning of January) and timing of maximum in *O. redekei*. Redrawn from Gibson and Fitzsimons (1982).

in the Lough water begin to decrease. SRP is quickly removed both by the diatoms and the blue-green algae, the relative amount taken up by the two groups depends on the size of the overwintering populations of blue-green algae. When SRP is available in the water column, it is taken up in excess of immediate requirements so that much of the SRP is removed before the growth of *O. redekei* begins. *O. redekei* has therefore mostly to grow at the expense of stored phosphorus and the larger the inoculum of *O. redekei* the greater the storage capacity for phosphorus. Fig. 10.10 shows the way in which the phosphorus in the seston (predominantly algae) changes during this period in different years. From the beginning of March, the phosphorus content of the seston (% ash-free dry weight) declines because the quantity of phosphorus remains constant whilst the biomass increases. The 4 years' data shown illustrate three different ways in which the *O. redekei* crop may behave. In 1977 *O. redekei* doubled only once between March and June and the phosphorus was exhausted by a very large diatom crop and a spring crop of *O. agardhii*. In 1978 and 1979 the *O. redekei* maximum was in June after 4 doublings; in 1980 a larger crop also doubled 4 times and produced a final crop of 22 $mm^3 l^{-1}$ at the end of July.

Because the seston is a mixture of algal species and a random quantity of detritus, it is not possible to ascertain directly how much phosphorus is contained in the *O. redekei*. However, by selecting data from periods when the ratio of species varied widely, multivariate analysis of the data enabled an estimate to be made of the ash-free dry weight (AFDW) of *O. redekei* and *O. agardhii*. 1.0 mm^3 volume represented 0.14 mg AFDW of *O. redeki*

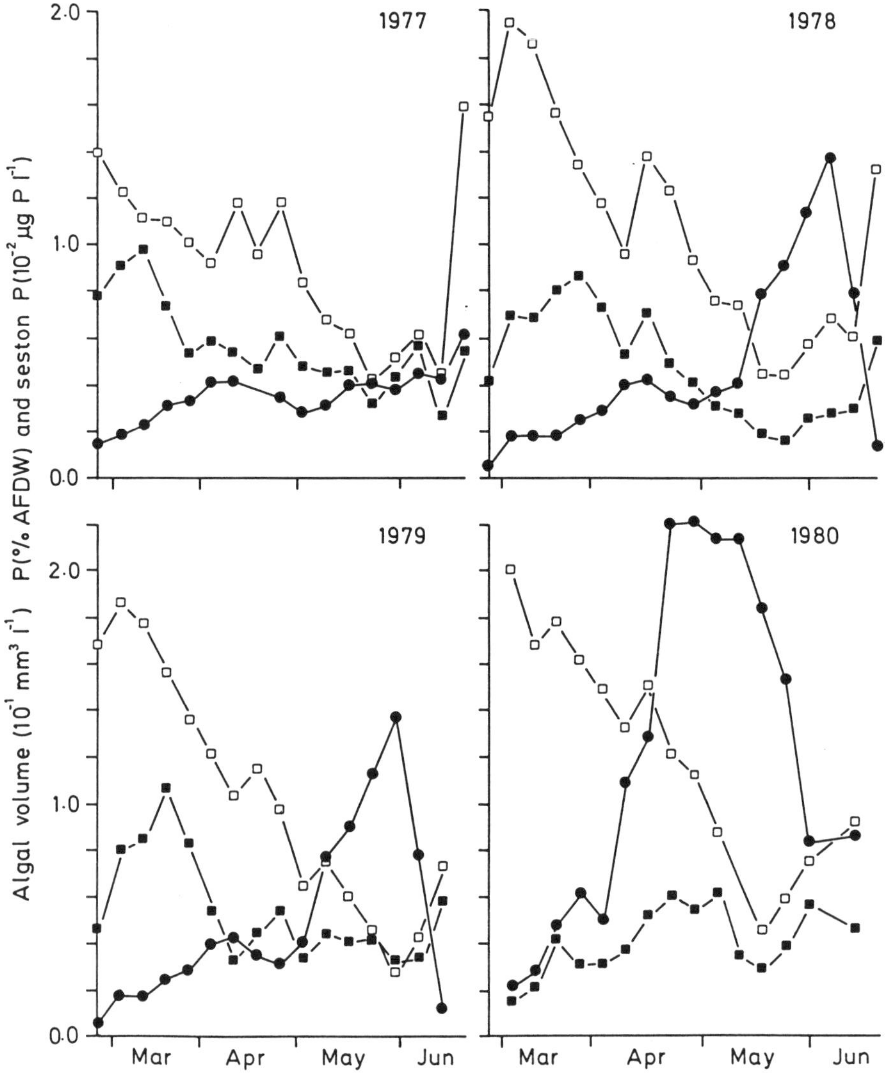

Figure 10.10. O. redekei crop ●—●; P in seston ■—■; and P in seston % □—□ ash-free dry weight in growth period of *O. redekei* 1977–1980. Redrawn from Gibson and Fitzsimons (1982).

and 0.18 mg AFDW of *O. agardhii*. When SRP is present in the water column, the seston contains a maximum of 2% phosphorus, so that the *O. redekei* crop contains approximately 3 μg P mm^{-3} but the phosphorus content may decline to 0.4% AFDW so that 1 μg P can perhaps produce a maximum of 1.7 mm^3. The uptake of 1 mg SiO_2 by diatoms is accompanied by the uptake of say 11 μg SRP (see Section 10.4.3) and since the concentration of SiO_2 in Lough Neagh at the beginning of the year is approximately 6.5 mg l^{-1},

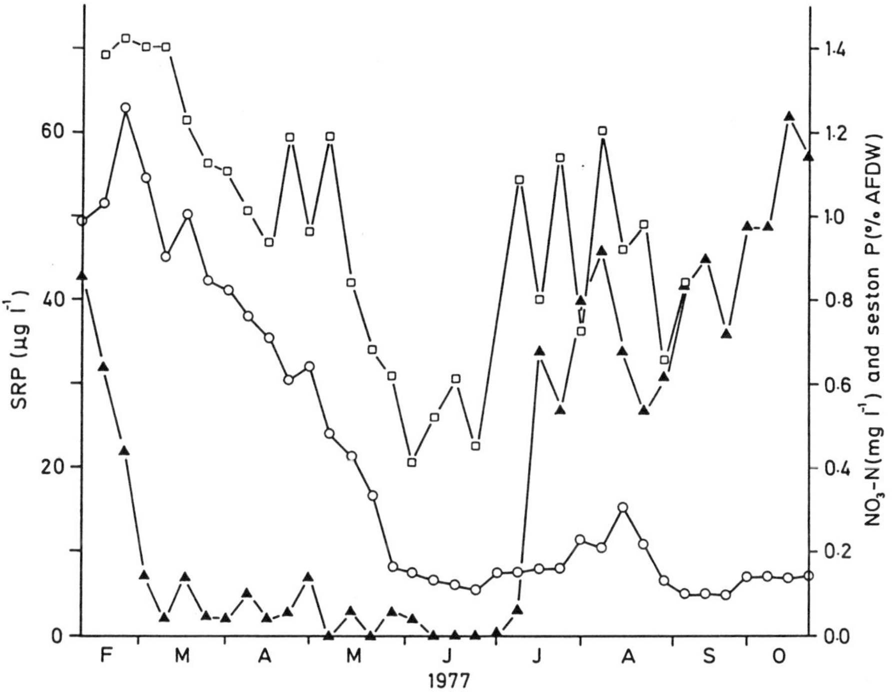

Figure 10.11. Nitrate-nitrogen ○—○ and SRP ▲—▲ in water column and seston P % □—□ ash-free dry weight 1977. Redrawn from Gibson and Fitzsimons (1982).

this could account for the removal of 70 μg SRP l^{-1}. Since SRP values in the early spring never exceed 80 μg P l^{-1}, the SRP remaining may only be sufficient to produce 17 $mm^3 l^{-1}$ of *O. redekei* and in some years less. Although further sources of P may arise from the soluble organic fraction, from the contemporary river inputs or from the bottom mud, it is clear that phosphorus supply is critical to the attainment of the *O. redekei* maximum. The intervention of the diatom crop plays a crucial role in determining the phosphorus supply until phosphorus is released from the mud at the end of June.

Nitrogen does not seem to be limiting during the spring growth period and nitrate is detectable in the water column until phosphorus is released from the sediment. Fig. 10.11 plots the dissolved nitrate, SRP and seston P in 1977 and shows that the uptake of nitrate proceeded smoothly until the phosphorus in the seston reached its base value, when nitrate uptake ceased. Calculating the uptake of nitrogen and phosphorus over the growth period (uncorrected for catchment inputs) gives a molar uptake ratio of 44:1 N:P. Similar calculations for other years give molar ratios of 73:1, 46:1 and 35:1 for 1978, 1979 and 1980 respectively which suggest that the algae are likely to be limited by phosphorus rather than by nitrogen. The situation changes when phosphorus is released from the sediment since little nitrate appears in the water column, and the possibility of nitrogen limitation increases. The competitive advantage of the nitrogen-fixing species seems to be greatest at this time, and Riddolls (1985b) has reported laboratory studies consistent with this.

In the early years of the study, *Aphanizomenon flos-aquae* fa *gracile* (Lemm.) Elenkin appeared regularly in May and formed a maximum which varied in size from 1.0 $mm^2 l^{-1}$ (1976) to 5.7 $mm l^{-1}$ (1972) in July (Fig. 10.8). After 1976, *A. flos-aquae* was only observed as occasional filaments in the plankton. Although there was a large crop of *Anabaena flos-aquae* in 1967, it did not reappear in the lake over the next 6 years. In the summers of 1974 and 1975, the waters of Lough Neagh were unusually clear, because the *Oscillatoria argardhii* crop failed and the euphotic zone extended to 4.5 m, i.e. half the water column; in both years *Anabaena flos-aquae* reappeared during the time of clear water (Fig. 10.8).

10.5.2.2. *Losses*

Observations on the sedimentation rate of the blue-green algae in Lough Neagh (Gibson, 1975, 1984) suggest that the loss of filaments from the water column by settlement is unlikely to be significant and few filaments are caught in sediment traps (Flower, 1980). Direct observation of the sediment surface suggests that some filaments do accumulate there (Jewson, pers. comm.) but this mode of loss is not as significant as it is for the diatom crops. Occasional chytrid parasites have been observed on *Oscillatoria agardhii* but after examination of thousands of samples no parasite has ever been observed on *Oscillatoria redekei*, and it is unlikely that parasitism plays a significant role in regulating the crops of blue-green algae in Lough Neagh. Cell death is much more difficult to assess since dead blue-green algae leave no obvious traces in the plankton. Meffert (1972) has suggested that carbon dioxide depletion can cause lysis of *O. redekei* filaments but there is no evidence that this happens in Lough Neagh. The loss down the outflow can be estimated accurately and it seems that only exceptionally high flows, as in December 1978, cause a significant loss. By elimination, therefore, grazing seems the most likely cause of crop decline. Calculation of the rate of disappearance of the *O. redekei* crop from the water column shows that the crop can decline by up to 34% per day. This is likely to arise from grazing by *Daphnia hyalina* Leydig *D. longispina*, *O.F. Mull* and *D. hyalina galeata* (Sars). The number of *Daphnia* increased yearly over the period 1972–1977 (see Chapter 14), then reverted to their former numbers in 1978. Most of the increase in numbers was due to the large species *D. longispina* and when the *D. longispina* began to increase, there were smaller *Oscillatoria* crops and clearer water during the summer. In the summers of 1974 and 1975 numbers of *Daphnia* and algae oscillated in sequence; *Daphnia* first increased then algal crops decreased and *Daphnia* finally 'crashed' in numbers.

10.6. Conclusions

The controlling variable at the beginning of the year is light and growth begins in the lake when the diatom populations receive a sufficient light dose. The size of the light dose is dependent primarily upon daylength, but light interception by other factors – principally water colour, detritis and other algae – also modifies the light climate in significant ways (see Chapter 4). In most years, *M. italica* and *S. astraea* begin growth together in January or February but if the water column is particularly clear, growth may start earlier in the season, even in the preceding autumn. If this is the case, growth of *S. astraea* is favoured. The relative success of the two species in spring growth seems to depend on the size of the overwintering population of *M. italica*. The evidence points to poor survival of cells on the sediment surface and it is probable that most of the cells which form the new inoculum have survived in the water column. *M. italica* grows more slowly than *S. astraea* and therefore requires a large inoculum of overwintering cells to compete with it. As the diatoms increase, silica is depleted rapidly, but SRP disappears from the water column whilst some soluble silica still remains. The SRP is shared between the diatoms and *Oscillatoria redekei* and SRP is taken up in excess of immediate requirements so that phosphorus represents up to 2% AFDW. As daylength increases, it exerts the major effect on light dose and differences in light attenuation have a smaller influence. *O. redekei* begins to grow whilst the diatoms are still increasing and growth depends on the phosphorus stored in the cells and the relatively small contemporary input from the catchment. Diatoms continue growing until all the soluble silica has been removed from the water column, but as soon as this happens, the cells begin to settle out to the bottom sediment. There is evidence (Gibson, 1984) that the sinking rate of the diatom cells increases when starved of silica in the lake. *M. italica* begins to sink before *S. astraea*. *O. redekei* is not constrained by a lack of silica and continues to grow at the expense of its stored phosphorus. A certain amount of phosphorus is released by the diatoms soon after they settle out, but the growth of *O. redekei* depends for the most part on the phosphorus taken up earlier in the year. When the stored phosphorus has been reduced to approximately 0.3–0.4% ash-free dry weight, growth is halted. During this time, significant changes are occurring on the sediment surface. Diatoms that have settled to the bottom and entered a resting state are consumed in large numbers by chironomid larvae and subsequently there is a release of silica, phosphorus, manganese and iron into the water column. If growth is limited by nutrients at all at this time, it is most likely to be limited by nitrogen. *O. redekei* may form a second late peak stimulated by the release of phosphorus, but zooplakton grazing seems to reduce the crop greatly and the population size becomes unstable. The place of *O. redekei* is usually taken by *Oscillatoria agardhii* which forms a midsummer peak and dominates the plankton for much of the rest of the year. If for any reason the *O. agardhii* crop fails, there is a period of clear water after the collapse of the *O. redekei* crop and the heterocystous species *Anabaena* or *Aphanizomenon* appear. It is likely that these species are favoured by nitrogen-limiting conditions, but the heavy energy expenditure of heterocyst production and nitrogen fixation requires a favourable light regime, i.e. a relatively clear water column. The two conditions needed for a crop of nitrogen-fixers are therefore likely to be (a) a low nitrogen budget – either through low river flows or through lower agricultural activity (see Chapter 8) and (b) the collapse of preceding crops caused by zooplankton grazing or by nitrogen limitation.

The collapse of the *Oscillatoria* crops also favours an autumn growth of diatoms and if this growth is large, much of the silica is taken up. River inputs of silica over the winter only supply approximately half of the amount normally available for the early spring growth so that if the silica released from the mud in the summer is taken up in the autumn, the diatom crop fails next spring.

10.7. References

Battarbee, R.W. (1976) *Coscinodiscus lacustris* in Lough Neagh – a case of mistaken identify? Br. Phycol. J. 11: 305–307.

Battarbee, R.W. (1979) Early algological records: help or hindrance to palaeolimnology. Nova Hedwigia Beih. 64: 379–394.

Dakin, W.J. and Latarche, M. (1913) The plankton of Lough Neagh. Proc. R. Irish Acad. B 30: 20–96.

Flower, R.J.L. (1980) A study of sediment formation, transport and deposition in Lough Neagh, Northern Ireland, with special reference to diatoms. D.Phil. Thesis, New University of Ulster.

Foy, R.H. (1980) The influence of surface to volume ratio on the growth rates of planktonic blue-breen algae. Br. Phycol. J. 11: 151–163.

Foy, R.H., Gibson, C.E. and Smith, R.V. (1976) The influence of daylength, light intensity and temperature on the growth of planktonic blue-green algae. Br. Phycol. J. 11: 151–163.

Gibson, C.E. (1975) Cyclomorphosis in natural populations of *Oscillatora redekei* van Goor. Freshwat. Biol. 5: 279–286.

Gibson, C.E. (1981) Silica budgets and the ecology of planktonic diatoms in an unstratified lake (Lough Neagh, N. Ireland). Int. Revue ges. Hyrobiol. 66: 641–664.

Gibson, C.E. (1984) Sinking rates of planktonic diatoms in an unstratified lake: a comparison of field and laboratory observations. Freshwat. Biol. 14: 631–638.

Gibson, C.E. and Fitzsimons, A.G. (1982) Periodicity and morphology of planktonic blue-green algae in an unstratified lake (Lough Neagh, Northern Ireland). Int. Revue ges Hydrobiol. 67: 459–476.

Gibson, C.E., Wood, R.B., Dickson, E.L. and Jewson, D.H. (1971) The succession of phytoplankton in Lough Neagh 1968–70. Mitt. Internat. verein Limnol. 19: 146–160.

Jewson, D.H., Rippey, B.H. and Gilmore, W.K. (1981) Loss rates from sedimentation, parasitism and grazing during the growth, nutrient limitation and dormancy of a diatom crop. Limnol. Oceanogr. 26: 1045–1056.

Lund, J.W.G. (1954) The seasonal cycle of the plankton diatom *Melosira italica* (Ehr.) Kütz subsp. *subarctica* O. Mull. J. Ecol. 42: 151–179.

Lund, J.W.G. (1955) Further observations on the seasonal cycle of *Melosira italica* (Ehr.) Kütz. subsp. *subarctica* O. Mull. J. Ecol. 43: 90–102.

Macan, T.T. and Lund, J.W.G. (1954) Records from some Irish lakes, 2: Phytoplankton. Proc. R. Ir. Acad. 56B: 135–157.

Meffert, M. E. (1972) Kultur und Wachstum von *Oscillatoria redekei* van Goor. Zur Bedeutung der Bakterien. Verh. int. Verein. theor. angew. Limnol. 18: 1359–1366.

Riddolls, A. (1985a) Aspects of nitrogen fixation in Lough Neagh. I. Acetylene reduction and the frequency of *Aphanizomenon flos-aquae* heterocysts. Freshwat. Biol. 15: 289–297.

Riddolls, A. (1985b) Aspects of nitrogen fixation in Lough Neagh. II. Competition between *Aphanizomenon flos-aquae*, *Oscillatoria redekei* and *Oscillatoria agardhii*. Freshwat. Biol. 15: 299–306.

West, W. and West, G.S. (1902) A contribution to the freshwater algae of the North of Ireland. Trans. R. Irish Acad. B 32: 1–100.

West, W. and West, G.S. (1906) A comparative study of the plankton of some Irish lakes. Trans. R. Irish Acad. B 33: 77–116.

11. Planktonic primary production in Lough Neagh

D.H. JEWSON

11.1. Introduction

Phytoplankton are the major primary producers in Lough Neagh (cf. Chapter 12) and during the 1970s a series of studies were carried out to estimate the amount of energy fixed by them in photosynthesis. Most of the work was based on oxygen evolution (Jewson, 1975, 1976; Jewson and Wood, 1975; Jewson and Taylor, 1978; Jones, 1977b, 1978) but ^{14}C-uptake (Jewson, 1977b; Dring and Jewson, 1982) and carbohydrate accumulation (Gibson, 1975, 1977) have also been measured. The majority of the material summarised in this chapter refers to the open lake but results are also included for a shallow area, Kinnego Bay (Jewson, 1976; Jones, 1977a,b,c).

11.2. Depth distribution of photosynthetic activity

Over the last decade the depth of the photosynthetic zone has varied from 1.2 to 4.6 m (Fig. 11.1). Compared to most aquatic environments these changes are relatively small and the reasons for this have been discussed in Chapter 4. In this section, we are mainly concerned with the response of the phytoplankton cells to the resulting steep light gradient. For instance, in the space of a few metres, cells can pass from the dark into limiting, saturating, or even inhibitory light irradiances (see Jewson and Wood, 1975).

Fig. 11.1 illustrates photosynthesis depth profiles typical for the open lake. Sub-surface irradiance values are usually sufficient to give light saturation, except during overcast weather in mid-winter (December/January). At other times inhibition is found. This frequently occurs down to 0.5 m and occasionally down to 1 m. However, it is possible that the use of bottles suspended at the same depth throughout one experiment may overestimate this (Jewson and Wood, 1975), particularly as water movements make it unlikely that freely circulating phytoplankton cells will remain at the surface for long periods. Even if the bottle figures are correct the effects on integral rates are limited. For instance, in 50 mid-day experiments over 2 years the average

R.B. Wood and R.V. Smith (eds.), Lough Neagh, 225–237.

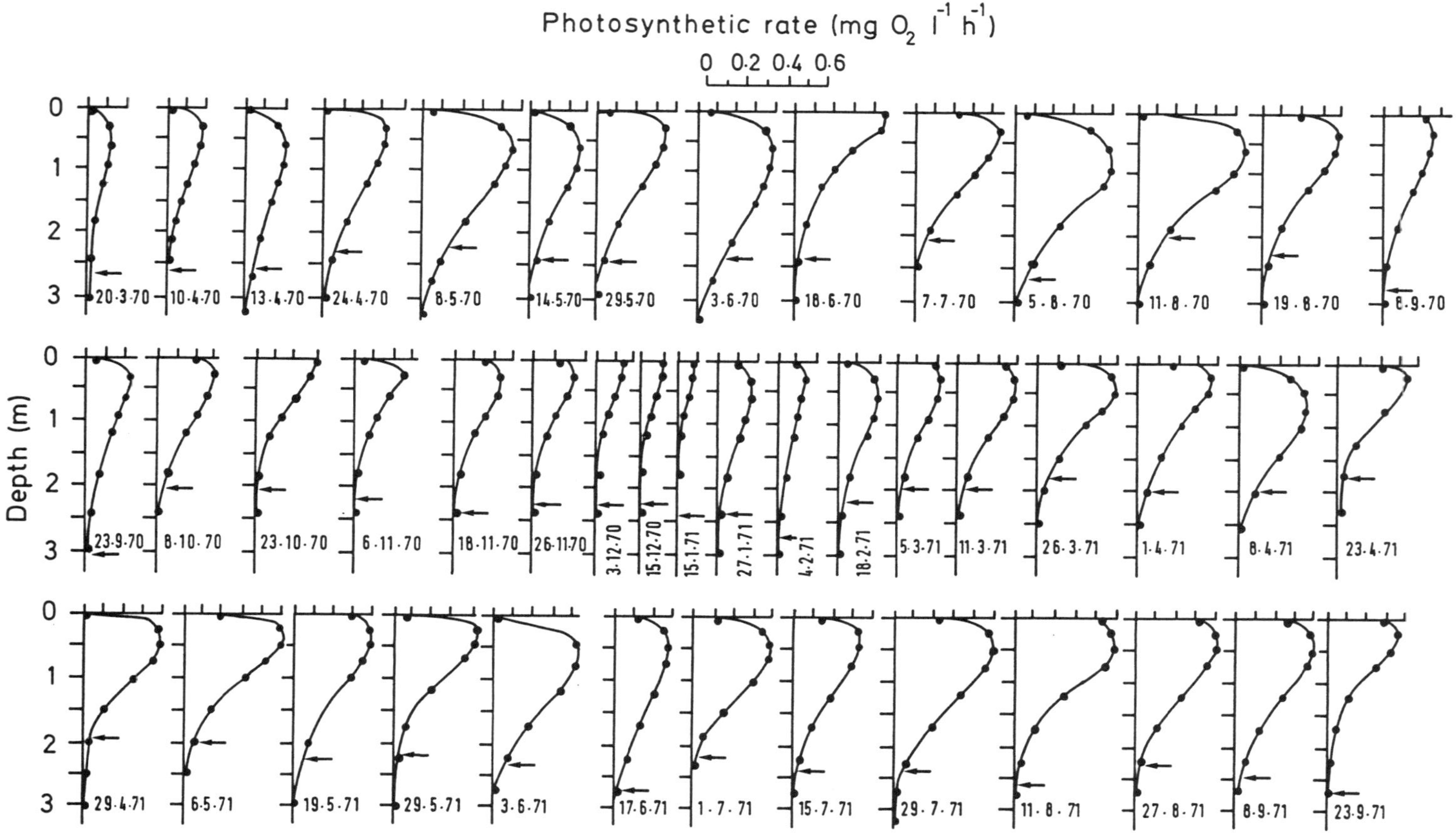

Figure 11.1. Depth profiles of photosynthetic rates per unit water volume (mg $O_2\,l^{-1}\,h^{-1}$) for the open lake, Lough Neagh. The arrows indicate the depth of the 1% level of available surface radiation, 400–700 nm (Redrawn from Jewson, 1976).

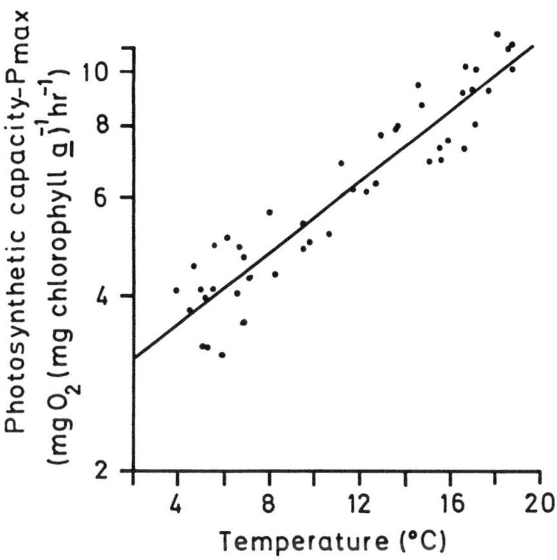

Figure 11.2. Variation of the photosynthetic capacity (P_{max}), plotted on a logarithmic scale, with temperature (Redrawn from Jewson, 1976).

inhibition at the surface was 48% of the maximum light saturated rate. When expressed per unit area this represents a drop of only 5% in the integral photosynthesis (Jewson, 1976; Jones, 1977b). It is probable, therefore, that inhibition is important only during very calm weather.

Maximum light saturated rates may occasionally occur at the surface during overcast weather but normally are found between 40 and 70 cm depth (Fig. 11.1). The highest rate recorded in the open lake was 1.0 mg O_2 l^{-1} h^{-1} (0.38 gC l^{-1} h^{-1}) in May 1979 at a time of high population density of the blue-green alga, *O. redekei*. Figures up to 2.0 mg O_2 l^{-1} h^{-1} have been recorded in Kinnego Bay (see Jewson, 1976; Jones, 1977a, b c). The magnitude of this light saturated value is dependent on both the concentration and specific activity of the algae. The latter, usually called photosynthetic capacity, is often expressed as mg O_2 evolved (or mg C uptake) per unit biomass per unit time (i.e. mg O_2 (mg chl *a*)$^{-1}$ h^{-1}). Its value varies with temperature (Fig. 11.2). However, if environmental conditions are unfavourable, particularly during nutrient restraint, the expected value may not always be reached. One example is discussed in the next section on diurnal rates.

11.3. Diurnal changes

Over most of the year the diurnal distribution of photosynthetic rates per unit area can be related to the amount of incoming solar radiation (Fig. 11.3

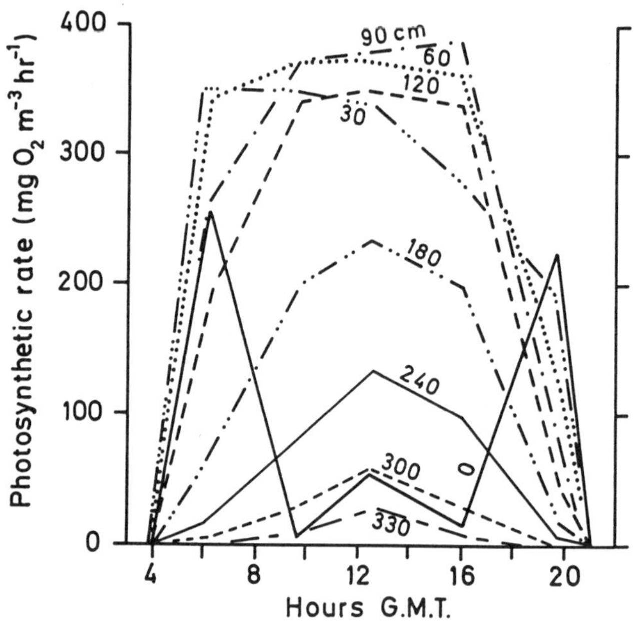

Figure 11.3. The diurnal variation of photosynthetic rates measured at various depths in the euphotic zone of Lough Neagh on 3.6.70.

and Jewson, 1975). This is because water turbulence rarely allows the build-up of adverse conditions such as high O_2 or low CO_2 concentrations and also limits the time any given algal cell stays in the euphotic zone. For instance, at times of reduced light penetration (Z_{eu} = 1.2 m) and homogeneous biomass distribution, any one cell may spend less than 15% of the daylight hours in the illuminated zone. Even in the most favourable conditions (Z_{eu} = 4.6 m) this will be increased to only 45%.

However during periods of calm weather, which rarely last more than a few days, the reduction in water circulation means that some cells will remain in the illuminated zone while the majority of the population is trapped in the aphotic (dark) regions. In these circumstances the classical afternoon depression of rates may occur (see e.g. the rates of photosynthesis at 0 cm, Fig. 11.3, and also Vollenweider, 1965). This is especially pronounced at times of nutrient restraint. Two examples illustrating the effects of water circulation are shown in Fig. 11.4. In these examples the shaded area shows the gross photosynthesis measured as oxygen changes in bottles fixed at individual depths throughout the light gradient. The profiles of chemical composition are for samples taken from open water. The first example was during the spring growth of diatoms when there was complete mixing. Photosynthesis was restricted to the top 2 m although biomass had a uniform vertical distribution throughout the water column (see protein profile Fig.

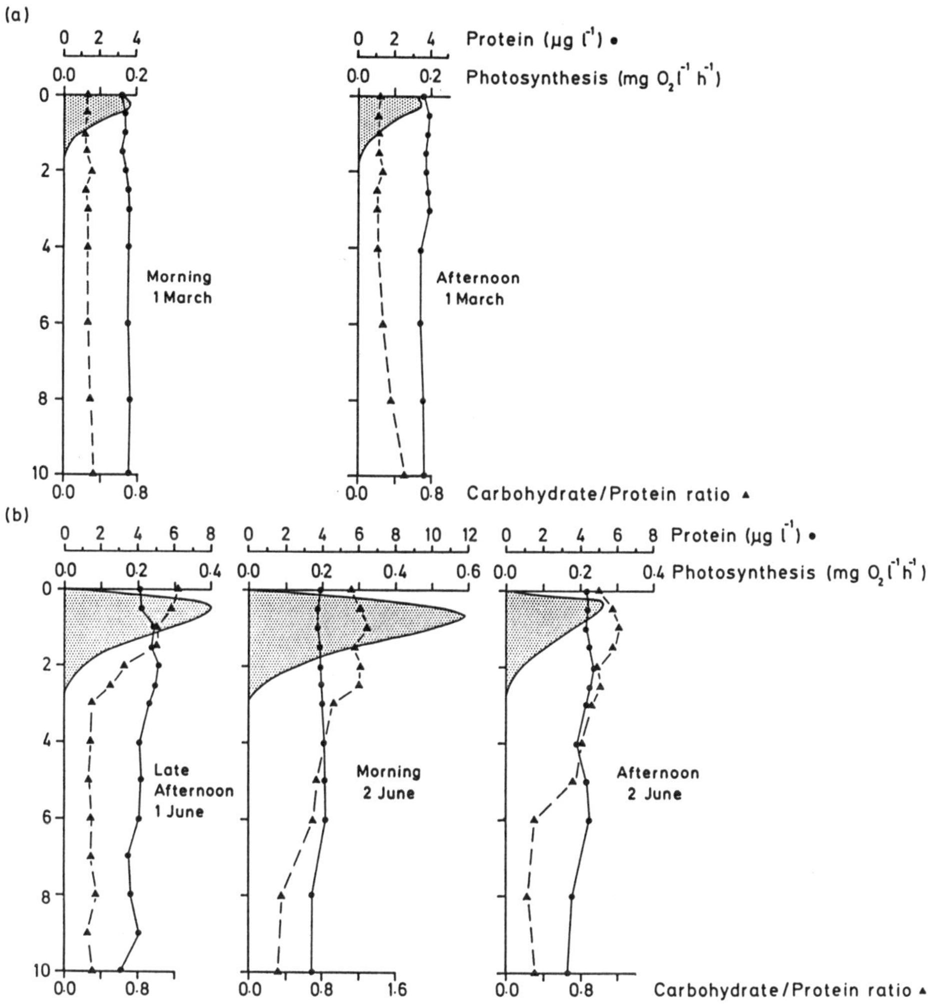

Figure 11.4. Two photosynthetic diurnal studies in Lough Neagh during (a) March 1976 and (b) June 1977. The shaded areas represent photosynthetic oxygen evolution (mg $O_2 \, l^{-1} \, h^{-1}$) in closed bottles (unpublished data). The protein (solid lines) and carbohydrate/protein ratio (broken line) were free-water samples (see Gibson, 1977) taken during the exposures.

11.4a). Photosynthetic rates were similar morning and afternoon. This was also true for cells sampled at 10 m and exposed at the surface. The second example, shown in Fig. 11.4b, was taken at a time when the blue-green algal population had passed its peak and phosphate was undetectable. The biomass (as protein) still remained evenly distributed with depth but a temperature gradient of 5° C difference between the surface and bottom developed (see Gibson, 1977). The light-saturated rate of photosynthesis in the morning

(9.8 mg O_2 (mg chl *a*)$^{-1}$ h^{-1}) was similar to that expected for a temperature of 16.5° C (see Fig. 11.2) but in the afternoon it was reduced by 45% (see Table 11.2). On the previous afternoon the reduction was even greater (67%). The difference in the physiological state of the cells in the two seasonal examples is shown by the much larger accumulation of carbohydrate relative to protein (Fig. 11.4). A fuller discussion of the carbohydrate content of the cells in this example is given in Gibson (1978) and the wider implications of chemical composition in Chapter 10. It is also possible that other factors were important in this study. For example supersaturation of oxygen in the surface waters develops during calm weather as a result of photosynthesis. Enclosing cells in bottles under these circumstances is likely to exaggerate any stress and make results difficult to interpret.

11.4. Range of photosynthetic carbon fixation and efficiency

During the period 1970–1980 the daily rates for gross photosynthesis ranged between 0.08 and 4.95 g C m^{-2} d^{-1} (0.21 to 13.2 g O_2 m^{-2} d^{-1}) in the open lake (Fig. 11.5) and between 0.15 and 5.85 g C m^{-2} d^{-1} (0.4 to 15.6 g O_2 m^{-2} d^{-1}) in Kinnego Bay (Jewson, 1976; Jones, 1977b). The annual figures were between 410 and 560 g C m^{-2} yr^{-1} in the open lake and between 675 and 854 g C m^{-2} yr^{-1} in Kinnego Bay (Table 11.1). These figures are below the highest values obtained for some tropical lakes (see Talling *et al.*, 1973) but amongst the highest recorded for temperate regions (see Westlake *et al.*, 1980). One lake which is higher is L. Leven in Scotland and Bindloss (1974) has reported values of between 597 and 971 g C m^{-2} yr^{-1}. One of the reasons for the higher values in L. Leven is that the algae intercept a greater proportion of the light, because background (i.e. non-algal) light absorption is much less. This can also be illustrated by comparing the annual efficiencies for energy conversion. In L. Leven Bindloss (1974) found an average of 0.9% (range 0.6 to 1.0%) of the energy in the total annual incoming solar radiation was converted in photosynthesis. In Lough Neagh the average for the open lake in the last decade was 0.6% (Table 11.1). During this time it is the years with the lowest standing crops that had the lowest efficiencies (e.g. 1974 and 1975) but as biomass increases so does the proportion of light intercepted by the algae (see Chapter 7A and Fig. 4.8). So in a shallow bay area, such as Kinnego Bay, which consistently maintains higher standing crops (see Jewson, 1976; Jones, 1977a) the conversion efficiency is up to 1.2% (see Jones, 1977b).

Results for individual days are obviously more varied. The lowest efficiency of 0.11% recorded was in the open lake during July 1974 when the phytoplankton concentration was very low (4 mg chl *a* m^{-3}). The low figure was largely due to the relatively small portion of the light absorbed by the algae but another contributory factor was that the light at high irradiances was not used because of inhibition of photosynthesis. Therefore, the highest

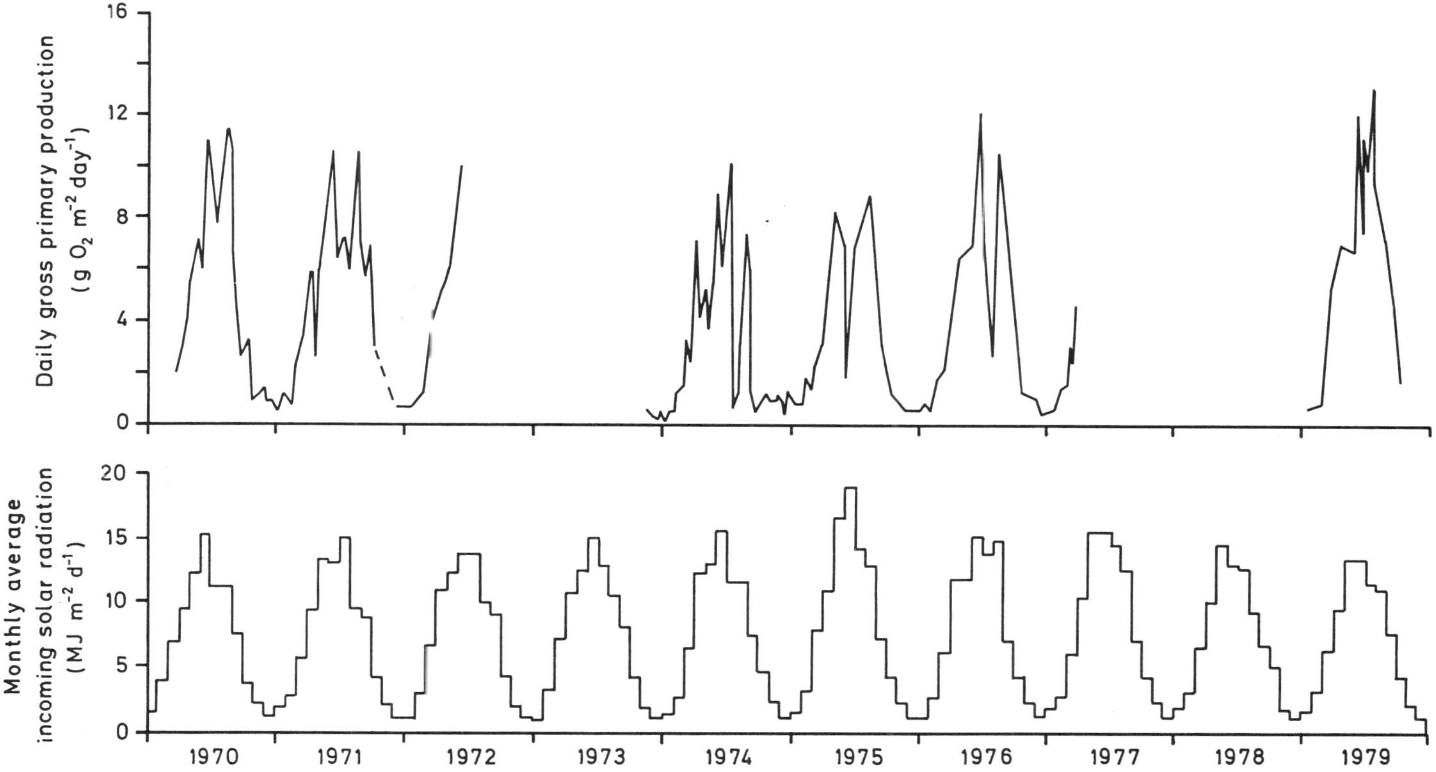

Figure 11.5. (a) The daily rates of gross primary production per unit area between 1970–1979 in the open lake of Lough Neagh and (b) The monthly average of the daily incoming solar radiation measured at Aldergrove Airport on the east side of the lake.

Table 11.1. Total solar radiation for 1970–1979 with the percentage deviation from the long term average for each year. The annual gross production by phytoplankton expressed both in terms of carbon uptake ($g\,C\,m^{-2}\,yr^{-1}$) and in energy equivalent ($J\,m^{-2}\,yr^{-1}$). In the two right-hand columns the utilisation efficiency of solar radiation by phytoplankton photosynthesis has been calculated for total incoming irradiance and also for sub-surface photosynthetically available radiation (i.e., Ph.A.R., 400–700 nm). Data for Kinnego Bay (1973/4) are from Jones (1977b) and the remainder is from Jewson (1976) and unpublished data

Year	Total solar radiation ($J\,m^{-2}\,yr^{-1} \times 10^7$)	Deviation from average (%)	Annual gross production		Utilization efficiency	
			($g\,C\,m^{-2}\,yr^{-1}$)	($J\,m^{-2}\,yr^{-1} \times 10^7$)	Total radiation (%)	Ph.A.R. sub-surface (%)
Open Lake						
1970	317	−3.6	537	2.19	0.7	1.7
1971	316	−4.0	560	2.28	0.7	1.7
1972	321	−2.5	–	–	–	–
1973	323	−1.8	558	2.28	0.7	1.7
1974	331	+0.6	410	1.67	0.5	1.2
1975	371	+12.8	439	1.79	0.5	1.2
1976	338	+2.7	477	1.95	0.6	1.4
1977	347	+5.5	–	–	–	–
1978	317	−3.6	–	–	–	–
1979	317	−3.6	541	2.21	0.7	1.7
Kinnego Bay						
1970/71	315	–	675	2.8	0.9	2.1
1973	327	–	854	3.5	1.2	2.6
1974	317	–	854	3.4	1.2	2.6

Table 11.2. The diurnal changes in photosynthetic capacity, integral photosynthesis and efficiency in the open lake of Lough Neagh in 1977 (see also Fig. 11.4)

	Afternoon 1 June	Morning 2 June	Afternoon 2 June
Total irradiance ($j\ m^{-2}$)	1.95×10^6	2.62×10^6	2.95×10^6
Photosynthetic capacity (mg O_2(mg chl *a*)$^{-1}$ h^{-1})	4.4	9.8	6.4
Integral photosynthesis			
in mg O_2 m^{-2} h^{-1}	304	712	449
in $J\ m^{-2}$	4.65×10^3	10.9×10^3	6.87×10^3
Efficiency (%)	0.24	0.42	0.23

conversion efficiences are found on occasions with no inhibition and high standing crops (e.g. 230 mg chl *a* m^{-3} in Kinnego Bay on 12 August 1971; Jewson, 1976) when efficiencies up to 1.5% were recorded.

Apart from the problems of light interception, the efficiency will also be reduced if there is a depression of the photosynthetic capacity (P_{max} (mg O_2 (mg chl *a*)$^{-1}$ h^{-1}), as discussed for the diurnal rates. For example, during the study shown in Fig. 11.4 the conversion efficiency dropped by half as the P_{max} dropped from 9.8 to 4.4 mg O_2 (mg chl *a*)$^{-1}$ h^{-1} (Table 11.2).

11.5. Seasonal changes in gross photosynthesis

The major factors controlling gross primary production at different seasons have been outlined for the main lake by Jewson (1976) and for Kinnego Bay by Jones (1977b). The events for the open lake can be summarised as follows. During the early part of the year the incoming solar radiation is insufficient for net population growth, with daily rates of gross photosynthesis of 0.08 to 0.2 g C m^{-2} day. Increasing solar elevation and irradiance (Fig. 11.5) would then usually shift the profile deeper and so extend the photosynthetic zone. However, at this time of year there are relatively frequent storms which resuspend benthic material. These can considerably reduce light penetration as shown in Chapter 4 and, on occasions, offset any advantage of increased irradiance (see Jewson, 1976; Jewson *et al.*, 1981). However, these stormy periods are usually of short duration and so, as daylength is also increasing, January to March is generally a time of improving light climate when the diatoms usually start growing (see Chapter 10). Their increasing biomass has a considerable effect on integral gross photosynthetic rates. This is partly due to a greater proportion of the light being intercepted (as more algal cells are packed into the euphotic zone; Jewson, 1977a) but there is also an increase in the photosynthetic rate per unit volume of lake water. This can be seen by the shift to the right of the photosynthesis depth profiles in Fig. 11.1. Rising temperature would also enhance this effect but at this time of the year (January to March) temperature is still only 3–5 °C.

By the end of March, integral gross photosynthetic rates are usually between 1 and 2 g C m^{-2} d^{-1} but begin to rise rapidly as temperature increases. This is largely due to the increases in the photosynthetic capacity (i.e. light saturated rate of photosynthesis per unit biomass) with temperature (Fig. 11.2), although the rates can still be affected by the size of the standing crop present. This varies from year to year depending on when growth starts and when nutrient limitation occurs (Chapter 10). Usually it is the diatoms that cease growth first, due to shortage of silica. The blue-greens continue to grow but then they become nutrient-limited by phosphorus during April or early May. Integral gross rates of carbon fixation are then up to 4 g C m^{-2} d^{-1} (Fig. 11.5). In summer, following the spring peaks, the algal biomass concentrations drop to between 40 and 60 mg chl *a* m^{-3} (except for the unusually low values in 1974 and 1975 resulting from zoo plankton grazing see Chapter 14) but the gradually increasing temperature keeps integral photosynthetic rates high. However, depression, by up to 50% of these high rates, may result from nutrient limitation (Jewson and Taylor, 1978; and Table 11.2). The rates return to nearer their expected values with the build-up of nutrients later in the year (see Chapter 9). There is usually a growth of the blue-green algal population, mainly *O. agardhii*, during the late summer and early autumn. The highest daily figures of over 5–6 g C m^{-2} d^{-1} have been recorded at this time due to the combination of high biomass, high temperatures and long daylength.

During September to November, with decreases in daylength and temperature, integral photosynthesis falls back to below 0.5 g C m^{-2} d^{-1} again. This is a time of generally decreasing biomass, as insufficient energy can be fixed to maintain the phytoplankton populations.

11.6. Net production and growth

One of the major features of a turbid well-mixed lake such as Lough Neagh is that during the day never less than half and sometimes up to 85% of the population is in the dark (or aphotic) regions of the lake. So, before net population increase can occur, sufficient energy must be fixed in the euphotic zone to cope with the metabolic demands of being circulated in the dark (see Jewson, 1976). This is in addition to any losses of cells through sedimentation, grazing, washout or parasitism (Jewson *et al.*, 1981). Jewson and Taylor (1978) discussed some of the implications of this for net photosynthesis. This latter term is frequently misunderstood particularly when respiration rates are referred to as 'losses' (e.g. see Westlake *et al.*, 1980). As well as maintaining cellular components, energy may be stored or diverted into building new photosynthetic or structural material. The balance between these processes varies with environmental conditions, so it should not be unexpected that respiration rates reflect these changes and vary with depth and time. This was not fully appreciated in productivity studies until work by Ganf (1974) on L. George (see also Talling, 1975). In Lough Neagh, Gibson (1975) has shown a decline in respiration with depth which is closely related to the carbohydrate content of cells and so to their previous light history.

Using data from an example in June 1976, when there were both depth and diurnal differences in respiration (Jewson and Taylor, 1978), the net photosynthesis was estimated as 90% of gross fixation in the photosynthetic zone at mid-day but about 78% if the whole water column to the mean depth of 8.9 m is considered. Over a full 24 hours it was reduced to only 6%. This was for an overcast day, whereas in brighter conditions that followed it increased up to 40%. Over the year the average is probably 10 to 15% (Jewson, 1976) with higher values of 40% being more common in shallow areas such as Kinnego Bay (Jones, 1977b). These figures, although low, are not unreasonable in view of the overall growth rate in spring. This shows an accumulation of new biomass of about 4% per day, with doubling times of between 7 and 10 days (Jewson *et al.*, 1981). The delicate energy balance in winter and early spring is shown by the delays and disruption to growth that may be caused by storms (Jewson, 1976; Jewson *et al.*, 1981).

Faster growth rates are possible in inshore areas such as Kinnego Bay which are not only physically but also optically shallower. Jones (1977c) has discussed how the differences in light interception and respiration rates, of the diatom *S. hantzschii* and the blue-green *O. agardhii*, interact to give each a different competitive advantage in different seasons. Essentially *S. hantzschii* is 'favoured' at low temperatures and blue-greens during summer.

11.7. Summary

Lough Neagh is a highly productive (eutrophic) lake with gross photosynthetic rates of over 500 g C m^{-2} yr^{-1}. The amount of primary production is ultimately controlled by the nutrient supply as shown in Chapter 9. However an important aspect for the ecology of the phytoplankton is the combination of rapid attenuation of light combined with a well-mixed water column which results in a uniform vertical distribution of biomass over most of the year. This means that the majority of the phytoplankton (50–85%), at any one instant during the day, are below the euphotic zone. This obviously has many implications for the energy balance of cells and their potential for growth. So Lough Neagh is a 'shade' environment in which sufficient energy has to be fixed in the light to cope with both the metabolic needs of being circulated for long periods in the dark, as well as overcoming any losses due to sedimentation, grazing, parasitism and wash-out (Jewson *et al.*, 1981). Net production per unit area is then relatively low, compared to gross rates, but in absolute terms this delays rather than limits the attainment of the very high standing crops that occur (up to 1300 mg chl *a* m^{-2}, see Chapter 10). The success of individual species in Lough Neagh is therefore often linked to their ability to cope with the 'losses' as well as optimising photosynthesis in a shade environment.

11.8. References

Bindloss, M.E. (1974) Primary productivity of phytoplankton in Loch Leven, Kinross. Proc. R. Soc. Edinb. (B) 74: 157–181.

Dring, M.J. and D.H. Jewson (1982) What does ^{14}C uptake by phytoplankton really measure? A theoretical modelling approach. Proc. R. Soc. Lond. B.: 351–368.

Ganf, G.G. (1974) Rates of oxygen uptake by the phytoplanktonic community of a shallow equatorial lake (Lake George, Uganda). Oecologia (Berlin) 15: 17–32.

Gibson, C.E. (1975) A field and laboratory study of oxygen uptake by planktonic blue-green algae. J. Ecol. 63: 867–880.

Gibson, C.E. (1977) Carbohydrate content as an ecological tool in the study of planktonic blue-green algae. Verh. Internat. Verein. Limnol. 20: 630–635.

Gibson, C.E. (1978) Field and laboratory observations on the temporal and spatial variation of carbohydrate content in planktonic blue-green algae in Lough Neagh, N. Ireland. J. Ecol. 66: 97–115.

Jewson, D.H. (1975) The relation of incident radiation to diurnal rates of photosynthesis in Lough Neagh. Int. Revue ges. Hydrobiol. 60: 759–767.

Jewson, D.H. (1976) The interaction of components controlling net phytoplankton photosynthesis in a well-mixed lake. Freshwat. Biol. 6: 551–576.

Jewson, D.H. (1977a) Light penetration in relation to phytoplanktonic content of the euphotic zone of L. Neagh. Oikos 28: 74–83.

Jewson, D.H. (1977b) A comparison between in situ photosynthetic rates determined using ^{14}C uptake and oxygen evolution method in Lough Neagh, Northern Ireland. Proc. R. Irish Acad. B77: 87–99.

Jewson, D.H. and Taylor, J. (1978) The influence of turbidity in net phytoplankton photosynthesis in some Irish lakes. Freshwat. Biol. 8: 573–584.

Jewson, D.H., Rippey, B. and Gilmore, W.K. (1981) Loss rates from sedimentation, parasitism and grazing during the growth nutrient limitation and dormancy of a diatom crop. Limnol. Oceanogr. 26: 1045–1110.

Jewson, D.H. and Wood, R.B. (1975) Some effects on integral photosynthesis of artificial circulation of phytoplankton through light gradients. Verein. theor. agnew. Limnol. 19: 1037–1044.

Jones, R. I. (1977a) Factors controlling phytoplankton production and succession in a highly eutrophic lake (Kinnego Bay, Lough Neagh). I. The phytoplantkon community and its environment. J. Ecol. 65: 547–559.

Jones, R. I. (1977b) Factors controlling phytoplankton production and succession in a highly eutrophic lake (Kinnego Bay, Lough Neagh). II. Phytoplankton production and its chief determinants. J. Ecol. 65: 561–577.

Jones, R. I. (1977c) Factors controlling phytoplankton production and succession in highly eutrophic lake (Kinnego Bay, Lough Neagh). III. Interspecific competition in relation to irradiance and temperature. J. Ecol. 65: 569–586.

Jones, R.I. (1978) Adaptations to fluctuating irradiance by natural phytoplankton communities. Limnol. Oceanogr. 23: 920–926.

Talling, J.F., Wood, R.B., Prosser, M.V. and Baxter, R.M. (1973) The upper limit of photosynthetic productivity by phytoplankton: evidence from Ethiopian soda lakes. Freshwat. Biol 3: 53–76.

Vollenweider, R.A. (1965) Calculation models of photosynthesis-depth curves and some implications regarding day rate estimates in primary production measurements. Mem. Ist. Ital. Idrobiol. Supp. 18: 425–457.

Westlake, D.F. et al. (1980) Primary production. In: E.D. LeCren and R.H. Lowe-McConnell (eds.) the Functioning of Freshwater Ecosystems. International Biological Programme 22: Cambridge Univ. Press, Cambridge, pp. 141–246.

12. Benthic algae in Lough Neagh

D.H. JEWSON and M. BRIGGS

12.1. Introduction

The availability of light in Lough Neagh restricts the growth of most benthic (bottom living) plants to a narrow coastal strip where the water depth does not exceed 3 m (Fig. 12.1). This represents an area of approximately 50 km^2, which is about 13% of the surface area of the lake. This inshore zone consists mainly of rocks and stones (Briggs, 1984) but there are also long stretches of sandy beaches. Both of these are erosive environments subject to considerable wave action, so fine sediments do not accumulate but are resuspended and transported out into deeper water. Rooted vegetation does not grow well in these conditions and is limited to a few sheltered areas. The major primary producers are microbial cells living on the surface of rocks and stones (epilithic) or attached to sand grains (epipsammic).

12.2. Epilithic flora

The depth distribution of algal biomass on stones is shown in Fig. 12.1. The stones were gathered by divers which is an inconvenient method of collection for routine work, so Briggs (1984) investigated the possibility of using artificial surfaces which could be placed in position from a boat and then resampled at regular intervals to follow seasonal patterns. She found that results were not reliable for green and blue-green algae but that either cleaned natural stone or 'rustic' bricks could be used to sample the diatom communities. This is because diatoms rapidly colonise bare areas. They form a mat over the surface of rocks and stones which is a complex of stalked, tube dwelling and mobile species. In spring this layer is dark red-brown and may grow up to several millimetres thick, frequently overgrowing the blue-green algae. During late spring and summer the cells are grazed down and the blue-greens again predominate.

The transect shown in Fig. 12.1 was carried out just before the time of the spring maxima of diatom growth in April, 1980. At this time, diatoms

R. B. Wood and R. V. Smith (eds.), Lough Neagh, 239–243.

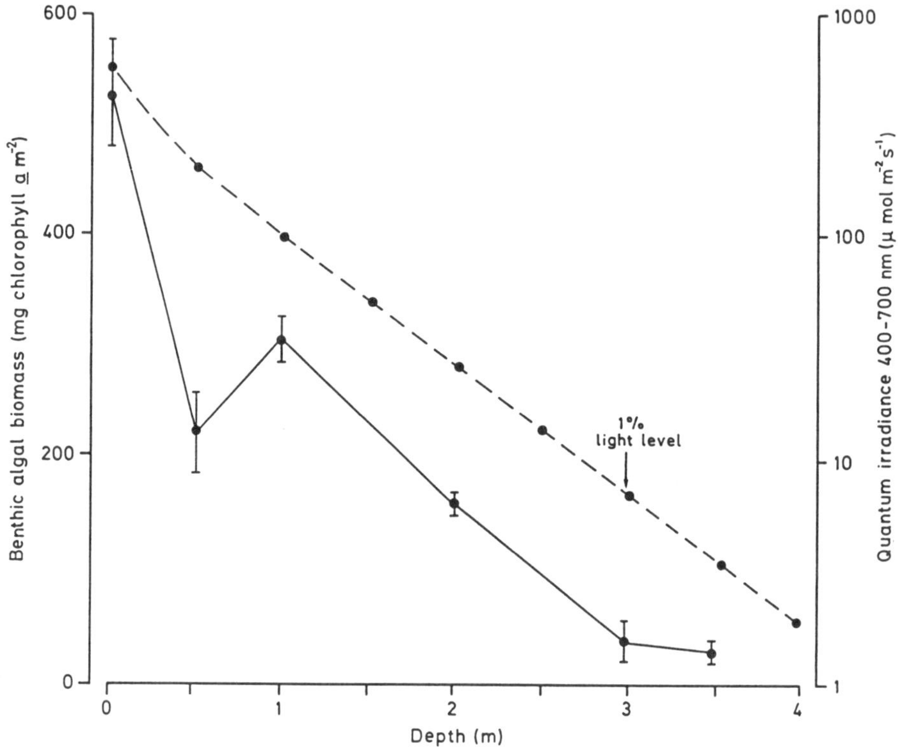

Figure 12.1. The depth distribution of chlorophyll *a* extracted from benthic algae living on the surface of rocks (April 1981) on the shore of Traad Point, Lough Neagh. Also shown is the light penetration of photosynthetically available radiation 400–700 nm).

formed approximately 40% by volume of the total epilithic flora while blue-green algae formed about 20%. Most of the remainder was due to filamentous green algae (mainly Cladophora) which grow in a narrow zone (0–10 cms water depth) at the water edge. However they are not always found there as this zone is very vulnerable to the changes in water level, which may be up to a metre during the year. The timing of the maxima of diatoms (April) and blue-greens (May) in the epilithic communities coincided with those of the representative groups in the phytoplankton during 1980. It is therefore probable that the benthic diatoms were also ultimately limited by silica and the benthic blue-greens by phosphorus. The maximum biomass concentrations recorded in 1980 were 567 mg chl *a* m^{-2} on rocks and 546 mg chl *a* m^{-2} for epiphytes on *Phragmites australis* (Cavanille) Trin. stems.

In contrast to the phytoplankton, where one or two species have dominated over the last decade, the epilithic flora is extremely varied. Some of the more common species are *Cocconeis placentula* Ehr., *Rhoicosphenia curvata* (Kütz) Grun., *Navicula gracilis* Ehr., *Navicula cryprocephala* Kütz., *Am-*

phora ovalis Kütz., *Amphora ovalis* var. *pediculus* Kütz., *Cymbella ventricisa* Kütz., *C. helvetica* Kütz., *Cymbelloritzschia diluviana* Hust., and *Nitzschia dissipata*. Planktonic species such as *Melosira italica* (Ehr.) Ralfs and *Stephanodiscus astraea* (Ehr.) Grun. also form a considerable portion of the community, especially during the period of sedimentation around the time of the peak growth in the water column. The major epiphytic species found associated with *P. australis* were *Rhoicosphenia curvata* (Kütz) Grun., and *Gomphonema divaceum* (Lyng) Kütz. The species assemblages are similar to those reported by Round (1959) for epipelic algae in Lough Neagh, which he classed as representative of calcarious lake types. The two major species forming respectively 20% and 17% of Round's counts were *Cocconeis placentula* Ehr., and *Cymbellonitzschia diluviana* Hust. This was the first record of the later species in the British Isles. In the epilithic flora sampled in May 1980 they were only 7% and 2% of the total diatom counts.

12.3. Epipsamic flora

Lough Neagh has many miles of wave-exposed sandy shores. It is a feature of this type of habitat that the depth of the euphotic zone (1% light level) is similar to the depth of bottom exposed to frequent wave action (i.e. lower limit of coarse sand). In such an unstable and abrasive environment macrophytes cannot establish themselves and the predominant plants are microbial cells which live attached to sand grains. Some motile diatoms do occur but they are much rarer than in the algal mats on the surfaces of rocks.

The depth distribution of algal biomass on the sand surface (Fig. 12.2a) is similar to that on stones (Fig. 12.1), with peak concentrations in the top metre and light limitation restricting the growth of most species to less than 3 m. However, at the water's edge there is a decline in biomass (filamentous green algae are absent) but an increase in phaeopigments (breakdown products of chlorophyll). It is not known whether the latter is an inwash of dead material or death of cells resulting from intermittent water coverage at the lake edge.

The depth distribution of biomass in Fig. 12.2a is only for the top centimetre of sand. However, in shallow waters live cells can be found attached to sand grains at depths of over 30 cm below the sediment surface with little reduction in concentration (e.g. Fig. 12.2b). As sufficient light for photosynthesis only penetrates a few millimetres into the sand, these cells must rely on being brought back to the surface to survive. Other evidence for such a turnover comes from a study of the sand bars (up to 5) which run parallel to the shore in 0.1 to 1.2 m depth (Carter and Balsillie, 1983; Carter, 1985). These bars are about 0.2 m in height and migratory, moving up to 0.3 m per day during storms. The distribution of animal communities within the bars and troughs has been related to the amount of wave action and sand disturbance they receive (Carter and Carter, 1983). A depth-profile of the

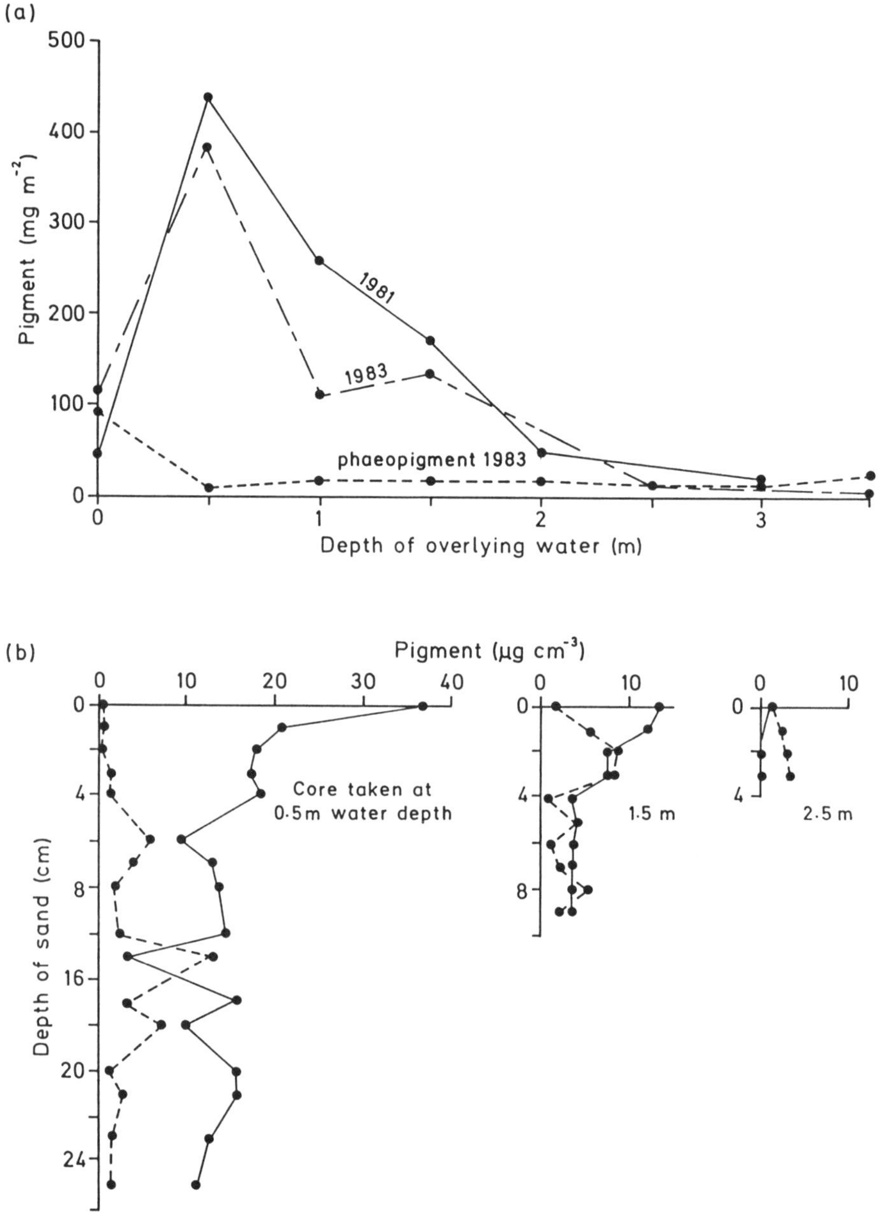

Figure 12.2. The depth distribution of chlorophyll *a* extracted from episammic algae in the sandy bay west of Traad Point, Lough Neagh. (a) surface sand samples of chlorophyll *a* (–, ---) (11 April 1981 and 1 April 1983) and phaeopigments (---) (1 April 1983), (b) chlorophyll *a* (----) and phaeopigments (---) sampled at water depths of 0.5, 1.5, and 2.5 m on 1 April 1983.

algal biomass distribution in one of these bars, at 0.5 m water depth, is shown in Fig. 12.2b. As a check on the chlorophyll *a* results, the numbers of live cells per sand grain were counted by autofluorescence. From the sediment surface down to 15 cm there were cells attached to over 90% of the sand grains. By 30 cm this was reduced to 15%. At the same time, the number of cells per grain declined from over 100 at the surface to 10 at 30 cm sand depth. In contrast, results from a trough behind the innermost bar, suggest much less disturbance, as a similar decrease occurred in only 5 cm.

Further offshore, as water depth increases there is also less sediment disturbance and so many fewer live cells are found below the sediment surface (Fig.12b). For example, in a water depth of 3.5 m only 68% of the sand grains on the sediment surface had algae attached, with less than 5 cells per grain. By 3 cm sand depth there were no live algae at all. The proportion of phaeopigments also increased, which suggests some accumulation of dead cells. The dominant species in this habitat are *C. diluviana*, *Opephora martyii* Heribaud, *Cocconeis disculus* Schumann and *Navicula scutilloides* Smith.

Other important factors affecting growth that have not been assessed are the changes in water level which may vary by over 1 m. As the majority of the biomass is contained in the top metre (Fig. 12.1 and 12.2), this must have considerable influence on the energetics of these shore populations.

During the 1980's the ecology of the benthic algae has been studied intensively and it becomes increasingly obvious that the inshore areas, although limited to less than 15% of the lough surface by light penetration and a physically hostile environment, are by no means unproductive. Further studies may well show that a highly significant fraction of the total energy flow enters the Lough Neagh ecosystem by this route.

12.4. References

Briggs, M. (1983) Algal sampling by use of artificial surfaces in Lough Neagh, Northern Ireland. Ir. Nat. J. 21: 151–155.

Carter, R.W.G. and Balsillie, J.H. (1983) A note on the amount of wave energy transmitted over nearshore sand bars. Earth Surface Processes and Landforms 8: 213–222.

Carter, C.E. and Carter, R.W.G. (1983) Factors influencing the chironomid community of a nearshore sand area. Mem. Amer. Ent. Soc. 34: 47–59.

Round, F.E. (1959) A comparative survey of the epipelic diatom flora of some Irish Loughs. Proc. R. Irish Acad. 60B: 193–215.

13. Physiological ecology of Lough Neagh phytoplankton

R.H. FOY and R.V. SMITH

13.1. Introduction

The annual cycle of phytoplankton in Lough Neagh has been described in Chapter 10 with two species of blue-green algae dominating the phytoplankton except for a short period during the spring diatom bloom. The regularity of this cycle, with *Oscillatoria redekei* dominant during the spring and early summer, giving way to *Oscillatoria aghardii* for the remainder of the year suggests that the cycle may be controlled by seasonal changes in light or temperature or a combination of the two. Because of high external nutrient loadings to Lough Neagh (Chapter 8), nutrient limitation may play only a subordinate role in the succession of these species. The aim of this chapter is to describe laboratory experiments which have been undertaken to investigate how the growth rates and cell composition of *Oscillatoria redekei* and *Oscillatoria agardhii* are determined by irradiance, day length and temperature. Additional studies have been carried out on *Aphanizomenon flos-aquae* and *Anabaena flos-aquae*, two potentially nitrogen fixing species which have been virtually absent from the phytoplankton since 1976 (Chapter 10). The results of these studies will be compared with changes in the species dominance observed in Lough Neagh.

13.2. Laboratory culture techniques

13.2.1. *Isolation and purification of blue-green algae*

Obtaining unialgal cultures of strains from the genera *Anabaena*, *Aphanizomenon* and *Oscillatoria* has proved to be relatively easy suggesting the absence of specialised nutrient requirements which have been found for other algal types (see e.g. Lindstrom and Rodhe, 1978). Cultures are routinely prepared by the transfer of single trichomes using micro-pipettes into a suitable liquid medium. For small species, such as *Oscillatoria redekei*, an intermediate stage may be employed where the microsamples are plated onto

R.B. Wood and R.V. Smith (eds.), Lough Neagh, 245–279.

agar and when the filaments have increased, a portion of the agar containing the unialgal sample is transferred into liquid medium.

In 1969 Heaney was the first to isolate two of the dominant algal species in Lough Neagh, *Aphanizomenon flos-aquae* and *Oscillatoria redekei*. The *Oscillatoria agardhii* strain which has been used in laboratory studies described in this chapter was isolated by Fitzsimons in 1971 from a small eutrophic lake, Loughgall, which lies to the south of Lough Neagh. This strain is similar in morphology and growth rates to a strain also obtained by Fitzsimons in 1974 from Lough Neagh (Foy, 1980). *Anabaena flos-aquae* has not been isolated from Lough Neagh and a strain from Windermere, England, has been used in the present studies. For physiological studies axenic cultures are desirable. However, attempts to achieve pure cultures of planktonic blue-green algae have frequently proved unsuccessful. Removing bacteria using antibiotics (Droop, 1967; Provasoli *et al.*, 1951) and radiation (Gerloff *et al.*, 1950) probably failed because of the similar prokaryotic response of bacteria and blue-green algal cells to these selection procedures. Purification has been successfully achieved by using washing and filtration which rely on the size difference between the alga and the small bacterium to produce a culture with a high algal filament:bacterium ratio followed by subsequent serial dilutions to isolate axenic algal filaments (Meffert, 1972; Heaney and Jaworski, 1977; Fitzsimons and Smith, 1983). In 1974 using this method on the Heaney isolate of *Oscillatoria redekei*, Fitzsimons increased the algal filament:bacterium ratio from 1:46 before filtration to 155:1 after filtration and by dilution obtained a pure culture of this alga (Fitzsimons and Smith, 1983). This method has been successfully used in purifying strains of *Anabaena flos-aquae*, *Aphanizomenon flos-aquae*, *Oscillatoria redekei* and *Oscillatoria agardhii* (Fitzsimons and Smith, 1983).

13.2.2. *Nutritional requirements of blue-green algae*

Autotrophic growth of Lough Neagh blue-green algal species (Smith and Foy, 1974; Foy *et al*, 1976; Foy and Smith, 1980; Foy and Gibson, 1982a and b) has been achieved using media based on the ASM-1 medium of Gorham *et al.* (1964). ASM-1 is a synthetic medium in which the only organic compound present is the chelating agent EDTA. To improve the capacity of this medium to support prolonged algal growth the nitrogen source, sodium nitrate was increased from $2\,\mathrm{mM\ l^{-1}}$ to $10\,\mathrm{mM\ l^{-1}}$. To eliminate the wide pH range observedc during growth, buffers in the pKa range 6.8–8.3 were tested and a hydrogen ion buffer *N*-2-hydroxyethypiperazine-*N'*-2-ethanesulphonic acid (HEPES) was selected because of the absence of inhibitory effects, favourable pKa of 7.5 and a low metal binding capacity (Smith and Foy, 1974). In subsequent studies (Foy and Smith, 1980; Foy and Gibson, 1982a) it was found that HEPES strongly interfered with the Lowry method of protein determination, the large quantities of organic nitrogen present made the direct determination of nitrogen impossible and the presence of HEPES caused errors in dissolved oxygen determinations using the Winkler Iodometric method. Because of these problems it was necessary to revert to ASM-1 medium to which some buffering capacity was provided by the addition of bicarbonate. By not allowing dense cultures to develop the majority of experiments took place in the pH range 8.0–9.0.

The axenic strain of *Oscillatorin redekei* grew at a rate comparable to the original isolate and has been used subsequently in physiological studies. However, the growth rate of the axenic *Oscillatoria agardhii* strain was considerably less than the bactericised isolate. When bacterial contamination of axenic cultures took place, growth rates rapidly improved (Fitzsimons and Smith, 1983). To date no satisfactory explanation for this phenomenon has been obtained as addition of vitamins, selenium and using filter sterilised medium which previously supported the contaminated culture, have all failed to relieve the growth inhibition. It may be that this alga secretes an autoinhibitory substance which can be metabolised by bacteria. Such substances have been found by Caldwell (1979) which, because of their strong chelating capacities, caused trace metal limitation of growth. Because of the poor growth of the axenic culture of *Oscillatoria agardhii*, it has not been possible to use it for experiments on the influence of light and temperature.

13.2.3. *Strain selection*

Although it seems likely that the original Heaney isolate of *Oscillatoria redekei* in 1969 was clonal, subsequently strains have developed under laboratory conditions with differing growth characteristics. Following isolation from Lough Neagh, this culture was maintained for 6 years under continuous light at a temperature in excess of 20 °C. In 1974 when experiments were carried out using a 6:18 light:dark (L:D) cycle growth was found to be strongly inhibited by irradiances in excess of 30 μE m^{-2} s^{-1} (Foy *et al.*, 1976). Subsequent experiments with the axenic strain showed similar inhibition. However, an axenic strain which had been maintained under a 6:18 L:D cycle from 1976 to 1978 was not inhibited by high irradiance. In 1981, the growth rates of 7 strains were determined at a high irradiance of 145 μE m^{-2} s^{-1} under a 6:18 L:D cycle at 15 °C. Two strains, including the original isolate grew slowly at 0.15 doublings day^{-1}, 4 strains grew at 0.20 doublings day^{-1}, one strain grew at 0.25 doublings day^{-1} and the final strain tested which was not inhibited by high irradiances grew at 0.30 doublings day^{-1}. Results presented by Gibson and Foy (1983) demonstrated that at subsaturating irradiances, there were no differences in the growth of different strains. However, under continuous light the strains which were inhibited by high light under a 6:18 L:D cycle showed no inhibition of growth and appeared to reach higher growth rates under continuous light than the strain which grew fastest under the 6:18 L:D cycle.

13.2.4. *Laboratory growth conditions*

Growth conditions are described in detail by Foy *et al.* (1976) and Foy (1980). In brief, batch cultures were grown in orbital shaking incubators with controlled temperature (±0.6 °C) and light provided by fluorescent tubes. Light measurement techniques varied as originally *lux* units were employed which are unsuitable for measuring photosynthetically active radiation. Quantum measurements of photosynthetically active radiation were subsequently obtained using a flat cosine corrected sensor, but such sensors are liable to serious error if there is a large component of scattered light which was the case with the incubators used. This has been overcome by the use of a 4π sensor placed inside the growth flask (Foy and Gibson, 1982a). All light values presented in this chapter are 4π quantum measurements ($\mu E\ m^{-2}\ s^{-1}$) and have been converted from the original units by direct comparison with the original culture conditions. Conversion factors are given in Gibson and Foy (1983).

13.3. Growth kinetics in continuous light

13.3.1. *Influence of cell size on growth*

Although physiological and metabolic factors can alter the growth rate of an alga, cell size is considered to be the dominant determinant of growth as it determines the surface area to volume ratio (S/V ratio) and hence indirectly, the number of nutrient uptake sites per unit volume (Fogg, 1975). A direct relationship between S/V ratio and the growth rate of 22 strains of *Anabaena*, *Aphanizomenon* and *Oscillatoria* grown under saturating continuous irradiance of approximately 60 $\mu E\ m^{-2}\ s^{-1}$ at 20 °C was demonstrated by Foy (1980). The fastest growing cultures were *Oscillatoria limnetica* and *Oscillatoria redekei* cultures which had cell diameters between 1.4 and 2.3 μm. The regression equation (equation 1) shows that although the relationship between the two variables was highly significant, approximately half the variation in growth rates was not associated with change in S/V ratio.

$$K = 0.624\ S/V + 0.374 \qquad r^2 = 0.46 \qquad p < 0.001 \tag{1}$$

where,
K = growth rate doublings day^{-1}
S/V = surface area to volume ratio μm^{-1}.

The causes of the remaining variation have not been determined but may be due to differing environmental optima between strains.

13.3.2. *Temperature*

The relationship between growth rate and temperature of *Oscillatoria redekei* and *Oscillatoria agardhii* has been examined by growing the algae at tempera-

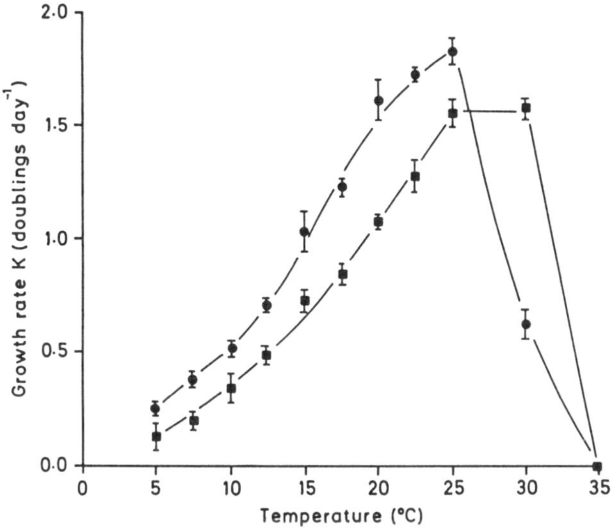

Figure 13.1. Growth rate: temperature relatiionship of *Oscillatoria agardhii* (-●-) and *Oscillatoria redekei* (-■-), grown under continuous light of 96 μE $m^{-2} s^{-1}$. Bars denote confidence limits at $p < 0.05$ level.

tures between 5 °C and 35 °C under continuous light using a saturating irradiance of 96 μE $m^{-2} s^{-1}$ (Foy *et al.*, 1976). It is apparent from Figure 13.1 that, between 5 °C and 20 °C, growth was highly temperature dependent with *Oscillatoria redekei* having a Q_{10} of 3.35 and *Oscillatoria agardhii* a Q_{10} of 4.11. A high Q_{10} value has been measured for *Oscillatoria agardhii* by Ahlgren (1978). The temperature optima of both species was at 25 °C and no growth was observed at 35 °C. At 25 °C and below, *Oscillatoria redekei* (S/V ratio 3.04 μm^{-1}) grew significantly faster than the larger *Oscillatoria agardhii* (S/V ratio 1.25 μm^{-1}).

In Lough Neagh, *Oscillatoria redekei* begins growth in February when water temperatures are close to 5 °C and although the day length is 9 hours the algae spend less than 3 hours in the photic zone because of rapid underwater light attenuation (see Chapters 4 and 10). When *in situ* growth rates of *Oscillatoria redekei* were calculated as doublings per 24 hours of light received (*Kcorr*) as in equation (2) these rates were greater than the corresponding laboratory continuous light growth rates, particularly at low temperatures (Table 13.1)

$$Kcorr - K \times \text{Zm} \times t^{-1} \times \text{Zeu}^{-1} \quad (2)$$

where,
Kcorr = growth rate calculated as doublings per 24 light hours received
K = measured *in situ* growth rate, doublings day^{-1},

Table 13.1. Comparison of Lough Neagh *in situ Oscillatoria redekei* growth rates (*K*) corrected for light hours received (*K corr*) and laboratory growth rates under continuous light

Growth period[a]	Temp (°C)	Algal day[b] (hours)	*in situ K* (doublings day^{-1})	*in situ K corr* (doublings day^{-1})	Laboratory *k* (doublings day^{-1})
9/4–28/4/69	5.3	3.24	0.106	0.786	0.274
11/10–10/11/71	11.5	2.78	0.029	0.250	0.622
1/12–22/12/71	5.4	2.06	0.030	0.350	0.276
27/9–25/10/72	11.6	3.60	0.051	0.340	0.652
14/2–5/3/73	3.8	3.07	0.061	0.477	0.202
3/4–8/5/74	9.4	5.32	0.108	0.487	0.480
6/5–12/6/74	12.7	5.06	0.191	0.906	0.720
16/10–20/11/74	8.1	3.32	0.107	0.773	0.402
5/3–2/4/75	6.1	4.19	0.085	0.486	0.312
2/4–30/4/75	8.0	4.09	0.110	0.645	0.400
14/5–4/6/75	12.3	4.81	0.138	0.688	0.696
23/7–13/8/75	17.6	5.18	0.480	1.99	1.22
10/9–24/9/75	14.7	4.09	0.376	2.20	0.930 .
9/2–8/3/76	4.5	3.18	0.078	0.588	0.230
15/3–12/4/76	6.3	3.73	0.087	0.560	0.310
28/2–21/3/77	5.6	2.23	0.045	0.484	0.288
9/5–13/6/78	13.9	4.41	0.058	0.316	0.856
1/5–15/5/79	7.7	3.72	0.097	0.625	0.384
3/3–24/3/80	4.1	4.09	0.073	0.428	0.216
1/4–28/4/80	6.5	3.01	0.100	0.845	0.328

[a]*In situ* growth rates selected from periods of not less than three consecutive algal counts showing exponential growth.
[b]Algal day length = photic depth (Zeu) ÷ mean depth (8.9 m) × day length (sunrise–sunset).

Zm = mean depth of Lough Neagh 8.9 m,
Zeu = photic zone depth, defined as level of 1% light penetration,
t = day length at surface as a proportion of 24 hours.

The possibility of strain differences between the growth rates of the wild *Oscillatoria redekei* and the laboratory strain under continuous light was tested, as was the presence of stimulatory substances in Lough Neagh water but both proved to be negative (Foy *et al.*, 1976). The disparity between *in situ* Lough Neagh growth rates and continuous light laboratory growth rates was largely removed by growing the algae under a L:D cycle. Using a 6:18 L:D cycle at 10 °C and 20 °C, *Oscillatoria redekei* and *Oscillatoria agardhii* showed improved growth efficiencies as evidenced by increased *Kcorr* values and also the Q_{10} values for 6:18 L:D growth were found to be close to 2.0, a value similar to that measured for the temperature dependence of maximum photosynthetic rates (Harris, 1978). A further important observation was that while under continuous light *Oscillatoria redekei* grew significantly faster than *Oscillatoria agardhii*, under a 6:18 L:D cycle *Oscillatoria agardhii* grew better than *Oscillatoria redekei* although at both temperatures the difference in *Kcorr* between the two species was not statistically significant (Table 13.2).

Table 13.2. Maximum growth rates and Q_{10} values of *Oscillatoria redekei* and *Oscillatoria agardhii* grown at 10 °C and 20 °C under continuous light and 6:18 lightdark cycle

	Oscillatoria rededei		*Oscillatoria agardhii*	
	LL *K*	6:18 LD *Kcorr*	LL *K*	6:18 LD *Kcorr*
10 °C	0.53 ± 0.06	0.84 ± 0.04	0.34 ± 0.03	0.93 ± 0.05
20 °C	1.61 ± 0.09	1.64 ± 0.07	1.17 ± 0.05	1.72 ± 0.07
Q_{10}	3.04	2.05	3.44	1.86

The improvement of growth efficiency under a L:D cycle did not appear to have been observed before for blue-green algae but had for other algal groups such as the observations on *Chorella* by Tamiya *et al.* (1953). They postulated that continuous light growth rates were limited by a dark reaction which had a high Q_{10}. When light:dark cycles were employed the limitation of the dark reaction was removed and growth was then governed by the photosynthetic rate of the alga which had a lower Q_{10}. Lorenzen and Hesse (1974) showed that *Chorella* grown under a 14:10 L:D cycle continued to synthesise protein during the dark phase while utilising carbohydrate as an energy source, a finding supported by observations on other algae by Werner (1966) and Cook (1963).

13.4. Growth of *Oscillatoria redekei* and *Oscillatoria agardhii* under light:dark cycles

13.4.1. *Photosynthetic characteristics of* Oscillatoria redekei

The high growth rates of L:D cultures in relation to continuous light cultures imply that L:D cells have either a higher photosynthetic capacity or efficiency. The pigmentation, photosynthesis and growth characteristics of the light tolerant strain of *Oscillatoria redekei* grown at 15 °C under a 6:18 L:D cycle and continuous light have been described by Foy and Gibson (1982b) and Gibson and Foy (1983). In common with other cyanophytes (Jones and Myers, 1965; Ghosh and Govindjee, 1966) *Oscillatoria redekei* responded to reduced light dose by an increase in chlorophyll *a* and phycocyanin (Table 13.3). Between high and low light treatments, the increase in pigments was greater for continuous light cells. Because of the variability of chlorophyll *a* content, changes in photosynthetic rates (*P*) normalised to chlorophyll *a* do not reflect changes in *P* (dry weight)$^{-1}$ or *P* (cell)$^{-1}$ as can be seen if photosynthetic rates normalised to chlorophyll *a*, dry weight and cell protein are plotted against irradiance. This has been done for *Oscillatoria redekei* in Figure 13.2 using the data from Table 13.3. Normalised to chlorophyll *a*, the maximum rate of photosynthesis (*P*max) was depressed when the algae were grown under reduced irradiance, markedly so under continuous light.

Table 13.3. Phycocyanin and chlorophyll *a* content of *Oscillatoria redekei* grown at 15 °C under continuous light and 6:18 LD cycle at a high, 250 μE m^{-2} s^{-1} and a low irradiance 13 μE m^{-2} s^{-1}

	LL		6:18 LD	
% Dry weight	13 μE m^{-2} s^{-1}	250 μE m^{-2} s^{-1}	13 μE m^{-2} s^{-1}	250 μE m^{-2} s^{-1}
Phycocyanin	11.41	0.75	12.62	7.35
Chlorophyll *a*	1.37	0.31	1.79	1.36
Phycocyanin:chlorophyll *a* ratio	8.33	2.42	7.05	5.41

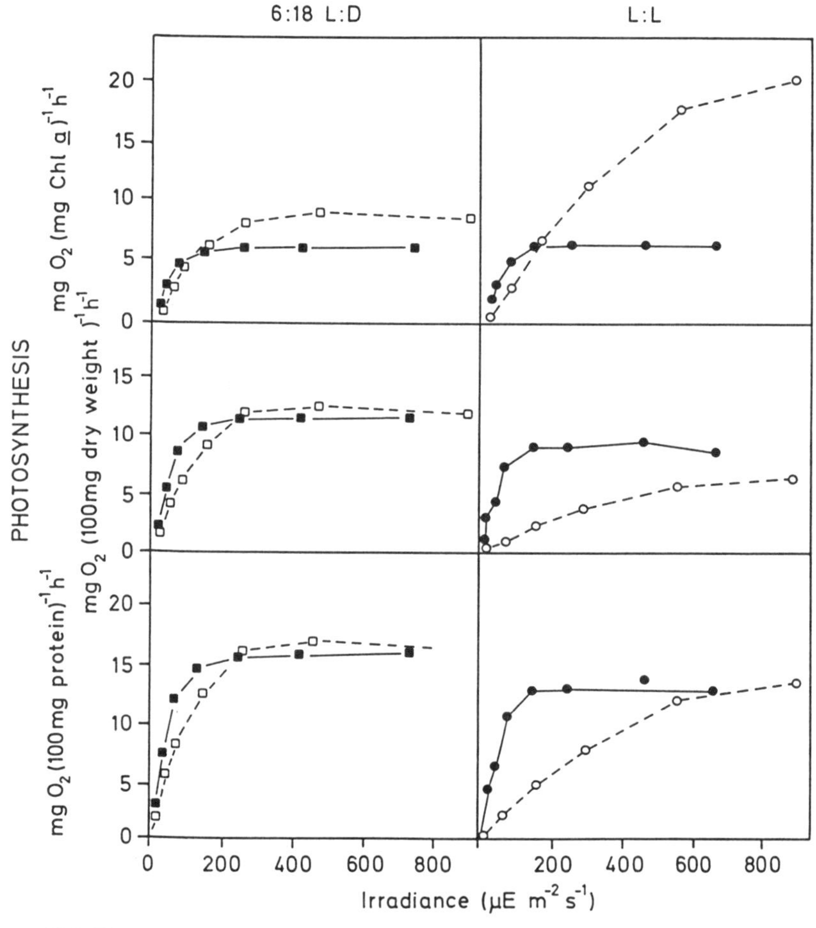

Figure 13.2. Photosynthesis: irradiance plots normalized to chlorophylla, dry weight and protein of *Oscillatoria redekei* cells grown at 13 μE m^{-2} s^{-1} (-■-) and 250 μE m^{-2} s^{-1} (-□-) under a 6:18 LD cycle and 13 μE m^{-2} s^{-1} (-●-) and 260 μE m^{-2} s^{-1} (-○-) under continuous light at 15 °C. After Foy and Gibson (1982a).

Table 13.4. The light saturation characteristic *I*k, of *Oscillatoria redekei* grown at 15 °C under various quantum irradiances (*I*) in continuous light and 6:18 L:D cycles. Values are $\mu E\ m^{-2}\ s^{-1}$

Continuous Light		6:18 light:dark	
I	*I*k	*I*	*I*k
13	78	14	78
29	120	30	93
36	120	42	96
95	250	77	113
117	254	109	128
183	304	167	139
259	471	250	152

Expressed on a cell protein basis, an estimate of cell number, *P*max was relatively constant, but algae which had received a reduced light dose showed a greater ability to harvest weak irradiances. The final base used, dry weight, produced similar results to *P* $(\text{protein})^{-1}$ except that in the high irradiance, continuous light culture, *P*max $(\text{dry weight})^{-1}$ was depressed. This depression corresponded to the high carbohydrate content of high irradiance continuous light cells.

Phycocyanin content was more variable than chlorophyll *a* content and cells grown under low irradiance had higher phycocyanin:chlorophyll *a* ratios than high irradiance cells (Table 13.3). Foy and Gibson (1982b) showed that there was a positive correlation between this ratio and the initial slope of the *P* $(\text{chlorophyll } a)^{-1}$ *vs* irradiance plot. This can be seen in Figure 13.2 where, as in the *P* $(\text{chlorophyll } a)^{-1}$ plot, the efficiency of light harvesting per unit of chlorophyll *a* was greater in the low irradiance cultures. The difference between high and low irradiance cultures was greater under continuous light corresponding to the large difference in the phycocyanin to chlorophyll *a* ratio. (Table 13.3).

The results obtained demonstrated that *Oscillatoria redekei* significantly improved its light harvesting ability in response to reduced light dose by increasing its photosynthetic pigment content rather than altering the photosynthetic capacity of the cell. A comparison of the growth irradiance under which the alga was cultured and the irradiance denoting the onset of light saturation of photosynthesis (*I*k) in Table 13.4 demonstrates that under continuous light, the growth irradiance was always considerably less than *I*k but under 6:18 L:D cycle at high irradiances, the growth irradiance exceeded *I*k. Therefore, under a 6:18 L:D cycle at the maximum growth irradiances, cells were fully utilising their maximum photosynthetic capacity. However, under continuous light, the growth requirements of the cell were met by carbon fixation rates considerably less than the cell's maximum potential. This result reflects the increased growth efficiency of light:dark cells and demonstrates that under the 6:18 L:D cycle the maximum growth rate was limited by the photosynthetic capacity of the algae.

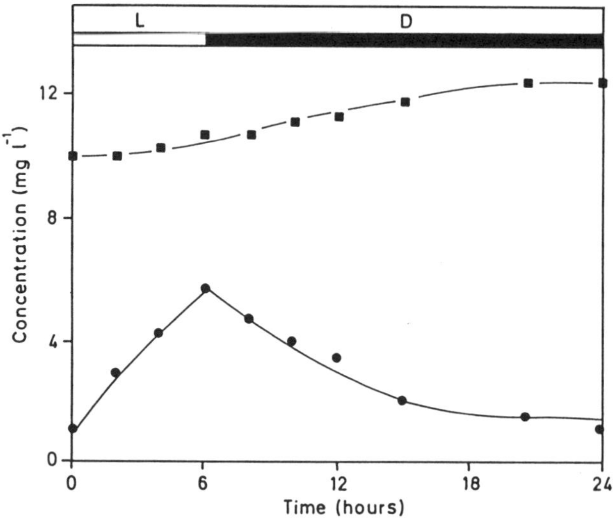

Figure 13.3. Diurnal changes in protein (-■-) and carbohydrate (-●-) of an *Oscillatoria redekei* culture grown at 15 °C under a 6:18 LD cycle and irradiance 47 μE m^{-2} s^{-1}. After Foy and Smith (1980).

13.4.2. *Role of carbohydrate accumulation*

Experiments described in this and in the subsequent section have been taken from the results of Foy and Smith (1980) which were carried out at 15 °C. Under a 6:18 L:D cycle, a light sensitive strain of *Oscillatoria redekei* had an optimum growth rate of 0.30 doublings day^{-1} at 47 μE m^{-2} s^{-1} compared to a continuous light growth rate of 0.92 doublings day^{-1}. *Oscillatoria agardhii* growth maximum, under a 6:18 L:D cycle of 0.37 doublings day^{-1}, occurred at 124 μE m^{-2} s^{-1} compared to a continuous light growth rate of 0.62 doublings day^{-1}. The most distinctive feature of the growth of both species grown under a 6:18 L:D cycle was found to be the rapid rise of carbohydrate during the light period followed by a decline in the subsequent dark period which is shown for *Oscillatoria redekei* in Figure 13.3. In contrast, protein only appeared to rise after a lag of 2–4 hours (Fig. 13.3) but continued to increase in the dark period only stopping when carbohydrate had been depleted to a basal level. A more detailed breakdown of the diurnal changes in cellular components of *Oscillatoria redekei* is given in Table 13.5. The dark loss of biomass was 13% of the dry weight at the start of the dark period but expressed as a percentage of the increase in biomass during the light period, dark loss was 44%. This was entirely due to the carbohydrate decline during the dark as RNA, DNA and chlorophyll *a* in addition to protein all increased. Dark production of protein, DNA and chlorophyll *a*

Table 13.5. Changes in cellular components of *Oscillatoria redekei* over a 24 h period when grown at 15 °C under a 6:18 LD cycle, irradiance 47 μE m^{-2} s^{-1}

Component mg l^{-1}	Start of light (a)	End of light (b)	End of dark (c)	Dark change as % of light increase $\frac{c-b}{b-a} \times 100$
Dry weight	100	142.7	123.71	−44.4
Protein	66.98	74.62	82.853	107.8
Carbohydrate	6.36	35.10	7.90	−94.6
RNA	7.70	9.04	9.51	20.1
DNA	1.94	2.08	2.34	200.6
Chlorophyll *a*	1.61	1.74	2.01	201.8

Table 13.6. Diurnal nitrogen budget of *Oscillatoria redekei* grown under a 6:18 LD cycle at 15 °C, irradiance 47 μE m^{-2} s^{-1}

Nitrogen fraction (mg N l^{-1})	Start of light	End of light	End of dark
Protein	1.600	1.782	1.979
RNA	0.161	0.189	0.199
DNA	0.040	0.044	0.049
Chlorophyll *a*	0.022	0.023	0.029
Soluble peptides + free amino acids	0.067	0.113	0.083
Sum of fractions	1.89	2.15	2.34
Measured total	1.90	2.16	2.35

was in excess of 50% of the daily total emphasising the important role of dark cell synthesis.

Dark cellular production created a cell nitrogen demand met largely by dark nitrate uptake which was observed to be 39% of the total nitrate uptake. Excretion of soluble organic nitrogen compounds by *Oscillatoria redekei* was not observed. However, the measured value of dark nitrate uptake was somewhat less than dark protein synthesis which was 52% of the total. This was due to an accumulation of a pool of free amino acids and peptides in the cell during the light period which declined in the following dark period. By including the changes in the amino acids and peptide pool, a diurnal nitrogen budget for *Oscillatoria redekei* showed good agreement between total organic nitrogen measured and the sum of the individual nitrogen fractions (Table 13.6).

The diurnal cycle of carbohydrate and protein of *Oscillatoria agardhii* grown under a 6:18 L:D cycle was similar to that of *Oscillatoria redekei*. Figure 13.4 shows the daily cycles of carbohydrate and protein of *Oscillatoria agardhii* at two growth rates of 0.31 and 0.37 doublings day^{-1}. The principal difference between the two growth rates was the greater carbohydrate ac-

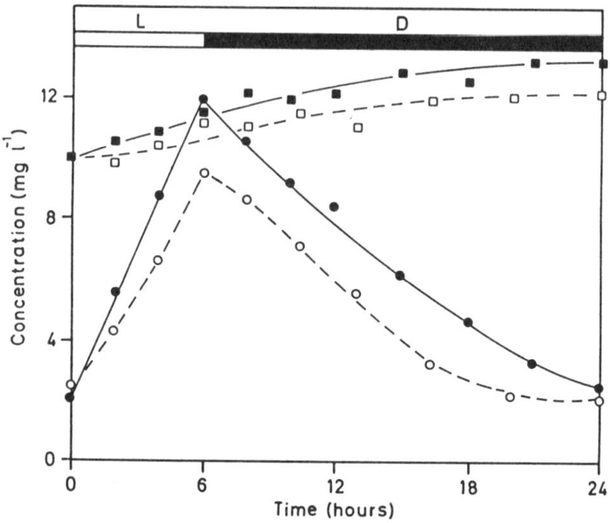

Figure 13.4. Diurnal changes in protein (-■-) and carbohydrate (-●-) of two *Oscillatoria agardhii* cultures grown at 15 °C under a 6:18 LD cycle. (i) closed symbols $K = 0.37$ doublings day^{-1} and (ii) open symbols $K = 0.31$ doublings day^{-1}. Modified after Foy and Smith (1990).

cumulation of the high growth rate culture which led to prolonged synthesis of protein in the dark period. The increase in growth rate under the L:D cycle was therefore dependent upon the ability of the alga to store surplus production as carbohydrate rather than enhanced rates of light protein synthesis.

The principal product of the increased photosynthetic rates under a 6:18 L:D cycle was shown to be carbohydrate which was used by the alga as an energy source to maintain cell synthesis in the dark period. As a result, growth rates expressed as *Kcorr* were higher under a 6:18 L:D cycle than under continuous light. The rapid accumulation of carbohydrate in the light period (Figs. 13.3 and 13.4) may have been stimulated by nitrogen limitation as evidenced by the lag in light mediated protein synthesis. Accumulation of metabolic intermediates such as 3–phosphoglycerate resulting from a nitrogen deficiency strongly activates ADP-glucose pyrophosphorylase, the enzyme most likely to be rate limiting carbohydrate production (Stanier and Cohen-Bazire, 1977; Lehmann and Wober, 1976).

13.4.3. *Photoperiod and growth efficiency*

The relationship between photoperiod and growth efficiency (*Kcorr*) of *Oscillatoria redekei* and *Oscillatoria agardhii* can be examined by comparison of *Kcorr* with the carbohydrate:protein ratio of the respective cultures at the

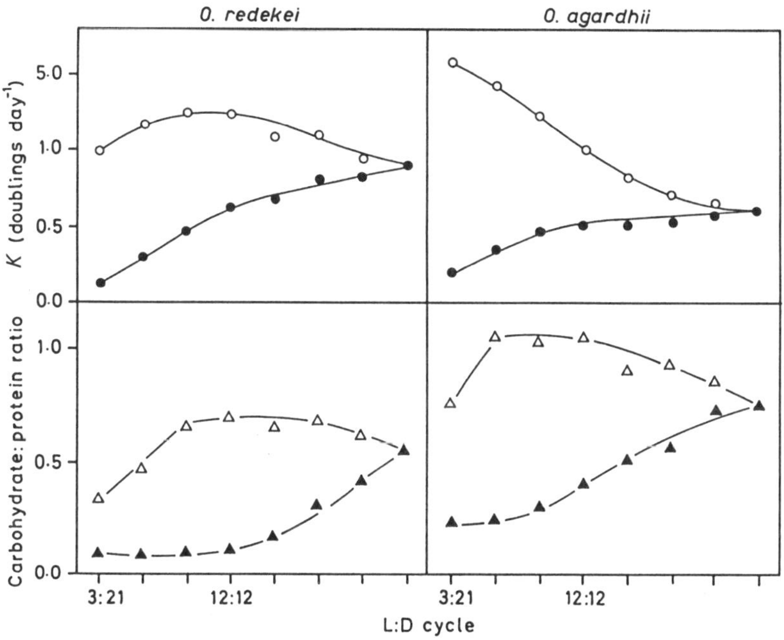

Figure 13.5. Effect of photoperiod on *K* (-●-), *K. corr* (-○-) and carbohydrate:protein ratios, (-▲-) start of light period and (-△-) end of light period of *Oscillatoria redekei* and *Oscillatoria agardhii* at 15 °C. After Foy and Smith (1980).

start and end of the light period (Fig. 13.5). The algae were grown under their optimum irradiances (see 13.4.2) at 15 °C. Comparison of *Kcorr* and carbohydrate:protein ratios shows that to obtain high values of *Kcorr* two conditions must be met:

(i) The carbohydrate:protein ratio at the start of the light must be low. If this is not the case, the cells will not have been able to convert all their stored carbohydrate into new cell protein because of an insufficiently long dark period.

(ii) The carbohydrate:protein ratio at the end of the light period must be below a certain maximum value which appears to be a characteristic of each species. If this value is reached, as for *Oscillatoria redekei* between 12:12 and 18:6 L:D cycles and *Oscillatoria agardhii* between 6:18 and 12:12 L:D cycles (Fig. 13.5), the implication is that the cells' capacity to store carbohydrate is less than the potentially available production of carbohydrate.

It is interesting to note that in the longer photoperiods when condition (i) was clearly not met, the cells exhibited reduced carbohydrate:protein ratios at the end of the light period (Fig. 13.5) suggesting that the cells do not solely store all potential surplus production as carbohydrate but attempt to balance light production with dark requirements as suggested by Cohen and Parnas (1976).

To achieve maximum values of *Kcorr* for *Oscillatoria redekei* and *Oscillatoria agardhii* the photoperiod regime is required to consist of short light and longer dark periods. For *Oscillatoria agardhii* the maximum growth effeciency occurred under a 3:21 L:D cycle which corresponds to the typical light experience of this algae within Lough Neagh (Table 13.1). The response of the *Kcorr* values of *Oscillatoria redekei* to photoperiod was flatter with a range of optimum photoperiods up to 12 hours (Fig. 13.5). This was due to the low irradiance optimum of the strain employed as higher growth rates were obtained under 3:21 L:D and 6:18 L:D cycles using the light tolerant strain of *Oscillatoria redekei* grown at 180–250 $\mu E\ m^{-2}\ s^{-1}$.

13.4.4. *Growth kinetics of* Oscillatoria agardhii *and* Oscillatoria redekei *grown under a 3:21 L:D cycle*

Comparison of the effects of light and temperature on the growth of these species grown under short photoperiods was initially hampered by light inhibition of *Oscillatoria redekei*, (Foy *et al.*, 1976; Foy and Smith, 1980). The fortuitous selection of a strain which was not light inhibited enabled experiments to be carried out comparing the effect of light and temperature on this strain and on *Oscillatoria agardhii* grown under a short photoperiod of 3:21 L:D cycle (Foy, 1983). As maximum growth rates under short photoperiod regimes had previously been shown to be determined by the maximum photosynthetic capacity of the cell, the temperature coefficient of growth under a 3:21 L:D cycle would have been expected to correspond to that of photosynthesis, i.e. $Q_{10} = 2.0$. The temperature:growth rate relationship for *Oscillatoria redekei* and *Oscillatoria agardhii* is plotted in Figure 13.6 with an exponential curve of $Q_{10} = 2.0$ drawn so that the measured growth rates lie on or below the curve. For each species, the relationship between temperature and growth was only exponential over limited temperature ranges, 14 °C–20 °C for *Oscillatoria agardhii* and 11 °C–20 °C for *Oscillatoria redekei*. Above and below these ranges, growth rates were less than predicted from the exponential relationship.

The rapid decline in growth rates at low temperatures observed for both species coincided with increases in the carbohydrate:protein ratio at the start of the light period (Fig. 13.7). At low temperatures, poor growth compared with the exponential relationship was due to the dark reactions being lowered by the reduction in temperature to the extent that the cells were unable to convert all stored carbohydrate into new cell material. Only above the temperature 11 °C for *Oscillatoria redekei* and 14 °C for *Oscillatoria agardhii* was the dark period sufficiently long for the cells to deplete their carbohydrate reserves and begin the light period with low carbohydrate:protein ratios (Fig. 13.7). At 23 °C, growth rates of the two species were lower than predicted from the exponential relationship (Fig. 13.6). The maximum

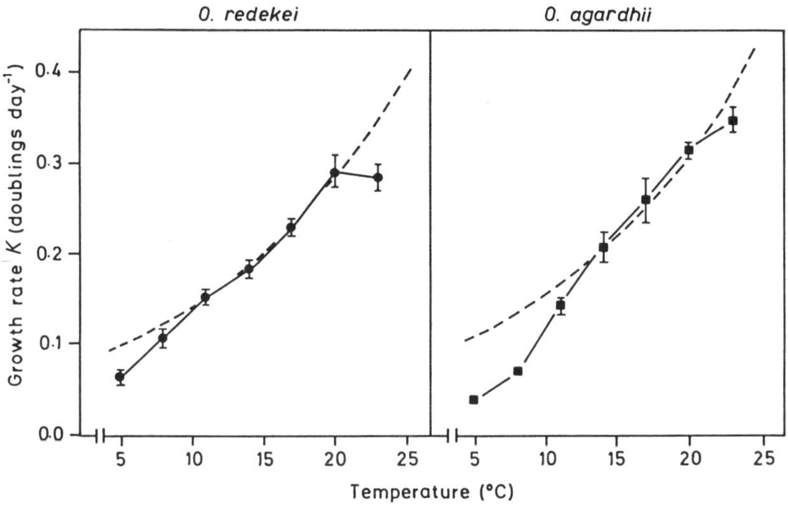

Figure 13.6. Effect of temperature on maximum growth rates of *Oscillatoria redekei* and *Oscillatoria agardhii* grown under a 3:21 L:D cycle. Dashed line-curve of $Q_{10} = 2.0$. Bars denote confidence limits at $p < 0.05$ level. After Foy (1983).

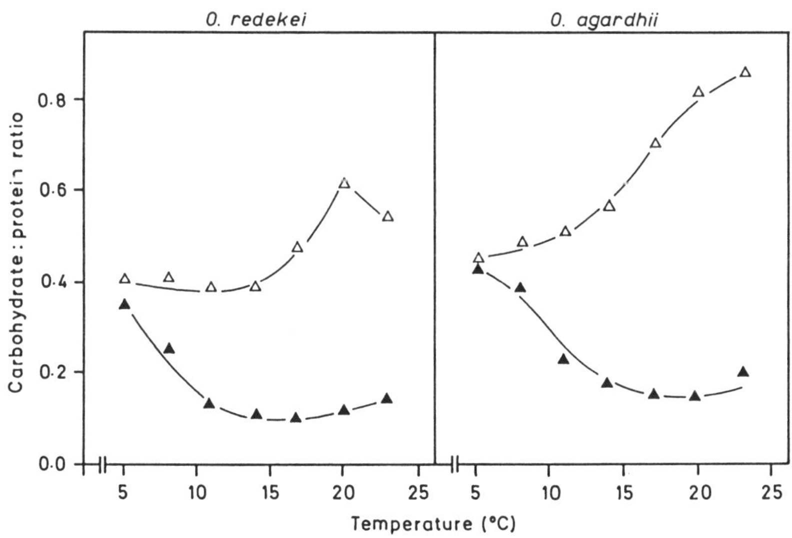

Figure 13.7. Effect of temperature on the carbohydrate:protein ratio at start (–▲–) and end (–△–) of the photoperiod for cells grown at maximum growth rate. After Foy (1983).

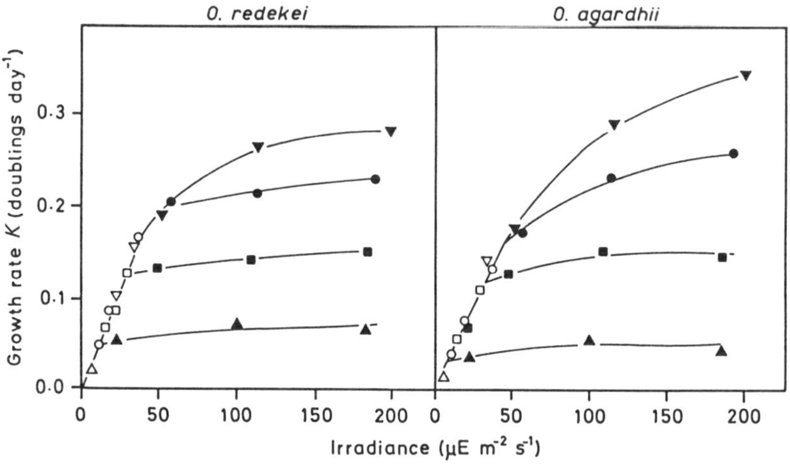

Figure 13.8. Growth rate:irradiance plots at 5 °C (–▲–), 11 °C (–■–), 17 °C (–●–) and 23 °C (–▼–). Open symbols – observations used to calculate initial slope. After Foy (1983).

irradiance of 200 μE m^{-2} s^{-1} was insufficient to saturate the growth of *Oscillatoria agardhii* (Fig. 13.7). The *Oscillatoria redekei* growth rate at 23 °C was less than at 20 °C and the temperature optimum of this species under a 3:21 L:D cycle was therefore lower than the 25 °C optimum of continuous light growth (Fig. 13.1).

At saturating irradiances, growth rate was temperature dependent but at sub-saturating irradiances, growth rates were found to become progressively temperature independent. Growth rate versus irradiance plots at 5 °C, 11 °C, 17 °C and 23 °C for *Oscillatoria agardhii* and *Oscillatoria redekei* under 3:21 L:D cycles in Figure 13.8 showed that at irradiances where the growth rate was approximately 50% or less of the maximum at that temperature, the relationship between irradiance and growth rate was linear, i.e. temperature independent for both species. Nine observations denoted by open symbols in Figure 13.8 together with growth rates from experiments at 14 °C and 20 °C and irradiance 20 μE m^{-2} s^{-1} were used to calculate the initial slope. *Oscillatoria redekei* grew faster than *Oscillatoria agardhii* in each of the 11 experiments but although the slope of the *Oscillatoria redekei* plot was greater than that of *Oscillatoria agardhii*, the difference between the slopes was not significant at the $p < 0.05$ level.

A comparison of the growth rates of the two species is shown in Figure 13.9 where open symbols denote that the difference between growth rates from an individual experiment was not significant at the $p < 0.05$ level. For a difference in growth rate to be statistically significant a difference in the region of 15% was required. Figure 13.9 demonstrates that only at temperatures in excess of 11 °C and irradiances greater than 80 μE m^{-2} s^{-1} did *Oscil-*

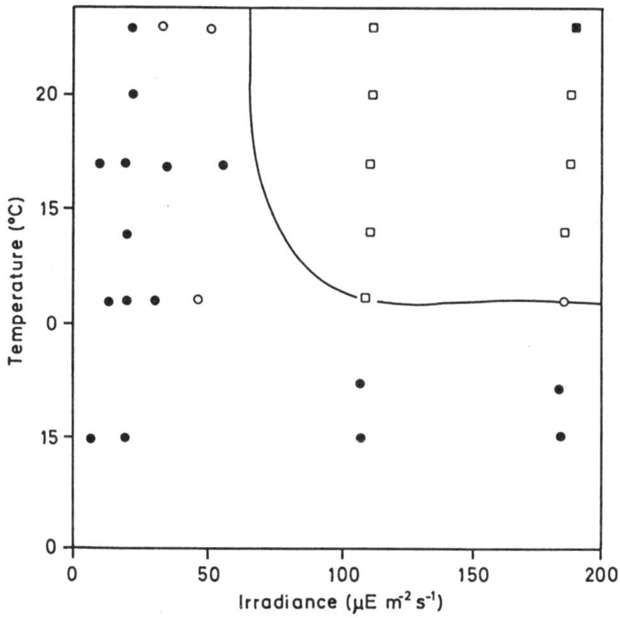

Figure 13.9. Light:temperature interaction on the growth of *Oscillatoria redekei* and *Oscillatoria agardhii*; (●) denotes experiment where *Oscillatoria redekei* grew faster and (■) experiment where *Oscillatoria agardhii* grew faster. Open symbols indicate that difference in growth rate was not significant at $p < 0.05$ level. After Foy (1983).

latoria agardhii grow faster than *Oscillatoria redekei* but in 7 of the 8 experiments, the individual difference in growth rates was not significant at the $p < 0.05$ level. Comparing the results from high light and temperatures as a group using a paired *t* test, showed that the difference between the two species was significant at the $0.01 > p > 0.001$ level.

A satisfactory explanation for the partial reversal in dominance observed under a 3:21 L:D cycle in comparison to continuous light growth rates has not been obtained. Shuter (1979) postulated that for a given light and temperature combination there was an optimum cell size. Under continuous light, small species would predominate but under reduced irradiances there would be a shift to larger species as he argued that small species have proportionally more carbon devoted to unproductive structural material. The maximum growth rates of *Oscillatoria redekei* and *Oscillatoria agardhii* above 11 °C under the 3:21 L:D cycle are consistent with this hypothesis as the pattern of carbohydrate:protein ratios suggested that growth was likely to be limited by carbon fixation at these temperatures. Below 11 °C, where the smaller species grew faster, growth was not limited by the photosynthetic capacity of the algae. At sub-saturating irradiances when growth rate was proportional to irradiance, the smaller *Oscillatoria redekei* grew at a faster rate which was at variance with the prediction of Shuter (1979). At sub-saturating irradiances, large species may experience greater internal self-shading of photosynthetic pigments and as a result be less efficient under low irradiances than smaller species (Kirk, 1975).

13.5. Nitrogen fixing blue-green algae

Anabaena flos-aquae and *Aphanizomenon flos-aquae* are capable of fixing atmospheric nitrogen and have in recent years decreased in abundance in Lough Neagh (Chapter 10). In the absence of combined nitrogen, they outgrow *Oscillatoria* species as the latter do not have the ability, at least under aerobic conditions, to fix nitrogen (Riddols, 1974, 1984a and b). The tendency to treat the presence of nitrogen fixing species as evidence of nitrogen limitation may be misleading as Foy (1980) and Foy *et al.* (1976) found that in the presence of nitrate, certain *Anabaena* and *Aphanizomenon* cultures grew at a faster rate under continuous light than some larger *Oscillatoria* cultures. Also under a 6:18 L:D cycle at high irradiances *Anabaena flos-aquae* and *Aphanizomenon flos-aquae* often grew faster than *Oscillatoria agardhii* and *Oscillatoria redekei* (Foy *et al.*, 1976). Therefore, success or otherwise of potentially nitrogen fixing species may not be explained simply on the presence or absence of combined nitrogen sources.

Foy and Gibson (1982a) undertook a comparison of the photosynthetic characteristics of 10 *Anabaena* and *Aphanizomenon* strains and 10 *Oscillatoria* strains. The algae were grown at high (150 $\mu E\ m^{-2}\ s^{-1}$) and low (30 $\mu E\ m^{-2}\ s^{-1}$) irradiance under a 6:18 L:D cycle at 20 °C. The response of the *Oscillatoria* cultures to low irradiance was broadly similar to that shown for *Oscillatoria redekei* in Figure 13.2 with an average reduction of *P*max chlorophyll a^{-1} of 42%. *Anabaena* and *Aphanizomenon* strains were more variable in their response as in only three cultures was *P*max (chlorophyll $a)^{-1}$ reduced by more than 20% and no significant differences between high and low irradiance treatments were found in four cultures. Variation in *P*max and, to a lesser extent, the slope of the initial portion of the photosynthesis irradiance plot (α) altered the light saturation characteristic *I*k (Talling, 1957) and mean values are listed in Table 13.7. Mean *I*k values of high and low light *Oscillatoria* cultures were significantly less than the corresponding *Anabaena* and *Aphanizomenon* means at the $p < 0.001$ level. As the *Oscillatoria* mean *P*max was depressed under low light, the mean *I*k value was similarly depressed. Despite the small change in *P*max values of *Anabaena* and *Aphanizomenon* strains, the mean *I*k value was reduced under low light because of a significant increase in the slope (α) (Table 13.7). However, despite this increase the mean α value of *Oscillatoria* cultures was greater than that of *Anabaena* and *Aphanizomenon*. The photosynthetic characteristics of the two algal groups suggested that on the basis of the lower *I*k

Table 13.7. Mean values of the photosynthetic characteristics of 10 *Anabaena* and *Aphanizomenon* strains and 10 *Oscillatoria* strains grown at irradiances of 30 μE m^{-2} s^{-1} and 150 μE m^{-2} s^{-1} at 20 °C under a 6:18 LD cycle. Maximum photosynthetic rate *P* max, initial slope of photosynthesis vs irradiance plot α, light saturation characteristic *I*k

	Anabaena and *Aphanizomenon* strains		*Oscillatoria* strains	
	30 μE m^{-2} s^{-1}	150 μE m^{-2} s^{-1}	30 μE m^{-2} s^{-1}	150 μE m^{-2} s^{-1}
P max mg O_2 (mg Chl*a*)$^{-1}$ h^{-1}	11.3	13.0	8.54	14.8
α mg O_2 (mg Chl*a*)$^{-1}$ E^{-1} m^{-2}	20.5	17.9	25.7	25.2
*I*k μE m^{-2} s^{-1}	164.1	214.6	100.1	174.8

value, *Oscillatoria* strains could be considered as low light organizms relative to the *Anabaena* and *Aphanizomenon* strains.

Further evidence was provided by comparing growth under a 3:21 L:D cycle of an *Aphanizomenon flos-aquae* strain (grown in nitrate medium at sub-saturating irradiances, i.e. on the linear portion of the growth vs irradiance plot, shown in Figure 13.8) with those of *Oscillatoria agardhii* and *Oscillatoria redekei* described in Section 13.4.4. In all instances, *Aphanizomenon flos-aquae* grew at lower rates than the *Oscillatoria* species (Figure 13.10). At the maximum irradiance used (180 μE m^{-2} s^{-1}) *Aphanizomenon flos-aquae* was the fastest growing species at temperatures below 14 °C although the difference was not statistically significant in each instance. Above 14 °C it grew at a faster rate than *Oscillatoria redekei* but slower than *Oscillatoria agardhii* but in these instances there was evidence that *Aphanizomenon flos-aquae* growth rates were not light saturated. Similar poor growth at low irradiances of *Aphanizomenon flos-aquae* in comparison with an *Oscillatoria agardhii* strain has been reported by Zevenboom *et al.*, (1981) which suggests that potentially nitrogen fixing blue-green algae are at disadvantage to *Oscillatoria* species if irradiance values are low. High light saturated growth rates obtained under light:dark cycles indicate that *Anabaena* and *Aphanizomenon* species could outgrow *Oscillatoria* species in the presence of nitrate if the irradiance is sufficiently high.

The energy requirements for growth under nitrogen fixing conditions in comparison to nitrate or ammonium media are not well known. A limited number of experiments have been undertaken to compare growth using different nitrogen sources (Table 13.8). Utilising atmospheric nitrogen under continuous light, growth of *Anabaena flos-aquae* and *Aphanizomenon flos-aquae* was less than in nitrate or ammonium media. The growth rates of these species again using atmospheric nitrogen under a 6:18 L:D cycle at 140 μE m^{-2} s^{-1} were also less than with the combined nitrogen sources but the difference between the nitrogen fixation and nitrate growth rates was reduced, especially for *Aphanizomenon* (Table 13.8). The marked stimula-

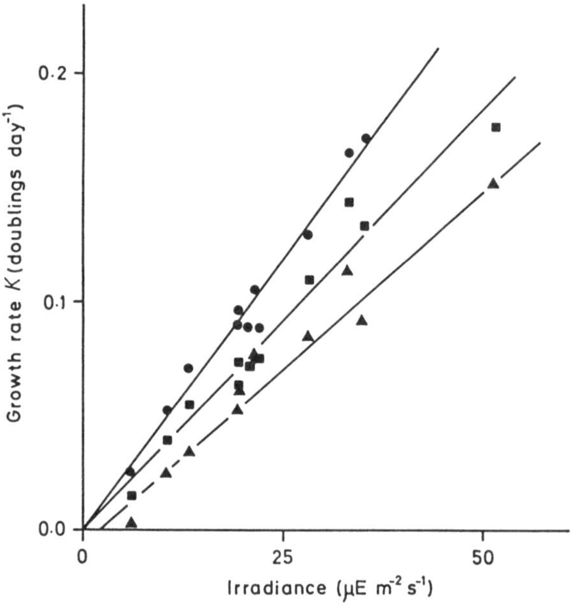

Figure 13.10. Relationship between irradiance and growth rate at subsaturating irradiances of *Oscillatoria agardhii* (–■–), *Oscillatoria redekei* (–●–) and *Aphanizomenon flos-aquae* (–▲–) grown under a 3:21 LD cycle at a range of temperatures 5 °C–23 °C. Modified after Foy (1983).

Table 13.8. Continuous light and 6:18 LD growth rates of *Anabaena flos-aquae* and *Aphanizomenon flos-aquae* grown in (a) medium with ammonium as the nitrogen source, (b) nitrate medium and (c) medium containing no combined nitrogen. Temperature 20 °C, irradiance 140 μE m^{-2} s^{-1}. Values in parenthesis – 95% confidence limits

Organism	Nitrogen Source	Growth rate (doublings day^{-1})	
		Continuous light	6:18 LD
Anabaena flos-aquae	a ammonium	1.447 (±0.061)	0.653 (±0.041)
	b nitrate	1.031 (±0.037)	0.416 (±0.020)
	c atmospheric nitrogen	0.629 (±0.035)	0.345 (±0.016)
Aphanizomenon flos-aquae	a ammonium	1.450 (±0.072)	0.525 (±0.020)
	b nitrate	1.297 (±0.048)	0.433 (±0.011)
	c atmospheric nitrogen	0.962 (±0.028)	0.397 (±0.005)

tion of the growth rate of the *Anabaena flos-aquae* culture using ammonium under both continuous light and the 6:18 L:D cycle is noteworthy.

13.6. Carbohydrate content of Lough Neagh phytoplankton

When grown under laboratory conditions, *Oscillatoria agardhii* and *Oscillatoria redekei* stored a large proportion of photosynthate as carbohydrate (Section 13.4.2). Furthermore under high irradiances the ability of each species to store carbohydrate and utilise it efficiently determined its growth rate for a given temperature and photoperiod. This section examines the evidence for carbohydrate accumulation in algae growing in Lough Neagh. The results were obtained by Gibson (1975, 1978a and b) not with the aim of verifying laboratory growth experiments but as a study of respiration losses in Lough Neagh.

Gibson (1975) described how annual changes in the respiration rate of Lough Neagh algae varied with depth. Moreover, this study demonstrated that depth gradients of respiration in Lough Neagh were of normal occurrence and that respiration was directly related to carbohydrate content. Although the lake was essentially homogeneous, in the majority of experiments the respiration rate of surface water was greater than that of water taken from 10 m. Further studies (Gibson, 1978a and b) examined in detail changes in carbohydrate in order to assess the use of carbohydrate content of seston as a tool for monitoring the previous light history of phytoplankton in Lough Neagh.

Enhanced levels of carbohydrate in the photic zone of Lough Neagh were shown by these studies to be a regular feature of the phytoplankton. However, the characteristics of each depth profile varied according to mixing, irradiance and the nutrient status of the phytoplankton. Successive depth profiles of carbohydrate:chlorophyll *a* ratios taken on 10.6.75 during calm conditions are plotted in Figure 13.11. In the first profile, at 07.05 hours, no distinct variation with depth was observed but when the second profile was taken at 10.15 hours there was a marked sub-surface maximum at 0.5–1.0 m. The shape of the second profile clearly resembled photosynthesis:depth profiles in Lough Neagh (Chapter 11) with evidence of surface inhibition and no increase in carbohydrate:chlorophyll *a* ratio below the 1% level of photosynthetically active radiation at 2.0 m. Enhanced carbohydrate:chlorophyll *a* ratios at depths below 2.0 m in the final profile taken at 18.05 hours indicated that downward circulation of cells from the photic zone had taken place. Comparison of maximum rates of carbohydrate increase at 0.5 m depth with maximum recorded rates of photosynthesis in Lough Neagh suggested that carbon accumulated as carbohydrate represented a major fraction of carbon fixed by photosynthesis (Gibson, 1978a).

Seasonal variations in carbohydrate content during 1976 and 1977 present a complex relationship but show that, with the exception of the period January–mid-March, the carbohydrate content of seston from surface samples was generally higher than samples taken from 10 m depth (Figure 13.12). In addition to there being little variation with depth between January and mid-March, carbohydrate values were also generally higher in this period

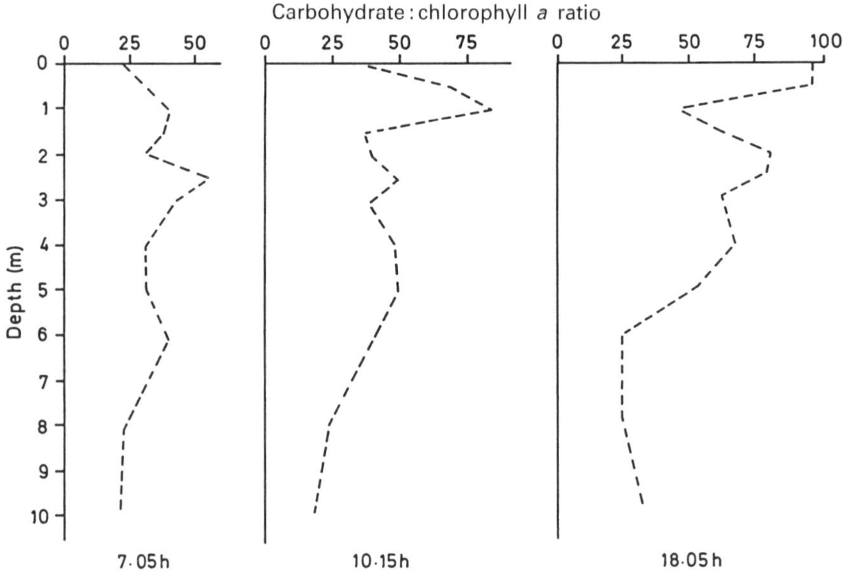

Figure 13.11. Variation in carbohydrate::chlorophyll *a* ratio with depth in Lough Neagh water column 10 June 1975. Modified after Gibson (1978a).

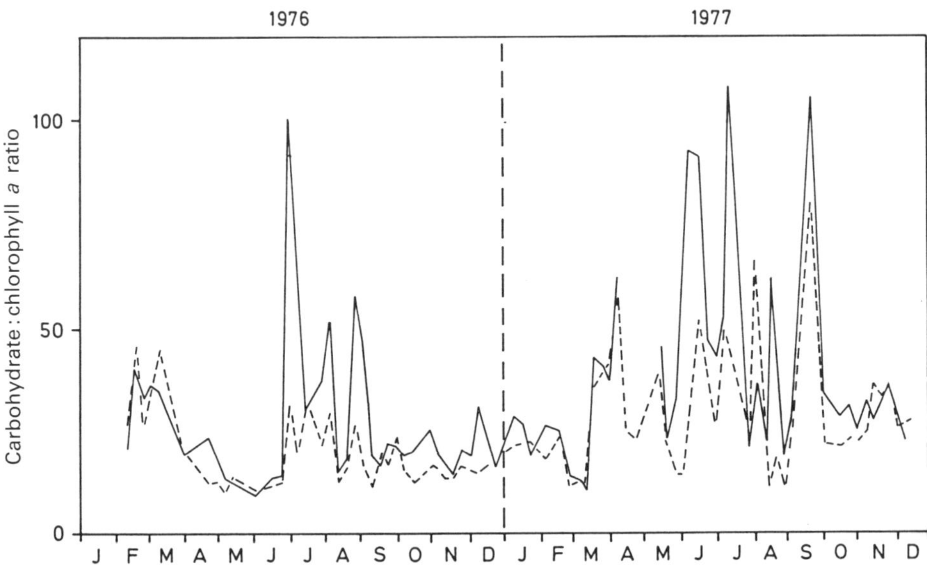

Figure 13.12. Carbohydrate:chlorophyll *a* ratio in Lough Neagh at 0 m (——) and 10 m (------) during 1976–77.

than those of the following weeks, March to June, 1976 and March, 1977. Higher carbohydrate levels could reflect the effect of low temperatures on carbohydrate content observed in laboratory cultures (see Figure 13.7). The rise in the field was not as distinct as found under laboratory conditions which may be due to the low light dose received by natural populations. The rapid rise of carbohydrate in mid-March to mid-April 1977 corresponded to a period when the diatom crop ceased to grow as a result of silica limitation (Chapters 9 and 10). The peak in carbohydrate content of surface samples observed during this period corresponded with high values from 10 m samples which would indicate that the algae were unable to utilise their stored carbohydrate for cell growth. While this evidence for high carbohydrate content being a result of silica limitation was only circumstantial, it is supported to a limited extent by observations made in 1976 when no such peak occurred; this corresponded to the failure of the diatom crop in that year (Chapter 10).

Gibson (1978a) further showed that, after the diatom crop had sedimented from the water column in mid-May, the carbohydrate content of the seston dropped briefly but the summer period, June-September, was characterised in both years by a complex series of peaks which only declined in the autumn months. Gibson and Stevens (1979) presented evidence that the first peak in June 1977 may have been a result of phosphorus deficiency as the peak coincided with a period when the phosphorus content of the plankton was at a minimum. For the remainder of the year sediment release of phosphorus ensured that phosphorus limitation did not occur. The high temperatures (>15 °C) during July to September allowed algae, given high irradiances, to achieve higher rates of photosynthesis than earlier in the year and so lead to higher carbohydrate levels. Therefore the variability observed in these months may have been a result of short term differences in irradiance and mixing patterns. There was a possibility that the high summer carbohydrate levels were due to short term nitrogen limitation because during this period nitrate levels were low (<0.20 g N l^{-1}) and, under nitrogen limitation, laboratory experiments have shown that blue green algae respond with increases in carbohydrate (Allen and Smith, 1969; Lehmann and Wober, 1976; Gibson, 1978a). It is difficult to assess the importance of nitrogen limitation as the algae may have been able to obtain sufficient nitrogen from zooplankton excretion products, microbial breakdown and river inputs. Gibson (1978a) showed that, when nitrogen deficient cells of *Oscillatoria redekei* were placed in the dark, carbohydrate levels did not decline as rapidly as cells which were given nitrate. Thus, if nitrogen limitation were occurring in Lough Neagh it would be expected that the surface maximum values would be accompanied by high values at 10 m. Apart from samples in mid-July and mid-September 1977 this was not the case.

Overall the results of Gibson (1975, 1978a and b) showed that carbohydrate plays an important role in the performance of Lough Neagh phytoplankton. The restraints that carbohydrate accumulation and decline place on the growth of *Oscillatoria agardhii* and *Oscillatoria redekei* under laboratory conditions may therefore also be of particular relevance to the situation in Lough Neagh itself.

13.7. Algal growth in Lough Neagh

Analysis of the effects of light and temperature on the growth rates of *Oscillatoria agardhii* and *Oscillatoria redekei* under natural conditions in Lough Neagh is made difficult as the populations are subjected to continual losses due to grazing, parasitism, sedimentation and washout. Sedimentation rates of *Oscillatoria* species are low as are the losses due to washout in Lough Neagh but the magnitude of grazing and parasitism losses are not known nor can they be easily determined. At first sight the controlled conditions available in the laboratory should make it possible to study the impact of selected environmental variables but, as the uniform laboratory environment by its very nature removes the selection pressures operating in natural populations this can lead to the development of laboratory strains or, equally misleading, to ecological interpretations divorced from the real world. Therefore the results of laboratory experimentation cannot be treated in isolation but must be considered in conjunction with observations from the field.

13.7.1. *In situ growth experiments*

An attempt to overcome some of the artificiality of the laboratory environment, especially in terms of light quality, but eliminate grazing, sedimentation and other losses has been to grow *Oscillatoria agardhii* and *Oscillatoria redekei*, separately, in clear plastic tissue culture flasks suspended over a range of depths in the photic zone, to 3.0 m. The experiments were conducted over a 15 month period, in 1980 and 1981, each experiment lasting 2–3 weeks. The seasonal pattern of growth of the two species was similar (Figure 13.13b and c). The observed range of maximum growth rates was almost 6-fold despite a temperature amplitude of only 12 °C so that Q_{10} values of 4.6 and 4.5 were calculated (Table 13.9). The high Q_{10} values were due to increasing daylength in the summer months. The Q_{10} value for maximum growth rate corrected for daylength was approximately 3.0 which was higher than laboratory 6:18 L:D experiments (Table 13.2) and reflects the long photoperiod (>8.0 hours) to which the trapped algae were subjected in relation to natural phytoplankton which would circulate throughout the water column. The temperature dependence of mean growth rates was lower than the maximum growth rate as growth at depths where sub-saturating irradiances occurred was less temperature sensitive. At 2.0 m growth rates, when corrected for light hours received, did not show a statistically significant relationship at the $p < 0.05$ level with temperature (Table 13.9). This result

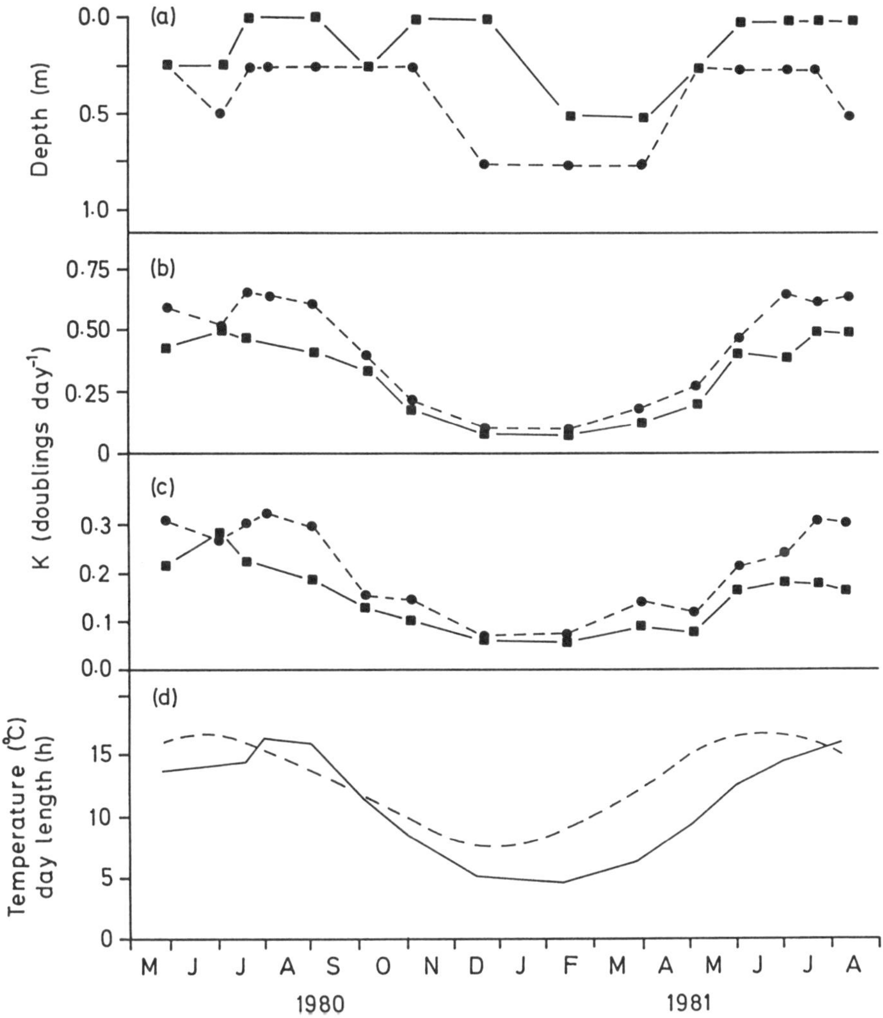

Figure 13.13. In situ growth experiments between May 1978 and August 1981 of *Oscillatoria agardhii* (–■–) and *Oscillatoria redekei* (---●---). (a) Depth at which maximum growth rate (*K max*) occurred; (b) *K max.*; (c) Mean growth rate (*K mean*) over 0–3 m water column; (d) Variation in day length (------) and temperature (——).

was in agreement with laboratory experiments (Figure 13.8) which showed that, given sufficiently low irradiances, growth was temperature independent.

The depth at which maximum growth occurred differed between seasons and species (Figure 13.13a). *Oscillatoria redekei* showed greater light inhibition but the ecological significance of this is not clear as a light sensitive strain was used. The increased light inhibition in the winter was paradoxical as it coincided with the lowest surface irradiances. Similar increased sensitiv-

Table 13.9. Temperature dependence of growth in Lough Neagh *in situ* experiments as measured by Q_{10} values; (a) maximum growth rate, K max.; (b) mean growth rate over 0–3 m water column, $\bar{K}$.; (c) growth at 2.0 Q_{10} values calculated for growth rate (K) and growth rates corrected for light hours received (K *corr*)

		a K max Q_{10}	b $\bar{K}$ Q_{10}	c K 2.0 m Q_{10}
Oscillatoria redekei	K	4.58	3.15	2.08
	K *corr*	2.85	1.91	1.26[a]
Oscillatoria agardhii	K	4.54	2.78	2.06
	K *corr*	2.69	1.65	1.22[a]

[a]Correlation coefficients of temperature: K *corr* relationships not significant at $p < 0.05$ level.

ity of blue-green algal growth at low temperatures was observed under laboratory conditions by Foy *et al.*, (1976). The comparative absence of surface inhibition of *Oscillatoria agardhii* was insufficient to compensate for growth rates lower than *Oscillatoria redekei* at greater depths. As a result, with one exception, the mean growth rate over 0–3 m water column of *Oscillatoria redekei* was faster than that of *Oscillatoria agardhii* (Figure 13.13c). This result was at variance with the observed cycle in Lough Neagh where *Oscillatoria agardhii* dominated from July and the changeover in dominance at 11 °C observed in the laboratory using a 3:21 L:D cycle (Section 13.4.4). This disparity was almost certainly a result of the long photoperiods that the *in situ* experiments were subjected to. The failure of the *in situ* experiments to reproduce the pattern of dominance in Lough Neagh highlight the importance of circulation in the ecology of these organisms.

13.7.2. *Growth of* Oscillatoria agardhii *and* Oscillatoria redekei *in Lough Neagh*

Although algae growing in Lough Neagh can achieve comparatively high net growth rates (Table 13.1), such occasions are confined to the spring and the general condition is for net growth rates to be low or negative. This is particularly so during the summer months when despite the high rate of carbon fixation during this period (Chapter 11) the algae must be subjected to loss rates similar in magnitude to potential growth rates. In this section, an attempt is made to compare the performance of *Oscillatoria agardhii* and *Oscillatoria redekei* relative to each other between 1976 and 1980 by examining changes in the ratio of the two species expressed as the natural logarithm ($ln\ \frac{Or}{Oa}$). The 1976–1980 period was chosen as both species were present throughout in measurable quantities and temperature and light data are available. The logarithm of the ratio was chosen because if two species differ in growth by a constant factor $ln\ \frac{Or}{Oa}$ will change at a constant rate whereas the simple ratio $\frac{Or}{Oa}$ will not. An upward trend of the $ln\ \frac{Or}{Oa}$ value implies that

Oscillatoria redekei was increasing relative to *Oscillatoria agardhii* which could have been because of one or two factors:

(1) *Oscillatoria redekei* possessed a higher growth rate than *Oscillatoria agardhii* and that the loss rate of the two species was similar.

(2) *Oscillatoria redekei* did not necessarily possess a higher growth rate but its loss rate was lower than *Oscillatoria agardhii* thus enabling it to increase relative to that species.

For a downward trend of $ln \frac{Or}{Oa}$, the opposite conditions in (1) and (2) would apply.

The annual variation in temperature, underwater light climate and $ln \frac{Or}{Oa}$ cycles between 1976 and 1980 are plotted in Figure 13.14. Underwater light data were available only in milliwatt hours and is suitable only for comparative purposes as the light values were obtained by multiplying incoming solar radiation by the photic depth expressed as a proportion of the water column. No attempt was made to correct for light attentuation, but as a percentage of sub-surface irradiance the mean irradiance in the photic zone is likely to be fairly constant. The annual underwater light cycle was different from the incoming radiation as high spring crops of diatoms and blue-green algae in Lough Neagh reduced the photic depth. As a result, the lag between incoming radiation and water temperature evident in Figure 13.13d was eliminated for the first half of each year (Figure 13.14). In the second half of the year underwater light declined at a faster rate than temperature. The bars on Figure 13.14b denote the periods when Lough temperature was at 11 °C or higher. The onset of temperatures above 11 °C coincided with the decline of the *Oscillatoria rededei* crop relative to *Oscillatoria agardhii*. In terms of algal crop, the ratio $ln \frac{Or}{Oa}$ can decline even if the increase in algal volume of *Oscillatoria redekei* is greater than that of *Oscillatoria agardhii* when the inoculum of the former is sufficiently greater (e.g. a 10% increase of an *Oscillatoria redekei* crop of 10 $mm^3\ l^{-1}$ would be greater in terms of volume than a 50% increase of an *Oscillatoria agardhii* crop of 1 $mm^3\ l^{-1}$ but the ratio $ln \frac{Or}{Oa}$ would decline from 2.30 to 1.99).

Prior to the onset of temperatures of 11 °C and above, the proportion of *Oscillatoria redekei* in the blue-green algal crop increased steadily in each year. Increases late in the year, after weeks 40–44, when the temperature was also below 11 °C were observed in 1978, 1979 and 1980, while in the remaining two years there was little change in the proportion of the two species during this period. The interaction of light and temperature on the two species was examined by plotting the temperature and light on the occasions when each species increased relevant to the other. To reduce variation in the $ln \frac{Or}{Oa}$ value due to counting errors, changes in the ratio were compared at 3 week intervals. If $ln \frac{Or}{Oa}$ altered by less than 10%, this was recorded as no change in dominance. The results of this analysis are shown in Figure 13.15 and clearly demonstrate the success of *Oscillatoria redekei* at low temperatures. Out of 120 observations below 11 °C, *Oscillatoria agardhii* was judged to be increasing relative to *Oscillatoria redekei* on only 16 oc-

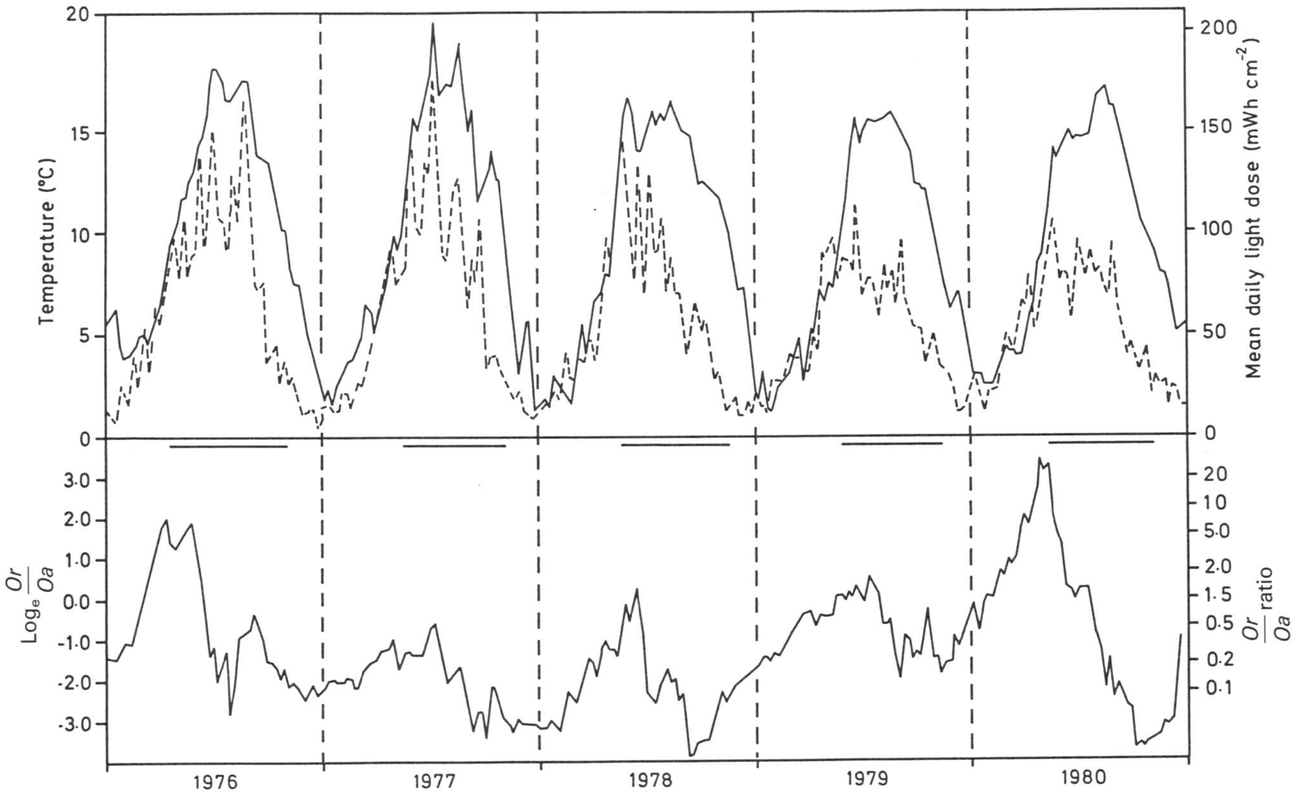

Figure 13.14. Lough Neagh *Oscillatoria agardhii* and *Oscillatoria rededei* 1976–1980. (a) Variation in under water light dose (------) and temperature (——); (b) Log$_e$ $\frac{Or}{Oa}$ ratio, bars denote periods when temperature was greater than 11° C.

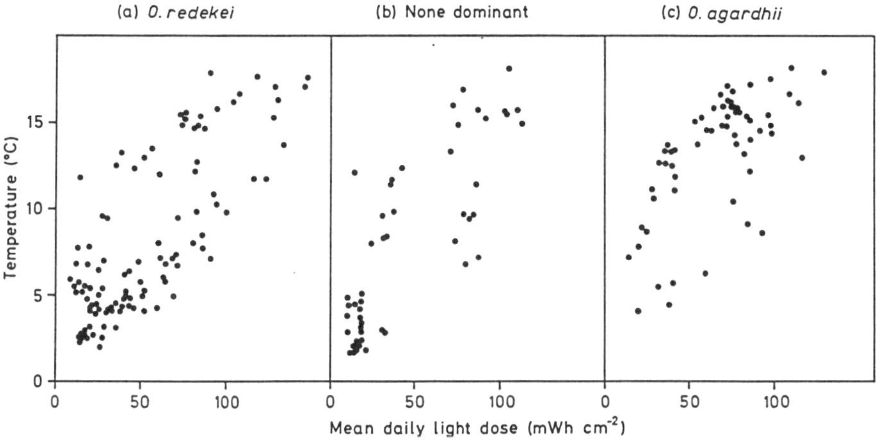

Figure 13.15. Interaction between temperature and light on *Oscillatoria redekei* and *Oscillatoria agardhii*. (a) *Oscillatoria redekei* increased relative to *Oscillatoria agardhii*; (b) $ln\ \frac{Or}{Oa}$ or altered by less than 10%; and (c) *Oscillatoria agardhii* increased relative to *Oscillatoria redekei*. Temperature and light data represent 3 week mean values.

casions, while *Oscillatoria redekei* was increasing on 76 occasions. No changes were recorded on the remaining 28 points which were clustered at the turn of the year when light was at a minimum and growth rates of both species were very low. Above 11 °C, the dominance of one species over the other was not as clear-cut, on 48 occasions, *Oscillatoria agardhii* did better compared to 29 for *Oscillatoria redekei* and 28 showed no difference. The distribution of points and the plot of the mean light-temperature cycle show that there was only a comparatively limited range of temperature and irradiance combinations available. Low irradiance and high temperatures and vice-versa, did not occur. There were significant numbers of *Oscillatoria redekei* observations at the highest irradiances; above 100 light units there were 12 *Oscillatoria redekei* points compared to 5 *Oscillatoria agardhii* points. This suggested that the strain or strains of *Oscillatoria redekei* present in Lough Neagh did not suffer selectively from light inhibition of growth during the summer months and resembled the light-tolerant laboratory strain. The appreciable number of observations when *Oscillatoria redekei* increased in relation to *Oscillatoria agardhii* above 11 °C was at variance with laboratory observations where under a 3:21 L:D cycle, *Oscillatoria agardhii* dominated at high irradiance above 11 °C (Figure 13.15). This may have been due to a number of factors:

(1) The high irradiance points were caused by daylengths in excess of 3 hours. An examination of daylength, light intensity relationship for points above 11 °C showed that daylength experienced by algae could approach 5 hours (but the overall distribution of observations was such that no firm conclusions could be made).

(2) The overlap of results above 11 °C reflects the overlapping confidence limits of individual growth rates of the two species above 11 °C at saturating irradiances and irradiances on the shoulder of the growth rate-irradiance relationship (Fig. 13.9). Whilst this is the case, it would have been expected however, that the algal daylength-irradiance plot of Figure 13.15 would have shown a clearer trend with *Oscillatoria redekei* dominant at the lower irradiances.

(3) The final factor considered arises in part from factor (2), because if growth rates above 11 °C were similar, then it would be predicted that the $ln \frac{Or}{Oa}$ ratio would change only slowly. However, this is not the case, as Figure 13.14 shows that the largest changes took place during the summer and were associated with decreases in the proportion of *Oscillatoria redekei* in the crop. Even allowing for the rapid potential growth rates of the summer months (0.3–0.4 doublings day^{-1}) the $ln \frac{Or}{Oa}$ ratio could not change by more than 0.3 units week^{-1} if the difference in growth rates was less than 20%. However, decreases of up to 0.6–1.0 units week^{-1} were observed in each year (Figure 13.14). Such changes could only have been caused by *Oscillatoria redekei* incurring greater loss rates than *Oscillatoria agardhii*. The effect of increasing the loss rate of *Oscillatoria redekei* on the crop composition of a hypothetical example where *Oscillatoria agardhii* grew at 0.35 doublings day^{-1} compared to 0.30 doublings day^{-1} for *Oscillatoria redekei* is shown in Table 13.10. This example shows that the comparatively small difference in loss rates combined with a similar difference in growth rates combine to give a large change in the ratio of the two species.

The cause of such an increased loss rate is not known. Selective grazing by *Daphnia* species on *Oscillatoria redekei* has been postulated by Gibson and Fitzsimons (1982), but a direct relationship between *Daphina* numbers and the proportion of the two species in Lough Neagh has yet to be shown. In the data of Figure 13.14 the $ln \frac{Or}{Oa}$ ratio does not decline indefinitely but reaches minimal values in the range −2.0 to −3.0. At these values the *Oscillatoria agardhii* crop was 5–20 times greater than that of *Oscillatoria redekei* but because the mean filament volume of *Oscillatoria redekei* is in the region of an order of magnitude less than *Oscillatoria agardhii*, the filament numbers of the two species would have been similar. It may be therefore, that selective grazing only takes place when there is an abundance of *Oscillatoria redekei* filaments. However, when the ratio of filaments approaches unity, it is no longer advantageous for the zooplankton to select for them in preference to the larger but perhaps less digestible *Oscillatoria agardhii* filaments. Such happenings have not been demonstrated in the field and because of the comparatively small differences in loss rates required to alter the algal species composition significantly, the technical problems in doing so are considerable.

The results presented in Figures 13.14 and 13.15 demonstrate the temperature dominance of *Oscillatoria redekei* below 11 °C. Above 11 °C, the situ-

Table 13.10. Impact of increasing loss rate of *Oscillatoria redekei* on the ratio of *Oscillatoria redekei* to *Oscillatoria agardhii*

EXAMPLE (1) LOSS RATES EQUAL				
	Growth rate (doublings day^{-1})		Loss rate (doublings day^{-1})	
Oscillatoria agardhii	0.35		0.33	
Oscillatoria redekei	0.30		0.33	
	O redekei crop	*O agardhii* crop	Total	$\log_e \frac{Or}{Oa}$
Time 0	10.0	5.0	15	0.693
Week 1	8.65	5.51	14.2	0.450
Week 2	7.47	6.07	13.5	0.208
Week 3	6.46	6.69	13.2	−0.034
Week 4	5.59	7.37	13.0	−0.277
			$\log_e \frac{Or}{Oa} = -0.24$ units week^{-}	
EXAMPLE (2) *OSCILLATORIA REDEKEI* LOSS RATE GREATER THAN *OSCILLATORIA AGARDHII*				
	Growth rate (doublings day^{-1})		Loss rate (doublings day^{-1})	
Oscillatoria agardhii	0.35		0.30	
Oscillatoria redekei	0.30		0.40	
	O redekei crop	*O agardhii* crop	Total	$\log_e \frac{Or}{Oa}$
Time 0	10.0	5	15	0.693
Week 1	6.16	6.37	12.5	−0.034
Week 2	3.79	8.12	11.9	−0.762
Week 3	2.33	10.35	12.7	−1.490
Week 4	1.44	13.20	14.6	−2.218
			$\log_e \frac{Or}{Oa} = -0.73$ units week^{-1}	

ation is complex but the results confirm the laboratory prediction that it is only above this temperature that *Oscillatoria agardhii* can compete with *Oscillatoria redekei*. However, the change in dominance of the two species in the June-July period observed between 1969–1980 (Chapter 10) in terms of alga volume was caused as much by *Oscillatoria redekei* experiencing higher loss rates as by *Oscillatoria agardhii* having a higher potential growth rate during the summer months.

13.8. Summary

The aim of the experiments described in this chapter was to elucidate the influence of light and temperature on the seasonal distribution of blue-green algae in Lough Neagh. They have demonstrated the problems of long-term laboratory studies on blue-green algae. The genotypic plasticity of *Oscillatoria rededei* observed in this study and elsewhere (Meffert and Krambeck, 1977; Kohl and Nicklish, 1981) places a constraint on interpreting results, especially when the algae are maintained in culture for a number of years. The variability of blue-green algae may of course be a factor in determining their success in aquatic environments. The importance of size and hence surface area:volume ratio in determining growth rates under continuous light has been demonstrated. The maximum growth rates observed under continuous light were, in comparison with those observed under light:dark cycles, relatively slow and could be sustained by considerably less than maximum carbon fixation rates. Under L:D cycles an obvious effect of size on growth rate was not apparent. If cultured under short photoperiods, less than 6 hours light per day, the maximum growth rates of *Oscillatoria redekei* were reached at irradiances at which the photosynthetic rate of the alga was saturated. The growth rate of an alga under a given L:D cycle appeared to be dependent not only on the cells' capacity to fix carbon but also the ability to store a large proportion of photosynthetic production as carbohydrate, subsequent dark conversion of carbohydrate to protein and the conservation of the metabolic machinery of the alga throughout the dark period. The ability of *Oscillatoria agardhii* in laboratory experiments to prolong dark protein synthesis over a long dark period appeared to give it a competitive advantage when growing at high irradiances and high temperatures. At low temperatures this became a disadvantage as the dark reaction was depressed to such an extent that the cell was unable to convert all accumulated carbohydrate to protein with the result that growth rate fell markedly and the small *Oscillatoria redekei* was at an advantage. At sub-saturating irradiances however, *Oscillatoria redekei* dominated at all temperatures indicating that at these low irradiances the efficiency with which an alga harvests incoming light is of overriding importance in determining growth rate.

The effects of temperature and light on the growth rates of *Oscillatoria redekei* and *Oscillatoria agardhii* were only partially observed in Lough Neagh where *Oscillatoria redekei* consistently grew better than *Oscillatoria agardhii* if the temperature was below 11 °C. Above 11 °C, although *Oscillatoria agardhii* was dominant, the speed at which it increased relevant to *Oscillatoria redekei* was much greater than would have been predicted from laboratory observation. Crucial to the success of *Oscillatoria agardhii* in the summer appears to be an as yet unidentified loss factor(s) discriminating against *Oscillatoria rededei* which enabled *Oscillatoria agardhii* to dominate the plankton. Selective grazing has been invoked as a probable loss mechanism but this has not been tested.

Comparing the growth kinetics and photosynthetic characteristics of *Oscillatoria* species with those of *Anabaena* and *Aphanizomenon* species suggested that *Oscillatoria* species were favoured by low light intensities but given high irradiances, *Anabaena flos-aquae* and *Aphanizomenon flos-aquae* could grow as fast as or faster than *Oscillatoria* species. The growth strategy of the 2 types appears to be different. *Oscillatoria* species are planktonic and spend the entire year in the water column. Their success under low light environments is an advantage during the winter and in turbid water when a potential for high growth rates is not necessary. *Anabaena* and *Aphanizomenon* species are not always entirely planktonic as they can form akinetes which sediment out of the water column during unfavourable conditions. Their high maximum growth rates would enable them rapidly to recolonise the water column when favourable high light conditions return. In Lough Neagh higher light conditions are likely to be associated with low nitrogen inputs so for an alga to utilise high light intensities the potential capacity to fix nitrogen would be an advantage. Increasing nutrient inputs to Lough Neagh during the 20th century and in particular the rapid increase of nutrient loading during the 1970s have enabled high, perpetual crops of *Oscillatoria* species to develop which has further led to severe light attenuation. The elimination of *Anabaena* and *Aphanizomenon* from the plankton may therefore be due to the low light environment made possible by the increased nutrient load acting upon an already naturally opaque water (see Chapter 4) rather than the removal of nitrogen limitation *per se* as paradoxically very low summer inorganic nitrogen levels occurred in the years when the plankton was dominated entirely by *Oscillatoria* species.

13.9. References

Ahlgren, G. (1978) Growth of *Oscillatoria agardhii* Gom. in chemostat culture. II Dependence of growth constants on temperature. Mitt. int. Verein. theor. angew. Limnol. 21: 88–102.

Allen, M.M. and Smith, A.J. (1969) Nitrogen chlorosis in blue-green algae. Arch. Mikrobiol. 69: 114–120.

Caldwell, D.E. (1979) Associations between photosynthetic and heterotrophic prokaryotes in plankton. In: J.M. Nichols (ed.) Third international symposium on photosynthetic prokaryotes, Oxford. p. A14 University, Liverpool, England.

Cohen, D. and Parnas, H. (1976) An optimal policy for the metabolism of storage materials in unicellular algae. J. theor. Biol. 56: 1–18.

Cook, J.R. (1963) Adaptation in growth and division in *Euglena* affected by energy supply. J. Protozool. 10: 436–444.

Droop, M.R. (1967) A procedure for routine purification of algal cultures with antibiotics. Br phycol. Bull. 3: 295–298.

Fitzsimons, A.G. and Smith, R.V. (1983) The isolation and growth of axenic cultures of planktonic blue-green algae. Br. phycol. J. 19: 157–162.

Fogg, G.E. (1975) Algal cultures and phytoplankton ecology. 2nd edition. University of Wisconsin Press, Madison and Milwaukee.

Foy, R.H. (1980) The influence of surface to volume ratio on the growth rates of planktonic blue-green algae. Br. phycol. J. 15: 279–289.

Foy, R.H. (1983) Interaction of temperature and light on the growth rates of two planktonic *Oscillatoria* species under a short photoperiod regime. Br. phycol. J. 18: 267–273.

Foy, R.H. and Gibson, C.E. (1982a) Photosynthetic characteristics of planktonic blue-green algae: the response of 20 strains grown under high and low light. Br. phycol. J. 17: 169–182.

Foy, R.H. and Gibson, C.E. (1982b) Photosynthetic characteristics of planktonic blue-green algae: Changes in the photosynthetic capacity and pigmentation of *Oscillatoria redekei* grown under high and low light. Br. phycol. J. 17: 183–193.

Foy, R.H., Gibson, C.E. and Smith, R.V. (1976) The influence of daylength, light intensity and temperature on the growth rates of blue-green algae. Br. phycol. J. 11: 151–163.

Foy, R.H. and Smith, R.V. (1980) The role of carbohydrate accumulation in the growth of planktonic *Oscillatoria* species. Br. phycol. J. 15: 139–150.

Gerloff, G.C., Fitzgerald, C.P. and Skoog, F. (1950) The isolation, purification and culture of blue-green algae. Am. J. Bot. 37: 216–218.
Ghosh, A.K. and Govindjee, (1966) Transfer of the excitation energy in *Anacystis nidulans* grown to obtain different pigment ratios. Biophys. J. 15: 139–150.
Gibson, C.E. (1975) A field and laboratory study of oxygen uptake by planktonic blue-green algae. J. Ecol. 63: 867–880.
Gibson, C.E. (1978a) Field and laboratory observations on the temporal and spatial variations of carbohydrate content in planktonic blue-green algae in Lough Neagh, Northern Ireland. J. Ecol. 66: 97–115.
Gibson, C.E. (1978b) Carbohydrate content as an ecological tool in the study of planktonic blue-green algae. Vehr. Internat. Verein. Limnol. 20: 630–635.
Gibson, C.E. and Fitzsimons, A.G. (1982) Periodicity and morphology of planktonic blue-green algae in an unstratified lake (Lough Neagh, Northern Ireland). Int. Rev. ges. Hydrobiol. 67: 459–476.
Gibson, C.E. and Foy, R.H. (1983) The photosynthesis and growth efficiency of a planktonic blue-green alga, *Oscillatoria redekei*. Br. phycol. J. 18: 39–45.
Gibson, C.E. and Stevens, R.J. (1979) Changes in phytoplankton physiology and morphology in response to dissolved nutrients in Lough Neagh, N. Ireland. Freshwater Biol. 9: 105–109.
Gorham, P.R., McLachan, J.L., Hammer, U.T. and Kim, W.K. (1964) Isolation and culture of toxic strains of *Anabaena flos-aquae* (Lyngb) de Breb. Vehr. int. Verein. theor. angew. Limnol. 15: 769–804.
Harris, G.P. (1978) Photosynthesis, productivity and growth: The physiological ecology of phytoplankton. Ergebnisse der Limnologie. 10: 1–171.
Heaney, S.I. and Jaworski, G.H.M. (1977) A simple sêparation technique for purifying micro algae. Br. phycol. J. 12: 171–174.
Jones, L.W. and Myers, J. (1965) Pigment variation in *Anacystis nidulans* induced by light of selected wavelengths. J. Phycol. 1: 6–13.
Kirk, J.T.O. (1975) A theoretical analysis of the contribution of algal cells to the attenuation of light within natural waters. II. Spherical cells. New Phytol. 75: 21–36.
Kohl, J-G. and Nicklish, A. (1981) Chromatic adaptation of the planktonic blue-green alga *Oscillatoria redekei* van Goor and its ecological significance. Int. Revue ges. Hydrobiol. 66: 83–94.
Lehmann, M. and Wober, G. (1976) Accumulation, mobilisation and turnover of glycogen in the blue-green bacterium *Anacystis nidulans*. Arch. Microbiol. 111: 93–97.
Lindstrom, K. and Rodhe, W. (1978) Selenium as a micronutrient for the dinoflagellate *Peridinium cinctum* fa *Westii*. Mitt. Intenst. Verein. Limnol. 21: 168–173.
Lorenzen, H. and Hesse, M. (1974) Synchronous cultures. In: W.D.P. Stewart (ed.) Algal physiology and biochemistry. Blackwell Scientific Publications, Oxford, pp. 894–908.
Meffert, M-E. (1972) Zur Isolierung von *Oscillatoria redekei* van Goor. Arch. Hyrobiol. Suppl. 41: 235–241.
Meffert, M-E. and Krambeck, H.J. (1977) Planktonic blue-green algae of the *Oscillatoria* redekei group. Arch. Hydrobiol. 79: 149–171.
Provasoli, L. Pintner, I.J. and Packer, L. (1951) Use of antibiotics in obtaining monoalgal bacteria free cultures. Proc. Am. Soc. Protozool. 2: 6.
Riddolls, A. (1974) Aspects of nitrogen fixation by blue-green algae in Lough Neagh. Ph. D. thesis, New University of Ulster.
Riddolls, A. (1985a) Aspects of nitrogen fixation in Lough Neagh. 1. Acetylene reduction and the frequency of *Aphanizomenon flox-aquae* heterocysts. Freshwat. Biol. 15: 289–298.
Riddolls, A. (1985b) Aspects of nitrogen fixation in Lough Neagh. 2. Competition between *Aphanizomenon flos-aquae*, *Oscillatoria redekei* and *Oscillatoria agardhii*. Freshwat. Biol. 15: 299–306.
Shuter, B. (1979) A model of physiological adaptation in unicellular algae. J. theor. Biol. 78: 519–552.

Smith, R.V. and Foy, R.H. (1974) Improved hydrogen ion buffering of media for the culture of freshwater algae. Br. phycol. J. 9: 239–245.

Stanier, R.Y. and Cohen-Bazire, G. (1977) Photorophic prokaryotes: the cyanobacteria. A. Rev. Microbiol. 31: 225–274.

Talling, J.F. (1957) The phytoplankton population as a compound photosynthetic system. New Phytol. 56: 133–149.

Tamiya, H., Hase, E., Shibata, K., Mituya, A., Iwamura, T., Nihei, T. and Sasa, T. (1953) Kinetics of growth of *Chorella*, with special reference to its dependence on quantity of available light and temperature. In: T.T. Burlew (ed.) Algal Culture from Laboratory to Pilot Plant. Carnegi Institution, Washington, pp. 204–234.

Werner, D. (1966) Die Kieselsaure im Stoffwechsel von *Cyctotella cryptica* Reimann, Lewin and Guillard. Arch. Mikrobiol. 55: 278–308.

Zevenboom, W., Van Der Does, J., Bruning, K. and Mur, L.R. (1981) A non-heterocystous mutant of *Aphanizomenon flos-aquae*, selected by competition in light limited continuous culture. FEMS Microbiol. Lett. 10: 11–16.

14. The seasonal succession of the zooplankton of Lough Neagh, 1968–1978

A.G. FITZSIMONS and T.E. ANDREW

14.1. Introduction

Since the first description of the zooplankton of Lough Neagh was published by Dakin and Latarche in 1913 the zooplankton of the largest lake in the British Isles has been neglected. A qualitative description was published by Graham (1970) and fleeting references are to be found in the reports of Macan and Lund (1954), Sutcliffe (1967) and Graham and Logan (1970). Since 1968 the zooplankton has been sampled on a regular quantitative basis and the results of this study are presented here. The results provide a basis for a qualitative comparison with earlier studies, particularly changes that have occurred over the last 60–70 years and the regular samples reveal detailed and subtle variations that occur on an annual basis over the last ten years.

Although many excellent and detailed studies of freshwater zooplankton have been reported (e.g. Edmondson, 1960; Hall, 1964; Wright, 1965; George and Edwards, 1974; Elliott, 1977; and de Bernardini and Soldavini, 1976) few include such extensive records of zooplankton populations. They also tend to concentrate on one element of the population, for example a single species or grouping such as the crustacean or rotiferan zooplankton. Herzig's (1979) report on Neusiedlersee is a welcome exception as it contains both long term descriptions and deals with all the species found. The work on Loch Leven, (Johnson and Walker 1974), provides a useful comparison.

Zooplankton, like other organisms, do not exist in isolation in the water-body and, to a greater or lesser extent, are controlled by, and control, many other elements of the system. Descriptions of the physical, chemical and algalogical characteristics are to be found in earlier chapters, but to interpret the zooplankton story it is convenient to reproduce in outline some of those results in this chapter. Attention may be drawn here to the size of the lake, 383 km^2, the mean depth 8.9 m and the extreme exposure to wind action. An outline map can be seen in Fig. 14.1, illustrating the relatively simple shape of the lake. The resultant effect of these features, as far as the zooplankton is concerned, is that the lake is extremely well mixed and has, except

R. B. Wood and R. V. Smith (eds.), Lough Neagh, 281–326.

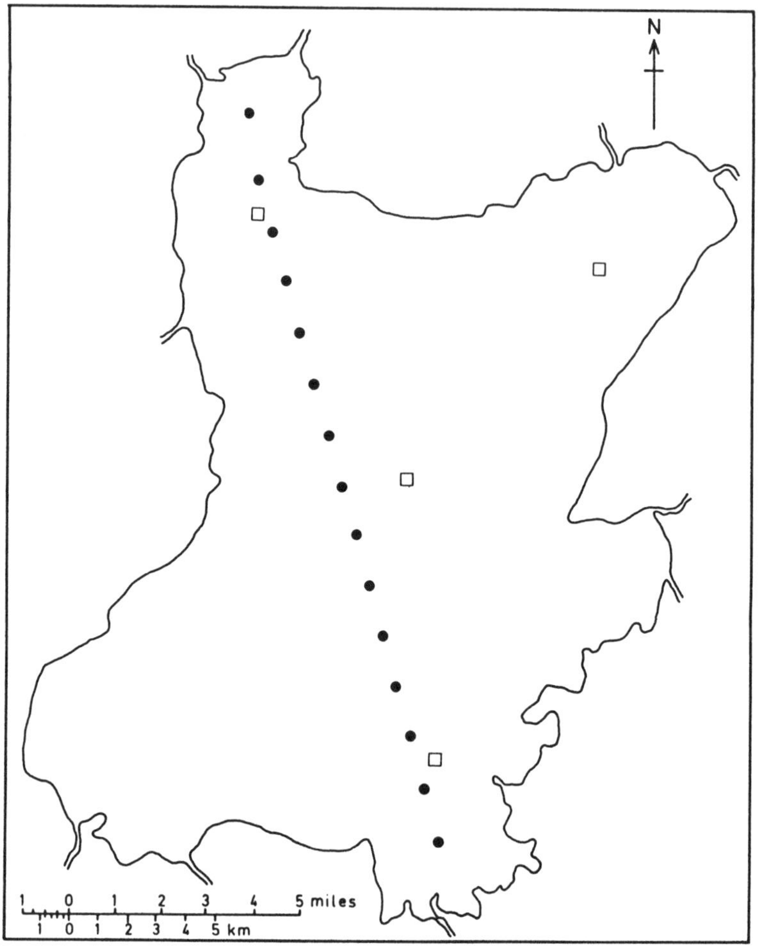

Figure 14.1. Map of Lough Neagh showing shape, sampling sites (□) and sampled transect (●).

in limited sheltered areas, a poorly developed littoral zone (see Chapter 19), an area often associated with the development of a particular type of zooplankton fauna (Herzig, 1979).

14.2. Sampling

Temperature was recorded with a platinum resistance thermometer by taking a depth profile at 1 m intervals. These have been integrated to give a mean column temperature.

Synoptic data covering chlorophyll *a* concentrations and phytoplankton counts are reproduced with the permission of R.V. Smith and C.E. Gibson (F.B.I.U., D.A.N.I.) (see also Chapter 10).

Cassie (1971) discusses logistic problems in relation to representative quantitative sampling but man-power and available time more often determine the frequency and number of stations sampled. Bottrell *et al.* (1976) point out that practical considerations determine the collection gear in different situations. Zooplankton samples were collected from four stations on the lough (Fig. 14.1) using vertical net hauls. Initially only fine nets (mesh aperture 0.05 mm) were used but in 1973 coarse nylon net (mesh aperture 0.19 mm) samples were taken in parallel. Although it has been demonstrated that more satisfactory quantitative techniques for collecting zooplankton are available (Bottrell *et al.*, 1976) it was judged expedient, because of the size of the lake, the meteorological conditions encountered and the available labour, to restrict sampling to this method. Samples were preserved in 5% formalin.

Samples were counted in the laboratory using 1 ml subsamples to estimate total numbers per vertical net haul following the recommendations of Lund *et al.* (1958). Species were identified wherever possible but copepod naupliar stages and copepodite stages were only identified as nauplii and copepodites respectively. Copepods were sexed and, since 1973, egg numbers recorded for all species. Subsampling variability for the normal four subsamples was found to be SE = 8.9% expressed as a percentage of the mean for the abundant species.

A number of sample testing programmes have been mounted to examine the variability inherent in the netting technique, due mainly to limited replicate sampling and restricted station sampling. Whether the results of these tests are applicable under all sampling conditions and at all times of the year is debatable as the execution of the tests themselves required reasonable working conditions. Groups of four replicate samples with fine nets showed an SE = 3.37% and coarse nets an SE = 5.18% for the species most commonly occurring in the net samples. For species common to both fine and coarse nets, in all except very high population densities, a students t test revealed no significant difference between means, $t_{27.04(7)}$ $p < 0.001$. When population densities become high, $>10^6\ m^{-2}$ individuals of one species, coarse nets catch more of the larger zooplankton. The effect of hauling speed on the zooplankton catch with fine nets was examined at two speeds, 0.67 and 0.41 $m\ s^{-1}$. Two-way analysis of variance reveals no significant difference between these hauling speeds. The results are presented in Table 14.1. The results show that both hauling speeds appear to be equally effective for catching the majority of species although comparisons of the means of the catch of individual species indicates some synergism between species. It may be safely assumed that nets do not catch all the animals in the water column sampled and it has been suggested that net loss factors be applied to correct for this (Bottrell *et al.*, 1976). These corrections are different for each species. Whether an individual is taken by the net or not is the resultant of two

Table 14.1. Anova table of the effect of hauling speed on species catch

Source	S.S.	d.f.	M.S.	F ratio
Groups	1.0371×10^{10}	19	5.458×10^{8}	76.37
Species	1.006×10^{10}	9	1.118×10^{9}	156.46***
Speeds	2.568×10^{7}	1	2.568×10^{7}	3.59 N.S.
Species × speeds	2.816×10^{8}	9	3.129×10^{7}	4.38
Within groups	8.576×10^{8}	120	7.147×10^{6}	
Total	1.123×10^{10}	139		
	$F_{0.001}$ (9,120) = 3.38		$F_{0.05}$ (1,120) = 3.92	

Table 14.2. Anova table for replicated samples along lake transect

Source	S.S.	d.f.	M.S.	F-ratio
Treatments	14852.93	14	1060.92	0.99
Error	64578.79	60	1076.31	
Total	79631.71	74		
			$F_{0.05}$ (14.60) = 1.84	

processes, assuming that it cannot pass through the mesh of the net. Firstly the ability of the species to escape from the path of the net, usually by activity, and secondly the back pressure caused by the resistance of the net to its passage through the water which effectively reduces the area of the column sampled. For the animals with the ability to swim out of the way, fast net hauls should be most effective but here the back pressure, proportional to haul speed, reduces the catch. Consequently it should not be surprising to find similar catches at different speeds as is indeed the case. For smaller animals that have limited escape potential only net pressure is effective in moderating the catch. For the speeds examined, the limits of the normal sampling procedure, there was no significant difference for the capture of the nauplii or rotifers.

There is evidence from other situations that the horizontal distribution of animals is likely to vary considerably (Herzig, 1979) but this may be more marked in lakes with highly developed littoral communities where, for example, species or stages of species might be found in reed beds. Examination of the samples from the three main sampling sites, north, mid and south as shown in Fig. 14.1 although revealing differences between stations on any individual sampling date, when each site is summed for the year, or ten year, period reveal totals for all three sites with differences of less than 1%. Net samples were also taken along a transect from NW to SE, the longest axis of the lake (see Fig. 14.1). Samples were dried (60 °C) and weighed and the weights transformed to unit volume figures, as the column depth along the transect varies considerably, 5.5 to 20 m. One-way analysis of variance showed no significant difference between samples (see Table 14.2).

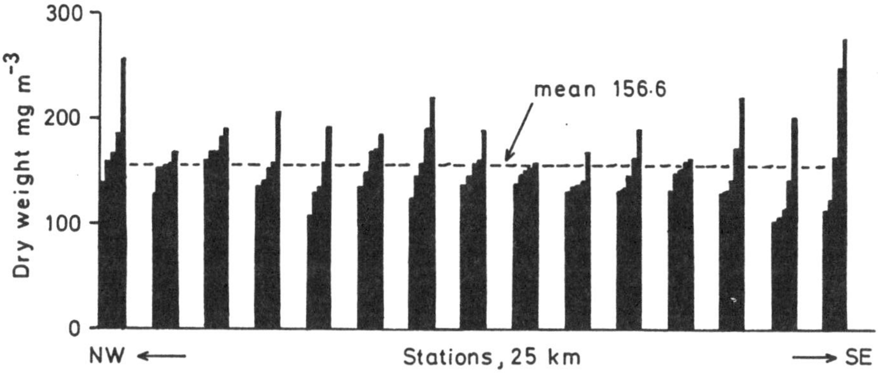

Figure 14.2. The distribution of zooplankton along the NW–SE transect shown in Fig. 14.1 of Lough Neagh (dry weight mg m^{-3}).

The distribution of samples along the transect is presented in Fig. 14.2 together with the sample values in histogram form. A similar exercise has been carried out in a limited area of the lake, Toome Bay in the NW corner, the results of which reinforce those described above. The selection of the three sampling stations for regular sampling as shown in Fig. 14.1 appears to be vindicated by the above results.

Diurnal variation in the vertical distribution of the zooplankton has been examined on several occasions and the results of one of these studies is examined below. The results are those of S. Pillar (unpubl.). Zooplankton were collected from a fixed sample site of 8 m depth in Toome Bay. Profiles were taken hourly for 24 h at depth intervals of 1 m from 0.5 m to 7.5 m using a Patalas sampler (Patalas, 1954). Counts were made of adult, copepodid and naupliar stages of the copepods. Oxygen and temperature profiles were taken each hour and phytoplankton was taken for analysis every four hours at 2 m depth intervals. Incident solar radiation and light attenuation (for wave lengths 460, 545 and 625 nm) were measured on five occasions during daylight. These light measurements were correlated with data from a spectroradiometer to give a detailed analysis of the light climate (Jewson, 1977).

At the time of this study (26 September 1975) the water column was well mixed with no detectable temperature (12.3 °C) or oxygen (100% saturation) gradients. The phytoplankton crop consisted of the blue-green alga *Anabaena flos-aquae* (Breb.), the diatoms *Stephanodiscus hantzii* (Grun.) and *S. astraea* (Ehr. Grun.) and flagellates of the *Cryptomonas* sp.. The results obtained for the calanoid copepod, *Eudiaptomus gracilis*, are shown in Fig. 14.3 and for the adult plus copepodite stages a clear pattern emerges. The migration pattern shows a congregation at the surface at night and a descent, during the day to below the 1.2×10^3 cal cm^{-2} min^{-1} irradiance contour. *Cyclops abyssorum* (not drawn) appears to exhibit a similar behaviour although the

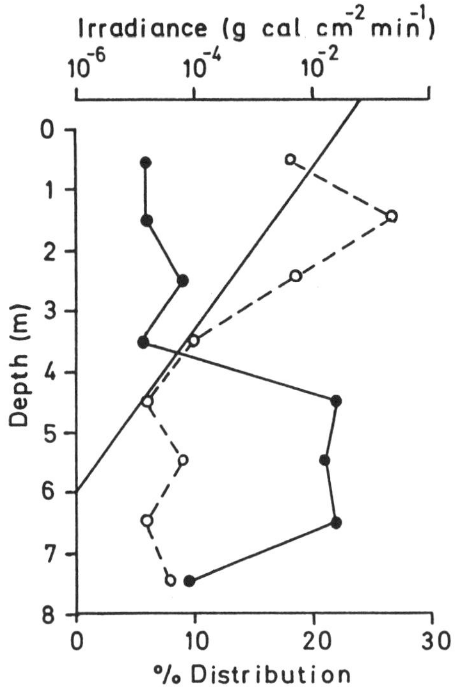

Figure 14.3. Vertical distribution of adult and copepodite *Eudiaptomus gracilis* in the water column at 12.00 h (●—●) and 23.00 h (○- -○), and the underwater irradiance at 12.00 h.

daytime distribution is more even through the water column. A problem encountered with the *C. abyssorum* samples is that the integration of the column samples produces very variable results for the sampling times which may be due to patchiness of the population as recorded by other workers (e.g. George and Edwards, 1974) or the migration of the population below the sampled depth of 7.5 m. Naupliar migration is also difficult to describe from the results obtained although the overall impression is of lesser migration with the nauplii tending to remain in the top two metres.

The phenomenon of vertical migration has been extensively reviewed in the literature (e.g. Cushing, 1951; McClaren, 1963; Longhurst, 1976) and the suggested reasons for its occurrence are numerous. An interesting feature of calanoid migrations in lakes with well-mixed water columns is that little advantage is to be gained from movement in terms of temperature, oxygen or food availability (e.g. Southern and Gardiner, 1932; Hart and Allanson, 1976; Zaret and Suffern, 1976). Zaret and Suffern (1976) suggest that predator avoidance might be an important operative and this might apply to L. Neagh. Fish, feeding by sight, will predate during the day whereas *Mysis relicta* seems to be active in the water column during the night.

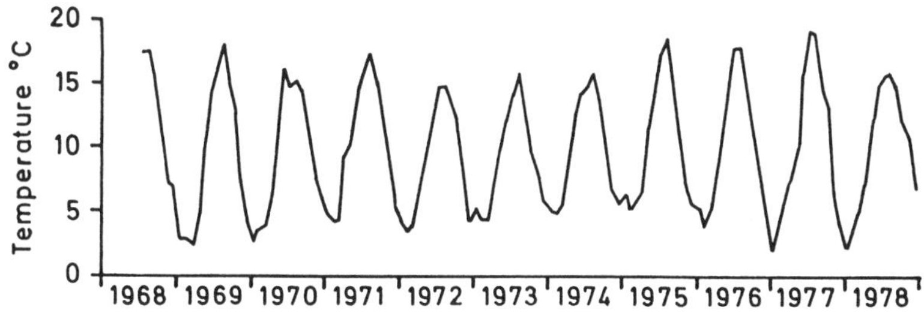

Figure 14.4. Seasonal variation in the mean column water temperature, °C.

14.3. The environment of the zooplankton

Although primarily concerned with the zooplankton, other features of the environment influence the succession, distribution and abundance of zooplanktonic species and some brief comments will be made on these features here.

14.3.1. *Physical conditions*

Lough Neagh is extremely exposed to winds blowing from any direction but has its longest fetch along its NW-SE axis. Complete mixing of the water column occurs at most times of the year. Even transitory thermal stratification has only rarely been recorded during the period of this study (see Chapter 9). Fig. 14.4 shows the seasonal fluctuation of temperature and of interest is that during 1972, 1973 and 1974 the extremes of temperature recorded were less than other years of the study. This has important implications for zooplankton survival and growth rates. The suspended solid content of the lake, potentially zooplankton food, is largely related to seasonal fluctuations in algal populations, the contribution of allochthonous material being small except when material is resuspended from the sediments (Chapter 7A and Flower, 1980).

14.3.2. *Chemical conditions*

Lough Neagh is a highly eutrophic lake with high concentrations of algal nutrients (see Chapters 8 and 9). Oxygen saturations rarely fall below 60% although the potential for lower oxygen saturations than this to occur undoubtedly exist (Wood and Gibson, 1973, and Chapter 9 section 7).

14.3.3. *Algal populations*

The reader is referred to Chapter 10 for a full description. Fluctuations in algal biomass, as chlorophyll *a* concentration, are shown in Fig. 14.5a, the maximum crops recorded may be very high, 100 mg m^{-3} chl. *a* in 1977. The maxima occur early in the year but during 1973, 1974 and 1975 the standing crop rarely exceeded 50 mg m^{-3} chl.*a*. The seasonal succession of the principal algal species is shown in Fig. 14.5b, where it is evident that considerable variation in amount and occurrence of different species occurs from year to year.

14.3.4. *Fish populations*

Chapter 18 lists the species of fish found in Lough Neagh. In Table 14.3 are listed the 8 species for which gut contents have been analysed. The larvae of these and other species are almost exclusively plankton feeders, gut content examination of larval pollan and perch revealing that rotifers, naupliar and copepodite copepods and daphnids are common items in the diet. This feeding pattern represents a significant predation pressure on the zooplankton particularly in the spring and summer months. The fraction of the zooplankton biomass taken by fish is difficult to estimate as there is little information yet available about the fish population size in the lake A secondary effect of fish feeding is caused by fish predating *Mysis relicta* which is itself, a significant feeder upon zooplankton (Goldman *et al.*, 1979).

14.3.5. *Benthic fauna*

An extensive benthic fauna, both littoral and sub-littoral exists in the lake and is described elsewhere in this volume (see Chapters 15 and 16). However the lake is noted for the presence of *Mysis relicta* L. (Dakin and Latarche 1913, Holmquist 1959) and this species migrates into the plankton particularly at night.

14.4. Zooplankton composition

14.4.1. *An historical perspective*

The zooplankton species identified from samples taken during this study are listed in Table 14.4 and compared with those recorded by Dakin and Latarche (1913). Not all the species listed are commonly found in the samples and *Mysis* is included as a frequent visitor to the plankton although more commonly considered a benthic type.

Many changes in nomenclature of rotifers have taken place since the study of Dakin and Latarche (1913) and we have ued the nomenclature of Ruttner-

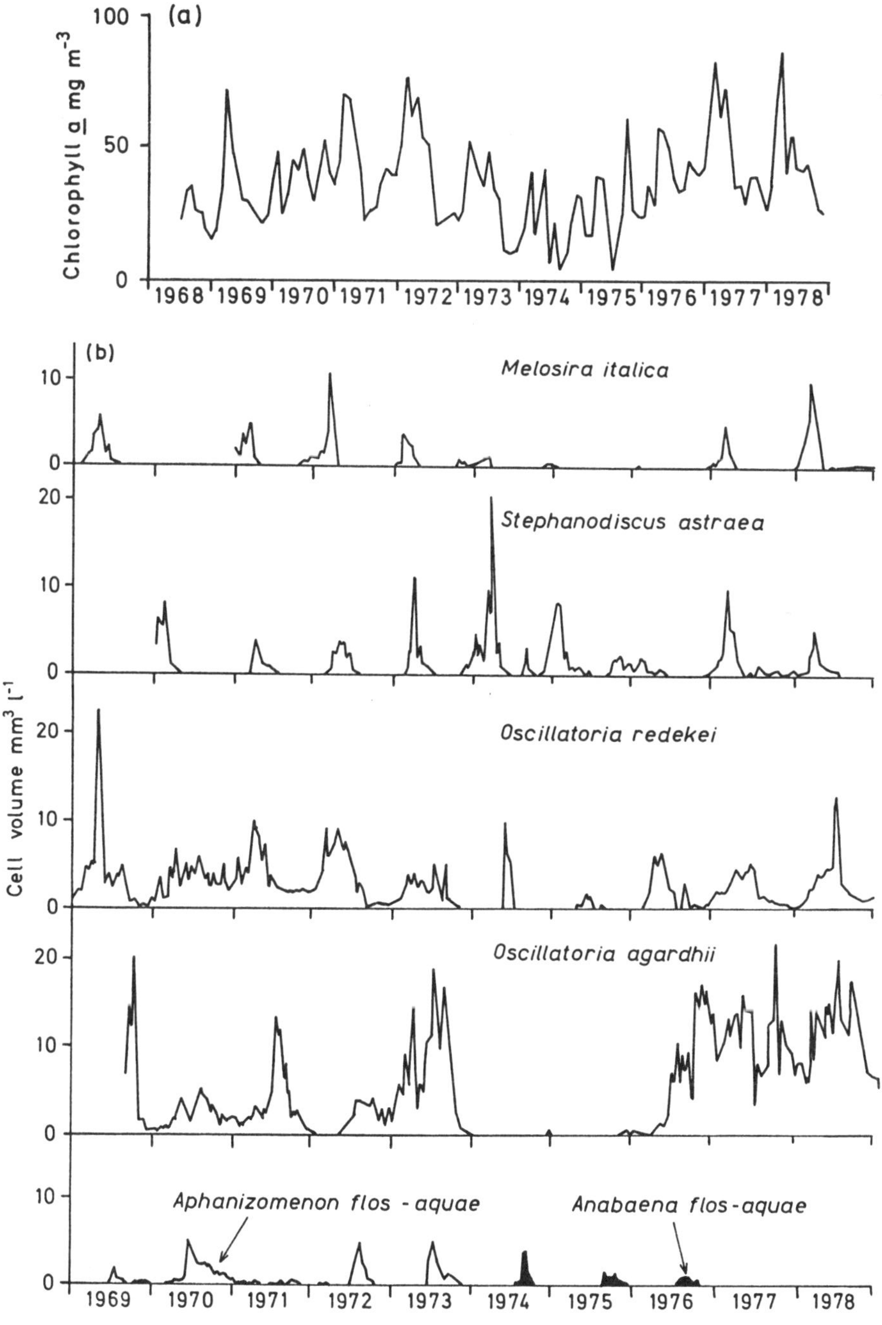

Figure 14.5. (a) Seasonal variation in the mean column chlorophyll *a* values, mg m^{-3}; (b) Seasonal occurrence of the principal algal species shown as cell volumes, $mm^3 l^{-1}$.

Table 14.3. The fish species found in Lough Neagh since 1968 for which diets recorded from gut analyses are available

Perca fluviatilis L.	F M C G Z
Rutilus rutilus L.	C G V Z I D
Anguilla anguilla L.	M C G A
Abramis brama L.	M C G A Z
Coregonus pollan (Thom.)	C Z G
Gobio gobio L.	M C G Z
Salmo trutta L.	G C Z
Esox lucius L.	M C G Z F

F – fish; M – *Mysis relicta*; C – chironomid larvae; G – *Gammarus* sp.; A – *Asellus* sp.; Z – zooplankton; I – various insects; V – vegetation D – detritus.

Table 14.4. The zooplankton species of Lough Neagh

	1910	1968–1978
Mysis relicta (Loven)	√	√
Eudiaptomus gracilis (Sars)	√	√
Cyclops abyssorum (Sars)	√	√
Cyclops vicinus (Ulianine)	×	√
Megacyclops viridis (Francois)	×	√
Bosmina longirostris (Sars)	√	×
Bosmina coregoni (Müller)	√	×
Daphnia hyalina (Leydig)	√	√
Daphnia longispina (Müller)	×	√
Daphnia pulex (de Geer)	×	√
Bythotrephes longimanus (Leydig)	×	√
Leptodora kindti (Focke)	×	√
Polyphemus pediculus (Linné)	×	√
Keratella cochlearis (Gosse)	√	√
Keratella quadrata (Müller)	√	√
Kellicottia longispina (Kellicott)	√	√
Filinia longiseta (Ehrenberg)	√	√
Polyarthra vulgaris (Carlin)	√	√
Cephalodella gibba (Müller)	×	√
Pompholyx sulcata (Hudson)	×	√
Notholca acuminata (Ehrenberg)	×	√

Kolisko (1974). Some doubt exists as to whether *Notholca striata* (Müller) was found in our 1968/1969 samples but re-examination of samples leads us to believe these to be a slightly striated form of *N. acuminata*.

Several major changes have occurred in the zooplankton since 1910. The most dramatic of these has been the disappearance of the two *Bosmina* spp., *B. longirostris* and *B. coregoni*, from the position of being the numerically most important element of the crustacean zooplankton. The position of dominant cladoceran was taken over by *Daphnia hyalina*. Since 1973 *Daphnia longispina* has been found in large numbers. The carnivorous cladocerans,

Leptodora kindti, *Bythotrephes longimanus* and *Polyphemus pediculus* make consistent appearances but were not reported as present in 1910. Their non-appearance in the earlier study may well be related to the sampling frequency of Dakin and Latarche rather than their absence at that time. The appearance of *Daphnia pulex* is numerically unimportant at present.

The copepods, *Eudiaptomus gracilis* and *Cyclops abyssorum* were both found in the earlier study but changes have occurred in their relative importance. *C. abyssorum* is currently the numerically most important member of the macro-zooplankton while *E. gracilis*, although consistently present, occurs as a smaller proportion. This is the reverse of the 1910 situation. Two other cyclopoid copepods have been found, *Cyclops vicinus* and *Megacyclops viridis* which occurred in very small numbers until 1974 when *C. vicinus* became established as a regular, significant fraction of the zooplankton. This coincides with the presence of *D. longispina*.

The rotifers formed an important constituent of the zooplankton in 1910 and retain that rôle now. They occur, unlike the situation in many other temperate lakes, in high numbers at all seasons of the year. *Keratella cochlearis* remains the most abundant species while *Keratella quadrata*, *Kellicottia longispina*, *Filinia longiseta* and *Polyarthra vulgaris* occur in about the same proportions as the earlier study. Three other species are now found, *Cephalodella gibba*, *Pompholyx sulcata* and *Notholca acuminata* but these species are not present every year. Nevertheless the continually large populations of rotifers always present in the zooplankton are a remarkable feature of Lough Neagh.

Major changes in the lake ecology must have occurred to cause the disapperance of the *Bosmina* spp. and the subsequent readustment of the relative importance of other cladoceran and copepod species. Until it is known when this change took place and what events accompanied the change, the reasons for it will remain speculation. These major changes may be attributed either to an increase in the eutrophic status of the lake, as chronicled by Battarbee (1978) and in Chapter 7B in relation to diatom crops, which has caused consequent changes in the composition and production of the phytoplankton – the prime source of food for most of the zooplankton, or to changes in the fish population composition and structure. The reasons for changes in eutrophic status are correlated with the introduction of sewage treatment, the more recent use of detergents and the increased use of agricultural fertilisers, all occurring during the last 100 years (see Chapter 8). The fish populations can be affected by eutrophic state, by fishing pressure, introductions (e.g. Roach, see Chapter 18) or by the quantity and composition of the zooplankton itself. *Mysis relicta* is another unknown in this relationship, an important predator of the zooplankton that can cause gross change in, particularly, daphnid abundance (Goldman *et al.*, 1979). *Mysis* is also an important element of the diet of some fish species and it is only possible to guess at what changes have taken place in its population structure and abundance. It has been suggested that the two *Bosmina* spp. may be used as an indication of increased eutrophic state if there is a shift from *B. coregoni* to *B. longirostris* (Edmondson *et al.*, 1956; Goulden, 1964). An alternative theory is that the presence or absence of *Bosmina* spp. is dependent upon the predation pressure on other cladoceran species (Brooks, 1968, 1969). Which of these two possibilities operates here is unclear except that the palaeolimnological evidence strongly points to increased eutrophication since 1910.

Another important change has been the establishment of *C. abyssorum* as the dominant crustacean in the zooplankton. This is an unusual but not unique phenomenon as it has been reported from Lake Hemjarden (Pejler, 1964) and Loch Leven (Johnson and Walker, 1974). *C. abyssorum* appears to be an algal feeder as a juvenile and omnivorous as an adult and consequently, in a lake with high standing crops of algae and zooplankton, is not short of suitable food items. The dominance of *C. abyssorum* does not preclude the occurrence of large crops of daphnids or rotifiers, in fact these may be essential for its occurrence. Pejler (1964) records *C. abyssorum* occurring together wlth the rotifer *K. cochlearis* in large numbers, a situation that exists in Lough Neagh and Loch Leven (Johnson and Walker, 1974). The success of *C. abyssorum* could be related to predation pressure; because of its large size and activity it is less predated by fish and invertebrate predators than the daphnid populations which are selectively predated by adult perch (J. Rogers, pers. comm.), by *M. relicta*, by the carnivorous Cladocera and *C. abyssorum* itself. The appearance of a second predatory copepod, *C. vicinus*, is also unusual (Smyly, 1968) but may be explained by the concurrent appearance of large numbers of potential prey in the form of *D. longispina*.

The rotiferan fauna shows relatively little long term change except that three more species have been recorded. The three additional species recorded do not appear every year and there may not, in reality, be very much difference between the 1910 situation and the present. That rotifers are found all year round and at high population densities is a feature of eutrophic situations. It has been suggested that extreme biotopes show a low species diversity but achieve high population densities (Thienemann, 1920; Saunders, 1969; Ruttner-Kolisko, 1974; and Herzig, 1979).

14.4.2. *Seasonal succession of crustacean zooplankton*

The visual presentation of such a long run of zooplankton population data causes diagrammatic problems so, except where otherwise stated, data are presented on logarithmic scales and, to reduce the number of points, shown as monthly mean numbers per metre-square column. This device results in a reduction of the, very variable, amplitude but the authors are convinced that these values adequately represent the situation in the lake. The practice of using monthly means is also used in the Tables in this section. Where numbers are reported in the text they are the actual values for a particular sampling date unless otherwise stated. A second device has been used to present a simplified annual picture of seasonality in figures of each species. The values here are the average of the monthly means for each species over the sampling period for the years in which these species occur in samples. Therefore they represent not any particular year but are presented to show features of the seasonality and how any one real year may depart from this pattern. A further point is that as the sampling programme progressed a more detailed analysis of the samples was undertaken and this detail has been included when it became available.

14.4.2.1. *Cyclops abyssorum*

This species is the most abundant member of the crustacean zooplankton of Lough Neagh. The seasonal changes in numbers of adults, copepodites, nauplii and eggs are presented in Fig. 14.6. *C. abyssorum* is present throughout the year and, although maximum and minimum population densities vary from year to year, the seasonal pattern is similar for each year, see Fig. 14.7. The population grows rapidly in the early part of the year from the overwintering population and the high egg production precedes the peak nauplii, copepodite and adult production which occurs four to six weeks later. The maximum population density usually occurs in May or June. The high biomass is generated by the larger stages, adults and copepodites, the copepodites reaching a maximum up to a month before the adults (this is not revealed in Fig. 14.7 because of the smoothing effect of using monthly means). The population then decreases throughout the rest of the year although a slight second peak may be seen in late summer or early autumn. Naupliar populations exhibit the same pattern as the older stages but slightly precede them chronologically. Egg production reaches a peak in early spring and may be high at other times of the year, in late summer there is a peak and winter levels can be quite high.

The actual population densities vary from year to year. The maximum population size, $4.2\times10^6\ m^{-2}$ adults and copepodites, was recorded in May 1975. This spring peak normally reaches 1 to $2\times10^6\ m^{-2}$ although from 1971 to 1973 only half this total was achieved. Winter minima are found in the order of 10^4 to $10^5\ m^{-2}$ although these values were higher from 1974 to 1976. Minimum populations are normally recorded in the early part of the year, January to March and often coincide with the lowest temperatures for the water column.

Table 14.5 shows the extent to which population size can vary from year to year in the *C. abyssorum* populations. High winter populations tend to be followed by high summer population densities which suggests that the population size achieved may be a partial reflection of the size of the initial or winter 'inoculum'. Several factors control population size and timing but it is often difficult to determine which, if any, factor is important at any one time. *C. abyssorum* has been demonstrated to be omnivorous (Johnson and Walker, 1974) as are other similar cyclopoid species (McQueen, 1969) and

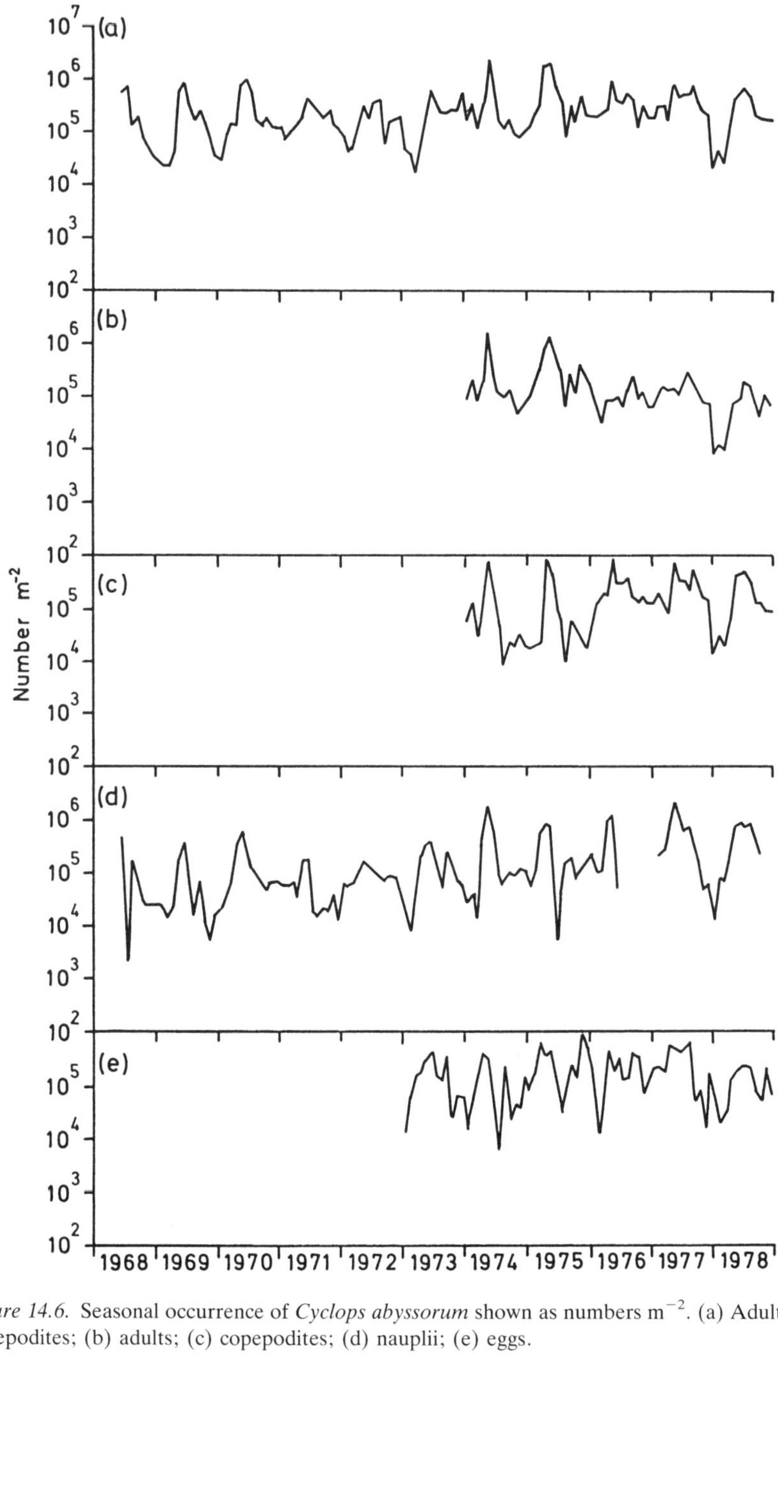

Figure 14.6. Seasonal occurrence of *Cyclops abyssorum* shown as numbers m^{-2}. (a) Adults plus copepodites; (b) adults; (c) copepodites; (d) nauplii; (e) eggs.

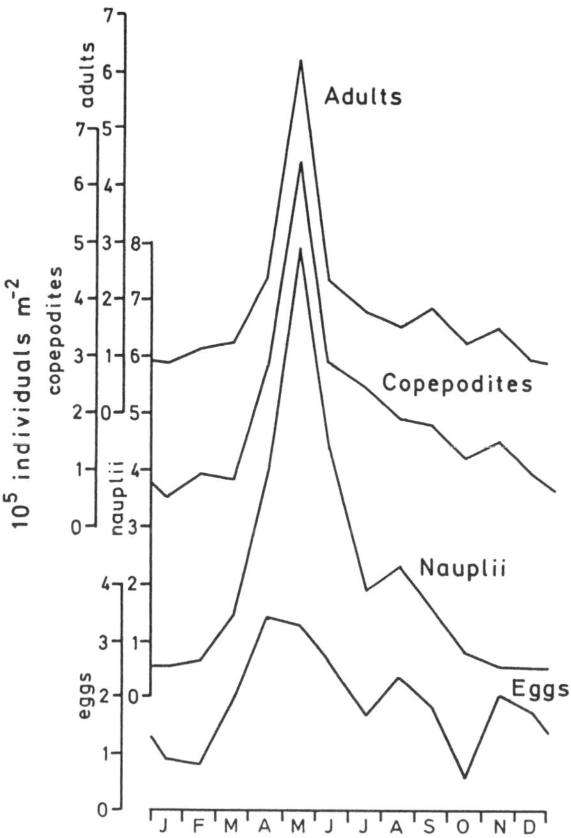

Figure 14.7. The mean annual population structure of *Cyclops abyssorum*.

this has been found to be the case in Lough Neagh (Jewson *et al.*, 1981). It can be deduced that food shortage in Lough Neagh is not necessarily limiting especially during the winter period as algal standing crops and suitable zooplankton food items are available all year round. The fecundity of adult *C. abyssorum* may, however, be controlled by food quality and as this improves with the early spring growth of algal, particularly diatom, populations, egg production reaches its peak in early spring. Population size achievement is the resultant of population growth limited by food availability and losses of individuals from the population. The losses can occur in a number of ways, for example by emigration through the natural overflow from the lake or by natural death. Diapause, well known in copepods (George, 1973), will be balanced by the emergence of already diapausing individuals but predation is an important loss to the populations. McQueen (1969) estimated that for *Cyclops bicuspidatus thomasi*, 31% of the nauplii were canabalised by their adults and the low winter levels of cyclopoid nauplii suggests that, despite fairly high egg production, a similar phenomenon occurs in Lough Neagh.

Table 14.5. Minimum and maximum population sizes of adult plus copepodite *C. abyssorum* presented as monthly mean numbers ($\times 10^3$ m^{-2})

	1968	1969	1970	1971	1972	1973	1974	1975	1976	1977	1978
Minimum	–	24.82	28.79	76.21	41.03	18.52	121.52	120.98	198.61	192.67	25.90
(month)	–	(Feb.)	(Feb.)	(Feb.)	(Mar.)	(Mar.)	(Mar.)	(Jan.)	(Feb.)	(Jan.)	(Jan.)
maximum	640.22	728.58	909.19	408.35	395.55	657.01	2276.93	1812.23	951.19	821.69	691.37
(month)	(June)	(June)	(June)	(Aug.)	(June)	(May)	(May)	(May)	(May)	(May)	(July)

C. abyssorum adults are consistently found as part of the diet of the plankton-feeding fish although not in proportion to other zooplankton species that are present. Temperature acts upon population development in two ways. Firstly temperature controls the rate at which individual stages develop (Bottrell *et al.*, 1976) and consequently controls how far a population will develop in a specific time. Secondly, in the colder winters the minimum population size is restricted further which results in a delayed population development and peak and lower population maximum. For example in 1969 the winter minimum temperature was 2 °C, population size $24.8 \times 10^3 \, m^{-2}$ and a summer peak of $728.6 \times 10^3 \, m^{-2}$. This compares with 1975, 6 °C, $121 \times 10^3 \, m^{-2}$ and $1812.2 \times 10^3 \, m^{-2}$ respectively. It is not the intention here to provide a simple explanation to cover all years but rather to point out the influence that specific factors might exert while accepting that these are only some factors from a complex situation.

14.4.2.2. *Cyclops vicinus*

This second species of cyclopoid copepod has been found in the zooplankton from 1974. The seasonal changes in numbers and population structure are presented in Fig. 14.8. Naupliar numbers are not shown as they are absorbed in the *C. abyssorum* counts and, except for 1976 and 1977, the adults and copepodites were lumped together. Fig. 14.9 shows the annual seasonal population structure. The maximum population density occurs in the spring, usually April or May and the population falls during the rest of the year. The peak in population numbers increased from 1974 to 1976 from 22.97 to $340.9 \times 10^3 \, m^{-2}$ but fell off in 1977, $125.9 \times 10^3 \, m^{-2}$ and were lower in 1978. Maximum egg production $1.4 \times 10^6 \, m^{-2}$ in April 1976, is coincident with the maximum population size and numbers can be high at other times of the year. This very high egg production does not manifest itself as large population increases and this must be ascribed to high mortality of the juvenile and naupliar stages or, which is more likely, to the juvenile stages being found living near or in the mud surface. Smyly (1968) reported most of the copepodite stages of *C. vicinus* as being bottom dwellers and these would be missed by the sampling technique in this study.

Numerically the appearance of *C. vicinus* is not as important as *C. abyssorum* but several interesting features are related to its occurrence. *C. vicinus* appears in the same years as *Daphnia longispina* apart from the first year of appearance of the cladoceran in 1973. The occurrence of these species may be related in some way as the years of maximum population densities of these two species coincide. A second point is that for the population of *C. vicinus* to survive with the more successful *C. abyssorum*, the niche width must be sufficiently broad (probably in food terms but possibly in a temperature related control as it occurs in the zooplankton in the years of low summer and high winter temperatures) for the two species to coexist. The coexistence of two open water forms of *Cyclops* spp. is not unknown but of

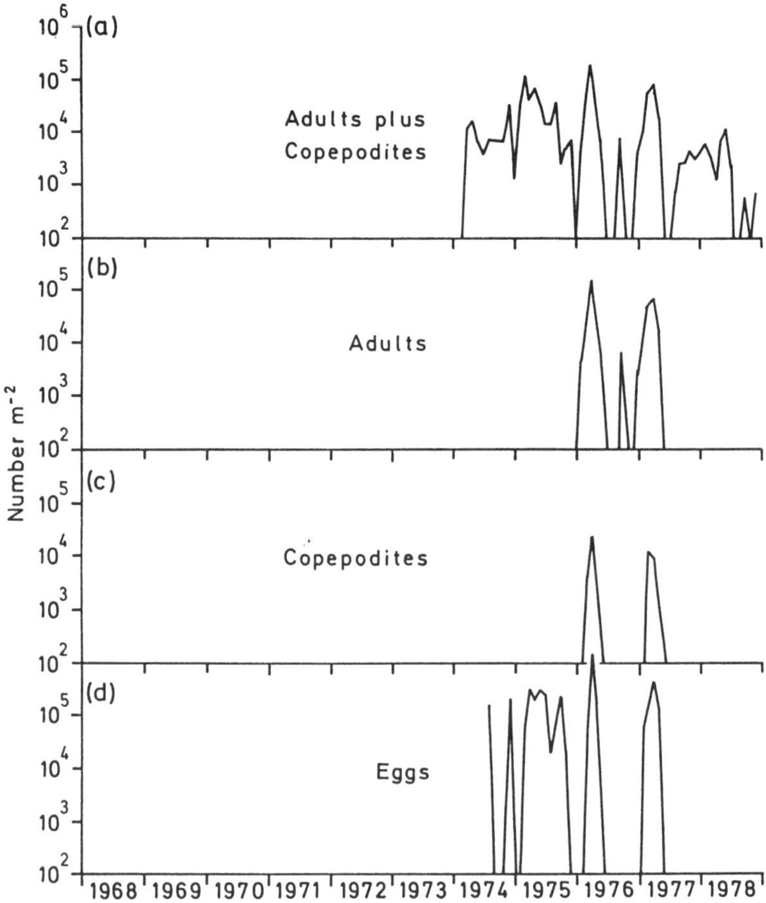

Figure 14.8. Seasonal occurrence of *Cyclops vicinus* shown as numbers m^{-2}. (a) Adults plus copepodites; (b) adults; (c) copepodites; (d) eggs.

18 lakes studied by Smyly (1968) in the Lake District only three shared this characteristic.

14.4.2.3. *Eudiaptomus gracilis*

This calanoid copepod, like *C. abyssorum*, is always present in the zooplankton throughout the year. The seasonal changes in numbers and population structure may be seen in Fig. 14.10 and the simplified structure for one year in Fig. 14.11. Naupliar numbers were not reported for 1968 as they were counted with the cyclopoid nauplii. As can be seen in Fig. 14.11 maximum egg production occurs very early in the year and is sustained close to this level through until the end of September. The peak of nauplii tends to occur in April and of the copepodites later, in May or July. Maximum adult and

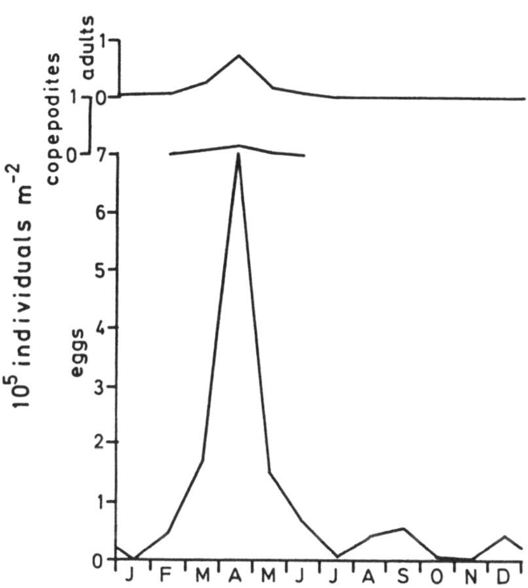

Figure 14.9. The mean annual population structure of *Cyclops vicinus*.

overall adult and copepodite populations are normally found in the summer, June to August. This seasonal pattern can be extremely variable from year to year and maximum population densities were recorded in November or December in 1972, 1973 and 1976. After the peak in 1972, the overwintering population remained low and the population increase did not commence until June/July and eventually reached its peak in December. Naupliar numbers had remained low from the preceding summer and again this suggests that some form of density dependence may operate here and cause delay as well as size regulation. In 1973 a summer peak can be seen but the winter levels exceeded this value. During 1976 the population remained fairly steady throughout the year and exhibited no marked seasonality.

The population densities varied considerably from year to year not only in timing but also quantitatively. Normally exceeding 10^5 m^{-2}, low values are found in 1971, most of 1972, 1973 and 1978. The maximum recorded population size occurred in July 1975, 819.3×10^3 m^{-2} adults and copepodites whilst the lowest population size was recorded for March 1978. Table 14.6 shows the variation in population density that can occur in different years and this may amount to a difference of an order of magnitude.

E. gracilis, a filter-feeding calanoid copepod, is normally a species described with a summer maximum but, as can be seen from the results presented here, this pattern can vary considerably. As with *C. abyssorum*, food is unlikely to be limiting on many occasions for *E. gracilis* and considerations of other environmental controls must be invoked to explain population fluc-

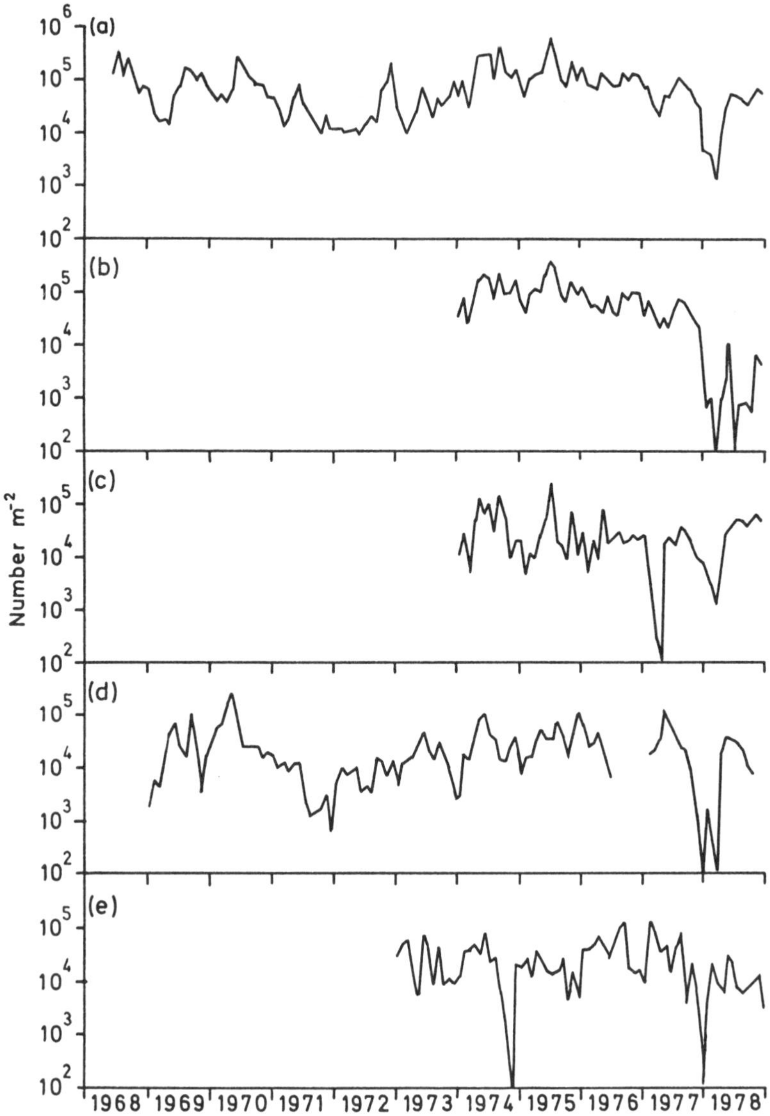

Figure 14.10. Seasonal occurrence of *Eudiaptomus gracilis* shown as numbers m^{-2}. (a) Adults plus copepodites; (b) adults; (c) copepodites; (d) nauplii; (e) eggs.

tuation and timing. Competition for food with daphnid, rotiferan and the juvenile stages of cyclopoid populations may take place but the peak of the population usually occurs when other populations are not so well represented. Predation will operate in the same way as *C. abyssorum* and *C. vicinus*, juvenile stages being susceptible to cyclopoid as well as fish predation. The later timing of the population peak, especially in the years with

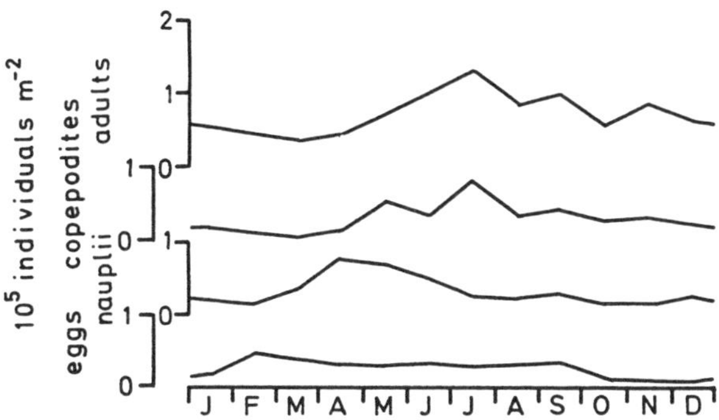

Figure 14.11. The mean annual population structure of *Eudiaptomus gracilis*.

very late peaks in September, November and December, suggests that the available degree-days for development, in this species with a long developmental period, may be an important consideration. The poor years of 1971 to 1974 coincided with the restricted amplitude in annual temperature cycle which reinforces this point.

14.4.2.4. *Daphnia hyalina*

Two separate varieties of different body shape were recognised in population samples but these have been considered together in this section. Seasonal changes in population numbers, structure and egg numbers are presented in Figure 14.12(a) the shaded portion representing the primiparous instars (instars 1 to 4) where distinguished. Figure 14.13(a) shows the generalised annual population structure. The population appears in spring, has a period of rapid growth culminating in a mid-summer maximum and disappears from the plankton during the winter.

The initial population input is from developing ephippial eggs and these reside in the benthic and littoral regions. These ephippial eggs are produced in an extremely synchronised fashion just before the disappearance of the autumn populations, except during 1975 when male daphnids were found in samples from mid-June until the end of July and females bearing ephippia were found in mid-August suggesting that the ephippial production phase occurred over this period. It is unclear what specific conditions generate this resting egg production. This period of intense ephippial egg production lasts for a short time, two to three days, and has not always been sampled. On the occasions when they have been sampled many of the ephippia are found floating on the surface of the water and from there are driven by the prevailing wind towards the margins of the lake. Although some of these ephippia sink and are found distributed in the benthic areas of the lake most of these resting eggs are concentrated in marginal areas. Preliminary results of samples taken from the margins show very high concentrations of ephippia,

Table 14.6. Minimum and maximum population sizes of adult plus copepodite *E. gracilis* presented as monthly mean numbers ($\times 10^3\ m^{-2}$)

	1968	1969	1970	1971	1972	1973	1974	1975	1976	1977	1978
Minimum	–	13.26	38.44	12.59	9.62	10.01	31.29	46.35	58.98	20.29	1.34
(month)	–	(May)	(Apr.)	(Mar.)	(Mar.)	(Mar.)	(Mar.)	(Feb.)	(Apr.)	(Apr.)	(Mar.)
maximum	319.71	145.67	256.44	76.14	198.39	86.04	329.28	593.80	121.51	104.28	49.76
(month)	(July)	(Aug.)	(June)	(Dec.)	(Dec.)	(Dec.)	(Sep.)	(July)	(Nov.)	(Aug.)	(June)

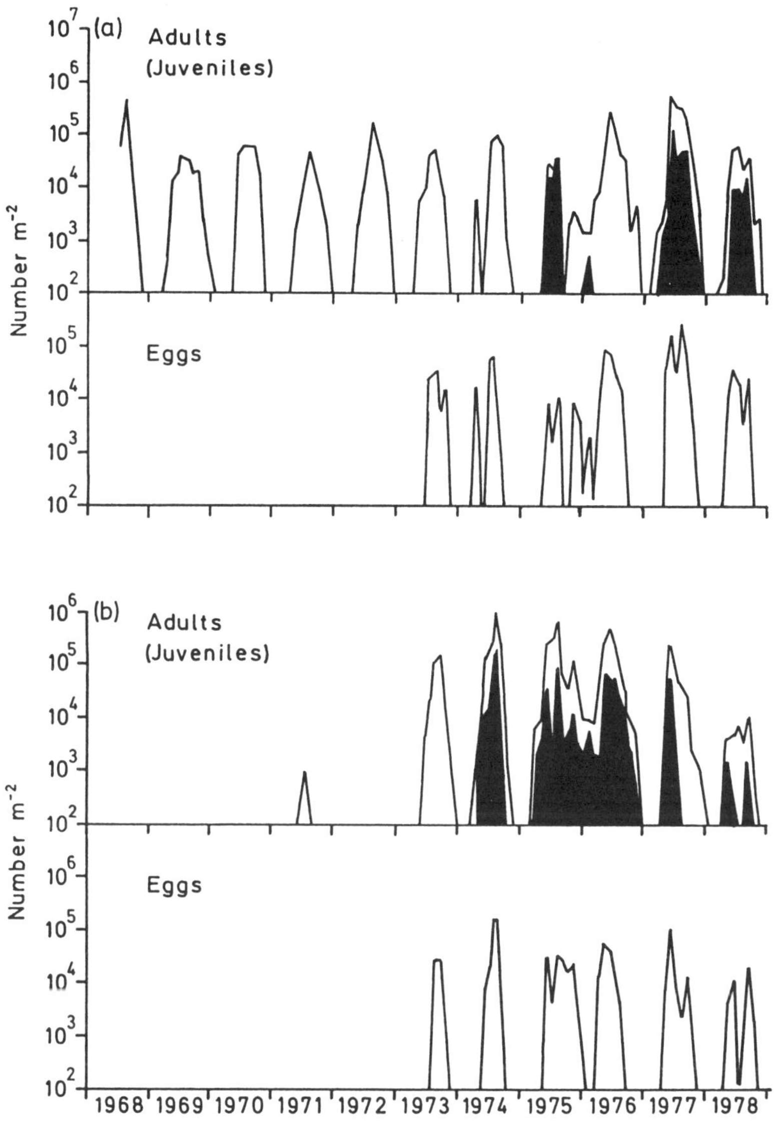

Figure 14.12. (a) Seasonal occurrence of *Daphnia hyalina* as numbers m^{-2}; (b) Seasonal occurrence of *Daphnia longispina* as numbers m^{-2}. Open figures = adults; shaded = juveniles.

10^6 to 10^7 m^{-2}, along some parts of the shore. Observations also show very high local concentrations of daphnids in sublittoral regions at the beginning of the seasonal population growth.

D. hyalina has been found in each year of the study but variations occur in the timing of the onset and disappearance of the populations and also the

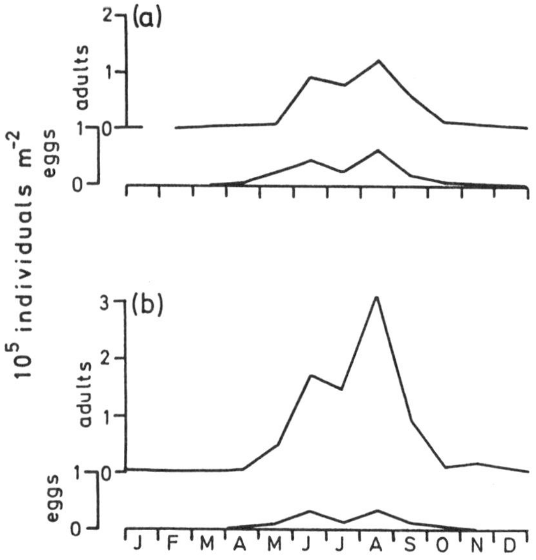

Figure 14.13. (a) The mean annual population structure of *Daphnia hyalina*; (b) The mean annual population structure of *Daphnia longispina*.

amplitude of the population peaks. High population densities were recorded in 1968, 527.6×10^3 m^{-2}, and 1976, 567.9×10^3 m^{-2} but the highest peak value was found in 1977, 937.0×10^3 m^{-2}. Population growth is usually exponential in form suggesting growth in unlimited food conditions until the maximum is reached. The years 1975–1976 show a different pattern in that the 1975 peak was lower than other years and the population collapsed and disappeared briefly in September. *D. hyalina* then re-established itself in October and produced a second, smaller peak in November. This population and its egg production overwintered and produced a very large population in 1976. Table 14.7 reveals the essentially summer nature of the *D. hyalina* populations and records the peak population size. An interesting feature is that an early start to the population growth, as in 1969, is not necessarily conducive to the production of large populations.

The filter feeding *D. hyalina* is a common inhabitant of the temperate lakes of the British Isles (Smyly, 1968; Johnson and Walker, 1974). The populations show a regular cyclical occurrence with absolute population size probably moderated by food availability and predation. A number of potential predators exist among the fish, *Mysis relicta*, the carnivorous Cladocera and the adult *Cyclops* spp. Temperature regulates individual and population growth rates and there probably exists a degree of competition for food from rotifers, *Eudiaptomus gracilis*, juvenile cyclopoids and other cladocerans when they occur.

Table 14.7. Peak population numbers (monthly mean 10^3 m^{-2}), time of peak occurrence, time of onset and time of disappearance of *D. hyalina* populations

	1968	1969	1970	1971	1972	1973	1974	1975	1976	1977	1978
Maximum (month)	435.64 (Aug.)	38.12 (July)	55.76 (Sep.)	44.79 (Aug.)	188.02 (Aug.)	57.13 (Aug.)	93.84 (Aug.)	33.17 (Aug.)	228.45 (June)	506.99 (June)	68.84 (July)
Onset	–	Apr.	June	May	May	May	May	June	Oct. 75	Mar.	Apr.
Disappearance	Nov.	Jan. 70	Nov.	Dec.	Dec.	Nov.	Nov.	Sep.	Dec.	Dec.	Dec.

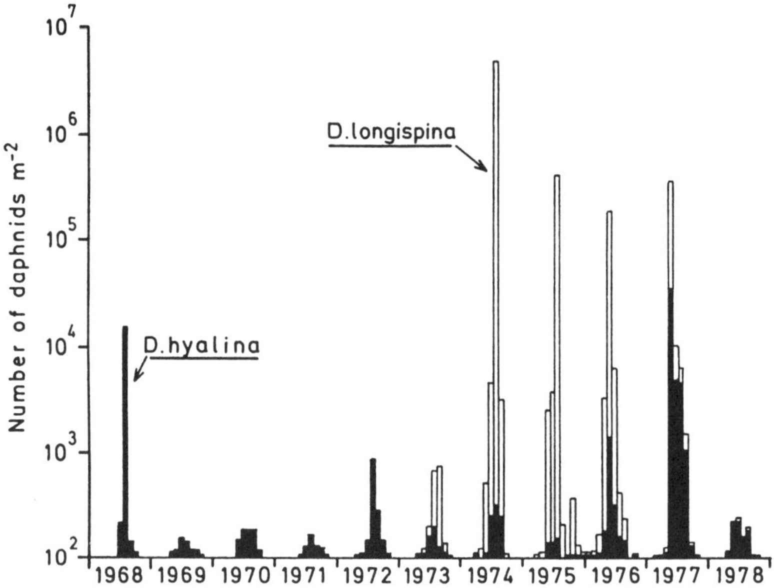

Figure 14.14. The composition of *Daphnia* populations to show the relative contribution of each species to total standing crop.

14.4.2.5. *Daphnia longispina*

First seen briefly in 1971, *D. longispina* became established as a regular member of the zooplankton community in 1973 (see Fig. 14.12(b)). This species shares the same seasonal pattern of occurrence as *D. hyalina* (Fig. 14.13(a and b)) but for several years became the dominant daphnid in the lake. Like *D. hyalina* the populations probably originate from ephippial eggs. The maximum population size was recorded in August 1975 at $2.2 \times 10^6\ m^{-2}$, although the mean monthly population size for August 1974, $937.0 \times 10^3\ m^{-2}$, was higher than August 1975, $722.6 \times 10^3\ m^{-2}$. As with *D. hyalina*, *D. longispina* overwintered in 1975/1976 but not the following winter.

The establishment of *D. longispina* as the dominant daphnid in the zooplankton between 1974 and 1976 does not appear to have seriously threatened the presence of *D. hyalina*. Except during 1975, when *D. hyalina* numbers were low compared with other years, but not significantly lower than 1969, the populations have maintained or exceeded the levels of other years (see Fig. 14.14). in 1977 *D. hyalina* overtook *D. longispina* as the dominant species and the populations of *D. hyalina* reached the highest recorded levels of this study. In 1978 the total population numbers of both species fell but *D. hyalina* remained dominant. The appearance and successful establishment of *D. longispina* is an interesting phenomenon. *D. longispina* is a more robust species than *D. hyalina* and as such is more susceptible to fish predation. It appeared at a time when, it is suspected, fishing pressure on the planktivorous fish was severe and the roach populations had not yet become established. This may have allowed the *D. longispina* populations to increase but, as the fish stocks improved, predation began to reduce the *D. longispina* numbers. The niche width is obviously sufficient to allow these two daphnids to coexist although without predation pressure *D. longispina* may be more successful. As a food organism, *Oscillatoria agardhii* appears to be more vulnerable to *D. longispina* grazing than that of *D. hyalina*. Fig. 14.14 shows clearly the changes in proportions and numbers of these two species.

14.4.2.6. *Daphnia pulex*
This species is not found frequently in the open lake, its distribution being restricted to Kinnego Bay at the SE end of the lake. This bay is morphologically isolated from the main lake, is very shallow and highly enriched with sewage works effluent. Biologically it exhibits different algalogical and faunal characteristics. *D. pulex* a larger species than *D. hyalina* and *D. longispina* is more adapted to these conditions. For a fuller description of Kinnego Bay see Jones (1977a,b,c).

14.4.2.7. *Leptodora kindti*
One of three species of carnivorous Cladocera found in Lough Neagh, this species was first observed in 1970. Apart from 1975, the numbers have increased during the course of this study (see Fig. 14.15a) until 1978. Normally this species appears in the early summer, in 1976 and 1977 it appeared during the spring and reached population peaks of $8.8\times10^3\,m^{-2}$ and $18.8\times10^3\,m^{-2}$ respectively. The early appearance in these years is probably related to the early appearance of daphnid populations on which it principally predates.

14.4.2.8. *Bythotrephes longimanus*
Another carnivorous cladoceran species with a summer occurrence that is found in every year except 1968, *B. longimanus* usually achieves a higher population density than *L. kindti* (Fig. 14.15b). Like *L. kindti* it has an extended season in 1976 and 1977. 1976 was the year of the largest population, $20.4\times10^3\,m^{-2}$ in September. This species was not observed in 1978 and as the factors controlling its occurrence are similar to *L. kindti* its loss from the zooplankton may be attributed to the reduced daphnid population.

14.4.2.9. *Polyphemus pediculus*
This third species of predatory cladoceran occurs in population numbers much lower than the other two species. Not found in 1973, 1974, 1975 and 1978 the records represent appearances rather than absolute numbers except in 1976 and 1977 when a population of $1.5\times10^3\,m^{-2}$ was recorded (see Fig. 14.15c).

For these three species of Cladocera Smyly's (1968) observation that 'records are scarce even when adequate sampling methods are used' is perti-

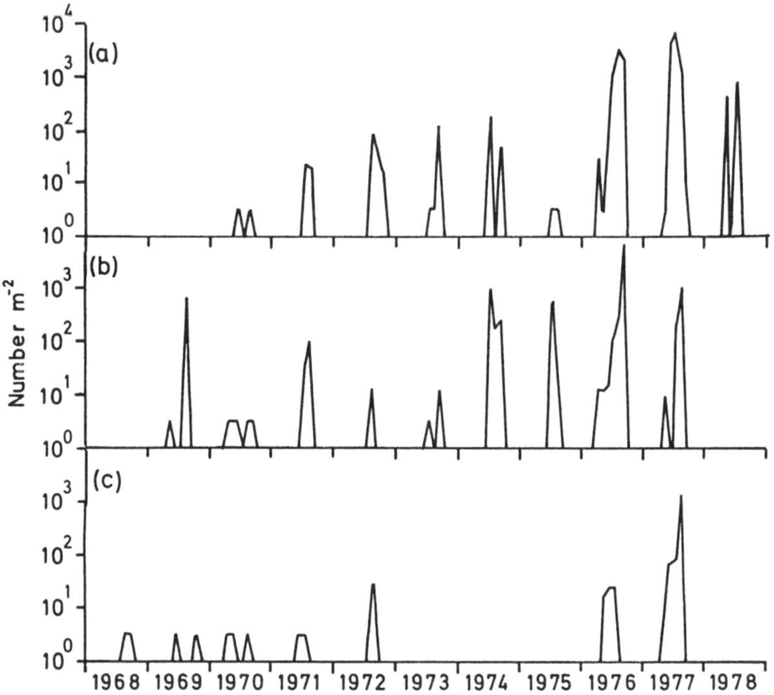

Figure 14.15. Seasonal occurrence of carnivorous Cladoceran, numbers m^{-2}. (a) *Leptodora kindti*; (b) *Bythotrephes longimanus*; (c) *Polyphemus pediculus*.

nent here. When numbers are reported for these species they certainly represent an underestimate. Their abundance is probably a response to the size of prey populations but they are probably competitors of both vertebrate and invertebrate predators.

14.4.3. *Seasonal succession of rotifer zooplankton*

Other workers describe rotifers in groups according to their seasonal occurrence (Ruttner-Kolisko, 1974; Elliott, 1977; Herzig, 1979). What is clear in this situation is that the timing of population peaks varies from year to year and no simple description is sufficient for any one year. Nevertheless the authors have persisted with the form of a 'mean' annual presentation in Fig. 14.16 so that departures from the norm can be seen. Two other relevant points are that no males were observed in any samples, sexual reproduction probably not occurring in overwintering species, and no marked seasonal variation in morphology was observed. Unfortunately samples of the smaller zooplankton were not collected during the latter half of 1976.

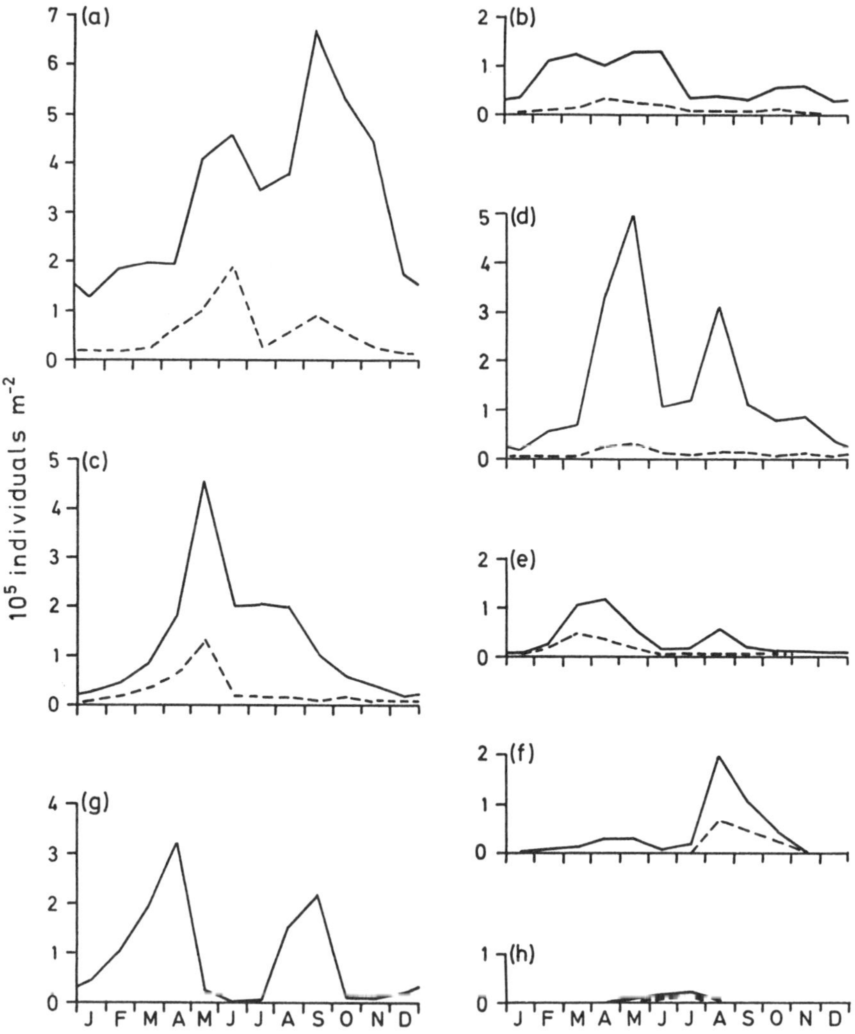

Figure 14.16. Mean annual population structure of Rotifera. (a) *Keratella cochlearis*; (b) *Keratella quadrata*; (c) *Kellicottia longispina*; (d) *Polyarthra vulgaris*; (e) *Filinia longiseta*; (f) *Pompholyx sulcata*; (g) *Notholca acuminata*; (h) *Cephalodella gibba*. (—— = total population, – – – = eggs).

14.4.3.1. *Keratella cochlearis*

This species was found on every sampling occasion in numbers usually exceeding 10^4 m^{-2} as can be seen in Fig. 14.17a. The annual population cycle usually showed two population peaks, the first in late spring and the second during the autumn. These peaks usually approached or exceeded 10^6 individuals m^{-2}, the maximum value being recorded in 1968 at 2.7×10^6 m^{-2} during

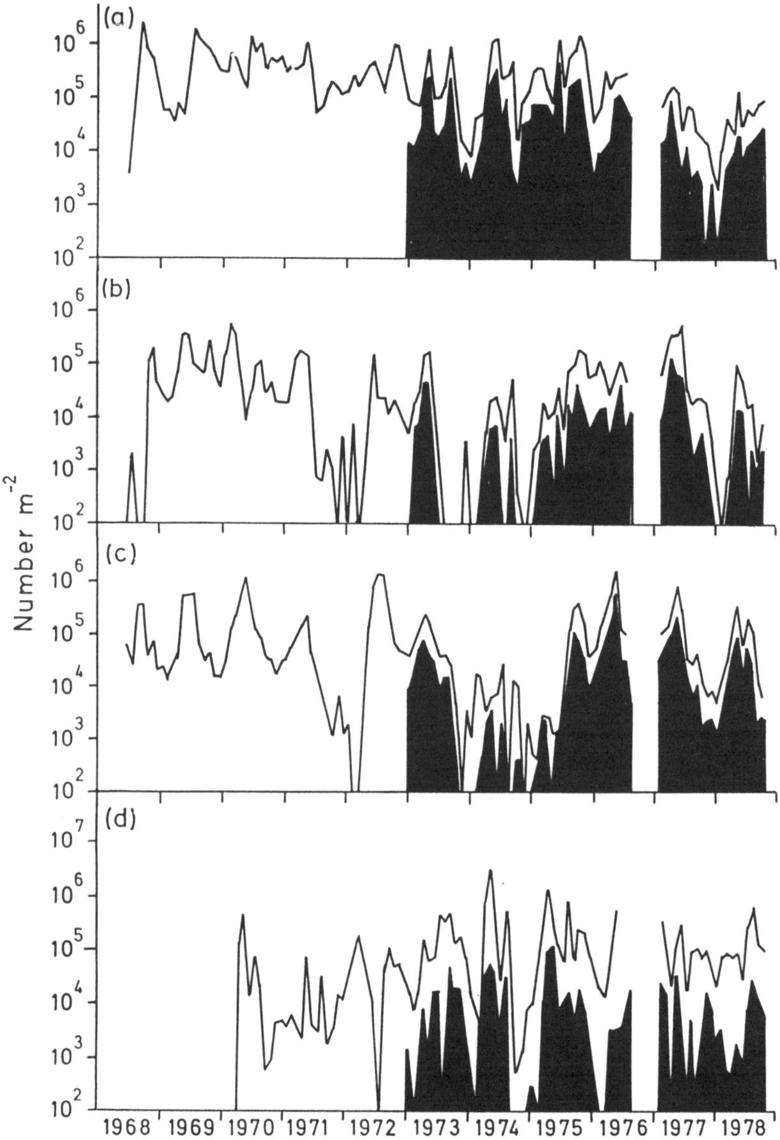

Figure 14.17. Seasonal occurrence of notifiers, numbers m^{-2}. Shaded portion represents egg numbers. (a) *Keratella cochlearis*; (b) *Keratella quadrata*; (c) *Kellicottia longispina*; (d) *Polyarthra vulgaris*.

the autumn. Several years depart noticeably from this pattern. In 1969 no proper spring or autumn 'peak' was observed but a single summer peak occurred at values similar to peak values of other years. In 1971 the autumn peak was low compared with other autumn maxima (274.9×10^3 m^{-2}) and in

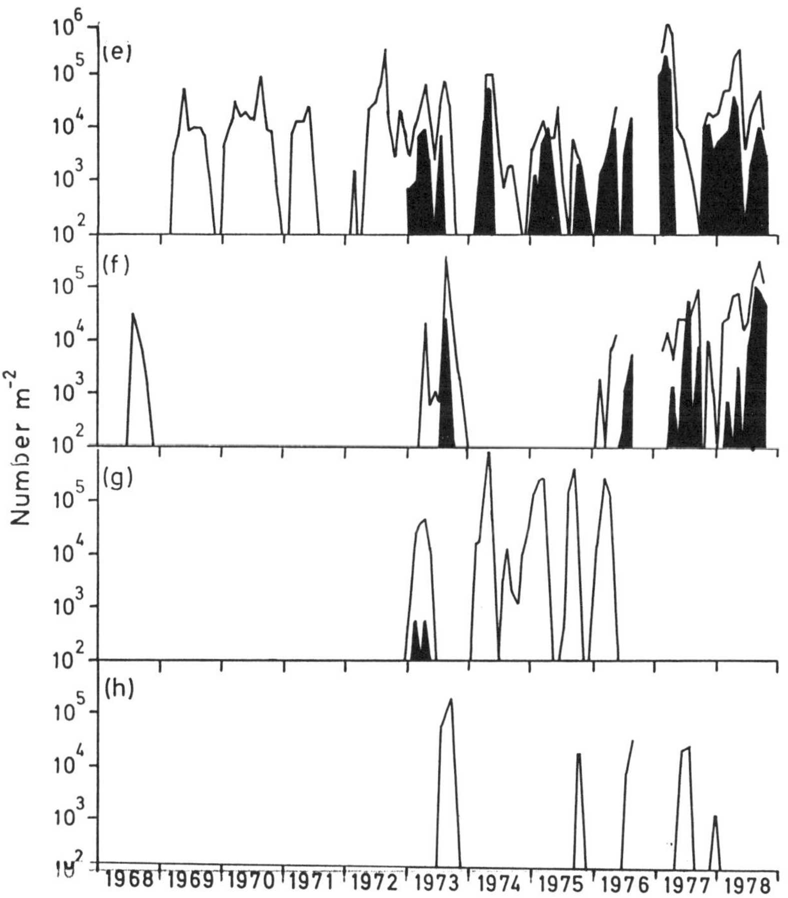

Figure 14.17. Seasonal occurrence of rotifers, numbers m^{-2}. Shaded portion represents egg numbers. (e) *Filinia longiseta*; (f) *Pompholyx sulcata*; (g) *Notholca acuminata*; (h) *Cephalodella gibba*.

1977 no autumn peak was found at all. The years 1977 and 1978 show populations an order of magnitude smaller than other years. Egg production appears to be a reflection of population size rather than having a marked independent seasonal occurrence. Peak egg numbers are synchronised with peak population size.

14.4.3.2. *Keratella quadrata*

The seasonal occurrence of *K. quadrata* varies from year to year. There is either a spring population (1971, 1973 and 1977), an autumn population (1968, 1972 and 1975) or populations found both in the spring and autumn (1969, 1970, 1974 and 1978) as shown in Fig. 14.17b. In some years there is an overwintering population (1968, 1969, 1970, 1972, 1975 and possibly 1976) the overwintering population size in the order of 10^4 individuals m^{-2}. The maximum recorded population size occurred in June 1977 at 1.3×10^6 m^{-2} but similar levels were observed in 1969 and 1970. The years 1971 to 1975 show a reduced population density both as maximum and mean population numbers. As with *K. cochlearis*, egg production appears to be a reflection of population size rather than seasons.

14.4.3.3. *Kellicottia longispina*

With exceptions between 1972 and 1974, *K. longispina* has always been found in samples. The population structure is normally unimodal the population peak occurring in spring (1970, 1971, 1973, 1976, 1977 and 1978), summer (1969 and 1972) or the early autumn (1968?, 1975). The timing of the population peak appears to depend upon the size of the overwintering population. If this is high, $>10^4$ m^{-2}, the peak occurs in spring or summer, if low, $<10^4$ m^{-2}, autumnal population maxima result. The exception is seen in 1978 but approaches the former rather than latter case. Population maxima approach or exceed 10^6 individuals m^{-2}, the maximum recorded size occurring in June 1972 at 1.4×10^6 m^{-2}. A low summer population density results in no real over-wintering population as occurred in 1971, 1973 and 1974. Seasonal changes can be seen in Fig. 14.17c. Egg production shows marked seasonality (see Fig. 14.16c) with maximum egg production occurring in spring but this is still tied to population size. Egg production is rather confused in 1974 but at this time the population did not achieve the levels of other years.

14.4.3.4. *Polyarthra vulgaris*

P. vulgaris was not found in the plankton until 1970 but since then has established itself as a consistent member. The populations are of a bimodal nature with a spring and autumn peak (Figs. 14.17d and 14.16d) reaching the highest population density in spring. The highest population density for any rotifer was observed for *P. vulgaris* in May 1974 at 4.4×10^6 m^{-2}. This large population coincided with the very large cryptophyte populations observed in 1974. This species has an overwintering population of between 10^3 and 10^4 individuals m^{-2}, although this was lower in the winter of 1974/75. Egg production appears to be seasonally modulated with higher production in the spring and autumn. 1978 proved to be exceptional again in that the peak adults and egg numbers were observed in the autumn. Egg production appears to be rather low to generate the observed populations suggesting that egg production may occur in individuals found in the benthic region.

14.4.3.5. *Filinia longiseta*

First recorded in the zooplankton in 1969. *F. longiseta* may have been present in the early part of 1968, see Fig. 14.17e. The population appears either as a single spring or summer population (1969, 1971, 1972, 1974, 1976? and

1977) or as spring and autumn populations (1970, 1973, 1975 and 1978). The populations usually disappear from the plankton during the winter, except for 1972/73 and the population maxima vary markedly from year to year, e.g., 1975 $26.0\times10^3\,m^{-2}$, 1977 $1.3\times10^6\,m^{-2}$. The egg production of *F. longiseta* is highest in the early spring and low during the summer (see Fig. 14.16e).

14.4.3.6. *Pompholyx sulcata*

Found in 1968 and again in 1973, *P. sulcata* was occasional in occurrence until it appeared successively in 1976, 1977 and 1978. When found in the zooplankton, see Fig. 14.17f, the populations are bimodal with a spring peak and an autumn peak as can be seen in fig. 14.16f. The maximum recorded population density was found in September 1978 at $778.8\times10^3\,m^{-2}$ but in 1973 the population density of $756.8\times10^3\,m^{-2}$ is not far short of this. Egg production is high in the autumn but is generally tied to the population size.

14.4.3.7. *Notholca acuminata*

This species first appeared in the zooplankton in 1973 and immediately formed significant populations but it has not been observed since 1976. In 1973 a spring peak was observed, the population disappearing by June. In 1974 and 1975 distinct spring and autumn peaks were observed (Figs. 14.17g and 14.16g) the maximum population density being recorded in April 1974 at $834.8\times10^3\,m^{-2}$. In 1976 only a spring population was found. The numbers of *N. acuminata* make an important contribution to the total rotiferan biomass in the years 1973 to 1976 and it is of interest that apart from 1973 its appearance coincides with the very low populations of the blue-green alga *Oscillatoria agardhii* (see Fig. 14.5b).

14.4.3.8. *Cephalodella gibba*

C. gibba is found as isolated summer peaks in 1973, 1975, 1976 and 1977. These populations do not form a very large part of the total rotiferan community, occurring over a short period and the maximum population of $194.0\times10^3\,m^{-2}$ was found in late summer 1973 (Figs. 14.17h and 14.16h).

14.4.4. *Rotifers and eutrophication*

Difficulty is encountered when trying to explain the seasonal occurrence and abundance of individual rotifer species. This is compounded by previous attempts to classify the eutrophic state of particular lakes by presence or absence of particular species (Pejler, 1957; Pourriot, 1965; Ruttner-Kolisko, 1974; Elliott, 1977). For example, oligotrophic waters have variously been described as containing (from among Lough Neagh's species):
Kellicottia longispina (Pejler, 1957; Pourriot, 1965; Ruttner-Kolisko, 1974)
Polyarthra vulgaris (Pejler, 1957; Ruttner-Kolisko, 1974)
Keratella cochlearis (Ruttner-Kolisko, 1974)

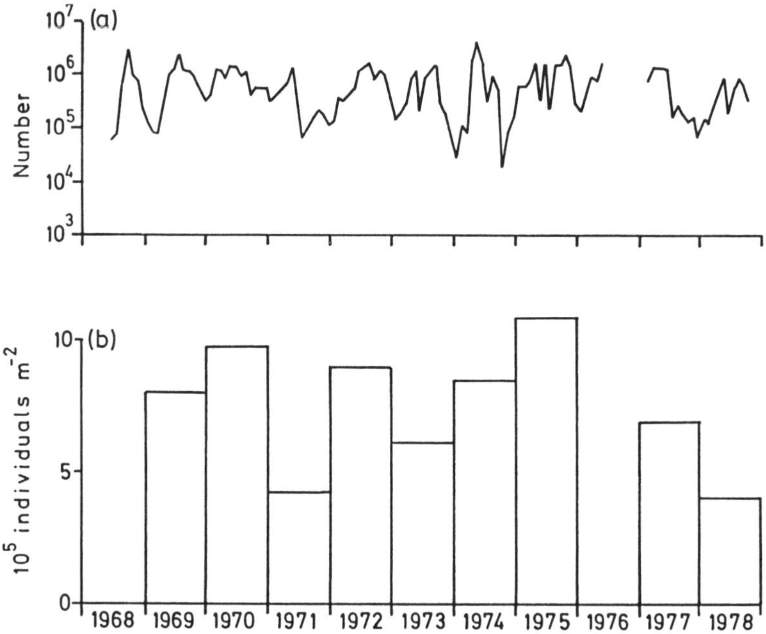

Figure 14.18. (a) Population density of all species of rotifer, numbers m^{-2}; (b) Mean annual population size of rotifers, numbers m^{-2}.

whereas Ruttner-Kolisko (1974) recognises that some of these species occur with *Keratella quadrata*, *Pompholyx sulcata* and *Filinia longiseta* in eutrophic situations. Herzig (1979) diplomatically ignores this debate and considers abundance as a measure of eutrophic state while recognising that this may be moderated by large populations of other zooplankton. Neusiedlersee has a larger number of species than Lough Neagh, although not all of these are present in one year, six of the eight species from Lough Neagh are common with Neusiedlersee which is considered to be a mesotrophic lake. The evidence of other studies (Chapter 8) points to Lough Neagh being eutrophic so perhaps other considerations are more important to the occurrence of particular rotifer species. It is suspected that the availability of suitable food items, (all Lough Neagh rotifers except *P. vulgaris* are filter feeders and, it is suggested, feed on items $<20\ \mu$) together with their own seasonal preference dictates whether or not the species survives. The abundance is moderated by growth rate, quantity of food, inter-specific competition and predation. Observations show that rotifers form a part of the diet of larval fish and cyclopoid copepods.

Abundance varies on a seasonal basis and Fig. 14.18a shows the population density for all species of rotifers from 1968 to 1978 as monthly means. This reveals that population densities reach or exceed 10^6 individuals m^{-2} in every year of this study. It can also be seen that these levels may be sustained as in 1970 and 1975 or only reach this level for a short period as in 1971 and 1978. Which situation pertains will have a very different effect on the zooplankton community in terms of grazing pressure or as potential food for other species. The mean annual population size is presented as a histogram in Fig. 14.18b showing more clearly the annual variation in standing crop. The mean for the whole study is 748.3×10^3 m^{-2} but, for example, the 1978 crop amounted to only 38% of the very high 1975 levels. These changes may be of a cyclical nature and controlled by an environmental factor such as temperature but further analysis is necessary to establish this.

14.5. Zooplankton feeding

14.5.1. *Grazing on phytoplankton*

Detailed studies of zooplankton grazing have not been undertaken except for work on *Cyclops abyssorum* feeding behaviour (Jewson *et al.*, 1981). However, examination of phytoplankton, chlorophyll *a* and zooplankton population data reveals trends which deserve further analysis and comment.

The rotifers, calanoid copepods and daphnids comprise the filtering element of the zooplankton and of these the daphnids show the closest association with the phytoplankton changes. The daphnids exhibit a marked seasonal periodicity, the populations occurring normally between April and November although this varies from year to year (see Fig. 14.12a and b). The species composition also varies in different years.

Phytoplankton biomass during the daphnid occurrence is composed mainly of three blue-green algal species, *Oscillatoria redekei* (van Goor) *O. agardhii* (Gomont.), and to a lesser extent *Aphanizomenon flos-aquae* (Ralfs.). A fuller description of phytoplankton seasonality may be found in Chapter 10 but features relevant to this discussion are described below. In the early years of the study *O. redekei* was the most abundant species reaching a maximum in late April but remaining in the plankton at a high level throughout the year. *O. agardhii* peaked in late summer or early autumn but again was present, in significant quantities, for most of the year. *A. flos-aquae* had summer peaks but was not as abundant as the other blue-green species. During 1974/1975 a major change took place, *O. agardhii* was found but at very low levels throughout the year. *A. flos-aquae* disappeared leaving *O. redekei* the major contributor to the standing crop biomass. During 1976 *O. agardhii* populations increased during the summer and this has been the dominant blue-green algal species since then (see Fig. 10.8, Chapter 10).

During the period 1969–1971 daphnid populations occurred at low population densities and gross grazing effects upon the blue-green crop were not apparent as the algal crops remain relatively constant during these summers. In 1972 peak numbers of daphnids were three times the previous level (mid-

August) and the peak collapsed in late September. However, between 22 August 1972 and 6 September 1972 the *O. redekei* crop fell from 4.3 mm^3 l^{-1} to 0.86 $mm^3 l^{-1}$, the fall coinciding with the peak period of daphnid numbers. Immediately preceding this event the *A. flos-aquae* populations fell, their minimum size also coinciding with the daphnid maximum population size. Similar events occurred in 1973. The daphnid numbers and the size of the *O. redekei* crop were almost identical with the 1972 situation and the *O. redekei* crop peaked and fell coincidently with the daphnid population rise and fall. Up to 1973 the daphnid species of the lough was *Daphnia hyalina* but during 1973 *Daphnia longispina* first appeared in the zooplankton and this species persisted into the autumn and population fluctuations were closely associated with a marked decrease in the *O. agardhii* crop subsequent to the decline in the *O. redekei* crop.

The years 1974 and 1975 are characterised by a marked increase in the population density of *D. longispina* and this species had two peaks of the population, the first in early July and the second in mid to late August, in both years (Fig. 14.19a and b). As described earlier, the bulk of the blue-green biomass in these years was composed of *O. redekei* and the maximum standing crop occurred later than in other years coinciding with the early *D. longispina* peak and there is a strong suggestion that the daphnids grazed out this algal crop. The disappearance of *O. redekei* heralded the appearance of large populations of cryptophytes and small green algae, which accounts for the high chlorophyll *a* concentrations during the first half of August, and this was followed by a very large and rapid increase in the daphnid populations and subsequent grazing out of the algal crops. There was a very marked decrease in daphnid population density during the interval between the chlorophyll *a* peaks. The second peak also partly consisted of the diatom *Stephanodiscus astraea* and this also disappeared during the second daphnid maximum. Daphnid populations, *O. redekei* populations and total algal standing crop biomass as chlorophyll *a* for the summer of 1974 and 1975 are shown in Fig. 14.19a and b.

During 1976 the blue-green algae re-establish their normal dominant position in the phytoplankton. The daphnid populations continued to oscillate synchronously with the *O. redekei* populations but, during the decline of the *O. redekei* crop in June, the *O. agardhii* crop increased and comprised the bulk of the phytoplankton biomass. Nevertheless the *O. agardhii* population fluctuated markedly appearing to respond to changes in the daphnid populations which remained established even through the winter. In 1977 *D. longispina* ceased to be the most abundant daphnid as *D. hyalina* re-established its pre-1973 place in the zooplankton. A similar algal growth pattern occured in this year except that the crops of *O. agardhii*, which had persisted throughout the first half of the year, decreased following the May growth of the daphnid populations. *O. redekei* exhibited an increase in population size during this period but declined rapidly when *O. agardhii* became scarce, presumably as the daphnids grazed it. The *O. agardhii* populations recovered

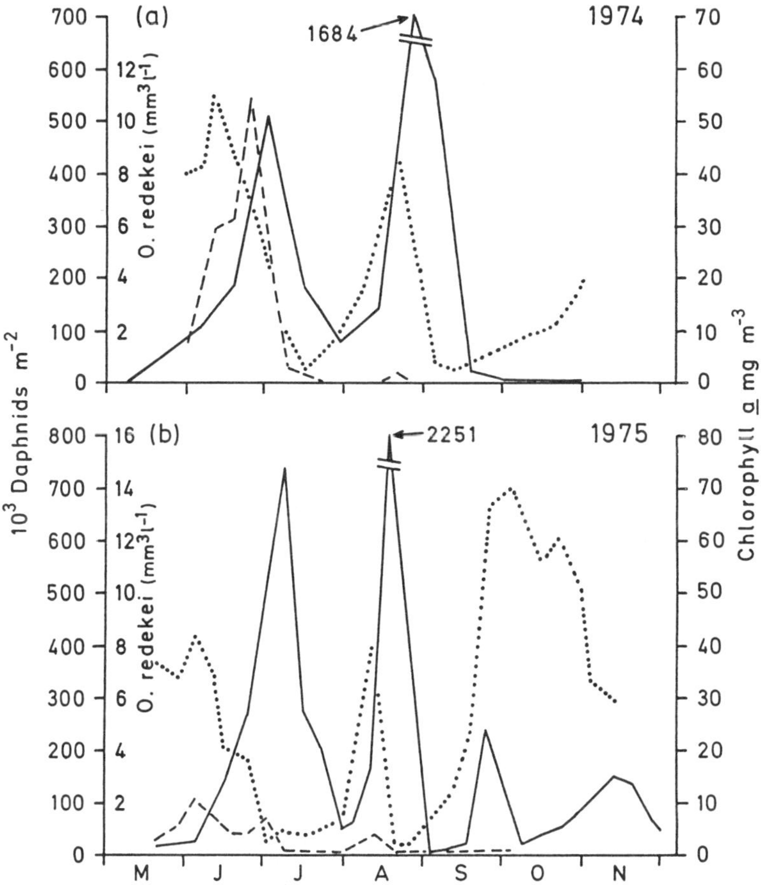

Figure 14.19. The relationship between *Daphnia* sp. (——), *Oscillatoria redekei* (– – –) and chlorophyll *a* (· · · · ·) during daphnid population occurrence for the years (a) 1974 and (b) 1975.

as the daphnid populations declined. During mid-September 1977 an increase in the daphnid standing crop, although smaller than other peaks, coincided with a halt in the growth of the *O. agardhii* population which, up until that point, appeared to be rapidly increasing. In 1978 the daphnid populations were much less abundant, population densities being equivalent to pre-1974 values. The *O. redekei* populations decreased during the periods of maximum population density but not in such a marked fashion as the years described above.

The conclusions to be drawn from these observations of the interaction of the daphnid and blue-green standing crops are of interest. There is a strong suggestion that the daphnids use *O. redekei* as a major source of food and this manifests itself in a marked reduction in the *O. redekei* standing crop especially during high population densities of the daphnids. The 1974 season illustrates this point as the *O. redekei* population peaked in June and was probably still in its active growth phase when grazed down by the daphnid population which was dependent upon it as its prime source of food. The switch to cryptophytes, green algae and diatoms in the August of that year suggests that the daphnids are flexible in their dietary requirements. Much unresolved debate in the literature suggests that blue-green algae may or may not provide a good food source for Daphnia (e.g. Schindler, 1971; Arnold, 1971; Porter, 1975; Porter and Orcutt, 1980) but clearly, from the observations here, *O. redekei* provides a food source of sufficient quantity and quality for the daphnid populations to grow and reproduce. The other principal herbivores found in the zooplankton at this time were *Eudiaptomus gracilis* and *Polyarthra vulgaris* but these species do not contribute significantly enough to the zooplankton biomass to cause the observed effects although they doubtless do contribute. Whether *O. agardhii* is grazed so freely by daphnids is not certain but there is a suggestion that the daphnids, particularly *D. longispina* switch to *O. agardhii* when *O. redekei* is unavailable. The situation is slightly complicated by the fact that the *O. agardhii* populations chronologically follow the *O. redekei* populations in the late summer which is normally towards the end of the daphnid growing season. However, in 1975/6 when daphnids overwintered *O. agardhii* was the principal algal species found in the lough. Decreases in the *O. agardhii* populations, when they occur, are linear implying a grazing function with a constant consumption of filaments per unit time. The larger size of *D. longispina* may be able to cope better with the larger filament size of *O. agardhii*.

The marked changes in the algal populations occur in the years of high daphnid population densities, which in itself suggests that over-grazing might be occurring; in turn this supports the idea that the daphnids can survive and grow on these blue-green species. One interesting feature of the years 1974 to 1976 is that probably in the absence of *O. agardhii*, *Aphanizomenon flos-aquae* establishes itself among the blue-green phytoplankton and this species may not be grazable or digestible by these daphnid species.

Trends such as those described above are not so easily described for other species although detailed multivariate analysis might reveal interesting correlations but the wide range of feeding types as classified by Morgan (1980) suggests that, particularly with the complex rotiferan fauna, a considerable grazing impact is created by the zooplankton on the phytoplankton.

Jewson *et al.* (1981) drew attention to the potential grazing impact of *Cyclops abyssorum* when they described adults and copepodites of this species from field samples having guts full of diatoms. Experimental work to measure feeding rates on *Melosira italica* (Ehr.) which, together with *Stephanodiscus astraea*, comprises the important elements of the diatom crops of the lough, gave estimates of consumption rates in excess of $1.5.10^3$ cells ind^{-1} day^{-1}. Although *C. abyssorum* tended to feed on the shorter cell filament (and up to 33% of the cells in the gut were unopened) their findings suggest that the cyclopoid populations may have an important grazing regulatory effect on diatom population growth.

14.5.2. *Predation*

The importance of predation in the regulation of community structure as well as population density has been recognised in a wealth of recent publications in the literature (see Zaret, 1980, for a review). Nevertheless the exact and quantitative nature of predatory relationships remain firmly in the realms of speculation as basic information as to population sizes of the principal predators are not known. This is true of Lough Neagh where three classes of predators upon the zooplankton may be recognised. Firstly zooplankton predating upon other zooplankton, secondly other invertebrates predating upon zooplankton and thirdly the planktivorous fish.

Two main groups represent predators from within the zooplankton, the carnivorous copepods and the carnivorous cladocera. A third group, carnivorous rotifers such as *Asplanchna* spp., are not found in this lake although found in comparable lakes such as Loch Leven (Johnson and Walker, 1974) and Neusiedlersee (Herzig, 1979). *Cyclops abyssorum* and *C. vicinus* population numbers are well known but the feeding behaviour of these species is difficult to interpret and to interpolate into field data. Records from the lake exist of adult cyclopoids feeding on a range of zooplankton and algal material including rotifers, daphnids, juvenile cyclopoid stages and adult calanoids (Gilmore, pers. comm.). The impact of this feeding on the zooplankton has not been assessed but must *per se* be considerable to maintain the standing crops of cyclopoids observed. The carnivorous cladocerans, *Leptodora kindti*, *Bythotrephes longimanus* and *Polyphemus pediculus*, do not occur in very large population numbers but, nevertheless, may have an important effect on daphnid populations if, as has been suggested, they select the smaller stages of the Daphnia spp. populations (De Bernardi and Giussani, 1978; Hall, 1964). These species however are vulnerable to, particularly, coregonid predation (De Bernardi and Giussani, 1978) which may in Lough Neagh restrict their own population development as suggested by observations of Wilson (1979, 1984) when in August 1977 *B. longimanus* was, jointly with *Daphnia longispina*, the commonest item in the diet of pollan fry.

The feeding of *Mysis relicta* on zooplankton is discussed in Chapter 15. Various authors have recently drawn attention to the significance of *M. relicta* predation particularly on daphnid populations (Richards *et al.*, 1975; Goldman *et al.*, 1979).

Some limited information is available concerning fish predation of the zooplankton. Gut analysis of young perch (*Perca fluviatilis*) reveals that the young fish feed only on zooplankton until they reach a fork length of 60–62 mm (Fig. 14.20). The diet consisted solely of *E. gracilis* and *C. abyssorum* copepodite and adult stages but this does not imply that they actively avoid feeding on cladocerans, rather that the cladoceran populations are not estab-

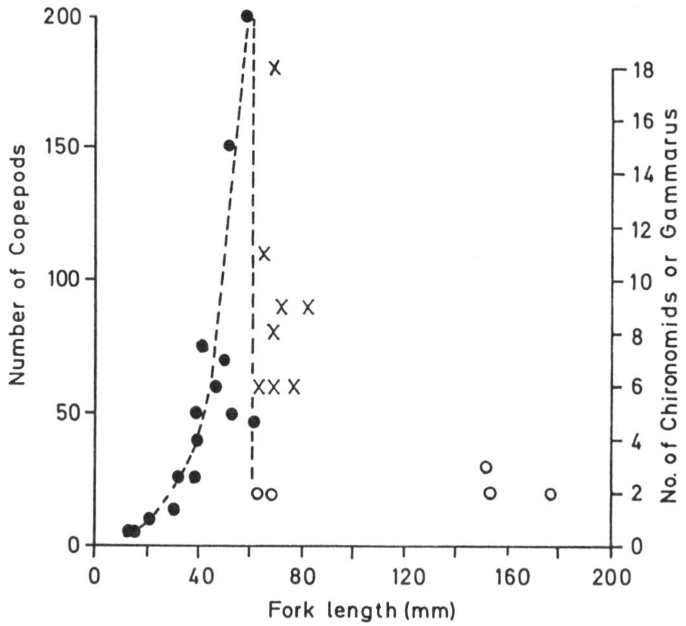

Figure 14.20. The relationship between larval fish size and gut content during the early part of the year for juvenile perch (N = 37, 4 empty guts).

lished when the young fish are in the planktivorous stage or that they are separated spatially. Larger fish were found to be feeding on benthic types, *Gammarus duebeni* and various chironomid larvae species which compose most of the benthic fauna. Adult fish, however, are not obligate feeders on one item of diet but switch from benthos to zooplankton and may become canabalistic on their own fry depending upon availability (Thorpe, 1974). It has been observed in Lough Neagh, that even large adult perch, will feed exclusively on *Daphnia* spp. if available at high population densities (E.J. Rogers, pers. comm.).

Pollan (*Coregonus pollan*) fry and adults have very similar diets (Wilson, 1984). The smaller fry feed upon zooplankton, mainly *Daphnia* spp., but will take *C. abyssorum* and *E. gracilis*, together with smaller benthic chironomids. The larger fry also feed upon daphnids but also take *Bythotrephes longimanus* when available. Adult fish feed on a variety of benthic organisms until the daphnid populations become established when they feed exclusively upon *Daphnia* spp. the impact of pollan upon the daphnid populations may be very marked, Wilson (1984) records an exceptional individual pollan gut containing 19.5×10^3 individual daphnids when the daphnid population density was 393.6×10^3 m^{-2}. Figure 14.21 shows the seasonality of zooplankton feeding of adult pollan.

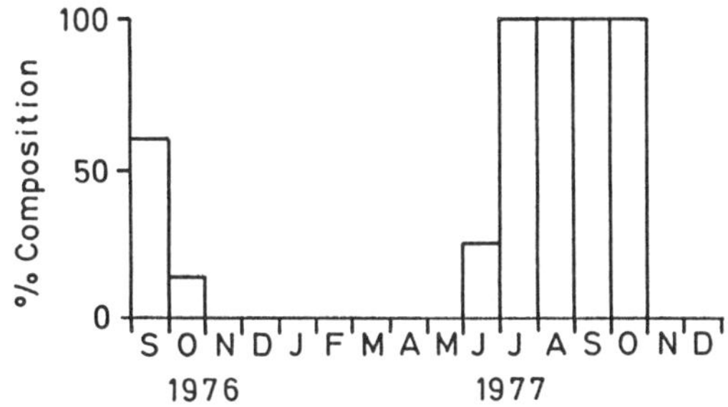

Figure 14.21. Seasonal occurrence of daphnids in pollan guts (after Wilson, 1984 modified).

Little is known about feeding by other species of fish in Lough Neagh except that roach (*Rutilus rutilus*) have similar feeding habits to perch. Thorpe (1974) has shown that trout are capable of feeding on all the crustacean zooplankton.

Morgan (1980) reviews the influences of fish upon zooplankton as does Zaret (1980) and they conclude that fish predation may be a powerful influence in moderating species composition of the zooplankton. Reasonable estimates of the total fish populations are necessary before quantitative effects on zooplankton can be seriously considered.

14.6. Summary

Smyly (1968), in his comparison of the crustacean fauna of the lakes of the English Lake District, concluded that most lakes have some peculiarity that distinguishes them from all others. He is referring to the species composition of a particular lake and suggests that the occurrence of a species, despite the catholic distribution of many zooplankton species, is determined by *local* environmental conditions. He further suggests that it is changes in these conditions, especially within the biological or chemical environment, that are likely to create a new species composition within a particular lake. He also concurs with the suggestions of Carl (1940) and Baldi (1951) that the species composition of the zooplankton will tend to remain unchanged in successive years. As stability appears to be the general rule, any changes in composition and their controlling factors are of ecological interest.

The situation in Lough Neagh is one of change in the species composition and this may be explained in terms of biological, chemical and physical variations in the environment. In the longer term (since the early 1900's) the change from a bosminid-diaptomid-rotiferan to a cyclopoid-daphnid-rotiferan situation may be best explained as a resultant of the increased eutrophic status of the lake (Chapters 7 and 8) and changes in the predation pressures (fish and *Mysis relicta*) on the zooplankton. This represents a combination of chemical and biological change and it would be natural to invoke the change in eutrophic status as the dominant factor as most evidence is available to describe that change. However, a physical change has taken place since the study of Dakin and Latarche (1913) in that the Lough level has been lowered, changing, as suggested in Chapter 2 the nature and extent of the shoreline. There is evidence, Herzig (1979), that a well established littoral region has a marked effect on the quality of the zooplankton, particularly *Eudiaptomus* sp. and *Bosmina* sp. Palaeoecological studies of the zooplankton sub-fossils of the sediments should define when and possibly why these changes have occurred.

In the shorter term, changes in the species composition, the occurrence of species not previously found – *Daphnia longispina*, *Cyclops vicinus*, *Pompholyx sulcata*, *Cephalodella gibba* and *Notholca acuminata* – and quantitative changes in the population densities, provide substantial evidence of lack of stability in the system. The standing crop of daphnids in 1974 was more than 20 times the 1969 level. Admittedly the higher numbers included two daphnid species, *D. hyalina* and *D. longispina* but, for example, the *D. hyalina* populations in 1977 were more than ten times the 1969 level. Changes of similar magnitude were not seen in the phytoplankton crops but (see Sections 9.8 and 10.2.2) individual species changed dramatically. Temperature probably plays an important part in this increase in population density. The maximum summer temperature in 1977 (20 °C) was high compared with 1969 (17 °C) and this resulted in faster growth and egg development rates which produced a much larger population level. During the winter of 1973/74 the minimum winter temperature was high (5.5 °C) compared with the winter of 1968/69 (2 °C). Similar situations were found in other years and the higher winter temperatures enhanced the growth rate at the beginning of the season and so produced larger populations. During the winter of 1975/76, the daphnid populations were even able to survive the winter and high winter temperatures were probably a major contributory factor to this event. During this period, especially from 1972 onwards, changes in the fish population structure may have relieved some predation pressures on the zooplankton. This has allowed the establishment of a previously unrecorded daphnid species, *D. longispina*, and also shows up as an all round increase in daphnid production. Although little information is available about *Mysis relicta*, other predatory crustaceans, particularly the carnivorous Cladocera, increased in numerical abundance which supports the view that fish predation of the zooplankton abate at this time. This release of the restraints on daphnid population growth result in the almost complete removal of the *Oscillatoria agardhii* crops (by grazing) and a significant reduction in the *O. redekei* populations. In turn these blue-green algal species were replaced by another, *Anabaena flos-aquae*.

Cyclops vicinus, a second species of cyclopoid copepod, was found, with the exception of 1973, in the same years as *D. longispina* and their appearances may be associated in some way. *C. vicinus* may be dependent upon *D. longispina* for food or may not be subject to fish predation as with *D. longispina* but its successful establishment may be related to other factors as well. *C. vicinus*, for example, has a stage inhabiting the benthic region, and the success of its introduction depends upon its survival here probably acting through a reduced predation pressure from *M. relicta* which is not found during daylight in this region. *C. vicinus* also needs to find sufficient room to survive in competition with *C. abyssorum* but the large rotifer populations together with the presence of *D. longispina* provide sufficient prey species.

The rotiferan fauna of Lough Neagh shows a lack of stability similar to the crustacea if it is being considered in Smyly's (1968) terms although the same basic assemblage of *Keratella cochlearis*, *K. quadrata*, *Kellicottia longispina*, *Filinia longiseta* and *Polyarthra vulgaris* is found at the beginning of the century and the present. How, if at all, the longer term changes described for the crustacean zooplankton affect the Rotifera is not known. What is seen is a variability occurring in successive years which suggests that the rotiferan composition is, for some species, less than stable. Even *K. cochlearis*, consistently found in the plankton, shows a downward tendency in population density from 1968 to 1978. *P. vulgaris*, not seen in 1968 or 1969, becomes an important element, while other species, *Pompholyx sulcata*, *Notholca acuminata* and *Cephalodella gibba* characteristically show scattered occurrences. Particularly interesting is the appearance of *N. acuminata* occurring during the years 1973 to 1976 a contemporary of *D. longispina* and *C. vicinus* suggesting perhaps an association with these species. Herzig (1979) suggests that there is a reciprocal relationship between the abundance of rotifers and crustaceans and this is mildly reinforced by the observation that the 'good' years for the cladocerans, 1973 to 1976, are the 'bad' years for the rotifers. The control of this situation may be sought in their feeding regimes which markedly overlap (see review in Morgan, 1980). Some species seem to show a degree of density-dependence, explaining, perhaps, the great variation in numerical abundance observed in different years. Again, the underlying control might be found in the environmental temperature and predation pressures exerted on the rotiferan populations.

In conclusion the authors would like to suggest that studies of this time span prove useful in several ways. Firstly they provide a more satisfactory description of specific variation than shorter term studies and incidently provide an example of a less stable situation than is normally assumed. Secondly they show that population abundance for individual species is very variable from year to year and also that the total population biomass can be very variable. Thirdly it becomes apparent that interactions between species at the same and different trophic levels are complex and are further complicated by physical and chemical variations in the environment. This area lends itself to both experimental investigation, through feeding and predation experiments for example, and statistical analysis, and sufficient information is available for multivariate and time series analyses to be applied. These latter techniques might identify underlying control mechanisms and also quantify the links between different parameters.

14.7. References

Arnold, D.E. (1971) Ingestion, assimilation, survival and reproduction by *Daphnia pulex* fed seven species of blue-green algae. Limnol. Oceanogr. 16: 906–920.

Baldi, E. (1951) Stabilita delle biocenosi limnetiche e sue probabili condizioni. Boll. Zool. 18: 65–78.

Battarbee, R.W. (1978) Observations on the recent history of Lough Neagh and its drainage basin. Phil. trans. R. Soc. 28: 303–344.

De Bernadini, R. and Giussani, G. (1978) Effects of mass fish mortality on zooplankton structure and dynamics in a small Italian lake (Lago di Annone). Verh. Internat. Verein. Limnol. 20(2): 1045–1048.

De Bernadini, R. and Soldavini, E. (1976) Long term fluctuations of zooplankton in Lake Mergozzo, Northern Italy. Mem. Ist. Ital. Idrobiol. 33: 345–375.

Bottrell, H.H., Duncan, A., Gliwicz, Z.M., Grygierek, E., Herzig, A., Hillbricht-Ilkowska, A., Kurasawa, H., Larsson, P. and Weglenska, T. (1976) A review of some problems in zooplankton production studies. Norw. J. Zool. 24: 419–456.

Brooks, J.L. (1968) The effects of prey size selection by lake planktivores. Syst. Zool. 17: 272–291.

Brooks, J.L. (1969) Eutrophication and changes in the composition of the zooplankton. In: NAS sponsored symposium: Eutrophication: causes, consequences, correctives. NAS, Washington, D.C. pp. 236–255.

Carl, G.C. (1940) The distribution of some Cladocera and free-living Copepoda in British Columbia. Ecol. Monogr 10: 55–110.

Cassie R.M. (1971) Sampling and subsampling: Theory and practice. In: W.T. Edmondson and G.G. Winberg (eds.), A manual on methods for the assessment of secondary productivity in fresh waters. Blackwell, London, pp. 174–209.

Cushing, D.H. (1951) The vertical migration of planktonic Crustacea. Biol. Rev. 26: 158–192.

Dakin, W.J. and Latarche, M. (1913) The plankton of Lough Neagh. Proc. Roy. Ir. Acad. 30(B): 20–96.

Edmondson, W.T. (1960) Reproductive rates of rotifers in natural populations. Mem. Ist Ital. Idrobiol. 12: 21–77.

Edmondson, W.T., Anderson, G.C. and Peterson, D.R. (1956) Artificial eutrophication of Lake Washington. Limnol. Oceanogr. 1: 47–53.

Elliott, J.I. (1977) Seasonal changes in the abundance and distribution of planktonic rotifers in Grasmere (English Lake District). Freshwat. Biol. 7: 147–166.

Flower, R.J. (1980) A study of sediment formation, transport and deposition in Lough Neagh, Northern Ireland, with special reference to diatoms. D. Phil. Thesis, New University of Ulster, pp. 212.

George, D.G. (1973) Diapause in *Cyclops vicinus*. Oikos, 24: 136–142.

George, D.G. and Edwards, R.W. (1974) Population dynamics and production of *Daphnia hyalina* in a eutrophic reservoir. Freshwat. Biol. 4: 445–465.

Goldman, C.R., Morgan, M.D., Threkeld, S.T. and Angeli, N. (1979) A population dynamics analysis of the cladoceran disappearance from Lake Tahoe, California-Nevada. Limnol. Oceanogr. 24(2): 289–297.

Goulden, C.E. (1964) The history of the cladoceran fauna of Esthwaite water (England) and its limnological significance. Arch. Hydrobiol. 60: 1–52.

Graham, T.R. (1970) The zooplankton of Lough Neagh, 1968–1969. Ir. Nat. J. 16: 265–271.

Graham, T.R. and Logan, K.G. (1970) Preliminary studies of Lough Neagh ostracods. Ir. Nat. J. 16: 326–328.

Hall, D.J. (1964) An experimental approach to the dynamics of a natural population of *Daphnia galeata mendotae*. Ecology 45: 94–112.

Hart, R.C. and Alanson, B.R. (1976) The distribution and diel vertical migration of *Pseudodiaptomus hessei* (Mrázek) (Calanoida: Copepoda) in a sub-tropical lake in southern Africa. Freshwat. Biol. 6: 813–819.

Herzig, A. (1979) The zooplankton of the open lake. In: H. Löffler (ed.), Neusiedlersee: the limnology of a shallow lake in Central Europe. Monographie Biologicae 37, pp. 281–335.

Holmquist, C. (1959) Problems on marine-glacial relicts, an account of investigations on the genus *Mysis*. Lund, Berlingska Boktryckereit, 270 pp.

Jewson, D.H. (1977) Light penetration in relation to phytoplankton content of the euphotic zone of Lough Neagh. Oikos 28: 75–83.

Jewson, D.H., Rippey, B. and Gilmore, W.K. (1981) Loss rates from sedimentation, parasitism and grazing during the growth, nutrient limitation and dormancy of a diatom crop. Limnol. Oceanogr. 26(6): 1045–1056.

Johnson, D. and Walker, A.F. (1974) The Zooplankton of Loch Leven, Kinross. Proc. R.S. Edin. (B) 74: 285–294.

Jones, R.I. (1977a) Factors controlling phytoplankton production and succession in a highly eutrophic lake: I. The phytoplankton community and its environment. J. Ecol. 65(2): 531–547.

Jones, R.I. (1977b) Factors controlling phytoplankton production and succession in a highly eutrophic lake: II. The phytoplankton production and its chief determinants. J. Ecol. 65(2): 547–561.

Jones, R.I. (1977c) Factors controlling phytoplankton production and succession in a highly eutrophic lake: III. Interspecific competition in relation to irradiance and temperature. J. Ecol. 65(2): 561–579.

Longhurst, A.R. (1976) Vertical Migration. In: D.H. Cushing and J.J. Walsh (eds.) Ecology of the seas. Saunders, Philadelphia pp. 116–140.

Lund, J.W.G., Kipling, C. and LeCren, E.D. (1958) The inverted microscope method of estimating algal numbers and the statistical basis of estimations by counting. Hydrobiologia 11: 143–170.

Macan, T.T. and Lund, J.W.G. (1954) Records from some Irish lakes. Proc. Roy. Ir. Acad. 56(B): 135–157.

McClaren, I.A. (1963) Effects of temperature on growth of zooplankton and the adaptive value of vertical migration. J. Fish. Res. Bd. Can. 20: 685–727.

McQueen, D.J. (1969) Reduction of zooplankton standing stocks by predaceous *Cyclops bicuspidatus thomasi* in Marion Lake, British Columbia. J. Fish. Res. Bd. Can. 26: 1605–1618.

Morgan, N.C. (1980) Secondary production. In: D. Westlake and R.H. Lowe-McConnell (eds.) The functioning of freshwater ecosystems. Cambridge University Press, 588 pp.

Patalas, K. (1954) Porównawcze badania nad nowyn typem samoczynnego czerpacza planktonowego; hydrochemicznego. Ekol. Pol. Ser. A. 2: 231–242.

Pejler, B. (1957) Taxonomical and ecological studies on planktonic Rotatoria from Central Sweden. K. svenska vetensk. Acad. Handl. 6: 1–52.

Pejler, B. (1964) Regional ecological studies of Swedish freshwater zooplankton. Zool. Bidrag Uppsala 36: 407–515.

Porter, K.G. (1975) Viable gut passage of gelatinous green algae ingested by *Daphnia*. Verh. Internat. Verein. Limnol. 19: 2840–2850.

Porter, K.G. and Orcutt, J.D. (1980) Nutritional adequacy, manageability and toxicity as factors that determine food quality of green and blue-green algae for *Daphnia*. In W.C. Kerfoot (ed.) Evolution and ecology of zooplankton communities. Univ. Press of New England, pp. 268–281.

Pourriot, R. (1965) Recherches sur l'écologie des rotifères. Vie et Milieu, Suppl. 21, 224 pp.

Richards, R.C., Goldman, C.R., Frantz, T.C. and Wickwire, R. (1975) Where have all the

Daphnia gone? The decline of a major Cladoceran in Lake Tahoe, California-Nevada. Verh. Internat. Verein Limnol. 19: 835–842.

Ruttner-Kolisko, A. (1974) Plankton rotifers (biology and taxonomy). In: Die Binnengewässer Vol. XXVI/I Suppl. Stuttgart, E. Schweizerbart'sche Verlagsbuchhandlung (Nägele & Obermiller), 146 pp.

Saunders, G.W. (1969) Some aspect of feeding in zooplankton. In: NAS sponsored Symposium: Eutrophication. Causes, consequences and correctives. NAS Washington D.C., pp. 556–573.

Schindler, J.E. (1971) Food quality and zooplankton nutrition. J. Anim. Ecol. 40: 589–595.

Smyly, W.J.P. (1968) Observations on the planktonic and profundal Crustacea of the lakes of the English Lake District. J. Anim. Ecol. 37(3): 693–708.

Southern, R. and Gardiner, A.C. (1932) Reports from the limnological laboratory II. The diurnal migrations of the Crustacea of the plankton in Lough Derg. Proc. Roy. Ir. Acad. (B) 40: 121–159.

Sutcliffe, D.W. (1967) A re-examination of observations on the distribution of *Gammarus duebeni* Lilljeborg in relation to the salt content of freshwater. J. Anim. Ecol. 36: 579–597.

Thienemann, A. (1920) Die Grundlagen der Biozönotik und Monards faunistische Prinzipien. Festschrift für Zschokke, Nr. 4, Basel.

Thorpe, J.E. (1974) Trout and Perch Populations at Loch Leven, Kinross. Proc. R.S.Edinb. (B) 74: 20, 295–331.

Wilson, J.P.F. (1979) The biology and population ecology of the Pollan, *Coregonus autumnalis pollan* Thompson, of Lough Neagh. D. Phil. Thesis, New University of Ulster, pp. 123.

Wilson, J.P.F. (1984) The food of the pollan, *Coregonus autumnalis pollan* Thompson, in Lough Neagh, Northern Ireland. J. Fish. Biol. 24: 253–261.

Wood, R.B. and Gibson, C.E. (1973) Eutrophication and Lough Neagh. Wat. Res. 17: 173–187.

Wright, J.C. (1965) The population dynamics and production of *Daphnia* in Canyon Ferry Reservoir, Montana. Limnol. Oceanogr. 10: 583–590.

Zaret, T.M. (1980) Predation and freshwater communities. Yale University Press pp. 208.

Zaret, T.M. and Suffern, J.S. (1976) Vertical migration in zooplankton as a predator avoidance mechanism. Limnol. Oceanogr. 21: 804–813.

15. Some observations on the populations of *Mysis relicta* in Lough Neagh

T.E. ANDREW and E. WOODWARD

15.1. Introduction

Within the British Isles, the occurrence of *Mysis relicta* Lovén is restricted to relatively few lakes and rivers (Tattersall and Tattersall, 1951). Two large lakes, Lough Neagh and Lough Erne, containing this species are to be found in Northern Ireland. The presence of *Mysis* in Lough Neagh has been known for a long time and a description of its collection and diurnal behaviour and speculation as to its origins in freshwater ecosystems appear in the early description of the lough by Dakin and Latarche (1913). They suggest that *Mysis*, often considered to be a benthic form, 'paradoxically is as important a plankton form as any other species of plant or animal caught in the lake'. Since then its presence in the lough has been referred to by a variety of workers including Tattersall and Tattersall (1951), Holmquist (1959) and Gledhill *et al.* (1976).

While much of the earlier interest in *Mysis relicta* had been stimulated by a description of the species as a glacial relict (Segerstråle, 1957; Holmquist, 1959) more recent studies in other lakes have been directed towards more quantitative aspects of its biology. *Mysis relicta* is generally considered to be a species whose breeding season is markedly restricted by the environmental temperature and, depending upon the lake, has a varying number of generations annually (see Mauchline, 1980, for a review). It is recognised as a species which often inhabits the benthic region but is capable of migration into the free waters and as such is important to both regions of the lake. Its occurrence in Lough Neagh is of interest then, because this lake, unlike many of the others that have received attention, is well mixed and of potentially high environmental temperature for this species. As a member of the lake community, the available evidence suggests that it not only serves as an important item of the diet of several commercially important fish but it also predates upon zooplankton and may be a competitor for the food of several zooplankton and benthic species.

R.B. Wood and R.V. Smith (eds.), Lough Neagh, 327–338.

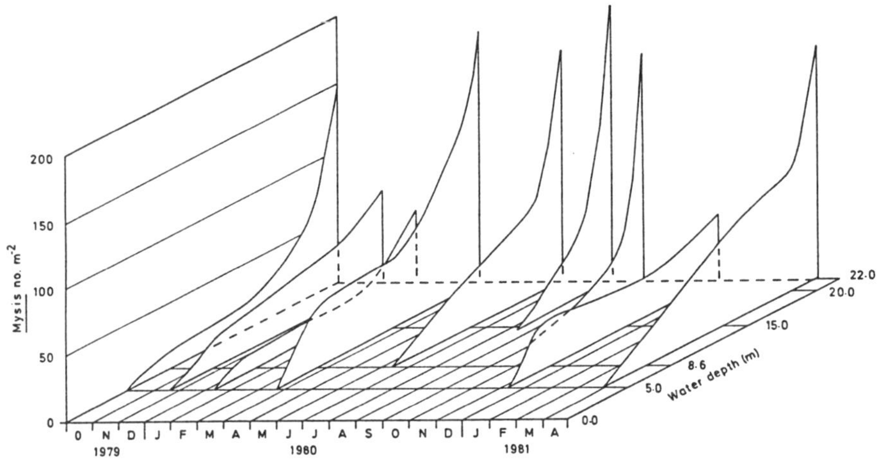

Figure 15.1. Horizontal-depth distribution of *Mysis* along transects on different sampling occasions.

15.2. Distribution

The vertical and horizontal distribution of *Mysis* in the lake is complex. *Mysis* is frequently captured in vertical net hauls used for routine zooplankton sampling and a combination of vertical and horizontal trawls were used for the population element of this study. No successful measurement of diurnal variation in vertical distribution was made although the routine sampling programme at 11.00 h indicated *Mysis* to be distributed throughout the bottom 2 m of the water column. *Mysis* is frequently captured in core samples taken during the day, i.e. at the surface of the mud but casual observations suggest that individuals may be captured in net or volume samples at different times of the day and at different depths. This appears to be at variance with the findings of Beeton (1960) and Fürst (1972) who find distinct diurnal patterns in their lakes but conforms with the observations of Hakala (1978). The underwater light climate has been advanced as a controlling mechanism for mysid migrations (Beeton, 1960) and Chapter 4, Fig. 4.4 includes a light level in Lough Neagh appropriate to this.

Horizontal distributions were mapped by transect sampling on several occasions and at different times of the year. The results may be seen in Fig. 15.1. Maximum population densities, on each sampling occasion, were recorded from the greatest sampling depth and animals have been collected from the deepest part of the lake, 27 m, although it is not known whether these animals were benthic or pelagic in origin. On these sampling occasions, around mid-day no animals were caught in less than 5 m water depth. However, this isobath closely corresponds with the change, with increasing depth, from sandy to muddy sediment types in this part of the lough (see Chapter

3), and the distribution may therefore be related to bottom type. During the summer months, June to September, the populations appear to retreat to waters with a column depth of 10 m or more. At this time the population is composed almost entirely of immature individuals. The reasons for and the mechanism that controls this migration are unclear. Hakala (1978), observing a similar migration pattern in Lake Pääjärvi, related *Mysis* distribution to the upper limit of the 7 °C isotherm, which occured at the 20 m isobath, and suggested that mysids prefer these colder waters. In Lough Neagh, the well mixed nature of the water column, allowing only transient stratification of temperature and oxygen levels, precludes temperature from this argument. The control may well be light induced, either a light intensity reaction or daylength reaction but alternative explanations must be sought to explain the phenomenon itself. Two more or less attractive possibilities exist. Firstly that the migration of the immature population is a response to light climate, as may indeed be the case in Lake Pääjärvi, but this retains no functional significance in Lough Neagh. Secondly, the size structure of the population and the feeding habits of the fish populations may render them more susceptible to fish predation in the shallower waters, particularly benthic feeders such as perch and eels, and it may therefore be a predator avoidance reaction again with a light climate control. These fish species, particularly the developing juveniles, are frequently found in the littoral and sub-littoral regions in the summer months and move into deeper water during winter (Tobin and Winfield, pers comm). *Mysis* appears to follow the reverse pattern, and more detailed study of any relationship might be illuminating.

15.3. Population growth

In Lough Neagh, *Mysis* is a univoltine species which breeds in the late autumn when water temperatures fall below 7 or 8 °C, corresponding with Hakala's (1978) observations. The population density, Fig. 15.2, varies considerably with season; low numbers, 10 to 30 m^{-2} are present in the winter months and maximum population densities occur in the spring or summer. In 1980 and 1981 maximum values of 58 m^{-2} and 62 m^{-2} respectively were observed although 398 m^{-2} immature individuals were recorded in 1978. Estimates of population density vary throughout the summer more than inherent sampling variation will allow. The explanation lies in the fact that the standard sampling point was at the 8.6 m isobath and the suitability of this as a regular sampling point may be questioned on the basis of results presented in Fig. 15.1. Its use for a population census certainly distorts the summer figures and probably underestimates the total population density throughout the year. Real summer values should be less than, but in the same order of magnitude as, the spring peak to be consistent with a single cohort population.

Population standing crop biomass, for the 8.6 m isobath, is also shown in

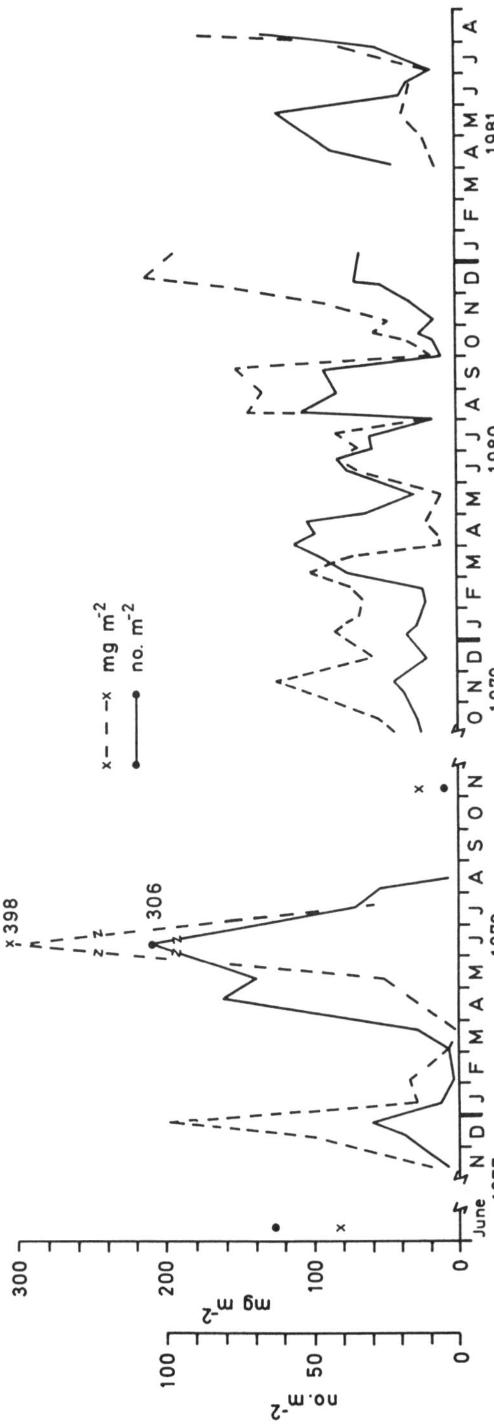

Figure 15.2. Population density and standing crop biomass of *Mysis*.

Fig. 15.2. Biomass does not directly follow the pattern seen for population numbers and two factors contribute to the variability of this estimate. Firstly, as with population density estimates, the distribution of *Mysis* is not homogenous and 8.6 m is not a representative sampling point, especially in early summer. Secondly, although adult *Mysis* may not have the same numerical importance in the population structure, they make a large contribution to the total biomass when present even in small numbers. Thus, during the reproductive season, October to March, standing crop biomass may be high, $>0.2\ g\ m^{-2}$ in 1977 and 1980 whereas in early spring, high population numbers correspond with low standing crop biomasses. However, when high population numbers and high individual weights coincide in summer situations, high standing crops obviously occur, $0.4\ gm^{-2}$ in 1978, $0.15\ gm^{-2}$ in 1980 and $0.18\ gm^{-2}$ in 1981.

The percentage composition of different elements of the population, classified as juveniles, sexually differentiated males and females, and females with or without broods, is shown in Fig. 15.3. The pattern that emerges is remarkably similar for each year. The single cohort nature of the population is clearly revealed with the juveniles being released from the marsupium in February or March synchronised with the disappearance of most of the females of the previous cohort. These juveniles, numerically representing >90% of the population, grow and become sexually mature during late August and September. Breeding occurs and females are observed to be carrying embryos from November on. At this point, the males begin to be lost from the population eventually becoming completely absent before the appearance of the next generation. The proportion of the females having broods increases, reaching a peak in February when, coincident with the release of the broods to the next generation, they largely disappear from the population. A low proportion, 2 to 10%, however, survive for up to six months after this time, until the next generation becomes sexually mature. The reasons for the survival of some females, but no males, is not clear. They do not go on to produce a second cohort, but although this may be energetically senseless in Lough Neagh they may be homologous with *Mysis* of other lakes that may go on to produce a second cohort in the same (Hakala, 1978) or subsequent season in more appropriate environmental conditions. The absence of males makes this impossible in Lough Neagh and it is possible that it is the lack of a suitable temperature refuge which does not allow the males to survive. The fate of the males is that large numbers appear to be washed out from the lake into the River Bann in a spent condition but, although fragile, they are not dead. Observations have been made of gulls feeding on these males in the surface waters of the river.

As noted above, embryos were found in the marsupium from late November, 1980/81, or early December 1977/78 and 1979/80, undergoing morphogenesis there until released as juveniles. The mean number of embryos per brood increased during this period from approximately 12 to 13 at the first observations to 18–19 towards the end of the breeding season. The

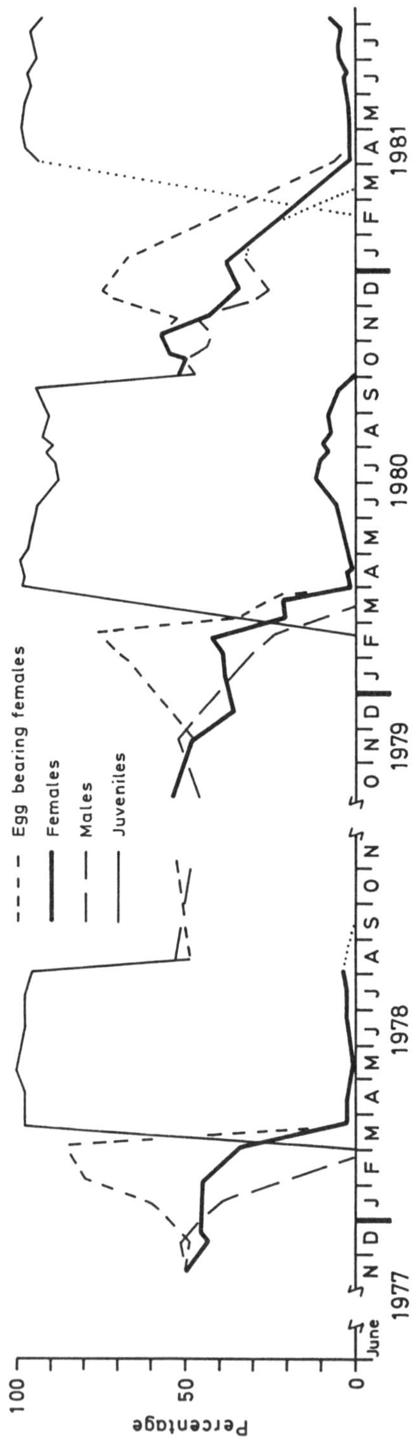

Figure 15.3. Seasonal variation of population structure of *Mysis* during study period.

variation in egg numbers per brood increased with the progress of the breeding season. The proportion of egg-bearing females also increased, reaching a peak in February or March, and at this peak period 35 to 50% of all females have a brood. The 1980/81 figures are distorted by missing samples. The time it took for the embryos to develop, taken in this case as the time of the first observation of eggs in population samples until the appearance of juvenile stages, varied between 124 and 135 days. This variation, although the smaller estimate might be artifactually short because of sampling inadequacies, is usually ascribed to temperature functions, for example the degree-days encountered by the eggs. Table 15.1 shows the mean environmental temperature, the minimum temperature and the degree-days encountered by the embryos during development. It is apparent that, in this case, temperature functions provided an inadequate description of developmental rate and the total time in the brood pouch appeared to be the controlling factor.

Developmental stages of the embryonic development were recorded from field populations. The following stages were recognised:

1. A ball of cells
2. An elongated embryo with no distinguishing features
3. The start of eye development
4. Prominent eyes
5. Embryo ready for release from marsupium

The growth of *Mysis relicta* has been described most frequently as an increase in length although this might be related to instar development (Morgan and Beeton, 1978). Hakala (1978) described growth in terms of increase in weight, the approach adopted here. This appears to be appropriate as the Pääjärvi and Lough Neagh populations conform in size, and Hakala (1978) presents a length-weight relationship including individuals of up to 20 mg but mainly falling below 10 mg ash free dry weight (equivalent to approximately 11 mg dry weight). It may be inferred from Carpenter *et al.* (1974) and Morgan and Beeton (1978) that in North American lakes significant numbers of *Mysis relicta* reach the 16 mm, >20 mg dry weight, categories, a size range rarely encountered in Lough Neagh but one that corresponds with instar IV of the Great Lakes mysids.

The seasonal individual growth rates, a composite of all years of this study, for a mean individual of one cohort, are presented in Fig. 15.4. Maximum embryo weights of the 1979/80 cohort were 0.124 mg dry weight falling to 0.065 mg just before release from the marsupium. The mean weight of the smallest individuals captured on the first sampling occasion after their release was 0.051 mg. The 60% drop in weight of eggs from breeding until the new cohort begins feeding occurred during the winter period. The corresponding figures for 1980/81 are 0.199 to 0.063 mg whereas Hakala (1978) reported embryo weights of 0.112 and 0.090, i.e. larger at the time of release from the marsupium. The release of young from the marsupium is highly synchronised occurring within two weeks in March. The mysids grow to a weight of 3.5 mg (2.5 to 3.00 mm) when sexes can be easily distinguished.

Table 15.1. State of embryonic development and brood size of *Mysis* in Lough Neagh (derived from field samples)

Date of sample	% Embryos in stage of development					Brood size (with SD)	Development time (days)	Temperature (°C)	Degree days above 0 °C
	1	2	3	4	5				
1977/78	–	–	–	–	–	–	130	mean 2.1 minimum 1.4	278
6.12.79	100	–	–	–	–	13.2 ± 3.2			
13.12.79	85	15	–	–	–	12.6 ± 2.9			
11.1.80	–	90	10	–	–	14.5 ± 3.6			
14.1.80	–	85	15	–	–	15.6 ± 4.8	124	mean 2.7	340
24.1.80	–	15	80	5	–	17.1 ± 5.2		minimum 1.7	
7.2.80	–	5	70	25	–	16.3 ± 4.5			
21.2.80	–	–	40	35	5	15.2 ± 3.4			
7.3.80	–	–	–	10	90	16.4 ± 8.6			
20.11.80	90	10	–	–	–	11.3 ± 4.6			
9.12.80	50	35	15	–	–	12.6 ± 3.5			
11.12.80	45	45	10	–	–	13.4 ± 2.4	135	mean 5.5	746
9.1.81	–	10	65	25	–	16.2 ± 5.1		minimum 4.0	
3.4.81	–	–	–	15	85	18.7 ± 3.0			
16.4.81	–	–	–	5	95	17.5 ± 5.8			

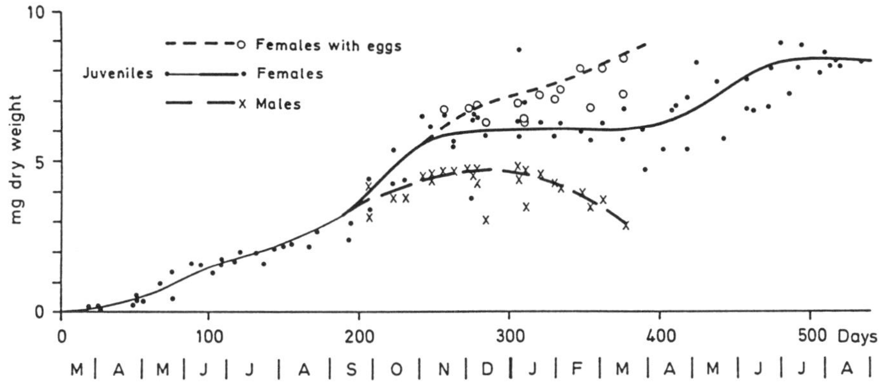

Figure 15.4. Mean growth of *Mysis* in Lough Neagh, derived from all data obtained during this study.

Beginning in September, male and female growth rates diverge markedly, with males growing more slowly than females and reaching a maximum weight of 4 to 6 mg dry weight which was maintained between November and January. Females grew rapidly becoming sexually mature at about 6 mg, a weight that is fairly constant during the breeding season, higher weights are a result of the brood weight. After the release of the brood, growth of the surviving females was resumed and these grew to a weight of 8.0 to 8.5 mg at which level they remained until they finally died the following August. This weight was slightly lower than Hakala's (1978) 10 mg level of the second broods in Pääjärvi.

15.4. Population production

Any estimate of production made from the population estimates available, bearing in mind the heterogenous nature of the distribution, must be treated with due caution. The values presented apply to the 8.6 m isobath and not the whole lake area. Using the Allen curve method (Allen, 1950), estimates were made for the 1977/78 cohort at 0.496 g m^{-2} yr^{-1} and for 1979/80 at 0.456 g m^{-2} yr^{-1}. The 1979/80 cohort, despite not reaching 1978 population biomass levels, still achieved 90% of that year's production. This may be ascribed to the adult production. Unestimated is the element of the production lost as exuviae during moulting which may occur between 9 and 16 times in the Lough Neagh population cycle. Hakala (1978) suggested that 5 to 10% of the dry weight is lost in cast skins at each moult. The annual production to biomass ratio for 1980/81 was calculated at 5.07, higher than the 3.0 to 3.8 for Pääjärvi.

15.5. A note of the genetic variation in Lough Neagh *Mysis relicta* (by R.D. Ward and R.I. Jones)

In 1980, a preliminary survey into the extent of genetically determined enzyme variation in the Lough Neagh population of *Mysis relicta* was carried out. Standard starch gel procedures were adopted, and buffer systems followed those given in Ward and Beardmore (1977).

A total of 19 loci was screened. Three were found to be polymorphic, and allele frequencies for these are given in Table 15.2. Sixteen loci were found to be monomorphic, although sample sizes for several of the loci are small. These loci are: *Pgm* ($N = 14$), *Idh*-1 (4), *Idh*-2 (4), *Xdh* (8), *Mdh* (17), *Ldh* (2), *Me* (10), *Ap*-2 (5), *Aat*-1 (25), *Aat*-2 (23), *Lap*-1 (16), *Sod* (14), *Alp*-1 (18), *Mpi* (12), *Hk* (3) and *Ada*-1 (5).

Table 15.2. Allele frequencies of polymorphic loci of Lough Neagh *Mysis relicta*

			Allele frequency	
Enzyme	Locus	*N**	*A*	*B*
Phosphoglucose isomerase	*Pgi*	47	0.979	0.021
Aminopeptidase – 1	*Ap*-1	12	0.208	0.792
Esterase – 2	*Est*-2	18	0.500	0.500

**N* is the number of individuals screened.

The percentage of polymorphic loci is 15.8%, and the average heterozygosity per locus, assuming Hardy-Weinberg equilibria, is 4.5%. These figures are rather low for invertebrates, where the average proportion of polymorphic loci is around 40% and the average heterozygosity per locus around 10% (Nevo, 1978). Preliminary studies on a Finnish population in Lake Pääjärvi have also given low estimates of variability. These low estimates may be the consequence of genetic drift in more or less isolated populations.

It is interesting to note that while some enzymes are monomorphic for the same alleles in the two populations thus far examined, the Lake Pääjärvi population is polymorphic at the *Ap*-1 locus for two slower alleles than those found in Lough Neagh. Thus for this particular locus, the genetic identity between the two populations is zero, and clearly the populations are genetically distinct from one another. Whether the variation at this locus represents an adaptive response to two different environmental regimes (see Section 15.6, this Chapter) or whether it results from founder effects and/or genetic drift cannot be ascertained at present.

15.6. The occurrence of *Mysis relicta* in Lough Neagh

The origins of *Mysis relicta* as an element of freshwater lake faunas is still the subject of unresolved debate (Segerstråle, 1957; Holmquist, 1959) but its occurrence in Lough Neagh is, in several respects, more of a puzzle when compared with its occurrence in Scandanavian and North American lakes. The lake morphometry, the physical and chemical environment and the highly eutrophic nature of Lough Neagh each contribute to this puzzle and modify the *Mysis* life history and ecology. The univoltine nature of its life history is not unique (Mauchline, 1980) but this appears to be controlled by relatively high water temperatures and the well-mixed nature of the water column encountered in Lough Neagh. This limits production to one breeding effort which occurs during the winter. The death of the males during the breeding season precludes the production of a second cohort despite the survival of a small proportion of females, which resume growth after the breeding season, suggesting potential for a second cohort. This may represent the close affinities with other European populations of *Mysis* as, apart from the death of the males, they conform in terms of size, growth rates, population densities, egg production and timing to those reported for Scandanavian lakes (Holmquist, 1959; Hakala, 1978). The specific environmental characteristics of Lough Neagh suggest, in relation to the *Mysis* population growth, avenues of study that might be profitably explored to explain the occurrence of this species in this lake. These include more detailed examination of the distribution and factors affecting the distribution of the species, feeding studies, fish predation rates and the physiological attributes of Mysis that enable it to tolerate the environment encountered in Lough Neagh.

15.7. References

Allen, K.R. (1950) The computation of production of fish populations. N.Z. Sci. Rev. 8: 89.

Beeton, A.M. (1960). The vertical migration of *Mysis relicta* in lakes Huron and Michigan. J. Fish. Res. Board Can. 17: 517–539.

Carpenter, G.F., Mansey, E.L. and Watson, N.H.F. (1974) Abundance and life history of *Mysis relicta* in the St. Lawrence Great Lakes. J. Fish. Res. Bd. Can. 31: 763–782.

Dakin, W. J. and Latarche, M. (1913) The plankton of Lough Neagh. Proc. R. Ir. Academy 30 (B) 3: 20–96.

Fürst, M. (1972) On the biology and introduction of the opossum shrimp *Mysis relicta* Loven. Acta. Univ. Upsalensis 207: 1–7.

Gledhill, T., Sutcliffe, D.W., and Williams, W. D. (1976) Key to British Freshwater Crustacea: Malacostraca. Freshwater Biological Association Scientific Publication No. 32, pp. 72.

Hakala, I. (1978) Distribution, population dynamics and roduction of *Mysis relicta* (Loven) in southern Finland. Ann. Zool. Fennici. 15: 243–258.

Holmquist, C. (1959) Problems on marine glacial relicts, an account of investigations on the genus Mysis. Lund, Sweden, pp. 270.

Mauchline, J. (1980). The biology of Mysids and Euphausids. In: J.H.S. Blaxter, F.S. Russell and M. Yonge. Advances in marine biology 18. Academic Press, pp. 681.

Morgan, M.D. and Beeton, A.M. (1978) Life history and abundance of *Mysis relicta* in Lake Michigan. J. Fish. Res. Bd. Can. 35: 1165–1170.

Nevo, E. (1978) Genetic variation in natural populations: patterns and theory. Theor. Pop. Biol. 13: 121–177.

Segerstråle, S. G. (1957) On the immigration of the glacial relicts of northern Europe, with remarks on their prehistory. Commentat. biol. 16: 16, 1–117.

Tattersall, W.M. and Tattersall, O.S. (1951) The British Mysidacea. London (Ray Society), pp. 460.

Ward, R.D. and Beardmore, J.A. (1977) Protein variation in the plaice, *Pleuronectes platessa* L. Genet. Res. 30: 45–62.

16. The macroinvertebrate fauna of Lough Neagh

C. E. CARTER and P.M. MURPHY

16.1. Introduction

Work on the macroinvertebrates of Lough Neagh has largely been concentrated on the profundal (deeper than *c.* 6 m) and the bulk of this chapter is devoted to that (Section 16.2). However, some work on the littoral fauna has been carried out and preliminary results from this are included (Section 16.3). The intermediate zone (2 to 6 m depth) is poorly known although some information is included in the profundal section.

The sediments of the lough are described in Chapters 3 and 7A, being very roughly 52% mud, 40% sand and 8% rocky. The profundal zone is entirely fine mud, the intermediate zone mainly sand and the littoral both sandy and rocky.

16.2. The profundal macroinvertebrate fauna

This section is devoted to the macroinvertebrates inhabiting the deeper area of the lough where fine mud (mean grain size 10 μm) is the predominant substrate (see Chapter 3). The water depth is generally greater than *c.* 6 m. The two main groups of macroinvertebrates found are Chironomidae (Diptera) and Oligochaeta (mainly Tubificidae) and most work has concentrated on these; *Helobdella stagnalis* (L.) (Hirudinea) and *Polycelis nigra* (Muller) (Platyhelminthes) are also found together with Sphaeriidae (Bivalvia) and *Gammarus* (Amphipoda).

16.2.1. *Chironomidae*

16.2.1.1. *Community composition*

Marks and Henderson (1970) studied the fauna of Lennymore Bay (eastern shore) and included one deep water (14 m) site and Irwin (1976) included a brief description of the habitats of some of the larvae in his study of the biochemical and morphological features of chironomid larvae in Lough

R.B. Wood and R.V. Smith (eds.), Lough Neagh, 339–367.

Table 16.1. Species composition of the chironomid community (mud sediment) of Lough Neagh

Chironomus anthracinus Zett.
Chironomus plumosus L.
Procladius crassinervis Zett
Tanytarsus v.d. Wulp sp.
Cryptochironomus 'defectus' gr.
Glyptotendipes [mainly *G. paripes* (Edw.) but also *G. pallens* (Meig.)]
Polypedilum Kieff. sp.
Pentapedilum Kieff sp.
Microtendipes [mainly *M. pedellus* (Degeer) but also *M. chloris* (Meig.)]
Endochironomus albipennis (Meig.)
Limnochironomus Keiff. sp.
Pentaneurini sp.

Neagh. Carter (1976, 1978, 1980a, 1980b) has studied larval chironomid populations, chiefly in the deeper areas of the lough, with particular emphasis on life cycles and abundance and Ripley (1980) has done work on the respiratory physiology of some of the deep water species. The species list for the mud substrate (Table 16.1), compiled from these sources, is not very extensive but is comparable to that found elsewhere in similar sediment, particularly Loch Leven (Maitland and Hudspith, 1974; Maitland, 1979). Some of the species, e.g. *Chironomus anthracinus* and *Procladius crassinervis*, are widely distributed throughout the muddy areas, while others, e.g. *Endochironomus*, seem to have either a very restricted or clumped distribution. There appears to have been no major change in species during the period of the studies (1969–present) although, as described later, there have been fluctuations in abundance and possible changes in distribution. For instance, *Glyptotendipes* sp. which in 1969–1971 was persistently present at a water depth of 8 m, was only occasionally found there in 1974–1976, apparently being concentrated in the sandy sediment which had always appeared to be its preferred habitat.

It is difficult to compare the fauna of Lough Neagh with that of other Irish loughs because of a lack of published work, particularly on the profundal fauna. The majority of species found here are listed in Murray (1972) where the location at which he found them is given but no indication of habitat. McGarrigle (1980) has worked on chironomid communities in Lough Derravaragh, Co. Westmeath, with particular reference to their relationship to sediment characteristics. The genera he found were very similar to those in Lough Neagh and calculation of a correlation matrix of sediment parameters and taxa revealed a strong correlation between fine grain size, high organic content and a group of three taxa, *C. plumosus*, *C. anthracinus* and *Procladius*, the three groups most common in the mud in Lough Neagh.

16.2.1.2. *Distribution in relation to water depth*

Within the apparent sediment uniformity of the mud environment, differences in larval distribution with water depth can be seen for the more abundant species (Carter, 1976, 1978) (Fig. 16.1). The depths 0.5 to 6 m in the 1969–1971 graphs have sandy sediments and are included to show the influence of the major change in sediment type on the different species. *C. anthracinus* is generally present throughout the muddy area, though not in sand, but tends to be less abundant at greater depths in contrast to *C. plumosus* which is generally more abundant there. *Glyptotendipes* sp. was less abundant in mud than in sand, very rarely occurring beyond 12 m water depth in 1969–1971 and later becoming uncommon in the mud, although still present in sandy areas. *Procladius crassinervis* was found to a limited extent in sand, but otherwise showed a similar distribution pattern to *C. anthracinus*.

16.2.1.3. *Abundance and life cycles*

The abundance and life cycles of the Chironomidae have been most intensively studied at a depth of *c.* 8 m, close to the mean depth of the lough. Two different samplers were used during the studies: in 1969–1971 a Gilson corer with restricted flow through, and in 1974–1976 a multiple corer which afforded virtually free flow. Thus, differences in numbers between the two periods are not to be construed as significant. A comparison of the two corers indicated that the multiple corer estimates varied from 1× to 10× those of the Gilson, the size of the variation being dependent to some extent on the size of larva being sampled. The 1974–1976 population estimates have 95% confidence limits of approximately ±25% of the mean.

The *C. anthracinus* population (Fig. 16.2) varied from a peak for the 4th instar of 2,500 to 4,500 m^{-2} in August to <1,000 m^{-2} in March prior to emergence. This species generally has an annual life cycle in Lough Neagh with emergence in April–May (Carter, 1975), a pattern which can be seen in 1970, 1971, 1974 and 1976; however, in 1975 full emergence did not occur and some larvae spent a second year in the lough. The reason for this appeared to be poor growth of the larvae, perhaps due to a low phytoplankton crop and low mean temperature the previous autumn (Carter, 1980a). There was a poor recruitment of larvae to the population that year, perhaps due to predation by the remaining larvae on the new population, as suggested for Lake Esrom by Jonasson (1972).

C. plumosus (Fig. 16.3) has a population *c.* 10% of that of *C. anthracinus* (the figures shown are for a depth of 20–25 m where it was most abundant). It had a regular one-year life cycle during the perod of study with emergence in May–August.

Glyptotendipes was consistently present and fairly abundant in 1969–1971 (Fig 16.4) but was generally uncommon in 1974–1976; the reasons for this are unknown. The earlier study indicated a one year life cycle with emergence of *G. paripes* in May–September.

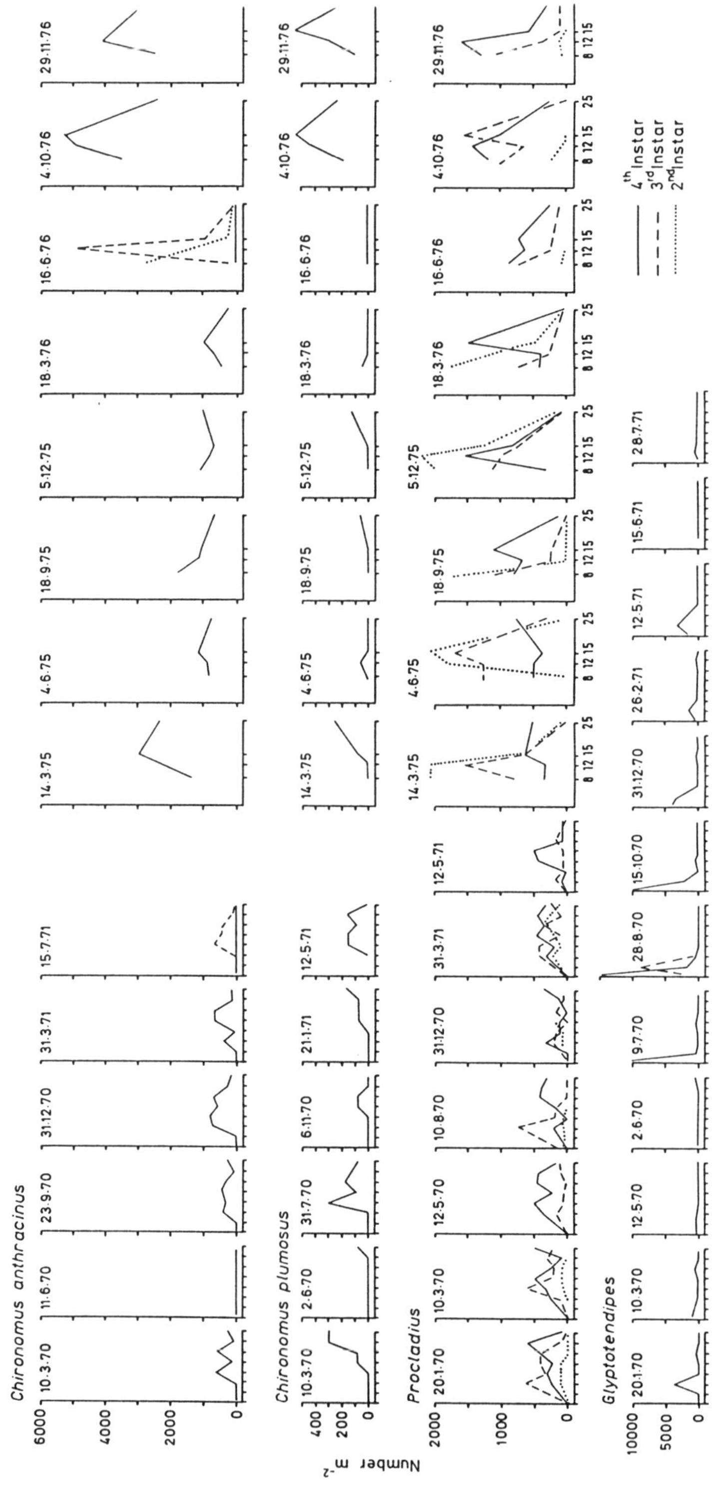

Figure 16.1. Distribution of chironomid species in relation to water depth.

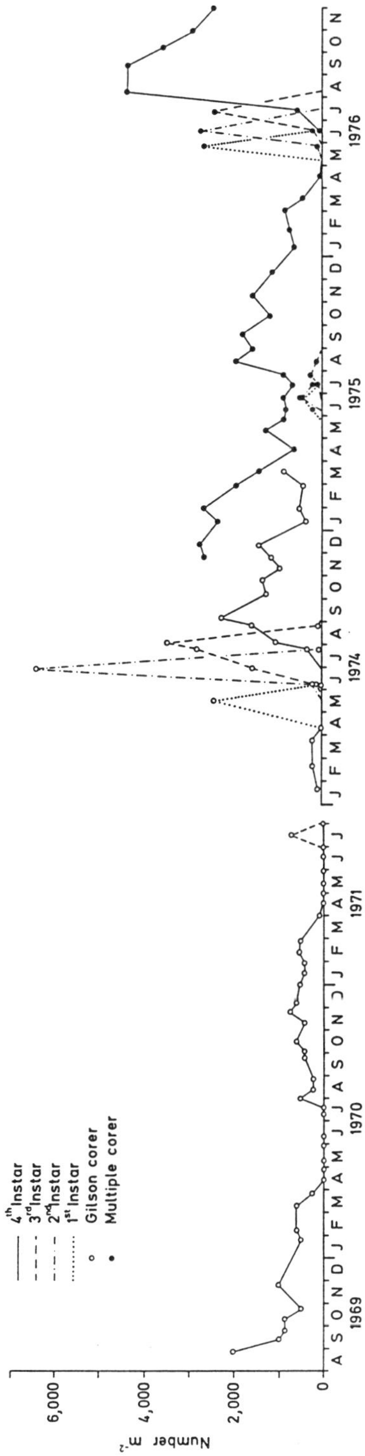

Figure 16.2. Larval population density of *Chironomus anthracinus*.

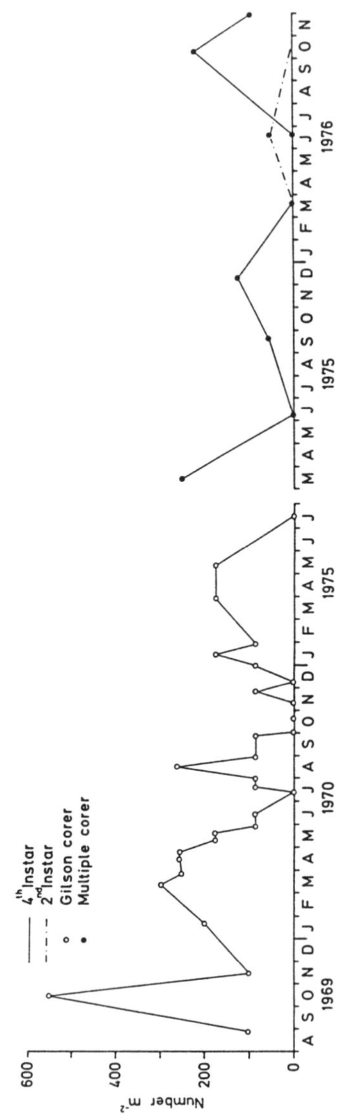

Figure 16.3. Larval population density of *Chironomus plumosus*.

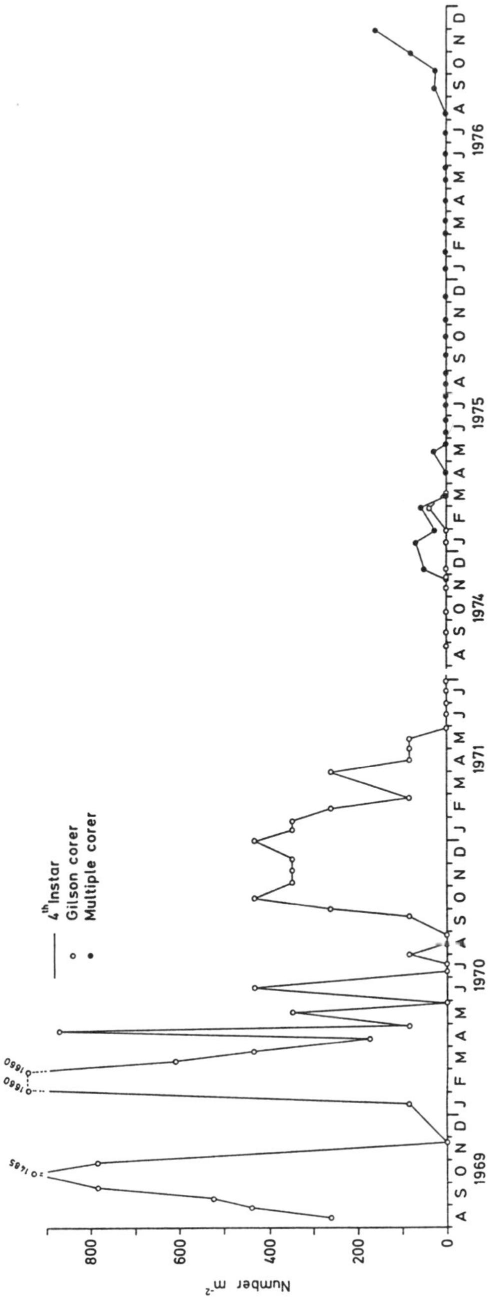

Figure 16.4. Larval population density of *Glyptotendipes*.

Procladius crassinervis larval numbers (4th instar) were generally less than 1,000 m^{-2}, although the earlier instars were more abundant (Fig. 16.5). Three instars of larvae were usually present at all times of year, with adult emergence from May to August. Scrutiny of the numbers and larval weights suggests that larvae which hatch early in the emergence period develop to the 3rd instar before the winter and emerge in late summer of the following year, a life cycle time of *c.* 15 months; those which hatch later, overwinter one year in the 2nd instar and a second one in the 4th instar, emerging early the next summer, a life cycle of 21 months (Carter, 1980b). *Procladius* was generally the second most abundant species, after *C. anthracinus*, at 8 m depth.

Tanytarsus larvae (Fig. 16.6) were abundant only in 1976 of the years studied, another example, like *Glyptotendipes*, of the rapid changes in chironomid populations that can occur. Assuming there was only one species (as the size distribution of the larvae suggests) there appeared to be three generations that year; May–June, July–September and October–April. Other species were found only irregularly and in small numbers ($<250\ m^{-2}$) so life cycles could not be determined.

16.2.1.4. *Biomass and production*

The biomass figures for the most abundant species (Fig. 16.7) show that in 1969–1971 *C. anthracinus* and *Glyptotendipes* were approximately equal contributors to biomass at 8 m while *C. plumosus* had the same biomass at 21 m. In 1974–1976, *C. anthracinus* had consistently the greater biomass, being about ten times more abundant than *Procladius* in biomass terms despite the similarity in numbers. (The apparent increase in *C. anthracinus* biomass between 1969–1971 and 1974–1976 is probably mainly due to increased sampling efficiency.) Assuming a P/B ratio of 4, about the middle of the range recorded for similar species (Charles *et al.*, 1974; Potter and Learner 1974; Titmus and Badcock, 1980), mean production of *C. anthracinus* in 1974–1976 was *c.* 10 g dry wt. $m^{-2}\ yr^{-1}$, of *Procladius c.* 1.2 g dry wt $m^{-2}\ yr^{-1}$. This is much less than the 34 g $m^{-2}\ yr^{-1}$ found by Charles *et al.* (1974) in Loch Leven, Scotland, but very similar to that in Lake Esrom (Jonasson, 1972) and several other lakes reported by Menzie (1981).

16.2.1.5. *The chironomid fauna of Kinnego Bay*

Kinnego Bay is a semi-enclosed bay at the south end of Lough Neagh, substantially enriched by sewage effluent (Jones, 1977; Carter and Rippey, 1982). There is an area of rich organic mud, water depth 3.5 m, inside the revetment which narrows the entrance, and samples have been taken from the mud to determine the possible changes in fauna which might occur if such extreme eutrophication took place in the open lough. The number of chironomids was less than at sites elsewhere in the lough, and only two genera were regularly present, *Chironomus* and *Procladius*. *Procladius* was

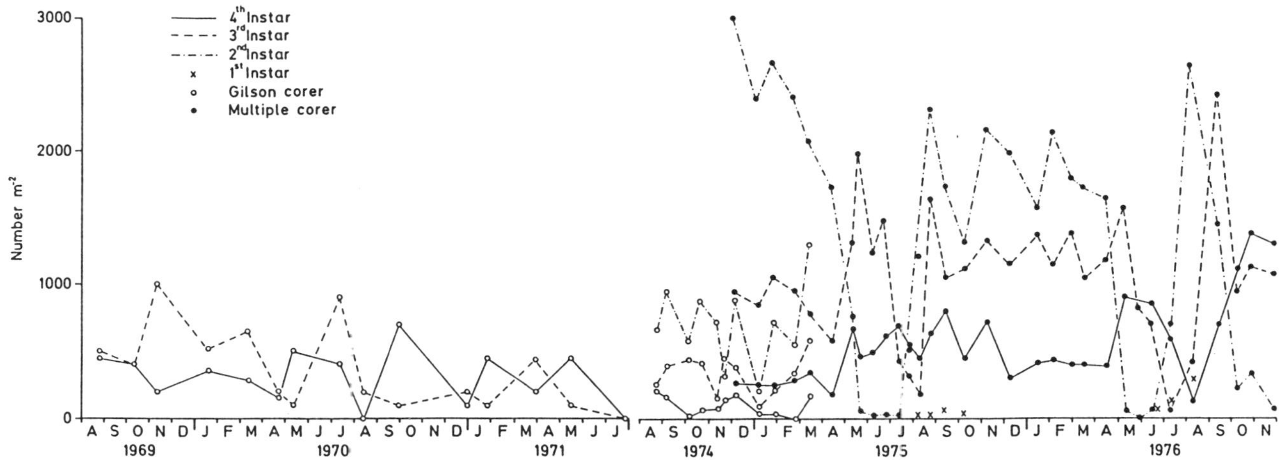

Figure 16.5. Larval population density of *Procladius crassinervis*.

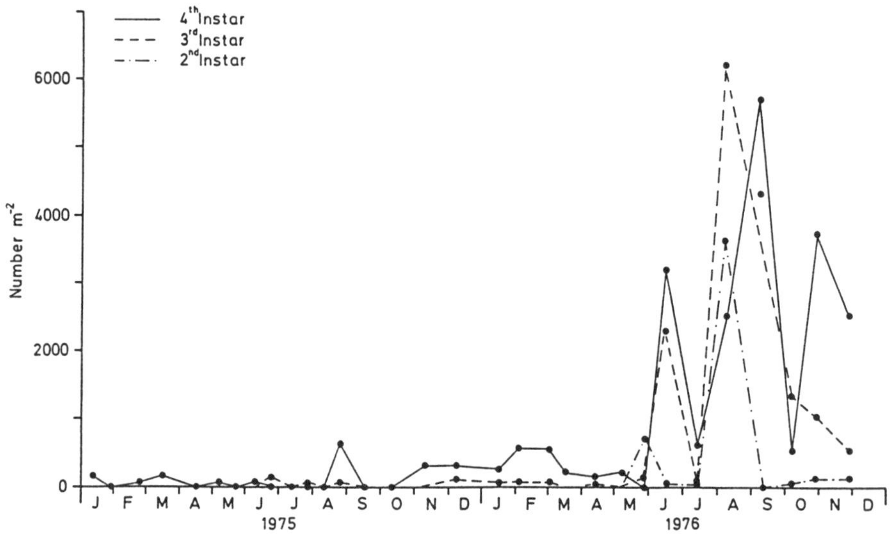

Figure 16.6. Larval population of *Tanytasus*.

the more abundant (Fig. 16.8), although the earlier instars appeared less able to tolerate conditions than the 4th instar; within *Chironomus*, *C. plumosus* was more common, at a density similar to that found in the open laugh at water depths of 20 m and more. Thus, this site had more affinities with the deepest areas of the lough than with comparable depths of similar sediment type.

16.2.1.6. *Respiratory physiology* (by M.P. Ripley)

Work has been done on the respiration of the main species of chironomids (Ripley, 1980) and the preliminary results from this are shown in Fig. 16.9. For all four species, there is reasonable agreement with the theoretical curve shown at the top, although the exact position on the axes varies. Zone A represents the winter range of temperatures to which the animals are adapted over which the Q_{10} varies from positive to negative; zone B is a non-acclimated area, representing the range of temperatures through which the lake passes fairly quickly while changing from the stable winter to the stable summer regime; zone C is the summer acclimation zone, when Q_{10} is approximately zero; and D is a zone in which the chironomids are not able to function properly. The temperature of onset of winter and summer acclimation can be derived from the curve for each species, and these are shown in Table 16.2. The two *Chironomus* species conform to the expected temperatures for their habitats, with *C. plumosus* acclimating at slightly lower temperatures because of being adapted to the temperature range of the deeper parts of the lough. *Procladius*, which is common over most of the lough, has a higher

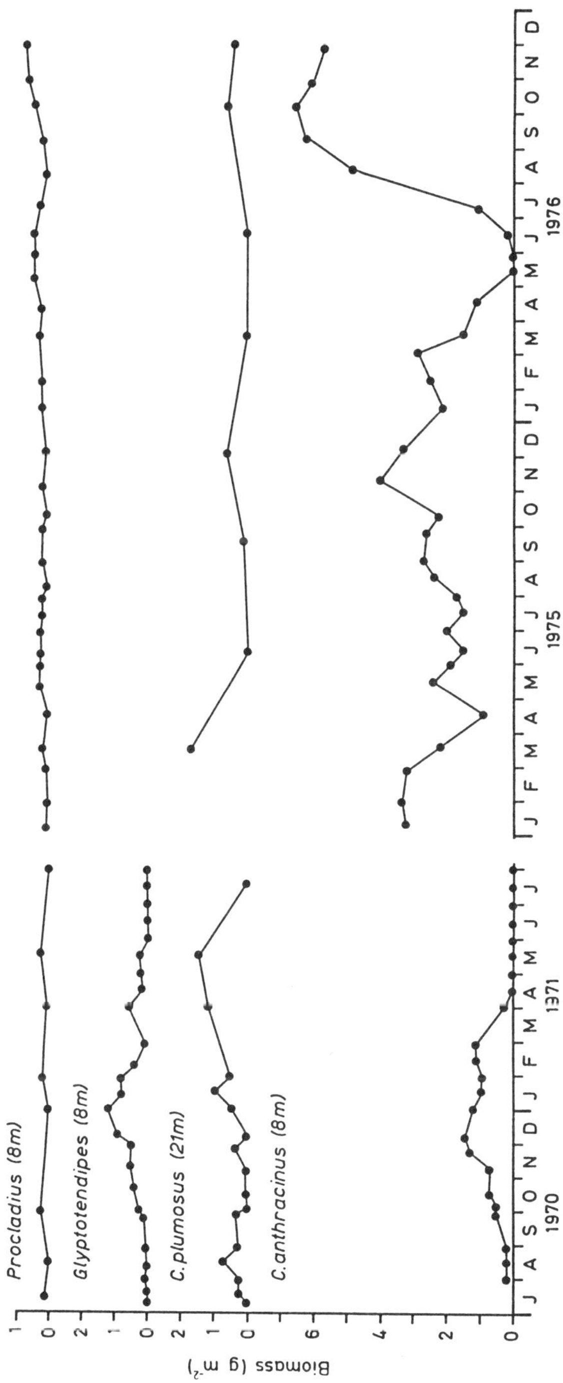

Figure 16.7. Biomass of the most abundant chironomid species.

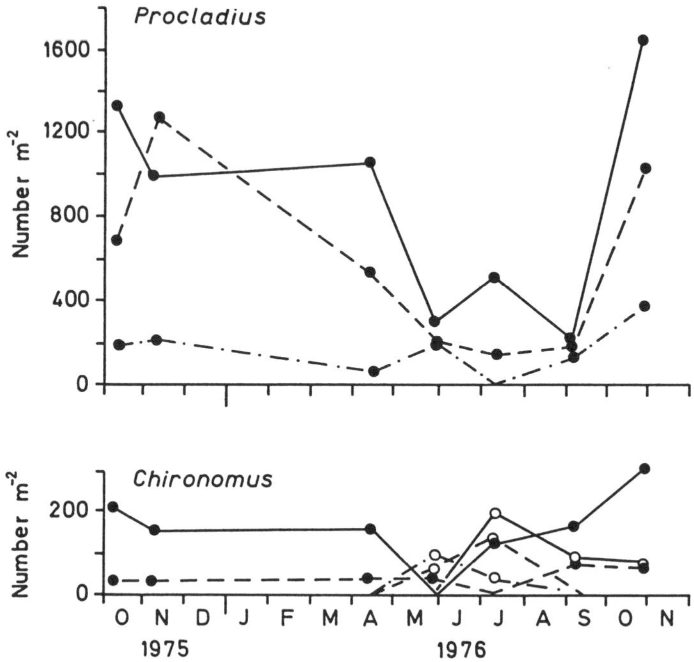

Figure 16.8. Density of principal chironomid species in Kinnego Bay. —— = 4th instar, - - - =3rd instar, -·-· = 2nd instar. In lower panel ● = *Chironomus plumosus*, ○ = *C authracinus*.

acclimation point and Glyptotendipes, confined to the shallower areas, has the highest of all.

16.2.1.7. *Summary*

The chironomid fauna of the muddy area of Lough Neagh shows similarities with that of a number of other lakes, particularly Loch Leven (Scotland), Lake Esrom (Denmark) and parts of Lake Erie (Brinkhurst *et al.*, 1968; Jonasson, 1972; Charles *et al.*, 1974; Maitland and Hudspith, 1974; Maitland, 1979). The closest correspondence in species terms is with Loch Leven, although Lough Neagh would not appear to be as productive, estimates giving a value only one-third of that of Loch Leven (see also Chapter 11 for comparisons of primary productivity), but very similar to Lake Esrom: in the latter, *C. anthracinus* is also the major contributor to chironomid production in the deeper areas. Wiederholm (1980) has calculated a benthic quality index based on chironomid indicator species and related this to eutrophication as indicated by total phosphorus concentration and mean depth. Calculating a similar index for Lough Neagh (Carter, 1980c) shows that the open lough falls into the eutrophic category and Kinnego Bay would be extremely eutrophic. The continuing presence of *Tanytarsus* (probably *T. lugens*) at 8 m depth is somewhat anomalous but the lack of stable summer stratification

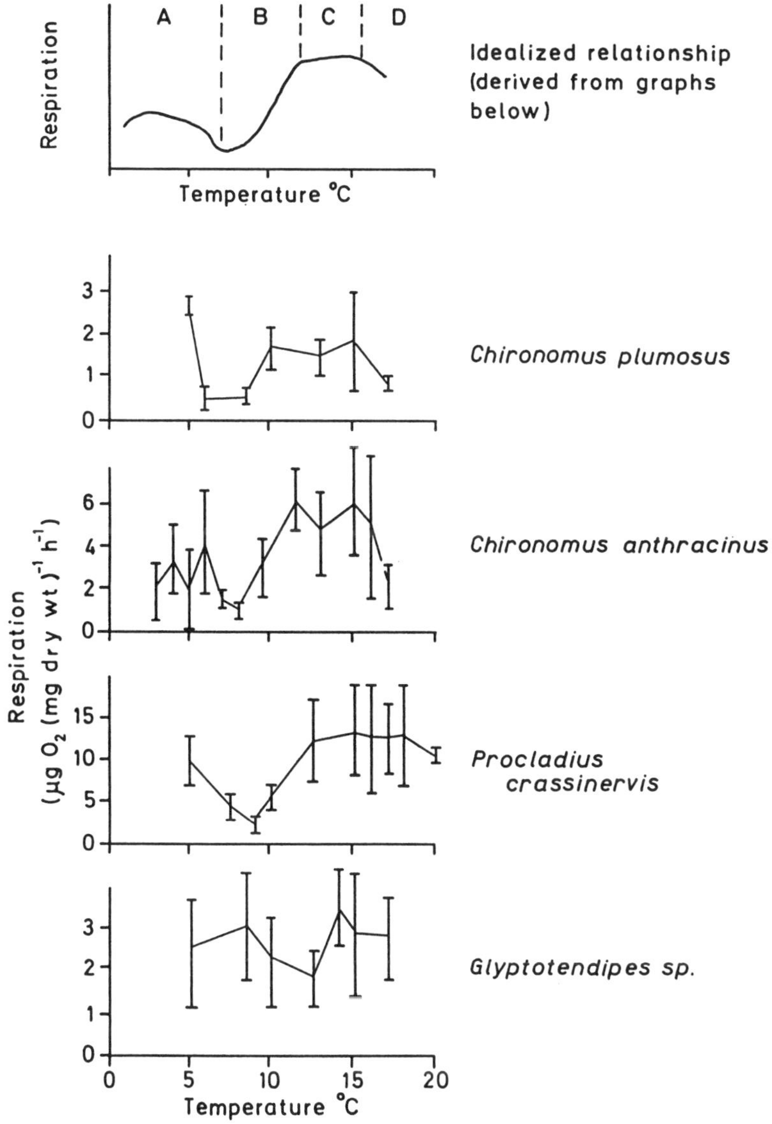

Figure 16.9. Respiration rate of chironomid species in relation to temperature. Redrawn from Ripley (1980).

does ameliorate conditions for the benthos. Studies of sub-fossil head capsules (Chapter 7B) have shown the relative change in the abundances of *Tanytarsus* and *Chironomus* over the last 100 years and, as the former is not found in any of the deeper areas, it would appear that it is living near the limits of its tolerance at 8 m; *Tanytarsus* in this context may therefore be of

Table 16.2. Acclimation temperatures for Lough Neagh Chironomidae

	Onset of summer acclimation (°C)	Onset of winter acclimation (°C)
C. plumosus	10.25	6.0
C. anthracinus	11.5	7.5
Procladius	13.0	8.5
Glyptotendipes	14.0	10.0

Table 16.3. Species composition of the oligochaet community (mud sediment) of Lough Neagh

Tubifex tubifex (Muller)
Potamothrix hammoniensis (Michaelsen)
Aulodrilus pluriseta (Piguet)
Limnodrilus hoffmeisteri Claparede
Nais sp.
Stylaria lacustris (L.)
Dero obtusa Udekem

value in indicating any further deterioration or recovery in conditons in the lough.

16.2.2. *Oligochaeta*

16.2.2.1. *Community composition*

The oligochaet community of Lough Neagh has been less extensively studied than the chironomid one, and the chief records come from the period 1974–1976. The population always contains large numbers of immature specimens and these are in most cases impossible to identify to species level. The species list for the mud (Table 16.3) will undoubtedly be added to when further studies are done, but it contains most of the expected common species. It is virtually identical to that from Loch Leven (Maitland and Hudspith, 1974) and very similar to that given by Potter and Learner (1974) for a shallow eutrophic reservoir in Wales. *Limnodrilus hoffmeisteri* could generally be distinguished from the other tubificids at all times, by the lack of hair chaetae, but the others could not be separated most of the time because they were immature.

16.2.2.2. *Distribution in relation to water depth*

L. hoffmeisteri (Fig. 16.10) showed a fairly consistent pattern of distribution in relation to depth, often being an order of magnitude more abundant at 25 m than 8 m. The other tubificids (mainly *Tubifex tubifex*, *Potamothrix hammoniensis* and *Aulodrilus pluriseta*) were less clearly influenced by depth, although generally least abundant at 8 m. The super-position of individual patterns of distribution may have led to the obscuring of them all. *Dero*

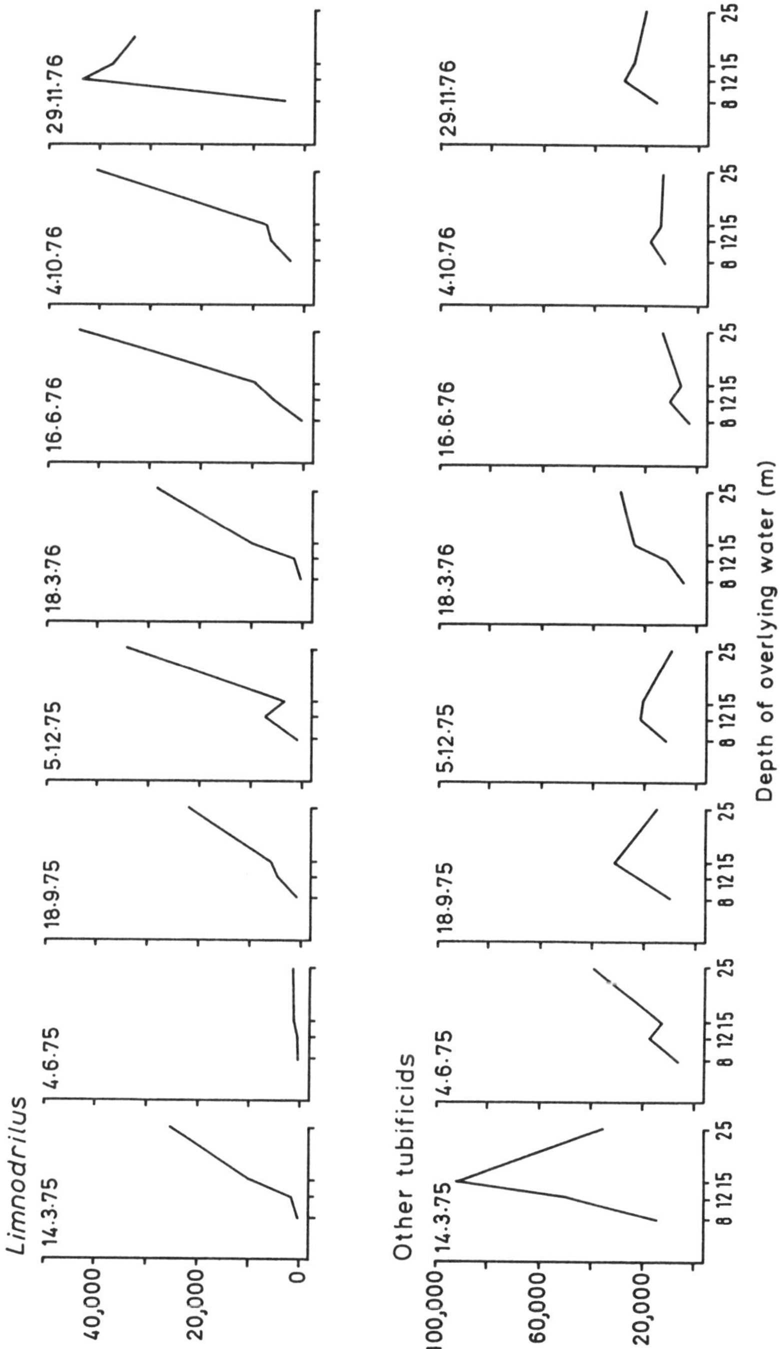

Figure 16.10. Distribution of Tubificidae in relation to water depth.

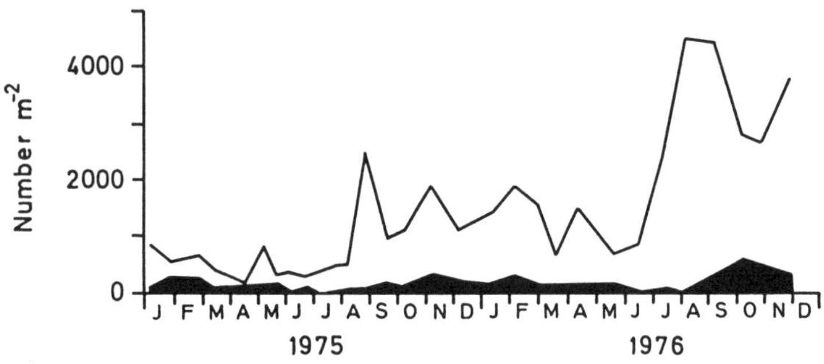

Figure 16.11. Density of *Limnodrilus* population. Shaded area = number of mature individuals.

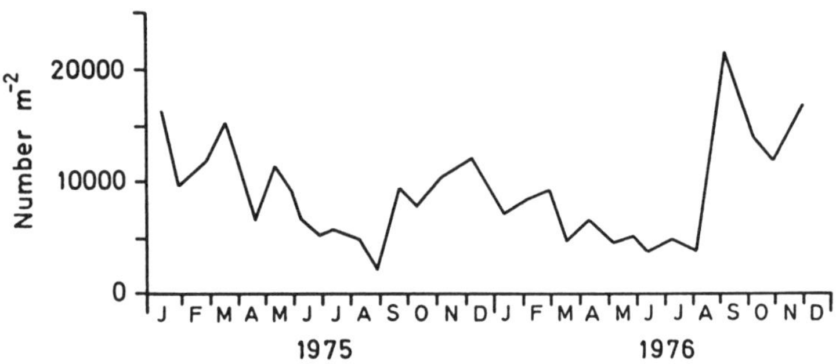

Figure 16.12. Density of Tubificidae, excluding *Limnodrilus*.

obtusa was occasionally found at 25 m, but was more common in Kinnego Bay (see Section 16.2.2.5).

16.2.2.3. *Abundance and life cycles*

Oligochaeta are, numerically, a very important component of the benthic fauna. At 8 m depth numbers of *Limnodrilus* averaged about 1,000 to 1,500 m^{-2} (Fig. 16.11) while the average of the other tubificids combined (Group A – *Tubifex*, *Potamothrix* and *Aulodrilus*) was *c.* 10,000 m^{-2} (Fig. 16.12). The numbers fluctuate greatly and one reason for this is the clumped distribution of these animals in the sediment. It can be seen from Fig. 16.10 that densities of 30 to 50,000 m^{-2} were common for *Limnodrilus* at 25 m depth. The fluctuation in numbers and the impossibility of distinguishing most species makes interpretation of life cycles very difficult. The proportion of mature *L. hoffmeisteri* in the population (shaded area of Fig. 16.11) does not vary greatly through the year except that it was at a minimum in June–July of both years studied. Numbers also increased at this time so this may be the breeding period in Lough Neagh, though breeding throughout the

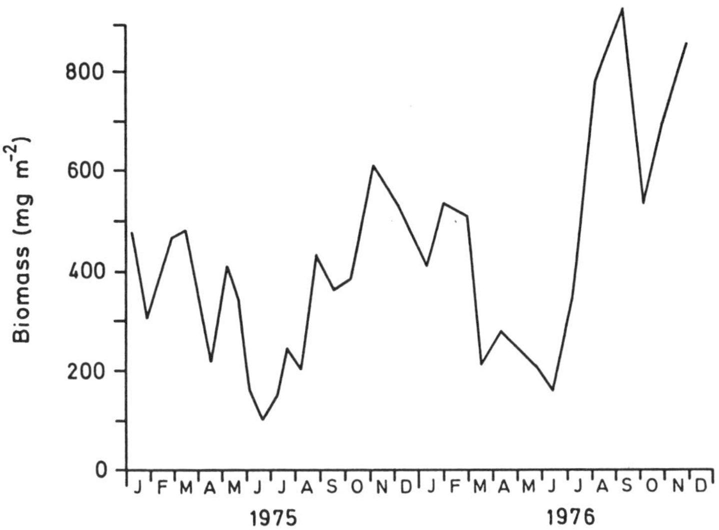

Figure 16.13. Biomass of Tubificidae.

year is also a possibility as has been found by other workers (e.g. Potter and Learner, 1974). The number of mature specimens of other tubificid species found was very small. *T. tubifex* was found in virtually all months except July–September inclusive, *P. hammoniensis* was found in March, October and November and *Aulodrilus pluriseta* in October and November.

16.2.2.4. *Biomass*

The biomass of tubificids (Fig. 16.13) averages *c.* 0.4 g dry wt. m^{-2} at 8 m depth, similar to that of *Procladius* despite the greater abundance of tubificids. *Limnodrilus* contributes about 50% of this biomass because of its larger individual size, mean dry weight 0.15 μg compared with 0.03 μg for other tubificids. At 25 m depth, the large numbers of *Limnodrilus* meant that, in 1975–1976, they contributed 74% of the average combined chironomid and tubificid biomass of 13.2 g m^{-2}, whereas at 8 m their contribution to the average combined biomass of 3.9 g dry wt. m^{-2} was only 10%. Potter and Learner (1974) found P/B ratios of 6.8 and 10.8 in successive years for *Limnodrilus* so, taking a conservative average of 8, production of *Limnodrilus* at 8 m would be *c.* 3 g dry wt. m^{-2} yr^{-1} and at 25 m *c.* 78 g m^{-2} yr^{-1}.

16.2.2.5. *The oligochaet fauna of Kinnego Bay*

The species of Oligochaeta present at 3.5 m depth in Kinnego Bay (muddy sediment) were generally the same as in the open lough. *Limnodrilus* (Fig. 16.14) is more abundant (4 to 12,000 m^{-2}) than at 8 m, though less so than at 25 m. Group A tubificids were slightly more abundant than at 8 m (6 to 7,000 m^{-2}) but the difference between the two sites was not significant.

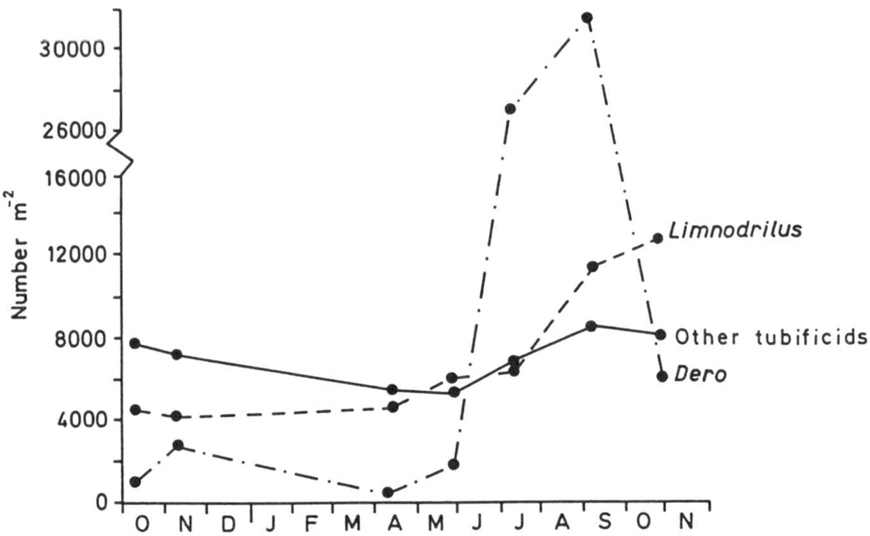

Figure 16.14. Density of Tubificidae in Kinnego Bay.

Dero obtusa (Naididae) was consistently found and at greater densities than elsewhere in the lough; the reasons for the large fluctuations in population density are unknown, they may be a seasonal pattern or a sampling artefact due to the very clumped distribution. In Kinnego Bay the combined average biomass in 1975–1976 at 3.5 m was 2.8 g dry wt. m^{-2}; *Limnodrilus* accounted for 40% of this, *C. plumosus* 28% and *Procladius* 20%.

16.2.2.6. *Summary*

There is still much work to be done on the Oligochaeta of Lough Neagh. Life cycles and patterns of reproduction need to be determined before the true importance of the oligochaets in benthic production can be assessed. The species in Lough Neagh are similar to those found by workers in other lakes of this type (Maitland and Hudspith, 1974; Potter and Learner, 1974) and represent an assemblage of species described by Brinkhurst (1969) as occurring in eutrophic, rather than organically polluted, regions. Cook and Johnson (1974), summarizing work on the St. Lawrence Great Lakes, found that *L. hoffmeisteri*, *T. tubifex* and *Dero digitata* can withstand highly enriched environments while *Potamothrix* and *Aulodrilus* belong to a mesotrophic association. This accords with the results here where *Limnodrilus* and *Dero obtusa* became more abundant at 25 m and in Kinnego Bay. The tubificids are not major contributors to the benthic biomass at 8 m depth and therefore are probably not as important as chironomids as biomass producers if the entire mud area of the lough is considered, but they account for three-quarters at 25 m depth. *Limnodrilus*, although not always the most abundant oligochaet, is the greatest contributor to biomass because of its larger individual weight.

16.3. The littoral macroinvertebrate fauna

Investigations of the littoral macroinvertebrate fauna have begun only recently and as yet remain largely of a descriptive nature.

The inshore areas of the lough can conveniently be divided into sandy and rocky shore types. The sandy littoral areas, found in sheltered bays, comprise approximately 30% of the lough shoreline. They typically support stands of *Typha* sp. or *Phragmites* sp. leading on to beds of submerged macrophytes usually *Elodea* sp. and/or *Potamogeton* sp. For details see Chapter 19. These macrophyte beds extend outwards to a water depth of about 1.5 m which in some localities occurs up to 500 m from the shore. Rocky shores comprise the remainder of the lough shoreline and in these localities macrophytes are invariably absent, the flora being comprised of benthic algae of which the macroscopic filamentous alga *Cladophora* sp. is often dominant. In the rocky areas the downward slope of the shoreline is more severe than in the sandy bays and a depth of 1.5 m would invariably be found within 50 m of the shore.

16.3.1. *Community composition and distribution*

In July 1981 ten quantitative replicate samples of the littoral macroinvertebrates, taken at depths between 10 and 50 cm, were removed from each of six inshore sampling sites (three rocky sites and three sandy sites). A summary of the available results from this brief survey, showing average species densities for rocky and sandy habitats, is given in Table 16.4.

A large proportion of the macroinvertebrate species, which were identified in this survey, were common to both the rocky and sandy littoral areas. However, certain groups, e.g., larval Trichoptera, and species, e.g., *Nais bretscheri*, *Gammarus duebeni* and *Ancylus fluviatilis* were apparently confined to the rocky areas whereas others, e.g., *Aulodrilus pluriseta*, *Limnodrilus udekemianus*, *Stylaria lacustris* and *Gammarus tigrinus* were found only in the sandy bays. Although the species lists for this survey are not comprehensive – some groups have yet to be identified or are incomplete, e.g., Hydracarina, Coleoptera (excluding Elminthidae) and Chironomidae – the available data have been published and presented in a tabular form which affords a visual separation between the faunas of the rocky and sandy shore lines, Table 16.5, taken from Murphy and Carter (1984).

16.3.2. *The gammarid fauna of Lough Neagh and its environs*

Gammarus duebeni Liljeborg and *G. lacustris* Sars are the indigenous freshwater gammarids in Ireland. The limnic populations of *G. duebeni*, which is

Table 16.4. The mean densities (m^{-2}) of littoral macroinvertebrate species on sandy and rocky shores in Lough Neagh as determined by a survey undertaken in July 1981. With the exception of the chironomid data, which are the values from only one rocky and one sandy sampling site, the densities given are mean values from three sampling sites on each of the two shore types

Platyhelminthes	Rocky	Sandy
Planariidae		
Polycelis tenuis Ijima	3	13
Dugesiidae		
Dugesia polychroa (O. Schmidt)	1	1
Dendrocoelidae		
Dendrocoelum lacteum (Muller)	20	5
Oligochaeta		
Tubificidae		
Peloscolex ferox (Eisen)	52	8
Aulodrilus pluriseta (Piguet)	–	10
Aulodrilus limnobius (Bretscher)	–	97
Psammoryctides barbatus (Grube)	3085	497
Tubifex ignotus (Stolc)	312	98
Rhyacodrilus coccineus (Vejedovsky)	15	1
Potamothrix hammoniensis (Michaelsen)	4	–
Potamothrix moldaviensis (Vejdovsky & Mrazek)	204	6
Limnodrilus hoffmeisteri Clapearede	104	217
Limnodrilus udekemianus Claparede	–	21
Naidiadae		
Chaetogaster diaphanus (Gruithuisen)	–	2
Ophidonais serpentina (Muller)	–	8
Uncinais uncinata (Ørstedt)	–	31
Homochaeta naidina Bretscher	–	2
Pristina idrensis Sperber	–	10
Stylaria lacustris (Linn.)	–	12
Nais barbata (Muller)	55	353
Nais bretscheri Michaelsen	157	–
Nais pardalis Piguet	7	379
Nais elinguis Muller	18	13
Nais communis Piguet	–	32
Lumbriculidae		
Stylodrilus heringianus Claparede	95	176
Lumbriculus variegatus (Muller)	47	849
Enchytraeidae sp.	1033	2585
Hirudinea		
Piscicolidae		
Piscicola geometra (Linn.)	4	–
Glossiphoniidae		
Theromyzon tessulatum (O.F. Muller)	1	5
Glossiphonia heteroclita (Linn.)	–	1
Glossiphonia complanata (Linn.)	176	6
Helobdella stagnalis (Linn.)	284	72
Erpobdellidae		
Erpobdella octoculata (Linn.)	16	27
Dina lineata (O.F. Muller)	2	–
Malacostraca (Crustacea)		
Asellidae		
Asellus aquaticus (Linn.)	289	952

Table 16.4. Continued

Platyhelminthes	Rocky	Sandy
Gammaridae		
Gammarus duebeni Liljeborg	171	–
Gammarus tigrinus Sexton	–	2206
Ephemeroptera (Insecta)		
Ephemerellidae		
Ephemerella ignita (Poda)	370	15
Trichoptera		
Polycentropidae		
Cyrnus trimaculatus (Curtis)	28	–
Polycentropus flavomaculatus (Pictet)	116	3
Psychomyiidae		
Psychomyia pusilla (Fabricius)	19	–
Tinodes waeneri (Linn.)	401	1
Limnephilidae		
Limnephilus lunatus Curtis	50	–
Allogamus auricollis sp. (Pictet)	2	
Sericostomatidae		
Sericostoma personatum (Spence)	108	–
Leptoceridae		
Arthripsodes cinereus (Curtis)	19	–
Mystacides azurea (Linn.)	1	
Hydroptilidae		
Hydroptila sp.	74	–
Coleoptera		
Elminthidae		
Limnius volkmari (Panzer)	1	–
Oulimnius tuberculatus (Muller)	3	1
Hemiptera		
Corixidae		
Arctocorisa germari (Fieb.)	–	17
Diptera (larvae)		
Empididae		
Wiedemannia sp.	44	–
Chelifera sp.	10	–
Muscidae		
Limnophora sp.	6	–
Chironomidae		
Natarsia sp. Fittkau	33	362
Procladius sp. Skuse	–	64
Prodiamesa sp. Kieffer	5	–
Potthastia sp. Kieffer	3	6
Monodiamesa sp. Kieffer	1	12
Synorthocladius sp. Thienemann	552	–
Psectrocladius sp. Kieffer	7	66
Limnophyes sp. Eaton	2	–
Cricotopus sylvestris v.d. Wulp	19	322
Cricotopus sp. A	–	2
Cricotopus sp. v.d. Wulp	–	242
Orthocladius sp. A.	231	–
B.	214	–
C.	140	–
D.	7	–

Table 16.4. Continued

Platyhelminthes	Rocky	Sandy
Orthocladius sp. v.d. Wulp	–	190
Microtendipes sp. Kieffer	2918	576
Glytotendipes sp. Kieffer	671	1628
Tanytarsus sp. v.d. Wulp	406	306
Polypedilum sp. Kieffer	17	–
Endochironomus sp. Kieffer	5	–
Micropsectra sp. Kieffer	1	–
Stictochironomus sp. Kieffer	7	2
Paratanytarsus sp. Baure	1	–
Dicrotendipes sp. Kieffer	2	–
Paracladopelma sp. Harnisch	–	44
Diptera		
Chironomidae		
Cladotanytarsus sp. Kieffer	45	74
Chironomus cingulatus Meigen	4	388
Cryptochironomus defectus Kieffer	–	30
Harnischia sp. Kieffer	19	–
Mollusca		
Valvatidae		
Valvata piscinalis (Muller)	1	11
Hydrobiidae		
Potamopyrgus jenkinsi (Smith)	–	42
Ancylidae		
Ancylus fluviatilis (Muller)	20	–
Physidae		
Physa fontinalis (Linn.)	6	6
Lymnaeidae		
Lymnea peregra (Muller)	2	7
Planorbidae		
Planorbis carinatus Muller	–	1
Planorbis albus Muller	7	11

the commoner of the two species, belong to the sub-species *G. duebeni celticus* (Gledhill *et al.*, 1976). *G. lacustris*, considered by Pinkster *et al.* (1977) to be a glacial relict species, is largely confined to low lying natural lakes in the north and west of the country and is not found in Lough Neagh (Strange and Glass, 1979).

Gammarus tigrinus Sexton is a North American immigrant which it has been suggested (Hynes, 1968) was introduced into Northern Ireland during 1914 to 1918. It is now present in several freshwater localities including Lough Neagh and the Lower Bann. The majority of the localities colonized by *G. tigrinus* during its rapid invasion of other European countries have been brackish or at least characterized by a raised ion content (Nijssen and Stock, 1966; Pinkster and Stock, 1967; Chambers, 1973; Holland, 1976; Pinkster *et al.*, 1980). However, Lough Neagh and the Lower Bann are indisputably fresh waters (see Chapter 5). Laboratory studies (Sutcliffe, 1968)

Table 16.5. The species/site data matrix arising from a survey of the littoral fauna of Lough Neagh arranged to effect visual distinction between the funas of rocky (R_1, R_2 and R_3) and sandy (S_1, S_2 and S_3) habitats. (Species densities given in nos per m^2.) Redrawn from Murphy and Carter (1984)

Species groups	Sites					
	R_1	R_2	R_3	S_1	S_2	S_3
1 *Homochaeta naidina* Bretscher						6
Aulodrilus limnobius Bretscher						292
Chaetogaster diaphanus (Gruithuisen)						6
Ophidonais serpentina (Muller)						24
Uncinais uncinata (Orsted)						94
Glossiphonia heteroclita (L.)						2
Planorbis carinatus Muller						2
2 *Aulodrilus pluriseta* Piguet					5	24
Stylaria lacustris (L.)					20	16
Pristina idrensis Sperber				24		6
Limnodrilus udekemianus Claparede				4	28	30
Nais communis Piguet				24	45	26
Gammarus tigrinus Sexton				5312	1156	152
Arctocorisa germari (Fieber)				2	24	24
Potamopyrgus jenkinsi (Smith)				2	98	26
3 *Polycelis tenuis* (Muller)			8		12	26
Oulimnius tuberculatus (Muller)	8			2	2	
Valvata piscinalis (Muller)	3		1		10	24
Theromyzon tessulatum (Muller)		1	2		6	10
Lymnea peregra (Muller)	2		5	4	18	
4 *Psammoryctides barbata* (Grube)	3103	2941	3210	684	648	104
Limnodrilus hoffmeisteri Claparede	261	12	39	30	75	546
Helobdella stagnalis (L.)	123	534	195	54	98	64
Erpobdella octoculata (L.)	23	4	20	14	52	14
Asellus aquaticus (L.)	127	116	623	346	2398	112
Dendrocoelom lacteum (Muller)	8	18	33	6	10	
Stylodrilus heringianus Claparede	44	65	177	46	483	
Lumbriculus variegatus (Muller)	49	18	75	694	1853	
Glossiphonia complanata (L.)	13	452	64	10	8	
Ephemerella ignita (Poda)	302	159	649	42	2	
Tubifex ignotus (Stoic)	102	473	360		288	6
Nais barbata Muller	42	26	96		483	576
5 *Dugesia polychroa* (Schmidt)		1	1		2	
Peloscolex ferox (Eisen)		34	125		25	
Rhyacodrilus coccineus (Vejdovsky)		38	7	2		
Physa fontinalis (L.)		13	6	18		
Nais pardalis Piguet	5	17				1138
Nais elinguis Muller	39	15				38
Gyraulus albus (Kerney)	3	7	11		34	
Potamothrix moldaviensis (Vejdovskyoc Mrazek)	453	71	87			18
Polycentropus flavomaculatus (Pictet)	11	14	323	8		
Tinodes waeneri (L.)	529	147	526	2		
Sericostoma personatum (Spence)	9	257	58	1		

Table 16.5. Continued

Species groups	Sites					
	R_1	R_2	R_3	S_1	S_2	S_3
6 *Nais bretscheri* Michaelson	9	285	176			
Gammarus duebeni Lilljeborg	233	87	194			
Psychomyia pusilla (Fabricius)	27	15	15			
Limnephilus lunatus Curtis	8	59	82			
Arthripsodes cinereus (Curtis)	5	37	16			
Ancylus fluviatilis Muller	11	16	33			
Piscicola geometra (L.)	4	2				
7 *Ecnomus tenellus* Rambur			1			
Mystacides azurea (L.)			2			
Potamothrix hammoniensis (Michaelsen)		11				
Limnius volkmari (Panzer)	1					
Dina lineata (Muller)	6					
Cyrnus trimaculatus (Curtis)	82					

indicate that *G. tigrinus* would be suffering some degree of osmoregulatory stress in such dilute media. However, in Lough Neagh, it appears to be competing successfully with the physiologically better adapted *G. duebeni celticus* with the latter being confined to rocky shores (see Section 16.3.3.1). The success of the invading species is undoubtedly due to its superior reproductive potential (Pinkster *et al.*, 1977). In accord with the recent survey, Hynes (1955) observed habitat partitioning between *G. duebeni celticus* and *G. tigrinus* in Lough Neagh similarly related to substrate type. Pinkster *et al.* (1977) concluded that *G. tigrinus* finds fresh, running waters unsuitable for colonization. It appears, therefore, that *G. tigrinus* is excluded from the rocky areas of the lough littoral by its inability to withstand strong wave action.

In the late 1950s *Gammarus pulex* (L.) was deliberately introduced into Northern Ireland where it is now found in several of the rivers and streams in the south and west of the Lough Neagh catchment (Strange and Glass, 1979). In those systems where *G. pulex* has been introduced it appears to have replaced *G. duebeni celticus* in the downstream reaches of the river systems. This in in accord with the conclusions of Dennert (1974) and Pinkster *et al.* (1970) that *G. pulex* could successfully compete with *G. duebeni celticus* in freshwater habitats. While *G. pulex* has not yet been recorded from Lough Neagh one might expect that it would eventually displace *G. duebeni celticus* from the rocky shores (Dennert 1974) although the evidence (Pinkster *et al.*, 1977) indicates that *G. tigrinus* would continue to predominate in the sandy littoral areas.

16.3.3. *Notes on corixids from the Lough Neagh district* (by A. Macfadyen)

The only justifications for the following notes are that study of Corixidae in Ulster has been very spasmodic so that even a few records may be of some future value and, secondly, in a more general context, because an indication of the habitat requirements of the less common species may be of wider interest to those who study the group in other countries.

The most important papers listing the Northern Ireland corixid fauna which are known to me are Macan (1954, 1965), Walton (1936) and in Flower (1982).

Most of my records below refer to the Traad Point area and date from the end of August 1978. These are supplemented by small collections in 1969 and 1982 from the vicinity of the Blackwater mouth and Maghery canal (south shore). Comparison is made with a few collections from Rathlin Island.

A total of 14 species of Corixinae have been found in the lake and its immediate vicinity and the habitats in which species occur usually agree well with the ecological notes given by Macan (1965) and Walton (1936). A single *Notonecta* (*N. glauca*) species was regularly collected but any *Gerris* found were not determined to species.

Collections in the vicinity of Traad Point were made along the shore of Lough Neagh for a distance of about 1 km, and in a complex of water-filled worked-out gravel pits which lie inland from the laboratory. These pits have been abandoned, partly overgrown and vary in depth to several metres. One of these pits, identified in Table 16.6 as 'Large gravel pit', is extensive and the West end is shallow, un-shaded, with a sandy bottom and with emergent vegetation limited to *Isoetes*. The water is brown and peaty. Limited habitat data were recorded at the time of collection. A further collection was made in the mouth of a drainage channel near Maghery (situated on the south-west shore of the lough at Map reference H93962B. The channel is fed by streams rising in hills to the south but also receives effluent from a peat-processing works about 250 m upstream. The water is brown but clear and supports a rich fauna of molluscs, and Odonata larvae among others.

The list of corixids given in Table 16.6, indicates that they can be associated in three main groups linked with habitat features:

1. The open lake and its shore where emergent vegetation is lacking supports *Arctocorisa germari* and *Callicorixa praeusta* which were also found in the Maghery channel a few yards from the open lake.
2. At the opposite extreme, never occurring in or close to the lake and confined to the smaller gravel pits, are *Hesperocorixa castanea*, *H. linnei*, *H. sahlbergi* and *Sigara scotti*.
3. An intermediate group occurred in the lake and the Maghery channel only in the presence of emergent vegetation, and also in the large shallow gravel pit above a sandy bottom. These are *Sigara distincta*, *S. dorsalis*, *S. falleni* and *S. fossarum*.

Of the remaining, less common species, *Corixa punctata* occurred only in the Maghery channel and a similar inlet to Lough Neagh near Traad Cottage. Both these localities contained a rich fauna including gastropod molluscs, water beetles and Odonata larvae and were choked with emergent macrophytes and floating *Potamogeton*.

Table 16.6. A summary of records of Corixinae from the Lough Neagh district. The table consolidates records from a number of collections at most sites. Species are arranged in the groupings indicated in the text. The final two rows refer to the Maghery Channel and to collections from Rathlin Island (see Flower, 1982)

Localities (L = Open to Lake)	*A. germari*	*C. praeusta*	*C. dentipes*	*S. concinna*	*C. punctata*	*S. distincta*	*S. dorsalis*	*S. fossarum*	*S. falleni*	*C. bonsdorffi*	*S. scotti*	*H. sahlbergi*	*H. linnei*	*H. castanea*
L Pier at Laboratory	√	√												
L Inlet, S. of laboratory	√	√	√											
L Sandy bay, S. of laboratory														
Open	√													
L Sandy bay, S. of laboratory														
Emergent vegetation					√	√	√	√						
L Inlet from Lake Traad cottage					√									
Large gravel pit sandy				√		√	√	√	√	√	√		√	
Open, some *Alnus*								√			√		√	
Gravel pit ponds											√	√	√	√
L Maghery channel	√	√	√	√	√		√		√					
Rathlin Island (various)			√			√	√	√		√	√	√	√	

Cymatia bonsdorffi occurred, together with *Notonecta glauca*, in the large gravel pit above sand and in the presence of limited emergent macrophytes.

Sigara concinna, generally considered a rarity, was found in the Maghery channel and also the large gravel pit, in both cases in peaty water (conforming with Macan's notes), and above a sandy bottom.

Corixa dentipes, also rather rare according to Macan, occurred, with *Callicorixa praeusta* in a small inlet to the lake, a few hundred yards south-west of the Traad Point Laboratory, overhung by *Alnus glutinosa* trees and also in the Maghery Channel, where it cohabited with *C. punctata*, an association to be expected from the literature.

Up to now the endemic, *Sigara fallenoidea* has not been found by me. Whilst Macan regards the middle leg measurements as diagnostic, I find great variability in this respect in specimens attributed to *S. falleni* and the only reliable character in females seems to be the chaetotaxy of the middle femur.

These collections hardly justify further analysis within the three main groups, between which, indeed, there is some overlap. In particular *S. scotti* and *H. linnei* extend to the large gravel pit where they coexist with the more lacustrine species.

A number of schemes has been published linking corixid species to habitat type and successional state e.g. by Macan (1938, 1963), Savage and Pratt (1976). Agreement between my data and these schemes is limited, partly in the latter case because only more alkaline localities were studied. In comparison with Macan's (1963) proposals my two groups from the open lake plus the large gravel pit occupy the centre of the diagram whilst the species from smaller pools are, according to him, unrelated and divergent. The series in his earlier paper seem easier to relate to mine. The Savage and Pratt diagram showing a eutrophic/oligotrophic gradient on one axis and shelter/open water on the other separates my species groups quite well but indicates better separation on the first of these axes whilst my data seem to relate more clearly to the shelter-open axis.

Clearly this brief survey constitutes little more than a species list. One aspect on which future collecting must concentrate is a much more precise characterization of micro habitats within each water body, using more and smaller sample units. In attempting to analyse habitats 'from the point of view of a corixid' it would be well to remember Walton's (1936) perceptive remarks about relations between anatomy and *physical* structure of the habitat which, in another medium can be linked with Duffey's (1963) emphasis on the importance of vegetation structure rather than taxonomy when studying spider distribution. Finally the time dimension, as revealed by Pajunen's (1977) discoveries about the fit between life cycle length and permanence of habitat, must always be remembered.

16.4. References

Brinkhurst, R.O. (1969) Distribution and abundance of tubificid (Oligochaeta) species in Toronto Harbour, Lake Ontario. J. Fish. Res. Bd. Can 27: 1961–1969.

Brinkhurst, R.O., Hamilton, A.L. and Herrington, H.B. (1968) Components of the bottom fauna of the St. Lawrence, Great Lakes, Univ. of Toronto No. PR 33.

Carter, C.E. (1975) Emergence periods of the main species of Chironomidae (Diptera) of Lough Neagh. Irish Nat. J. 18: 245–246.

Carter, C.E. (1976) A population study of the Chironomidae (Diptera) of Lough Neagh. Oikos 27: 346–354.

Carter, C.E. (1978) The fauna of the muddy sediments of Lough Neagh, with particular reference to eutrophication. Freshwat. Biol. 8: 547–559.

Carter, C.E. (1980a) The life cycle of *Chironomus anthracinus* in Lough Neagh. Holarctic Ecol. 3: 214–217.

Carter, C.E. (1980b) The population structure of *Procladius crassinervis* (Zett.) in Lough Neagh, Northern Ireland. Acta Univ. Carol. Biol. 1978: 27–33.

Carter, C.E. (1980c) Methods of Summarizing Survey Data from Lakes, illustrated by Reference to Lough Neagh, Northern Ireland. In: D.A. Murray (ed.), Chironomidae – Ecology, Systematics, Cytology and Physiology. Pergamon Press, p. 217–224.

Carter, C.E. and Rippey, B. (1982) Water quality in four bays in Lough Neagh, N.I. Irish J. Env. Sci. 2: 51–60.

Chambers, M.R. (1973) Notes on the gammarid fauna of the Friesian Lake District following the invasion of the alien amphipod *Gammarus tigrinus* Sexton. Bull. zool. Mus. Univ. Amsterdam 3: 1–5.

Charles, W.N., East, K., Brown, D., Gray,. C. and Murray, Y.D. (1974) The production of larval Chironomidae in the mud at Loch Leven, Kinross. Proc. R. Soc. Edinbr. B. 74: 241–258.

Cook, D.G. and Johnson,. G. (1974) Benthic macroinvertebrates of the St. Lawrence Great Lakes. J. Fish. Res. Bd. Can. 31: 763–782.

Dennert, H. G. (1974) Tolerance differences and interspecific competition in three members of the amphipod genus *Gammarus*. Bidjr. Dierk. 44: 83–99.

Duffey, E. (1963) Ecological studies on the spider fauna of the Malham Tarn area. Field Studies 1(5): 65–88.

Flower, R.J. (1982) A comparative limnological survey of Rathlin Island, Co. Antrim, with particular reference to diatoms. Proc. R. Ir. Acad. 82B(1), 1–20.

Gledhill, T., Sutcliffe, D.W. and Williams, W.D. (1976) Key to British Freshwater Crustacea: Malacostraca. Scient. Publ. of the Freshwater Biological Association No. 32.

Holland, D.G. (1976) The inland distribution of brackish water *Gammarus* species in the area of the Mersey and Weaver River Authority. Freshwat. Biol. 6: 277–285.

Hynes, H.B.N. (1955) Distribution of some freshwater amphipoda in Britain. Int. Verh. Ver. theor. angew. Limnol. 12: 620–628.

Hynes, H.B.N. (1968) Report on the invertebrate fauna of Lough Neagh, (Unpublished) Ministry of Finance, Northern Ireland.

Irwin, A.G. (1976) Biochemical and morphological studies of Chironomidae in Lough Neagh. Unpublished Ph.D. Thesis, Queen's Univ. Belfast.

Jonasson, P.M. (1972) Ecology and production of the profundal benthos in relation to phytoplankton in Lake Esrom. Oikos Suppl. 14: 1–148.

Jones, R.I. (1977) Factors controllllng phytoplankton production and succession in a highly eutrophic lake (Kinnego Bay, Lough Neagh). 1. The phytoplankton community and its environment. J. Ecol. 65: 547–559.

Macan, T.T. (1938) Evolution of aquatic habitats with special reference distribution of Corixidae. J. Animal Ecology 7: 1–19.

Macan, T.T. (1954) Records from some Irish Lakes Pt. 1. Proc. Ir. Acad. (B), 56: 135–143.

Macan, T.T. (1963) Freshwater Ecology. Longmans, London.

Macan, T.T. (1965) A revised key to the British Water Bugs. Sci. Pub. Freshwat. Biol. Assn. Windermere, 78 pp.

Maitland, P.S. (1979) The distribution of zoobenthos and sediments in Loch Leven, Kinross, Scotland. Arch. hydrobiol. 85: 98–125.

Maitland, P.S. and Hudspith, P.M.G. (1974) The zoobenthos of Loch Leven, Kinross and estimates of its production in the sandy littoral area during 1970 and 1971. Proc. R. Soc. Edin. B 74: 219–239.

Marks, R.J. and Henderson, A.E. (1970) An examination of the larval Chironomidae (Diptera, Nematocera) in Lennymore Bay, Lough Neagh. Irish Nat. J. 16: 328–334.

McGarrigle, M. (1980) The distribution of chironomid communities and controlling sediment parameters in Lough Derravaragh, Ireland. In: D.A. Murray (Ed.), Chironomidae – Ecology, Systematics, Cytology and Physiology. Pergamon Press, pp. 275–282.

Menzie, C.A. (1981) Production ecology of *Cricotopus sylvestris* (Fabricius) (Diptera: Chironomidae) in a shallow estuarine cove. Limnol. Oceanogr. 26: 467–481.

Murphy, P.M. and Carter, C.E. (1984) A summer survey of the littoral macroinvertebrate fauna (excluding Chironomidae) of Lough Neagh, N. Ireland. Proc. R. Ir. Acad. 84B: 103–108.

Murray, D.A. (1972) A list of the Chironomidae known to occur in Ireland with notes on their distribution. Proc. R. Ir. Acad. 72B: 275–293.

Nijssen, H. and Stock, J.H. (1966) The amphipod *Gammarus tigrinus* Sexton (Crustacea), introduced in the Netherlands. Beaufortia, 13: 197–206.

Pajunen, V.I. (1977) Population structure in rock-pool Corixids (Hemiptera, Corixidae) during the reproductive season. Ann. Zool. Fennici. 14: 26–47.

Pinkster, S. (1972) On members of the *G. pulex* group from West Europe. Bijdr. Dierk. 42: 164–191.

Pinkster, S., Dennert, A.L., Stock, B. and Stock, J.H. (1970) The problem of European freshwater populations of *G. duebeni*. Bijdr. Dierk. 40: 115–147.

Pinkster, S., Dielman, J. and Platvoet, D. (1980) The present position of *Gammarus tigrinus*, Sexton 1939, in the Netherlands with the description of a newly discovered amphipod species

Crangonyx pseudogracilis Bousfield, 1958 (Crustacea: Amphipoda). Bull. zool. Mus. Univ. Amsterdam 7: 33–45.

Pinkster, S., Smit, H. and Brandse-de Jong, N. (1977) The introduction of the alien amphipod *Gammarus tigrinus*, Sexton 1939, in the Netherlands and its competition with indigenous species. Crustaceana Suppl. 4: 91–105.

Pinkster, S. and Stock, J.H. (1967) Range extension in 1966 of the alien amphipod *Gammarus tigrinus*, Sexton 1939, in the Netherlands. Beaufortia 14: 81–6.

Potter, D.W.B. and Learner, M.A. (1974) A study of the benthic macro-invertebrates of a shallow eutrophic reservoir in South Wales with emphasis on the Chironomidae (Diptera): their life histories and production. Arch. Hydrobiol. 74: 186–226.

Ripley, M.P. (1980) The relation of dry weight and temperature to respiration in some benthic chironomid species in Lough Neagh. In: D.A. Murray (ed.), Chironomidae - Ecology, Systematics, Cytology and Physiology. Pergamon Press, pp. 81–88.

Savage, A.A. and Pratt, M.M. (1976) Corixidae (Water Boatmen) on the North-West Midland Meres. Field Studies 4(3): 465–476.

Strange, C.D. and Glass, G.B. (1979) The distribution of freshwater gammarids in Northern Ireland. Proc. R. Ir. Acad. 79B: 145–153.

Sutcliffe, D.W. (1968) Sodium regulation and adaptation to freshwater in gammarid crustaceans. J. Exp. Biol. 48: 359–380.

Titmus, G. and Badcock, Ruth M. (1980) Production and emergence of chiromids in a wet gravel pit. In: D.A. Murray (ed.), Chiromidae - Ecology, Systematics, Cytology and Physiology. Pergamon Press, pp. 299–305.

Walton, G.A. (1936) A new species of Corixidae, *Sigara pearcei* (Hemipt.) from Ireland; together with descriptions of its closely related species. Trans. Soc. Brit. Ent. 3: 33–47.

Wiederholm, T. (1980) Chironomids as indicators of water quality in Swedish lakes. Acta Univ. Carol. - Biol. 1978: 275–283.

17. Heterotrophic micro-organisms in the water column and sediments of Lough Neagh

J.P. QUINN

17.1. Introduction

The multitude of experimental difficulties which beset aquatic microbial ecology have made this perhaps the least advanced branch of limnology generally (not just in Lough Neagh) but an increasing body of evidence is accumulating (Overbeck, 1974; Berrie, 1976) which suggests that only a small proportion of freshwater primary production may pass directly to grazing zooplankton. Up to 90% has been found to enter the detrital food web as decaying phytoplankton and macrophyte remains, as dissolved organic matter subsequently scavenged by bacteria and as partially digested zooplankton faeces (Pomeroy, 1980). Clearly no lake ecosystem can be fully understood if the heterotrophic micro-organisms are ignored.

While this chapter is based on only one year's data (and those from an essentially exploratory study) it does serve to illustrate the potentially very great importance of the detrital food web and the likely value of deeper study.

Samples were taken weekly throughout 1982 at a station on the 10 m isobath and about 1000 m from the shoreline in the North-West of Lough Neagh (the 'profundal' site); it overlay deposits of the loose brown/black sediment which characterizes some 52% of the bed of the lake, and had previously been judged to represent adequately the body of the lake which is essentially homogeneous throughout the year. Less frequent 'littoral' samples were taken from a station of water depth 0.75 m, 20 m offshore in the extreme north-western corner of the lake; the substratum was a coarse sand which supported a thin cover of *Elodea* sp. in summer and autumn.

17.2. Populations of heterotrophic organisms

17.2.1. *Planktonic bacteria and yeasts*

Total numbers of planktonic bacteria present during the year at a depth of

R. B. Wood and R. V. Smith (eds.), Lough Neagh, 369–380.

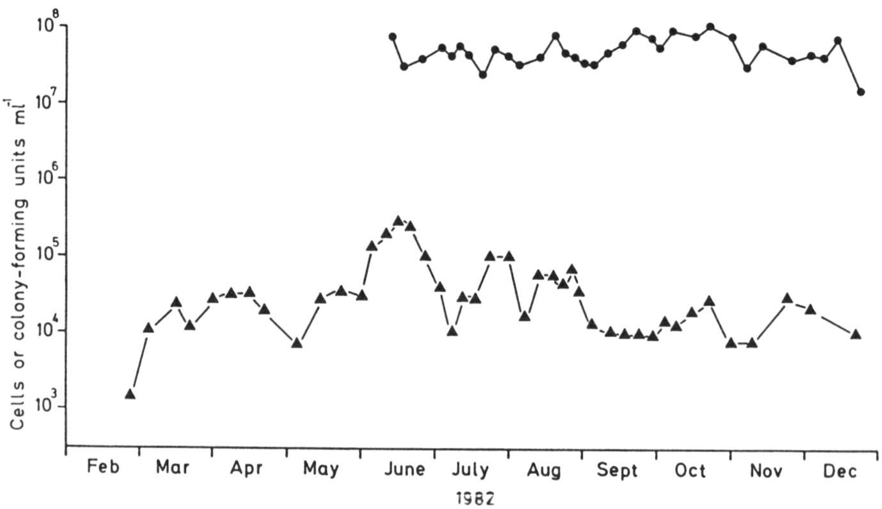

Figure 17.1. Estimates of total (●—●) and viable (▲—▲) planktonic bacterial populations at a profundal sampling site in Lough Neagh during 1982. Redrawn from Quinn *et al.* (1985).

1 m at the profundal site, as determined by the acridine orange/epifluorescence direct count procedure of Jones and Simon (1975) are shown in Fig. 17.1; numbers of 'viable' aerobic bacteria in the same samples are also shown as determined by an MPN modification (Harris and Somers, 1968) of the traditional plate count on CPS medium (Collins and Willoughby, 1962). Counts at 1 and 9 m rarely varied by more than 10%, i.e., there was no evidence of extensive or prolonged stratification. Direct counts of planktonic bacteria at the littoral site were also very similar to those further offshore, but plate count values were consistently much higher, reaching on 17 June a peak of 1.7×10^6 viable cells ml^{-1} which represented about 5% of the total count, and presumably reflected the surrounding presence of rooted macrophytes. Planktonic yeasts were enumerated at both sites after plating on the medium of Aaronson (1970), but numbers rarely exceeded 20 ml^{-1} and no obvious seasonality could be detected.

In all samples, the percentage of viable planktonic and benthic (see Section 17.2.2) bacteria capable of proteolysis and cellulolysis was also routinely estimated. It was found that virtually all viable bacteria were proteolytic throughout the year – in several instances their numbers exceeded those in control samples plated on CPS medium. By contrast, numbers of cellulolytic bacteria rarely exceeded 50% and frequently represented less than 10% of the viable population.

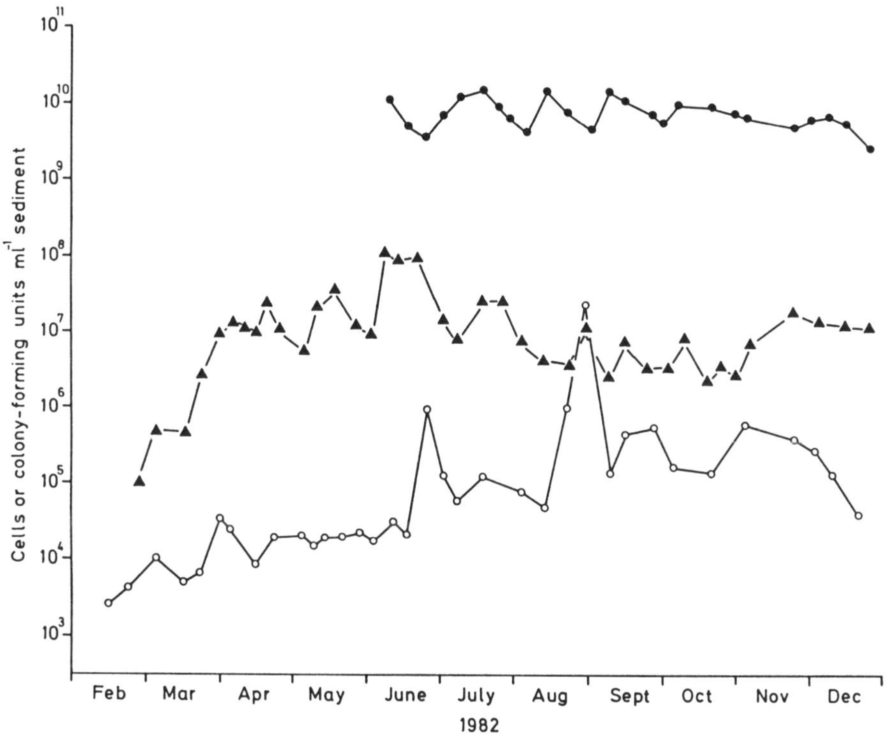

Figure 17.2. Estimates of total (●—●), viable aerobic (▲—▲) and viable anaerobic (○—○) benthic bacterial populations in the uppermost centimetre of sediment at a profundal sampling site in Lough Neagh during 1982. Redrawn from Quinn *et al.* (1985).

17.2.2. *Benthic bacteria and yeasts*

The mean results of total (epifluorescent) and viable aerobic and anaerobic MPN determinations of bacterial numbers in 0–2 cm surface sediment samples, taken in triplicate at the profundal site during 1982, are shown in Fig. 17.2. Numbers of yeasts in the same samples are illustrated in Fig. 17.3. At the littoral site, numbers of both total and viable bacteria and yeasts in the surface sediment were normally greater than those at the deepwater station by some 50%

17.2.3. *Epiphytic bacteria*

A study was carried out of the bacterial colonization of the surfaces of submerged vegetation on a sheltered sandy area of the north-west shoreline of Lough Neagh. The extent of epiphyte colonization was recorded fort-

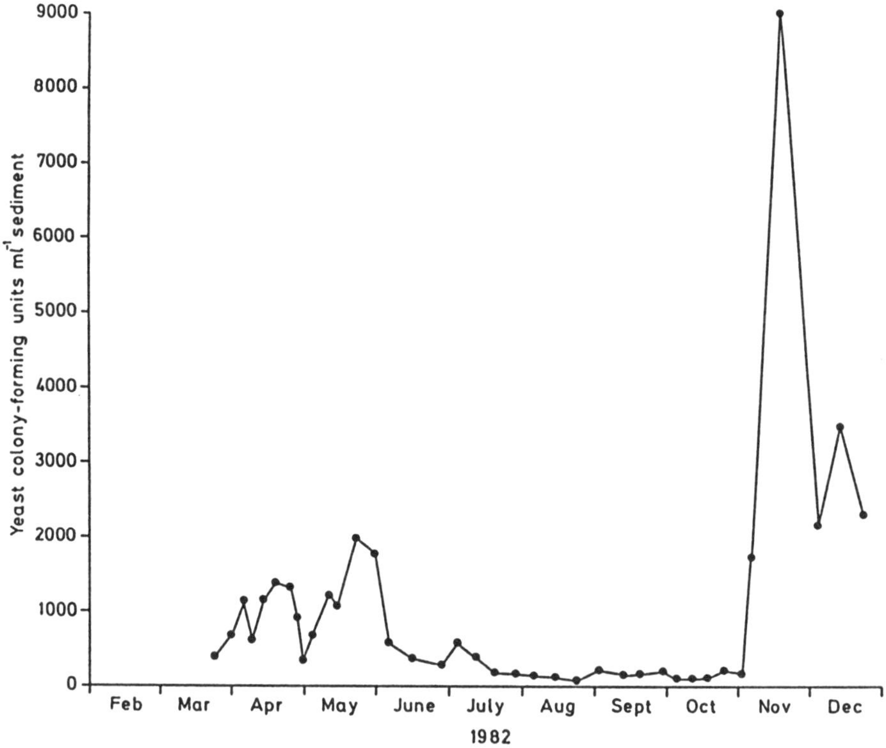

Figure 17.3. Seasonal distribution of yeasts in the uppermost centimetre of sediment at a profundal sampling site in Lough Neagh during 1982. Redrawn from Quinn (1984b).

nightly during the macrophyte growing season (June–November) but density did not vary appreciably throughout. Typical direct count values obtained were (for 29 September): On *Myriophyllum* sp. 8.7×10^8 cells cm^{-2}, on *Elodea* sp. 8.6×10^8 cells cm^{-2}, on *Ranunculus* sp. 9.4×10^8 cells cm^{-2}, on *Potomageton* (stems) 8.7×10^8 cells cm^{-2} and on *Potomageton* (leaves) 8.4×10^8 cells cm^{-2}. The sessile alga *Cladophora* was also densely colonized; it was found impossible to express results on a surface area basis in this instance but counts of 10^{11} bacterial cells g^{-1} fresh wt., were regularly obtained. By contrast, the filaments of *Spirogyra* supported only 4×10^9 epiphytic cells g^{-1} fresh wt. (this might be attributable to their highly mucilaginous outer cell walls). A survey was carried out in September 1982 of an atypically sheltered littoral site on the north-west shore of Lough Neagh, lying in 0.5 m of water with some 80% macrophyte 'ground cover' amounting, on average, to 5800 g fresh wt. plant material m^{-2}. It was found that, of the 1.4×10^{14} cells m^{-3} in the water column overlying the site, only some 7% were planktonic; the remainder were attached to plant surfaces. Nevertheless the fact that extensive macrophyte populations are confined to the most sheltered bays and inlets must severely diminish the impact of epiphytic bacteria on Lough Neagh as a whole, although their role in many smaller freshwater ecosystems is of undoubted significance (Fry and Humphrey, 1978).

17.2.4. *The characterisation of bacteria*

Attempts were made to identify – or at least characterise – the predominant heterotrophic bacterial groups encountered in routine sampling. The exercise was by no means definitive but it was found – confirming the findings of investigators in similar lentic environments – that the microflora consisted overwhelmingly of gram-negative, catalase-positive rods, frequently motile and at least weakly fermentative, with oxidase-positive and -negative organisms present in roughly equal proportions. An isolate, tentatively identified as *Aeromonas* sp., was found to be particularly common in all sampling locations throughout the year. Chromogens formed a large proportion of plate counts on samples from the water column and littoral sediments; profundal sediment samples yielded predominantly cream or colourless colonies. A possible explanation is the protection presumed to be afforded by pigmentation against solar radiation in more exposed situations.

A further interesting observation, made during an unusually hot, calm two-week spell in August 1982 was the occurrence in the water column of a filamentous bacterium, tentatively identified as *Leptothrix* sp. at concentrations of up to 3×10^5 ml^{-1}. This may represent a migration by these organisms from the sediment in response to a temporary period of anoxia (see Chapter 9) and resultant nutrient release at the sediment surface; it may be significant that only at this time (Fig. 17.2) did numbers of anaerobes exceed aerobes in plate counts carried out on samples of the uppermost sediment layer.

17.2.5. *Bacterial biomass and periodicity*

On the basis of the results given in Sections 2.1, 2.2 and 2.3 it is clear that the microbial biomass of Lough Neagh is certainly similar in magnitude to that of other eutrophic bodies of water (Quinn *et al.*, 1985). Jones (1977b), for example, reported that direct counts of planktonic bacteria from water bodies of varying trophic status in the English Lake District lay between 1×10^6 and 1×10^7 ml^{-1}; viable counts ranged from 1×10^3 to 2×10^4 cells ml^{-1}. In 1979 Jones *et al.* also reported on the microbiology of the sediment of 16 Cumbrian lakes previously ranked as a series ranging from oligotrophic to eutrophic (Jones, 1972). Maximum values were obtained from Blelham Tarn, the most productive of the series, and showed total counts for the top 1 cm of sediment of 1.5×10^{10} ml^{-1} and viable aerobic and anaerobic counts of 1.8×10^8 ml^{-1} and 3.9×10^7 ml^{-1} respectively.

These values are quite remarkably similar to those obtained in the present study (Fig. 17.2).

Seasonal fluctuations in lentic bacterial populations have proved notoriously difficult to interpret, even when extensive monitoring of background variables has made possible a sophisticated statistical treatment of results e.g. Jones (1977a). In his study of seasonal changes in the planktonic bacterial populations of Windermere, Esthwaite Water and Blelham Tarn, Jones concluded that the bulk of variability could be accounted for by the effects of stratification and deoxygenation; neither of these phenomena are generally applicable to Lough Neagh. It is tempting, however, to link the sharp spring maxima of viable planktonic bacteria (Fig. 17.1) and of viable bacteria and yeasts in the sediment (Figs. 17.2 and 17.3) to the spring diatom crop which in 1982 peaked in mid-April and 'crashed' during May. Components representing algal productivity could account for 10% to 20% of variability in Jones' study. The increase in numbers of viable bacteria and, more dramatically, of yeasts in the surface sediment of the lake during autumn might conversely be related to the increased input of allochthonous organic material to Lough Neagh at leaf-fall.

17.2.6. *Total vs. viable populations*

The discrepancy of some two to three orders of magnitude between total and viable bacterial counts in both the water column and sediments (Figs. 17.1 and 17.2) is an outstanding feature of the season's results, and is not unique to Lough Neagh (Overbeck, 1974). Those bacteria which respond to plate culture techniques have been variously described as 'saprophytic', 'saprotrophic', 'heterotrophic' or 'zymogenous'; the vast bulk of the freshwater microbial flora which cannot easily be cultured has been termed 'oligotrophic', 'autochthonous' or 'oligocarbophilic'. This larger group – which consistently forms at least 95% of the planktonic and benthic bacteria of Lough Neagh, and which did not show visibly improved growth even on lake water solidified with agar (Table 17.1) – has been described by Overbeck (1974) as 'possibly the most important, and from the metabolic point of view the most interesting, but nevertheless the most unknown part of the aquatic microflora'. The fluctuating relationship between total bacterial numbers and saprophyte counts in lentic environments has frequently been interpreted as an indication of trophic status; thus the relatively 'low' ratios of 140 : 1 and 39 : 1 in the water column and sediment respectively in June 1982 may indicate an enrichment resulting from phytoplanktonic extracellular production or decomposition products. Similarly the ratio of 25 : 1 among those bacteria epiphytic on *Elodea* (Table 17.1) possibly reflects the high nutrient status of this microniche even in late autumn.

Whilst many workers believe, with Overbeck, that the non-culturable aquatic bacteria do play an important ecological role – most probably that of the mineralization of the more recalcitrant higher molecular weight com-

Table 17..1. A comparison of techniques for the assessment of viable, metabolically-active bacteria in Lough Neagh. 95% confidence limits are ±10% of the reported values

	Planktonic bacteria					Benthic bacteria			Epiphytes on *Elodea*
Date of sample:	11 Nov.	15 Nov.	22 Nov.	23 Nov.	24 Nov.	19 Nov.	19 Nov.	22 Nov.	15 Nov.
Direct counts (Nos. ml^{-1})	8.3×10^7	6.4×10^7	11.0×10^7	5.0×10^7	8.5×10^7	3.2×10^9	1.5×10^9	4.1×10^9	8.7×10^7 cm^{-2}
Viable Count (as % of direct count) as determined by:									
(a) NPN enumeration on CPS agar	0.2	0.3	0.6	0.9	0.2	2.5	0.7	1.6	4.0
(b) MPN enumeration on lake water agar	0.1	0.1	0.3	0.4	0.2	1.0	0.6	0.9	1.7
(c) ONT reduction in absence of added substrate	no data	21.4	29.4	25.0	27.8	15.7	18.8	31.7	38.0
(d) INT reduction in presence of added NADH + NADPH (0.5 and 0.08 mg ml^{-1} respectively)	no data	35.6	39.1	34.8	37.0	no data	no data	no data	42.3
(e) Cell elongation in presence of yeast extract and nalidixic acid	20.7	22.6	18.8	30.7	16.6	17.9	9.6	24.2	30.9
(f) Hydrolysis of fluorescein diacetate	4.2	2.2	5.5	11.4	3.8	2.9	1.2	5.2	3.6

pounds including the ubiquitous and poorly-defined 'Gelbstoff' – it is sometimes held that it is the zymogenous fraction, despite its relatively low numbers, which is largely responsible for decomposition and nutrient turnover in lakes with a high nutrient status (Rosswall and Persson, 1982) and that many of the cells included in the total count are dormant and inactive, if not actually dead. In an effort to clarify this point a study was carried out (Quinn, 1984a) in which a number of water and sediment samples from Lough Neagh, and a sample of epiphytic bacteria, were enumerated using the traditional plate count and acridine orange/epifluorescence direct count techniques, and in addition examined for the presence of a number of cytochemical criteria of metabolic activity: (i) the ability to reduce the terminal electron acceptor 2-(p-iodophenyl)-3-(p-nitrophenyl)-5 phenyl tetrazolium chloride (INT), either with or without an exogenous substrate; this is believed to indicate possession of an intact electron transport system (Zimmermann *et al.*, 1978), (ii) the ability to respond to the presence of added yeast extract by cell elongation – this is visually enhanced by addition of nalidixic acid which inhibits cell division (Kogure *et al.*, 1979), (iii) the ability to cleave the fluorigenic substrate fluorescein diacetate, which is believed to provide an indication of the integrity of the cellular membrane and hence of viability (Rotman and Papermaster, 1966).

The results, shown in Table 17.1, indicate clearly that some 20% of total planktonic, benthic and epiphytic bacteria, as enumerated by the direct count technique, consistently fulfilled at least two of these criteria of metabolic activity. Furthermore, monitoring of all planktonic bacterial samples obtained at the profundal station from June to November 1982 indicated that the number of cells capable of reducing INT (in the absence of an added substrate) was never less than 16% and occasionally as high as 54%. A cell's ability to respond to added substrate by rapid elongation (criterion b) is surely significant and, just as the direct count procedure overestimates the metabolically-active population, so the conventional plate count technique seriously underestimates the number of cells which have at least the potential for metabolic activity.

17.2.7. *Fungi other than yeasts*

Finally, an attempt was also made to assess the distribution and possible significance of fungi in the sediments of the profundal and littoral sites (Quinn, 1984b). Propagule numbers were estimated by plate counts on the glucose-yeast extract-starch-tellurite (GYST) medium of Collins and Willoughby (1962); the seasonal variation in their numbers is shown in Fig. 17.4. Fungal counts on samples from the water column consistently showed less than 12 propagules ml^{-1}. The interesting seasonal pattern obtained from the sediment counts is interpreted as indicating that fungal spores – the majority presumably of terrestrial origin - do not remain viable for long periods in the sediments. Two short periods of oxygen depletion in the surface sediments of

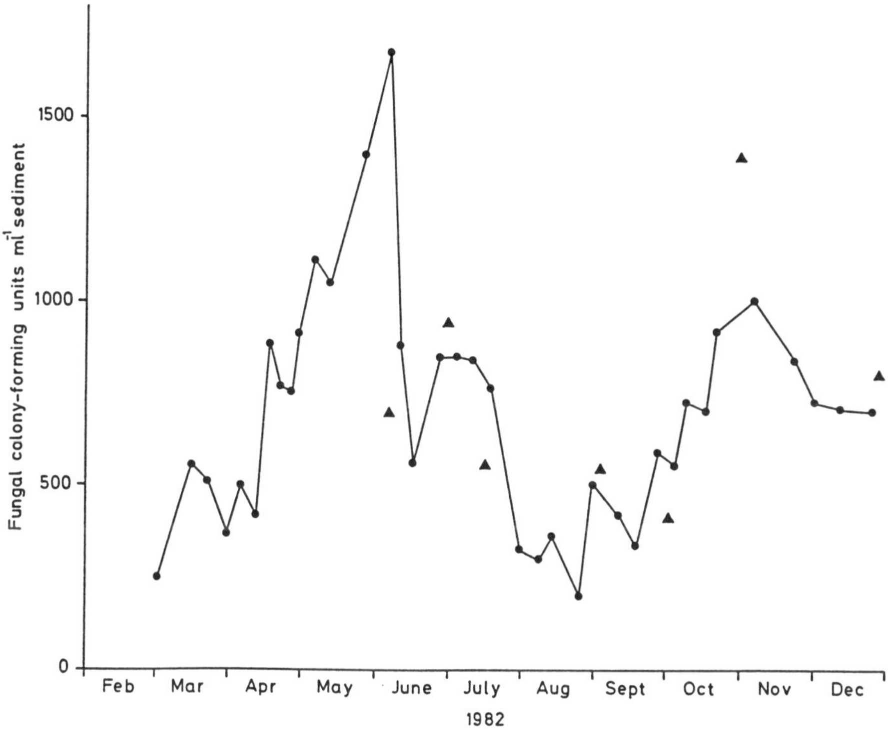

Figure 17.4. Seasonal distribution of fungal propagules in the uppermost centimetre of sediment at a profundal (●—●) and a littoral (▲—▲) sampling site in Lough Neagh during 1982. Redrawn from Quinn (1984b).

the profundal site which occurred in June and August may have been responsible for the particularly sharp decline in viable propagule numbers during mid-summer – only 54% of fungal isolates from the sediment proved facultatively anaerobic under laboratory conditions. Despite extensive examination of the sediments by the fluorescein diacetate staining technique of Soderstrom (1977), and the application of the homogenization technique successfully used by Johnson and Cross (1976) in their assessment of actinomycete activity in sediments, no evidence of active fungal mycelium was obtained, and their role in Lough Neagh sediments must tentatively be presumed to be passive.

17.3. Summary

On the basis of the preliminary, highly selective, and purely 'structural' data presented in this review it is clearly impossible to draw any definite conclusions as to the function of heterotrophic micro-organisms in Lough Neagh and their ecological significance.

It seems likely, however, that the total planktonic bacterial biomass of Lough Neagh is of the order of $1.3 \times 10^4\,m^{-3}$ or 1.5×10^4 tonnes fresh weight, assuming an average 'total' standing crop of 5×10^7 cells ml^{-1} and a mean cell volume of $0.08\,\mu^3$. With regard to this last assumption, Pedros-Alio and Brock (1982), using scanning electron microscope studies on the eutrophic freshwater Lake Mendota, obtained a mean cell volume of $0.16\,\mu^3$ but this was heavily influenced by larger attached bacteria. Straskrabova and Komarkova (1979) calculated mean cell volumes between 0.06 and $0.125\,\mu^3$ in their Czechoslovak reservoirs and, pending further SEM studies on Lough Neagh bacteria (and wishing to avoid wildly exaggerated claims for the possible magnitude of bacterial production), the more modest $0.08\,\mu^3$ value is taken. On several occasions during November 1982 bacterial growth rates in Lough Neagh were determined by the frequency-of-dividing-cells (FDC) technique of Hagstrom *et al.* (1975); on the basis that 25% of total cells were metabolically active (Table 17.1) a mean generation time of 19.8 h was obtained for the 'active' fraction. If this value applied to the whole year which is probably an overestimate, then using the values for bacterial cell density of $1.1\,g\,cm^{-3}$ and for carbon-to-wet-weight ratio of 0.15 (Hagstrom *et al.*, 1979) annual heterotrophic bacterial production in the water column of Lough Neagh in 1982 might be as high as $0.65 \times 10^3\,g\,C\,m{-}^2$. An assumed growth efficiency of 60% Payne, 1970) and specific maintenance energy requirement of $1\,g\,C\,g^{-1}$ wet cells yr^{-1} (Paul and Voroney, 1980) imply a bacterial carbon consumption of $1.1 \times 10^3\,g\,m^{-2}\,yr^{-1}$. A similar, but necessarily even more tentative, calculation which assumed a mean standing crop over the 0–2 cm Lough Neagh sediment layer of 6×10^9 cells ml^{-1} implies an annual carbon consumption by benthic micro-organisms of $3 \times 10^2\,g\,m^{-2}\,yr^{-1}$. Such a value amounts to a significant proportion of the estimated 4.1 to $5.6 \times 10^2\,g\,C\,m^{-2}\,yr^{-1}$ production by phytoplankton in Lough Neagh (see Chapter 11) recognising of course a variety of other organic inputs to the lake exist, benthic autochthonous and allochthonous. For a first crude estimate, the Lough Neagh value is in tolerable agreement with the findings of Pedros-Alio and Brock (1982) for Lake Mendota but it must be reemphasised that many of the assumptions made above may well be shown by further work on Lough Neagh to be inappropriate. Even so, there seem to be good reasons for anticipating a major influence of the very large heterotrophic microbial population on the functional ecology of Lough Neagh.

17.4. References

Aaronson, S. (1970) Experimental Microbial Ecology. New York, Academic Press.

Berrie, A.D. (1976) Detritus, micro-organisms and animals in fresh water. In: J.M. Anderson and A. Macfadyen (eds.), The Role of Terrestrial and Aquatic Organisms in Decomposition Processes. London, Blackwell.

Collins, V.G. and Willoughby, L.G. (1962) The distribution of bacteria and fungal spores in Blelham Tarn with particular reference to an experimental overturn. Archiv. fur Mikrobiologie 43: 294–307.

Fry, J.C. and Humphrey, N.C.B. (1978) Techniques for the study of bacteria epiphytic on aquatic macrophytes. In: D.W. Lovelock and R. Davies (eds.), Techniques for the Study of Mixed Populations. London, Academic Press.

Hagstrom, A., Larsson, U., Horstedt, P. and Normark, S. (1979) Frequency of dividing cells, a new approach to the determination of bacterial growth rates in aquatic environments. Appl. Environ. Microbiol. 37: 805–812.

Harris, R.E. and Sommers, L.E. (1968) Plate-dilution frequency technique for assay of microbial ecology. Appl. Microbiol. 16: 330–334.

Johnson, D.W. and Cross, T. (1976) Actinomycetes in lake muds: dormant spores or metabolically active mycelium? Freshwat. Biol. 6: 457–464.

Jones, J.G. (1972) Studies on freshwater microorganisms: phosphatase activity in lakes of differing degrees of eutrophication. J. Ecol. 60: 777–791.

Jones, J.G. (1977a) The effect of environmental factors on estimated viable and total populations of planktonic bacteria in lakes. Freshwat. Biol. 7: 67–91.

Jones, J.G. (1977b) Variations in bacterial populations in time and space. Annual Report Freshwater Biological Association 45, pp. 55–61.

Jones, J.G. and Simon, B.M. (1975) An investigation of errors in direct counts of aquatic bacteria by epifluorescence microscopy, with reference to a new method for dyeing membrane filters. J. Appl. Bact. 39: 317–329.

Jones, J.G., Orlandi, M.J.L.G. and Simon, B.M. (1979), A microbiological study of sediments from the Cumbrian Lakes. J. Gen. Microbiol. 115: 37–48.

Kogure, K., Simidu, U. and Taga, N. (1979) A tentative direct microscopic method for counting living marine bacteria. Can. J. of Microbiol. 25: 415–420.

Overbeck, J. (1974) Microbiology and biochemistry of lakes. Mitt. Internat. Verein. Limnol. 20: 198–228.

Paul, E.A. and Voroney, R.P. (1980) Nutrient and energy flow through soil microbial biomass. In: D.C. Ellwood, J.N. Hedger, M.J. Latham, J.M. Lynch and J.H. Slater (eds.), Contemporary Microbial Ecology. London, Academic Press.

Payne, W.J. (1970) Energy yields and growth of heterotrophs. Ann. Rev. Microbiol. 24: 17–52.

Pedros-Alio, C. and Brock, T.D. (1982) Assessing biomass and production of bacteria in eutrophic Lake Mendota, Wisconsin. Appl. Environ. Microbiol. 44: 203–218.

Pomeroy, L.R. (1980) Detritus and its role as a food source. In: R.S.K. Barnes and K.H. Mann (eds.), Fundamentals of Aquatic Ecosystems. London, Blackwell.

Quinn, J.P. (1984a) The modification and evaluation of some cytochemical techniques for the enumeration of metabolically active heterotrophic bacteria in the aquatic environment. J. Appl. Bact. 57: 51–57.

Quinn, J.P. (1984b) The seasonal occurrence of yeasts and fungi in the water column and sediments of a freshwater lake. Trans. Br. Mycol. Soc. 83: 53–58.

Quinn, J.P., Gillan, M.B. and McGrogan, H. (1985) The planktonic and benthic bacterial populations of Lough Neagh. J. Appl. Bact. 58: 87–98.

Rosswall, T. and Persson, I.B. (1982) Functional description of bacterial populations from seven Swedish lakes. Limnologica (Berlin) 14: 1–16.

Rotman, B. and Papermaster, B.W. (1966) Membrane properties of living mammalian cells as studied by enzymic hydrolysis of fluorogenic esters. Proc. National Academy of Science, U.S.A. 55: 134–141.

Soderstrom, B.E. (1977) Vital staining of fungi in pure cultures and in soil with fluorescein diacetate. Soil Biol. Biochem. 9: 69–63.

Straskrabova, V. and Komarkova, J. (1979) Seasonal changes of bacterial plankton in a reservoir related to algae. Int. Rev. ges. Hydrobiol. 64: 285–302.

Zimmermann, R.R., Iurriagan, R. and Becker-Birck, J. (1978) Simultaneous determination of the total number of aquatic bacteria and the number thereof involved involved in respiration. Appl. Environ. Microbiol. 36: 926–935.

18. The fish of Lough Neagh

Part A. A historical and taxonomic perspective of the fish fauna of the Lough Neagh catchment

G.J.A. KENNEDY and K.U. VICKERS

18A.1. Introduction

The fisheries of the Lough Neagh catchment have been the subject of a number of historical reviews (Donnelly, 1986; Mitchel, 1965; Thompson, 1856, see also Chapter 6) and recent species lists have been compiled by Kennedy (1978) and Wilson (1979). Much of the emphasis in both the historical and scientific literature has been placed on the commercial importance of two main species – the Atlantic salmon (*Salmo salar* L.) and the European eel (*Anguilla anguilla* L.) (Frost, 1950; Lowe, 1951, 1952; Menzies, 1936; Parsons *et al.*, 1977). Salmon are anadromous and eels are catadromous, and much of the fishing effort is therefore concentrated on the River Lower Bann, where their migratory habits have made these species vulnerable to capture at natural and man-made obstacles for perhaps millenia.

The fishing implements of prehistoric man have been found in abundance during dredging operations on the river (see Chaper 6), and radio carbon dating from one site near Toomebridge has indicated that harpoon-like fishing implements date from about 5725 BC (Mitchell, 1955). These bone harpoon points and flint instruments known as 'Bann flakes', also thought to be associated with fishing activities, were found to be particularly abundant at the sites of fords and natural weirs where fish might congregate (Frost, 1950; Mitchel, 1965). Mitchel (1965) also reports that although fishing was historically regarded as a valuable resource, and rights at weirs were defended with raths and castles, the fishing methods used were somewhat haphazard until more recent times. It is only in the last century that some of the weirs and fords of the lower River Bann have been adapted specifically for the purpose of catching salmon and eels. The older eel weirs at Movanagher and Portna (see Fig. 18A.1) are still constructed from woven hazel, larch and laurel 'skeaghs', or downstream vee traps, which direct the migrating silver eels into a hooped net trap known as a 'coghill'. The present eel weir at Toombridge was completed in 1946, replacng a traditional weir, and is a concrete and metal structure leading to four double hooped nets. Several

R. B. Wood and R. V. Smith (eds.), Lough Neagh, 381–395.

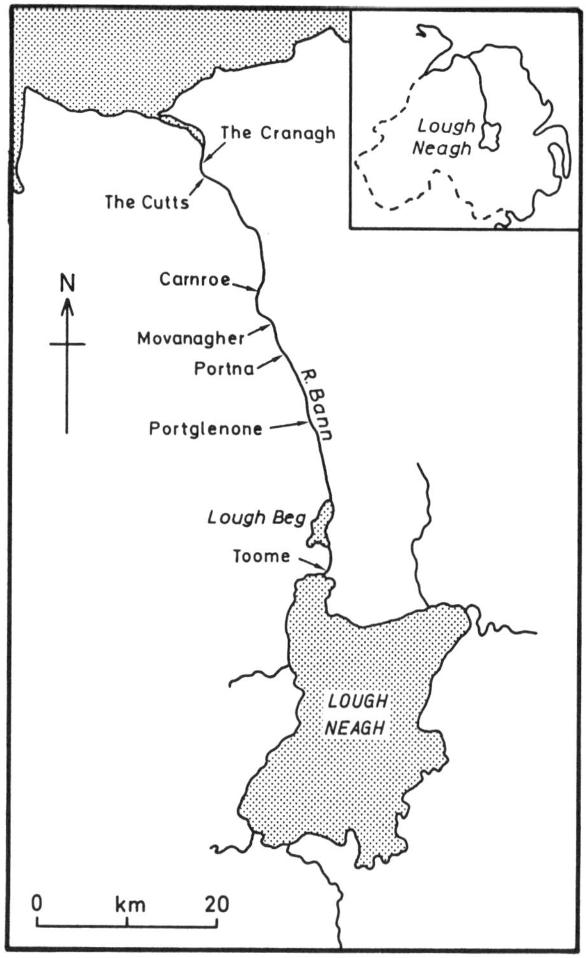

Figure 18A.1. Map of Lough Neagh and Lower River Bann, Northern Ireland, showing the position of the Toome silver eel fishery, and the sites of capture of elvers. Insert shows position of Lough Neagh and River Bann in Northern Ireland.

temporary weirs made from woven wood or even whins weighted with stones apparently existed right up to the time of the drainage scheme of the 1930s.

In contrast to the numerous eel fishing sites, only one permanent salmon fishing weir has been constructed on the River Bann. This is sited at the Cutts, about a mile upstream from the town of Coleraine (see Fig. 18A.1), and was first developed as a permanent fishing trap sometime in the 17th century (Mitchel, 1965). At present the weir is fished by means of four stone and metal 'in-scale' traps which were constructed in their present form in the mid-19th century. These traps have also been occasionally converted for the capture of eels after the close of the salmon season. Permanent traps are also incorporated in the structure at the Cutts for the capture of elvers during their upstream migration (Menzies, 1936; Lowe, 1951, 1952). Many draft net sites were also fished for salmon on the river in past years – among them the Cranagh, Loughen Island, Portglenone and Toomebridge. However, the only site now fished in this way is the Cranagh, which is situated on the tidal stretch of the lower Bann.

The eels of Lough Neagh have also been traditionally fished by long lines at the pre-migratory, brown eel, stage in the lough itself (see Chapter 6), along with many of the other species present. Details of this and the taxonomy of all the fish of the Lough Neagh catchment have not been fully reviewed until the present due to introductions and species changes brought about by environmental factors, as well as a variety of taxonomic interpretations in the past (Day, 1887; Regan 1911; Thompson, 1856; Yarrell, 1859).

A taxonomic review of the fish species of the Lough Neagh catchment, past and present, along with their life histories and indigenous details is outlined below.

18A.2. Salmonidae

18A.2.1. *Salmo salar* L. – Atlantic salmon

Atlantic salmon are anadromous and are captured seasonally in the River Bann and Lough Neagh on their migration to spawning areas in the tributary streams of the system, which the juveniles utilise as nursery areas (Cragg-Hine, 1972). The adults enter the River Bann from about the beginning of May onwards (Anon, 1975), although some 'spring' fish may enter earlier than this. About 90% of the adults have spent one winter at sea and are known as grilse (Went and Vickers, 1971). Most of the remainder have spent two winters at sea and, depending upon the time of their return to fresh water and on scale characteristics, are referred to as spring or summer fish. Occasionally fish which have spent three winters at sea have been found, and these are referred to as 'large spring or summer fish'. A proportion of the adult run is also made up of fish which have spawned one or more times previously and have returned to the sea as 'kelts' to feed for another winter, on average about 5% of the run in Irish rivers (Went, 1947). Spawning takes place from late November until mid January in gravelly areas of the tributary streams, where the eggs are laid in gravel beds or 'redds', in which they hatch during late March and April. At this stage the young fish are called alevins, and they only emerge from the gravel as fry after a further four or five weeks – when the yolk sac has been almost completely absorbed. In the Lough Neagh catchment the juveniles then remain in the nursery streams for one, two or occasionally three years as parr before turning into smolts and migrating to sea during the spring to complete the life cycle.

The River Bann has always had a large run of salmon, with Thompson (1856) reporting annual catches of up to 120 tons (i.e. about 30,000 fish), at the Cranagh in 1790. A decrease in the salmon runs in many rivers, including the River Bann, has been observed in Ireland in recent years. There is no doubt that the proportion of salmon as opposed to grilse in the Bann today is smaller than previously, with the last very large fish of 58 lbs (= 26 kg) captured at the Cutts having been photographed in 1928. A three sea-winter male salmon of 44 lbs (20 kg) was taken in the River Maine in 1955 and its cast is kept in the Ulster Museum. This disappearance of large fish is common to many rivers and is related to a number of factors. There appear to be cyclical changes in the proportion of salmon and grilse, and netting at Greenland has been selective for the large spring and summer fish as well as the repeat spawners. Large grilse are selectively captured by drift nets in home waters. In addition, the Bann drainage scheme in the 1930s probably removed many of the spawning sites favoured by the larger spring and summer fish.

18A.2.2. *Salmo trutta* L. – Brown trout

Brown trout are common in the Lough Neagh catchment, and their considerable propensity for phenotypic variation here as elsewhere has led to more taxonomic confusion for this species than probably any other. However, it is now considered that there is only one polytypic species of brown trout in Ireland despite the electrophoretic identification of varying allelic frequencies in separate breeding populations (Ferguson and Mason, 1981; Fleming and Ferguson (1983); Taggart *et al.*, 1981). Lough Neagh brown trout are now collectively referred to as dollaghan (Kennedy, 1975), although in the early records (Day, 1887; Thompson, 1856; Yarrell, 1859) dollaghan were divided into three distinct species: the river trout, *Salmo trutta* L.; the common trout, *Salmo fario* L. (of which the gillaroo trout, with thickened stomach walls and large spots, was conidered a distinct variety) and the great lake trout, sometimes called the buddagh of Lough Neagh, *Salmo ferox* L. Very large specimens of these 'species' were reported captured in the past in Lough Neagh. Thompson (1856) records that gillaroo were taken up to 12 lb weight (5.5 kg) and that the lake trout went up to over 30 lb (or even 50 lb (>22 kg) by traditional records (Regan, 1911)). A trout 13 years old and weighing 28 lb was taken by net in 1928. However, larger brown trout have now become very rare in Lough Neagh, and this has been attributed to heavy fishing mortality combined with the latent effects of the 'salmon disease', ulcerative dermal necrosis (UDN), which seriously depleted trout stocks in the late 1960s and particularly affected the larger fish (Kennedy, 1975).

The life cycle of the anadromous form of the brown trout i.e. sea trout, is similar to that of the salmon, and spawning of these and the lacustrine forms conform to the same patterns, with the juveniles occupying similar nursery areas in the tributaries. The Lough Neagh catchment is not noted for runs of very large sea trout, but considerable numbers of small sea trout known as finnock or whitling (Menzies, 1936) inhabit the estuary of the River Bann and the surrounding coastline. This situation has pertained in the Bann at least since the early 19th century, as Thompson (1856) has recorded that 'great numbers [of sea trout] are taken at the Cranagh, Coleraine; but they do not at any season average more than one pound weight'. However, large sea trout at this time were often regarded as a separate species – the grey trout, bull trout or roundtail, *Salmo eriox* L. (Thompson, 1856; Yarrell, 1859). Finnock have a sea life of only a few months duration after they go down as smolts, and they typically do not wander far from their native river estuary (Menzies, 1936). Fahy (1977) reports that the most probable spawning sites of the sea trout in the River Bann are in the lower tributaries, and that their is little penetration as far as Lough Neagh or beyond. He believes that poor feeding at sea is a major contributory factor to this downstream localisation of Irish sea trout stocks in general. However, we consider that the Bann estuary sea trout may originate in neighbouring rivers such as the Roe and Faughan (which are well known for their sea trout runs), since few are captured by anglers in Bann tributaries, such as the Agivey.

18A.2.3. *Coregonus pollan* Thompson – Pollan

This was first described as a separate species in Lough Neagh by Thompson (1856) as distinct from the vendace, powan or other coregonids of Britain. Regan (1911) was the first to recognise that this species had much in common with an Arctic species and Ferguson *et al.* (1978) have recently provided electrophoretic evidence that *C. pollan* and the Alaskan *C. autumnalis* are conspecific and separated only since the last ice age. The most recent studies of the Lough Neagh pollan are described more fully in Part D of this chapter.

18A.2.4. *Salvelinus alpinus* L. – Char

This was formerly abundant in Lough Neagh but is now extinct. Thompson (1856) reports that the last char were caught commercially about 1824, and a substantial reward offered in 1837 failed to produce a single specimen. In summer the Lough Neagh char were restricted to the deepest water – mainly in an area called the 'whiting hole' – and oxygen depletion of the deeper areas caused by eutrophication has been blamed for the disappearance of this species from Lough Neagh. However, as pointed out by Kennedy (1978) the disappearance of char occurred considerably prior to the main period of enrichment in Lough Neagh – as extrapolated from sedment analysis (see Chapter 7) – and it is therefore likely that some other factor was involved. Regan (1911) considered that overfishing was to blame and Thompson (1856) has noted that char disappeared from Lough Leven at about the same time.

18A.2.5. *Oncorhynchus mykiss* Walbaum – Rainbow trout

Small numbers of these are found in the Lough Neagh catchment as a result of escapes from fish farms. They are considered an exotic species throughout Ireland and are non-breeding under Lough Neagh conditions.

18A.3. Esocidae

18A.3.1. *Esox lucius* L. – *Pike*

Pike were originally introduced to England from mainland Europe and were still considered rare in the 16th century (Yarrell, 1859). From England they were brought to Ireland and probably introduced into ornamental lakes by land owners. They then spread throughout Ireland and their presence in the River Bann is first mentioned in 1682 (Went, 1957). By the 19th century Thompson (1856) noted them as 'common'. The largest he recorded from Lough Neagh was 36 lb (16 kg) which is only about half the reported size of the largest pike captured in Ireland.

Pike can apparently grow rapidly in conditions of a plentiful food supply, putting on up to 4 lb a year (Yarrell, 1859). They are cannibalistic and therefore tend to be solitary for much of the year, holding a feeding home range to which they will return if displaced (Anon, 1974). Spawning takes place during February to April in Irish lakes (Healy, 1956) when the eggs are deposited on weeds in shallow areas. These hatch in 14 to 21 days depending on water temperatures and the fry initially feed on small invertebrates. Pike become sexually mature at either two or three years of age and have a very high fecundity of about 12,000 eggs per lb. bodyweight.

Pike are common throughout the Lough Neagh catchment, where they have traditionally been commercially fished (Cragg-Hine, 1973), with Lough Beg just downstream from Lough Neagh producing some of the best specimens. They have reportedly declined in recent years in Lough Neagh due to excessive netting activities, although lowering of the lake level by 0.6 m in 1959 may have resulted in a reduction in the amount of available spawning weed beds. However, a specimen of over 29 lb (13 kg) was captured in the River Bann in 1980 (Anon, 1980).

18A.4. Percidae

18A.4.1. *Perca fluviatilis* L. – Perch

The historical status of the perch in Ireland is difficult to assess. Kennedy and Fitzmaurice (1974) point out that its distribution is extensive but that it is absent from several catchments where conditions are suitable for it and that its Irish name (péirse) seems to be derived from the English. It seems likely therefore that it is an introduced species. Thompson (1856) noted that great numbers were taken in Lough Neagh in the nineteenth century while fishing with pollan nets. Intensive commercial fishing has recently enjoyed a somewhat erratic revival and is described in Part D of this chapter.

18A.5. Cyprinidae

18A.5.1. *Abramis brama* L. – Bream

Kennedy and Fitzmaurice (1968) speculate that the bream is not a native of Ireland due to its patchy (although extensive) distribution prior to transplanting operations by the Inland Fisheries Trust from 1955 onwards. However, there is no record of this possible introduction in the literature, and although Thompson (1856) does not specifically mention its presence in Lough Neagh, it is likely that it has been there for many years. Cragg-Hine (1973) recorded that it is very common in the Lough Neagh catchment in present times and that there is a traditional bream fishery, the catch exported to mainland Europe.

In Ireland the bream is a long-lived fish, showing a range of growth rates in different systems. Its growth rate in Lough Neagh has not been investigated, but Kennedy and Fitzmaurice (1968) quote mean lengths of between 23.4 cm and 50.3 cm for ten year old fish in other Irish lakes. Spawning takes place among weeds in depths up to 3 m at temperatures over 15 °C (Kennedy and Fitzmaurice, 1968), and hybridisation with roach (*Rutilus rutilus* L.) and rudd (*Scardinius erthrohthalmus* L.) has taken place in the Lough Neagh catchment (Cragg-Hine, 1973).

18*A*.5.2. *Rutilus rutilus* L. – *Roach*

The roach is a relatively recent introduction to Ireland and is not included in Thompson's (1856) review. Went (1950) relates that an English angler using live bait first accidentally released it into the River Blackwater, Co. Cork, in 1889 (along with dace, *Leuciscus leuciscus* L., which have not yet spread throughout Ireland). From this fairly recent start in Ireland roach have now successfully invaded nearly every catchment – due mainly to transfer by anglers (Fitzmaurice, 1981), and have displaced rudd in some areas (Cragg-Hine, 1973). Cragg-Hine (1973) reports that the first recorded roach in the Lough Neagh catchment was taken in 1970 during electrofishing operations carried out by the Fisheries Research Laboratory staff on a tributary of the River Blackwater. In 1971 anglers reported catching roach in the Upper Bann, and they are now common throughout Lough Neagh and the Lower Bann.

Conditions are ideal in the Lough Neagh catchment for roach spawning, which takes place between early May and mid June when the water temperatures reach 15 °C. As well as the roach/bream hybrids which are present, roach/rudd have also been identified. Investigations into the biology of roach in Lough Neagh are outlined in Part D of this chapter.

18*A*.5.3. *Scardinius erythrophthalmus* L. – Rudd

Went (1950) asserted that the rudd is a species native to Ireland, but there is no factual historical evidence. Kennedy and Fitzmaurice (1974) point out that the species has an incomplete distribution throughout all the suitable habitats in Ireland and that this suggests that it is an introduced rather than an indigenous species, possibly from as far back as the time of the Norman invasion in the 12th century. Certainly by the mid-nineteenth century it was noted to be 'exceedingly common in the North of Ireland' by Thompson (1856), who identified specimens from Lough Neagh. However, Yarrell (1859) maintained that Lough Neagh rudd had some meristic characters slightly different from those in the English rudd. Cragg-Hine (1973) noted that rudd is very abundant in the Lough Neagh system and speculated on the possibility that the recent colonisation of the catchment by roach might soon displace it – as occurred in Lough Erne during the late 1960s. Part D of this chapter outlines the limited studies so far made of the interactions and hybridisation of these species in Lough Neagh as well as the general biology of the rudd.

18*A*.5.4. *Gobio gobio* L. – Gudgeon

There is no historical information available on the possible introduction of this species to Ireland. Kennedy and Fitzmaurice (1972a) pointed out that because its Irish distribution is imperfectly known its status cannot be fully assessed. However, in a later paper these authors (Kennedy and Fitzmaurice, 1974) considered that its known patchy occurrence was sufficient to suggest that it is not a native. Thompson (1856) noted the presence of gudgeon in Lough Neagh, but Regan (1911) and Yarrell (1859) consider that the species is more common in flowing water than still water habitats. Certainly, Fisheries Research Laboratory staff have captured specimens in small streams in the Lough Neagh catchment during electrofishing operations in recent years and it was found in Lough Neagh itself in 1984–5 by Tobin (pers. comm.).

The biology of the species has not been investigated in the Lough Neagh catchment, but Kennedy and Fitzmaurice (1972a, 1974) note that spawning at other sites in Ireland takes place in June in small shallow streams. These authors found that growth rates are similar to those recorded in England, with fish of three years of age from different sites ranging in mean length from 9.5 to 11.1 cm.

18*A*.5.5. *Phoxinus phoxinus* L. – Minnow

Thompson (1856) considered that the minnow is an introduced species to Ireland and relates several incidences of transplanting operations carried out by individuals on southern Irish rivers where minnows were previously not found. Kennedy & Fitzmaurice (1974) felt that because the range of this species in Ireland was still expanding it was likely that the minnow was not indigenous.

Kennedy (1977) studied the homing behaviour prior to and after a spawning migration in a small stream in the Lough Neagh catchment. He noted that spawning took place during late May and early June in shallow gravelly areas, and that minnows returned to downstream home ranges after this time. Thompson (1856) noted that three specimens from Lough Neagh were over 6 inches in length (152 mm), but their growth rate in this catchment has not been assessed since that time.

18*A*.5.6. *Tinca tinca* L. – Tench

Kennedy and Fitzmaurice (1970) assert that the tench is a fairly recently introduced species to Ireland, where it has a very restricted distribution. However, Went (1950) cites a reference to it in Co. Cork as early as 1750. It is not known whether there are tench in Lough Neagh itself, but it has been introduced to two lakes in the Lough Neagh catchment – Portmore and Loughgall – and in the latter the population is reputedly self sustaining (Cragg-Hine, 1973). A minimum temperature of 20 °C is required for spawning, which take place in late June and July, and the temperature in Lough Neagh rarely reaches that (see Chapter 3, Section 3.5).

18A.6. Cobitidae

18A.6.1. *Noemacheilus barbatulus* L. – Stoneloach

There are no historical data on stoneloach in Ireland and little is known about its distribution. Kennedy and Fitzmaurice (1974) consider that it is probably native since it is found throughout Ireland, often in small streams which do not form part of larger catchments, and it seems unlikely that it would be deliberately introduced into such areas. It is present in the Lough Neagh catchment, but is of no commercial value and little is known about its biology in this area.

18A.7. Anguillidae

18A.7.1. *Anguilla anguilla* L. – Eel

The eel is a native species to Ireland and common in every catchment. Thompson (1856) and Yarrell (1859) considered that there were three species, all of which could be found in Lough Neagh – the sharp nosed eel, *Anguilla acutirostris* Yarrell, the broad nosed eel, *Anguilla latirostris* Yarrell, and the snig eel, *Anguilla mediorostris*, Yarrell. However, Regan (1911) pointed out that the eel is subject to great phenotypic variation and that there is in fact only one species of European eel.

The life cycle of the eel has now been well documented (Bertin, 1956; Moriarty, 1978; Sinha and Jones, 1975; Tesch, 1977). After crossing the Atlantic as leptocephali, the young eels change into elvers and ascend into freshwater during the spring months. They then live in freshwater for at least 7 or 8 years during which time they are called yellow eels and grow to over 40 cm in length. After this time the eels then turn silvery in appearance and stop feeding prior to their seaward migration during the autumn and early winter to complete the life cycle somewhere in the Sargasso sea. As previously pointed out, it is because of this migratory habit that eels have been vulnerable to capture in the River Bann for many years. This silver eel fishery along with the yellow eel harvest on the lough itself combine to make the Lough Neagh eel fishery one of the largest and most commercially important fisheries for this species in Europe. Part B of this chapter details the studies undertaken of eels in the River Lower Bann.

18A.8. Gasterosteidae

18A.8.1. *Gasterosteus aculeatus* L. – Three spined stickleback

These are native to Ireland (Kennedy & Fitzmaurice, 1974). Previously Thompson (1856) had considered that there were several species of three spined stickleback in Ireland, but Regan (1911) noted that all three spined sticklebacks in the British Isles were the same species, showing various phenotypic variation in the number and spacing of the spines. They were recorded by Thompson (1856) in the Bann at Toomebridge and along the shores of Lough Neagh, and are common throughout the catchment today.

Their biology has not been studied in this area, but elsewhere they are known to spawn in the spring and summer (Regan, 1911). Although abundant they are of no commercial importance.

18A.8.2. *Gasterosteus pungitius* L. – Ten spined stickleback

This species was not recorded in Lough Neagh by Thompson (1856), who noted that it was rare in Ireland. However, it has more recently been recorded as being common in Lough Neagh (Anon, 1968) although its biology has not been studied in this area.

18A.9. Petromyzonidae

18A.9.1. *Petromyzon marinus* L. – Marine lamprey

This species commonly migrates into the rivers of Ireland to spawn, and Thompson (1856) notes that it is of regular occurrance in the Lough Neagh system. It can grow to over 5 lb (2.5 kg) in weight and Thompson (1856) recorded that specimens of this size have been captured in the River Ballinderry. However, nowadays it probably cannot ascend beyond Carnroe weir. It has not been investigated in the Lough Neagh catchment, but Regan (1911) recorded that the upstream spawning migration can take place from February to June in different English and Scottish rivers. Spawning takes place in the faster flowing gravelly areas of streams from June to August.

18A.9.2. *Lampetra fluviatilis* L. – River lamprey

This lamprey is also found in the Lough Neagh catchment, and Thompson (1856) identified one specimen which was captured adhering to a Lough Neagh trout. Although no investigations have been carried out on them in the Lough Neagh catchment, it was noted that 18.5% of pollan and 42.9% of trout captured during experimental fishings in 1966 showed wounds from lamprey bites (Moriarty, unpublished). This species may therefore be responsible for considerable mortalities in some Lough Neagh fish stocks. It is not known what proportion of the Lough Neagh population migrates to the sea. Spawning takes place in more upstream sites than the marine lamprey from April to June (Jenkins, 1946; Regan, 1911).

18A.9.3. *Lampetra planeri* Black – Brook lamprey

This species is widely distributed in Ireland (Regan, 1911) and is common in the tributaries of Lough Neagh and the River Bann. It was first recorded here by Thompson (1856) who wrongly assumed that a pride captured in the River Blackwater was a separate species known as *Ammocoetes branchialis*. These are now known to be immature lampreys, which exist in this larval form for three or four years (Regan, 1911). This lamprey never migrates to the sea and its spawning habits are similar to those of the river lamprey.

18A.10. Pleuronectidae

18A.10.1. *Platichthys flesus* L. – Flounder

Thompson (1856) notes that the flounder is the only flat-fish which has been captured in freshwater in Ireland. The first record of a specimen from Lough Neagh was by McCurdy (1977) who supposed that it had migrated at least 50 miles through the navigation system of the River Bann to its point of capture.

It is common in the estuary of the River Bann where it has been investigated by Wirjoatmodjo (1980). In a physico-chemical survey of the estuary he identified a permanent salt wedge containing a partially de-oxygenated zone in deeper water and large areas of shallow mud and sand flats which were available as foraging ground when covered by the tide. Feeding during the high tide period was found to be independent of time of day, and the main preferred food item was *Nereis* followed by *Corophium*, particularly in the warmer months. The diet was more diverse closer to the sea, and young flounder fed less selectively than older fish. Residents did not fast in the winter and the seasonal condition factors did not fluctuate greatly up to three years of age, when both sexes reached maturity. Migration to spawning grounds in the sea took place in winter followed by a return to the estuary for feeding in the summer. The young of the year entered the estuary in large numbers in May, and, after one year, annual mortality was about 52%. Adults up to six years of age were captured, by which stage they had reached a mean length of about 39.5 cm. An unusual ambicoloured specimen has also been recorded from this area (Briggs, 1977).

18A.11. Muglidae

18A.11.1. *Crenimugil labrosus* Risso – Thick lipped mullet

Two species of mullet have been recorded from the Irish Coast (Kennedy and Fitzmaurice, 1969, 1972b), but only the thick lipped variety is common, and it is this that is found in the Bann estuary. This species was originally referred to by Thompson (1856) and Regan (1911) as *Mugil chelo* Cuvier.

Work by O'Neill (unpublished) has shown that the mullet population in the River Bann is composed entirely of adult fish over five years old, and this indicates that the estuary lies close to the northern limit of the range for this species. Spawning takes place at sea during the winter, and although all the fish are absent from the estuary during this time only a proportion of the females spawn in any one year. The fish reappear in early spring (usually mid April), and recapture data from tagging studies have indicated that the same fish return to the River Bann each year. There is also evidence that small shoals fed regularly at the same place in the estuary. The fish move in with the flood tide to feed over the mud flats, and return to sea with each ebb tide. Food consists mainly of algae, detritus and invertebrates, and this is occasionally supplemented by dairy waste from a local discharge. Individuals are long lived, and fish of over fifteen years are not uncommon. Growth, however, is slow due to the poor food value of much of the diet, and fish of more than 2 kg are rare.

18A.12. Summary

The natural distribution of freshwater species in Lough Neagh and the River Bann was largely determined as the catchment area evolved upon the retreat of the ice cap following the last ice age. Certainly electrophoretic evidence from the work of Ferguson *et al.* (1978) on Lough Neagh pollan (see Part D of this chapter) suggests that these are relicts of a wider distribution some 10,000 years ago. According to Regan (1911) the present number of all species of fish in eastern England and continental Europe was determined by those fish which inhabited the archaetypal Rhine. Ireland was isolated from this species reservoir, and only anadromous, catadromous and a few euryhaline species were able to colonise the area. In addition, it was thought that a few freshwater species had colonised through a glaciological 'Shannon', which included the present Lough Neagh, but current Quaternary geology largely discounts this connection (see Chapter 2). Clearly the relatively large number of species either known or surmised to have been introduced or spread through man's activity has had a considerable effect on the fish fauna of Lough Neagh.

From this review of the fish to be found in the Lough Neagh catchment it is evident that most information is available about the most economically important species. These, both in the historic and present day context, are the salmon and eel and are summarised in Part B of this chapter. However, more recent investigations have concerned trout, pollan and cyprinids and these are described in Parts C and D.

18A.13. References

Anonymous (1968) Symposium on large lakes and impoundments. Lough Neagh, Northern Ireland. HMSO Spec. Publ., 78 pp.

Anonymous (1974) Annual report of the Inland Fisheries Trust Incorporated, 1973–74, pp. 8–10.

Anonymous (1975) Department of Agriculture for Northern Ireland. Ann. Rep. on Res. and Tech. Work, HMSO, pp. 107–114.

Anonymous (1980) Irish Specimen Fish Committee. Ann. Rep., 30 pp.

Bertin, L. (1956) Eels. A biological study. Cleaver-Hume, London, 192 pp.

Briggs, R.P. (1977) An ambicolorate flounder *Platichthys flesus* (L.) from the River Bann, Northern Ireland. Ir. Nat. J. 19: 70–71.

Cragg-Hine, D. (1972) The juvenile salmonid populations of some streams in the Lough Neagh and lower Bann catchment areas. Dept. Agric. for N.I., Fish. Res. Leaflet No. 4, 13 pp.

Cragg-Hine, D. (1973) Coarse fish and fishery management in Northern Ireland. Proc. 6th Brit. Coarse Fish. Conf. Univ. of Liverpool 1973, pp. 52–59.

Day, F. (1887) British and Irish Salmonidae. Williams and Norgate, London, 298 pp.

Donnelly, D.J. (1986) On Lough Neagh's Shores. Publ. The Donnelly Family, Galbally, Tyrone, 331 pp.

Fahy, E. (1977) Characteristics of the freshwater occurrence of sea trout *Salmo trutta* in Ireland. J. Fish Biol. 11: 635–646.

Ferguson, A., Himberg, K-J.M. and Svardson, G. (1978) Systematics of the Irish pollan (*Coregonus pollan* Thompson): An electrophonetic comparison with other holarctic coregoninae. J. Fish. Biol. 12: 221–233.

Ferguson, A. and Mason, F.M. (1981) Allozyme evidence for reproductively isolated sympatric populations of brown trout *Salmo trutta* L. in Lough Melvin, Ireland. J. Fish Biol. 18: 629–642.

Fitzmaurice, P. (1981) The spread of roach *Rutilus rutilus* (L.) in Irish waters. Proc. 2nd Brit Freshwat. Fish Conf. Univ. of Liverpool, pp. 154–169.

Fleming, C.C. and Ferguson, A. (1983) Genetic variation in Irish brown trout populations: Identification and conservation of stocks. Irish Fisheries Investigations, Ser A 23: 56–60.

Frost, W. (1950) The eel fisheries of the River Bann, Northern Ireland, and observations on the age of the silver eels. J. Cons. Perm. Int. Con. Explor. Mer., Inter. Expl. 16: 358–383.

Healy, A. (1956) Pike (*Esox lucius* L.) in three Irish Lakes. Sci. Proc. R. Dub. Soc. 27: 51–67.

Jenkins, J.T. (1946) The Fishes of the British Isles, both freshwater and salt. Warne & Co., London and New York, 408 pp.

Kennedy, G.J.A. (1975) Age, growth and feeding studies on Lough Neagh trout *Salmo trutta* (L.). Dept. Agric. for N.I., Fish. Res. leaflet No. 10, 14 pp.

Kennedy, G.J.A. (1977) Experiments on homing and home range behaviour in shoals of roach (*Rutilus rutilus* L.) and minnows (*Phoxinus phoxinus* L.). D. Phil. thesis, New University of Ulster, 267 pp.

Kennedy, G.J.A. (1978) Lough Neagh – The algal problem and its solution. Impact on fisheries. Proc. Inst. Biol. Symp. on L. Neagh, 1978, pp. 22–26.

Kennedy, M. and Fitzmaurice, P. (1968) The biology of the bream *Abramis brama* (L.) in Irish waters. Proc. R. Ir. Acad. 67: 95–157.

Kennedy, M. and Fitzmaurice, P. (1969) Age and growth of thick lipped grey mullet, *Crenimugil labrosus* in Irish waters. J. Mar. Biol. Ass. U.K. 49: 683–699.

Kennedy, M. and Fitzmaurice, P. (1970) The biology of the tench *Tinca tinca* (L.) in Irish waters. Proc. R. Ir. Acad. 69: 31–82.

Kennedy, M. and Fitzmaurice, P. (1972a) Some aspects of the biology of gudgeon *Gobio gobio* (L.) in Irish waters. J. Fish Biol. 4: 425–440.

Kennedy, M. and Fitzmaurice, P. (1972b) Occurrence of thin lipped grey mullet, *Liza ramada* (Risso) in County Mayo. Ir. Nat. J. 17: 242–243.

Kennedy, M. and Fitzmaurice, P. (1974) Biology of the rudd *Scardinius erythrophthalmus* (L.) in Irish waters. Proc. R. Ir. Acad. 74: 245–305.

Lowe, R.H. (1951) Factors influencing the runs of elvers in the River Bann, Northern Ireland. J. Cons. Perm. Int. Explor. Mer. 17: 299–315.

Lowe, R.H. (1952) Letter to the editor on 'Factors influencing the runs of elvers in the River Bann, Northern Ireland'. J. Cons. Perm. Int. Explor. Mer. 18: 241–244.

McCurdy, W.J. (1977) A record of *Platichthys flesus* L. flounder in Lough Neagh. Ir. Nat. J. 19(4): 1.

Menzies, W.J.M. (1936) The run of elvers in the River Bann, Northern Ireland. J. Cons. Perm. Int. Explor. Mer. 11: 249–259.

Mitchel, N.C. (1965) The lower Bann fisheries. Ulst. Folk. 11: 1–32.

Mitchell, G.F. (1955) The Mesolithic site at Toome Bay, Co. Londonderry. Ulst. J. Archaeol. 18: 1–16.

Moriarty, C. (1978) Eels. A natural and unnatural history. David & Charles, London, Vancouver, 192 pp.

Parsons, J., Vickers, K.U. and Warden, Y. (1977) Relationship between elver recruitment and changes in the sex ratio of silver eels *Anguilla anguilla* L. migrating from Lough Neagh, Northern Ireland. J. Fish Biol. 10: 211–229.

Regan, C.T. (1911) The freshwater fishes of the British Isles. Methuen, London, 287 pp.

Sinha, V.R.P. and Jones, J.W. (1975) The European freshwater eel. Liverpool University Press, 146 pp.

Taggart, J., Ferguson, A. and Mason, F.M. (1981) Genetic variation in Irish populations of brown trout (*Salmo trutta* L.): electrophoretic analysis of allozymes. Comp. Biochem. Physiol. 69B: 393–412.

Tesch, F.W. (1977) The eel. Biology and management of Anguillid eels. Chapman and Hall, London, 434 pp.

Thompson, W. (1856) The Natural History of Ireland. IV Mammalia, reptiles and fishes. H.G. Bohn, London, 516 pp.

Went, A.E.J. (1947) Value of the kelt. Some notes on previously spawned salmon in Ireland. Sal. and Trt. Mag., January 1947: 41–48.

Went, A.E.J. (1950) Notes on the introduction of some freshwater fish into Ireland. J. Ir. Dept. Agric. 47: 1–8.

Went, A.E.J. (1957) The pike in Ireland. Ir. Nat. J. 12: 177–181.

Went, A.E.J. and Vickers, K.U. (1971) A comparison between the salmon and grilse of the Foyle and Bann systems. Foyle Fish. Comm. Ann. Rep. 170–171, Appendix IV, 4 pp.

Wilson, J.P.F. (1979) The biology and population ecology of the pollan, *Coregonus autumnalis pollan* Thompson, of Lough Neagh. D. Phil. thesis, New University of Ulster, 123 pp.

Wirjoatmodjo, S. (1980) Growth, food and movement of flounder *Platichthys flesus* (L.) in the estuary. D. Phil. thesis, New University of Ulster, 306 pp.

Yarrell, W. (1859) A History of British Fishes. Volume I. Van Voorst, London, 675 pp.

18. The fish of Lough Neagh
Part B. Investigations on salmon (Salmo salar L.) and eels (Anguilla anguilla L.) in the River Lower Bann

G.J.A. KENNEDY and K.U. VICKERS

18B.1. Salmon investigations

18B.1.1. *Migration routes*

As stated in Part A, the salmon has long been exploited in Lough Neagh and, although the value of the commercial fishery on the River Lower Bann currently is perhaps as little as £70,000 per annum (Wood and Smith, 1987), concern for salmon remains high, particularly with sport angling and tourism in mind. Note that thoughout this section the River Bann is the River Lower Bann and that many of the sites referred to are located in Fig. 18A.1.

Salmon investigations, in the form of tagging studies, have been carried out on the River Bann since 1901 (Hillas, 1904; Holt, 1903). Up to 1922 over two and a half thousand fish had been marked as kelts in the Bann (Went, 1947). Of 117 recaptures only one was taken in another river system (the Bundrowes River in Co. Sligo). Five other recaptures were taken in fixed nets on the North Antrim coast, and the remainder were caught in the commercial fisheries of the Bann at the Cranagh and the Cutts. This was some of the first good evidence available that returning kelts had a very marked homing instinct for the river where they first spawned. The tagged fish recaptured at sea also gave the first clues as to the migration routes of salmon returning to the Bann to spawn. This work was not extended to the tagging of first time spawners until 1948, when a tagging programme of adult salmon at sea was initiated and recaptures were monitored in the River Bann and elsewhere (Went, 1964a and b; Went and Vickers, 1953, 1956; Vickers and Ferris, 1970).

These studies indicated that the majority of fish returning to the River Bann migrated along the North Antrim coast in a Westerly direction, and many fish which were originally tagged at Carnlough, Torr Head, Carrick-a-Rede, Port Bradden, Portballintrae and Portrush were subsequently recaptured in the River Bann (see Fig. 18B.1). Other fish heading for more Westerly rivers mainly the Foyle, but also for example the River Lennon, were found to share this migration route with River Bann salmon. However,

R. B. Wood and R. V. Smith (eds.), Lough Neagh, 397–417.

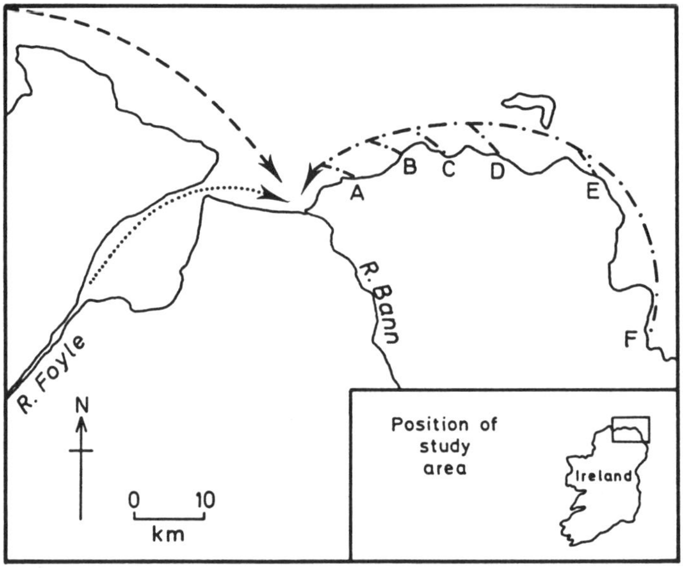

Figure 18B.1. Migration routes of tagged salmon to the River Bann. East to West salmon migration route to River Bann, described by Went (1964a), Went and Vickers (1953, 1956) and Vickers and Ferris (1970). Salmon tagged at coastal netting sites (A – Portrush; B – Portballintrae; C – Portbradden; D – Carrick-a-rede; E – Torr Head; F – Carnlough) (–·–·). Salmon migration route to River Bann from West to East, described by Went (1958, 1964a). (–––). Movement of salmon taggerd in River Foyle estuary to River Bann, described by Went (1963, 1964a and b). (· · · · ·).

some River Bann fish apparently approached the river from the opposite direction, and a few fish which were originally tagged off County Donegal were later recaptured in the Bann (Went, 1958). More recent investigations using micro-tags in River Rush salmon have shown that large numbers of these are taken in the Donegal drift net fishery, and that the importance of the west to east migration route may have been underestimated in earlier studies due to the non-returning of traditional tags by fishermen (Crozier and Kennedy, 1987). Tagging carried out by Went (1963) also indicated that some fish which were eventually captured in the River Bann had first entered the neighbouring estuary of the River Foyle. Similarly, tagging studies carried out in the estuary of the River Bann in 1970 and 1971 (Anon, 1971; Vickers and Parsons, 1971) indicated that 2.7% to 11.1% of the fish entering the River Bann were in the 'wrong' estuary, and subsequently left again to be recaptured elsewhere (Table 18B.1). Vickers and Parsons (1971) pointed out that of the 11.1% recaptured outside the Bann in 1970, most were taken in local shore nets, and mainly after one particular day's tagging. The authors believed that these fish dropped down due to the trauma of tagging and would subsequently have returned to the Bann.

Table 18B.1. Recaptures of salmon tagged in the estuary of the River Bann during 1970 and 1971

Year	No. Tagged	Percentage recaptured		
		At Cranagh	At Cutts	Elsewhere
1970	389	13.6	13.1	11.1*
1971	403	11.7	23.2	2.7

* Over two thirds of these were recaptured in shore nets in the locality of Portrush and Portstewart close to the mouth of the River Bann.

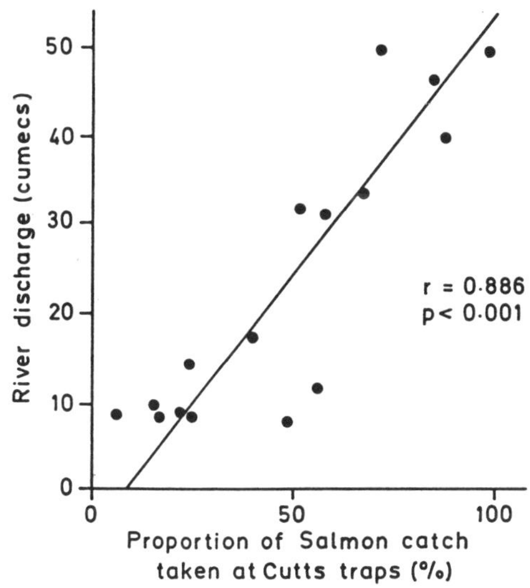

Figure 18B.2. Correlation of water discharge in the River Bann (measured at Movanagher) with the proportion of the salmon catch taken at the Cutts.

18B.1.2. *River discharge and salmon movements*

Of the salmon recaptured in the River Bann itself, the proportions taken by the traps at the Cutts and the nets at the Cranagh were found to be significantly correlated to water flows (Fig. 18B.2). On days of low discharge salmon linger at the mouth of the river and a high proportion of the total catch is made at the Cranagh, whereas on days of high flows the fish move upstream and the Cutts traps take the greater part of the catch (Parsons, 1971). Water discharge was approximately three times higher in 1971 than in 1970, and this is reflected in the greater percentage of tagged fish taken at the Cutts in 1971 (Table 18B.1). Also, in the high flow conditions of 1971 the tagged salmon moved more quickly the three miles to the Cutts, where most recaptures occurred two days after release at the Cranagh. In 1970 the

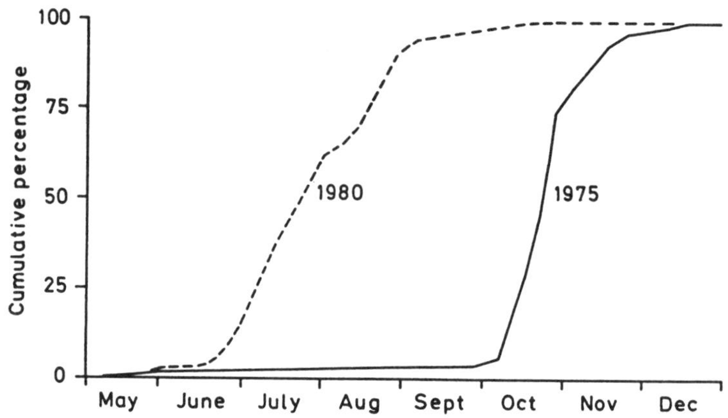

Figure 18B.3. Cumulative percentage salmon count at Portna on the River Bann during a year of low water conditions (1975) and year of high water conditions (1980).

majority of salmon took four to five days to move this distance. Due to this quicker movement through the estuary in higher flow conditions and the consequent lower efficiency of the Cranagh nets, the upriver escapement in years of high water discharge is likely to be greater and has been estimated at 48% to 55%, compared to 35% to 40% in a drier summer (Anon, 1971).

Since 1935 the flow rates of the River Bann have been regulated by manipulation of the sluices at Toomebridge. This has resulted in the ability to control the rate of flow to facilitate the migration of salmon up the river, and recommendations have been made that flows should not be less than 9.4 to 11.8 $m^3 s^{-1}$ at Movanagher (Bell, 1971). A fish counter installed at Portna showed that during periods when the flows were maintained at about 1.5 times this recommended rate, salmon did move upstream (Anon, 1973). However, the majority of movement occurred only when there were considerable and prolonged increases in flows, although smaller runs of salmon can be induced by small increases in water level (Anon, 1973; Parsons, 1971). During dry summers this leads to a fairly late run of salmon in the Bann, with the majority of fish not passing upstream until about October (Anon, 1975). In wet summers, however, when water discharge is high, the majority of fish have passed upstream by the end of August (Fig. 18B.3). Information on the diurnal movements of salmon was also provided by the fish counter at Portna, indicating that very few fish moved during the hours of darkness (see Fig. 18B.4(a)). The peak run occurred during the early morning, and as the season progressed there was a second peak in the evening (Fig. 18B.4(b)).

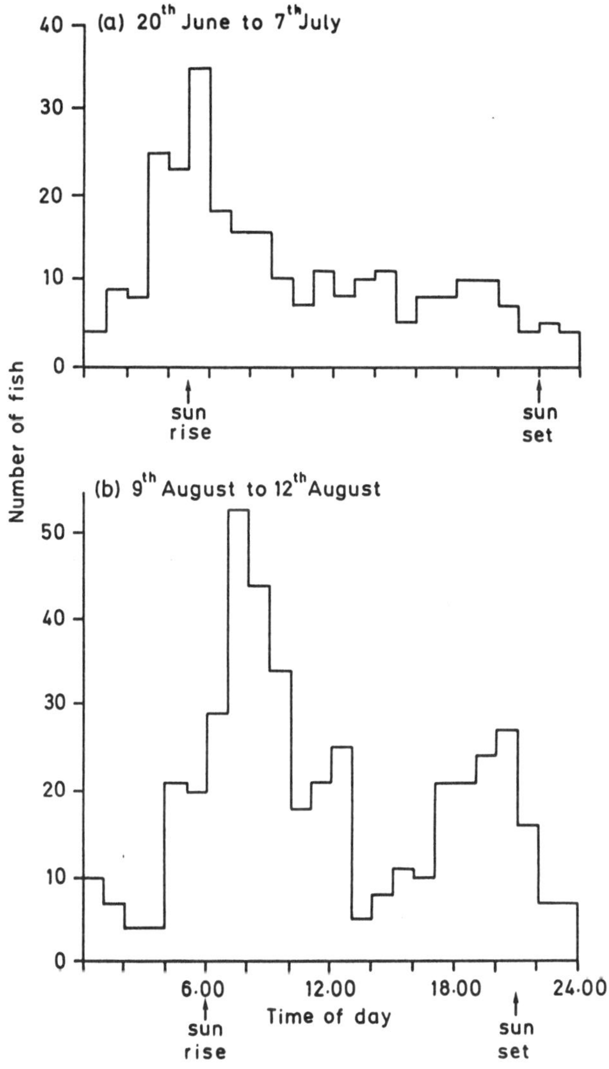

Figure 18B.4. Diurnal pattern of movement of salmon through Portna counter, 1973.

18B1.3. *Age and weight of salmon*

An investigation of the age distributions and weights of salmon and grilse entering the River Bann was undertaken in 1971 (Anon, 1971; Went and Vickers, 1971). The size distribution of these was found to range from 49 cm to 89 cm (1.4 kg to 8.4 kg) with a mode at 65 cm (3.3 kg) (see Fig. 18B.5).

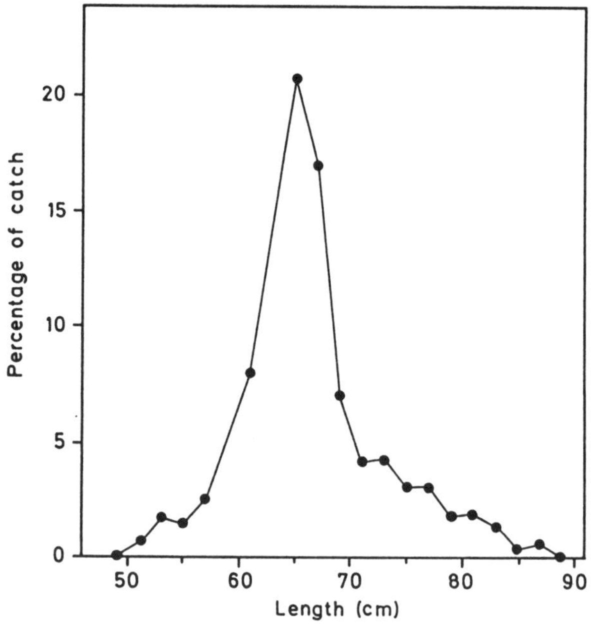

Figure 18B.5. Size distribution of commercially caught River Bann salmon in 1971.

Table 18B.2. Details of fish heavier than 3.63 kg (8 lb) sampled in the River Bann commercial catch during June, July and August 1971

Month	Fish classification by scale reading		Fish heavier than 3.63 kg in sample		
	Description	No of sea winters	No.	%	Mean wt (kg)
June	Grilse	1+	24	18.8	3.87
	Salmon	2 and 2+	103	80.5	5.04
	Repeat spawners	2+	1	0.7	3.86
July	Grilse	1+	237	88.1	4.24
	Salmon	2 and 2+	27	10.0	5.65
	Repeat spawners	2+	5	1.9	5.67
August	Grilse	1+	56	98.2	4.43
	Salmon	2 and 2+	1	1.8	6.12
	Repeat spawners	2+	0	0	–

One of the aims of this investigation was to determine the relevance of the 3.63 kg (8 lb) limit operated in marketing as the dividing line between salmon and grilse. Scale readings showed that as the season progressed an increasing proportion of the fish exceeding 3.63 kg were in fact grilse (Table 18B.2).

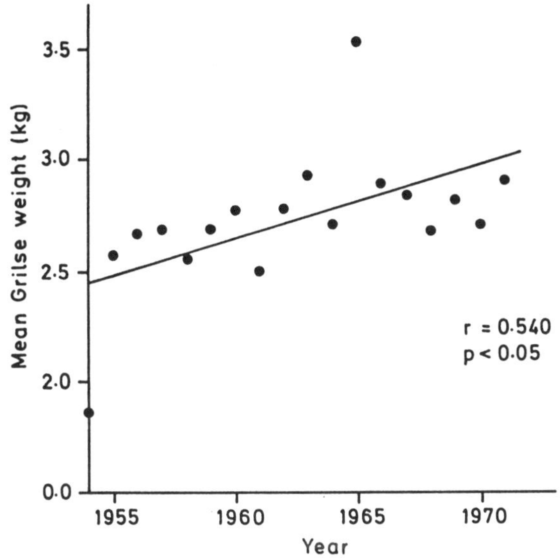

Figure 18B.6. Mean weight of grilse caught in the River Bann 1954–1971.

Over the whole season 69.6% of the 3.63 kg plus fish which were sampled were grilse. However, there is evidence that this has not always been the case, as the mean weight of grilse has been increasing steadily since the early 1950s (see Fig. 18B.6). It has been hypothesised that this is due to the late run of grilse on the Bann becoming progressively later during this period, and the extra time spent feeding at sea has produced the increase in mean weights. Certainly there has been a significant decline in the runs of the earlier running spring salmon (i.e. 2 sea winter fish) in the catch and therefore a concomitant increase in the proportion of the later running grilse over this period (see Fig. 18B.7).

18B.2. Eel investigations

Of all the fish in the Lough Neagh system eels are the most commercially important, with the annual catch estimated to be of the order of 950 tonnes in 1980, worth approximately £2,000,000 (Anon, 1980). In 1987 the total catch had fallen to around 600 tonnes (see Fig. 18B.8) although with price inflation the total value has probably risen.

18B.2.1. *Elver runs*

The first investigation of eels in this system was concerned with the very large run of elvers, which were monitored following the construction of elver traps at the Cutts, Coleraine, during the drainage scheme of the 1930s

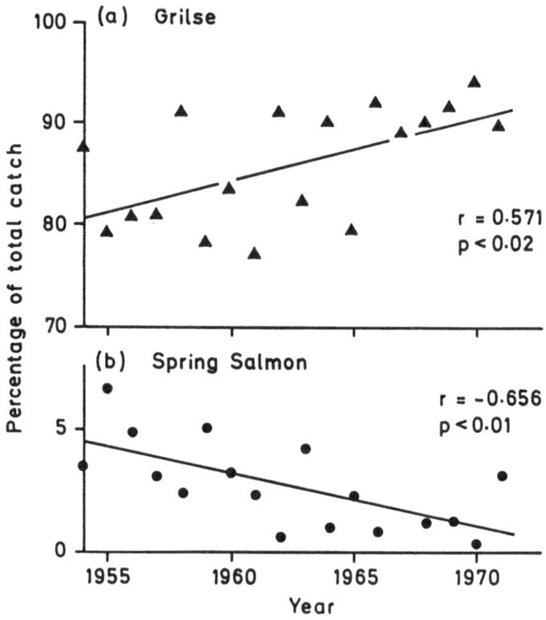

Figure 18B.7. Percentage of (a) grilse (▲) and (b) spring salmon (●) in the total River Bann catch 1954–1971.

(Menzies, 1936). These traps were constructed so that elvers could be captured and transported by road for release into the River Bann and Lough Neagh at sites beyond all the new navigational locks and weirs, which it was felt could have obstructed their migration to the lough. Menzies reported that the first elvers arrived at the Cutts traps in April, with most being captured in May and June and only small numbers in July. However, it was not until the late 1930 that the traps became fully operational, and Frost (1950) reported that elvers had arrived at the Cutts as early as 6 March. Frost also noted that although the total annual number of elvers trapped in the Bann from 1936–1946 varied from about 16 million to 30 million, this bore no relation to the length of season. Factors influencing the run of elvers in the Bann were investigated by Lowe (1951, 1952), who reported that elvers tended to run more often from spring tides only at the beginning and end of seasons, when relatively few elvers were present or in seasons when river discharge was very low. She hypothesised that the presence of large numbers of elvers during the mid-season in some way stimulated migration irrespective of tidal influences, and that floods masked the effects of tide due to high freshwater discharges attracting elvers into the estuary. However, later work by the Department of Agriculture (Anon, 1970) demonstrated that there is a significant inverse correlation between elver runs and water discharge (see Fig. 18B.9) i.e. the elvers require periods of low water dis-

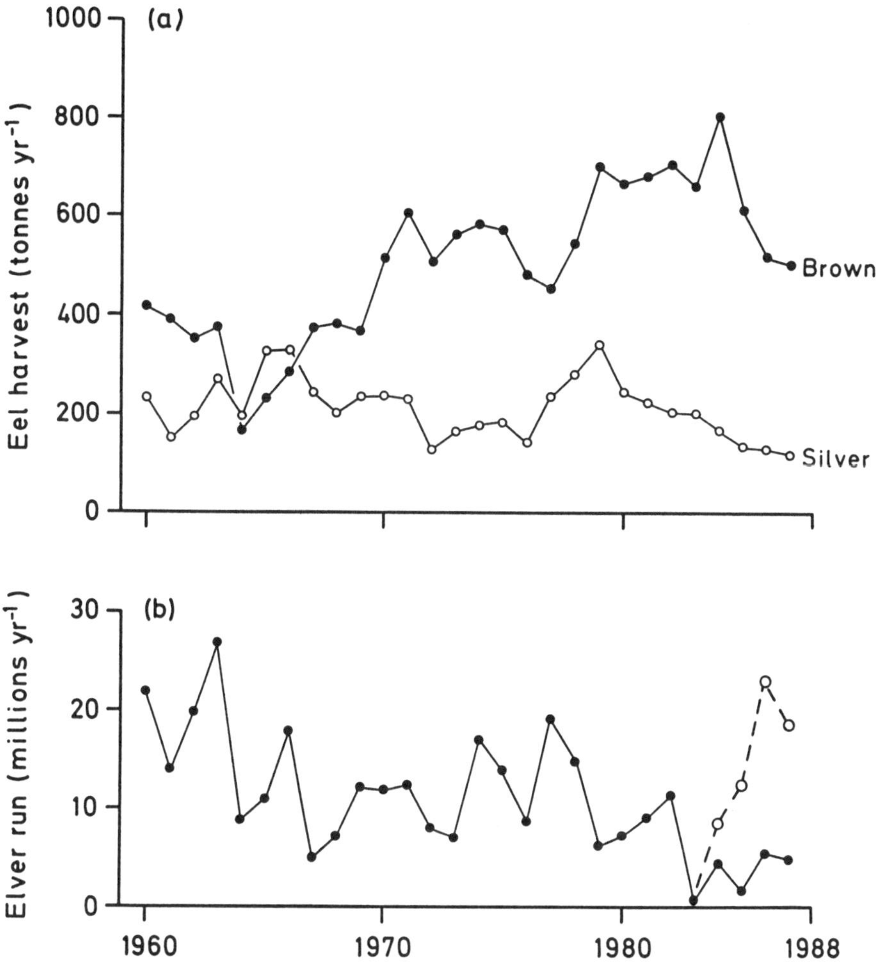

Figure 18B.8. The eel fishery of Lough Neagh since 1960. (a) brown and silver eel catches; and (b) elver runs in the River Lower Bann. Total elver recruitment since 1984 includes elver stocking (○--○). (Data from the Lough Neagh Fishermen's Cooperative.)

charge to reach the traps at the Cutts, and high water flows appear to impede their upriver migration. Lowe (1952) recognised that her conclusions were subject to inaccuracies arising from the irregular data collection on the Bann carried out by the commercial elver transporters. Locally, the influence of factors such as temperature, wind direction, time of day or state of the tide on elver migration have not been satisfactorily assessed on the Bann. There is also no obvious explanation of the sharp fall in elver recruitment in recent years, shown in Fig. 18B.8b. Moriarty (pers. comm.) believes this to be a phenomenon common to many Irish rivers.

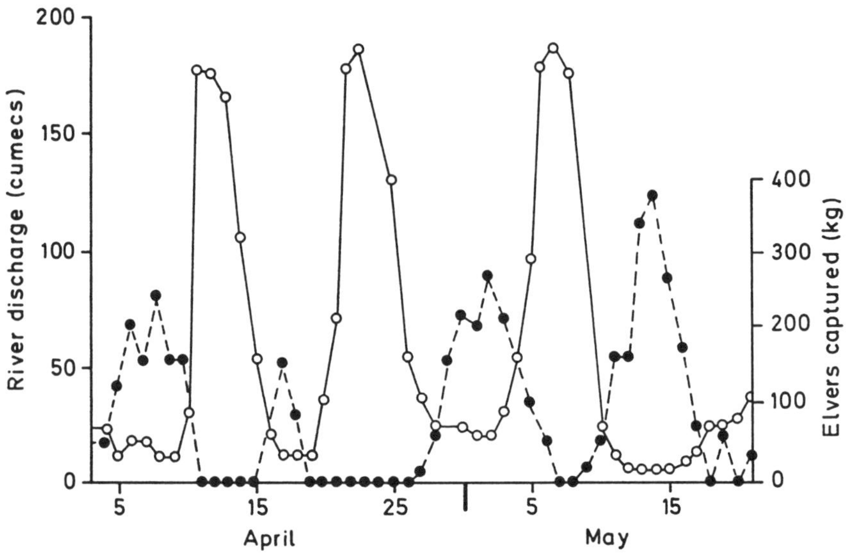

Figure 18B.9. Relationship between water flows (○—○) and elver runs as measured by the weight of elvers captured at the Cutts traps (●– –●) in the River Bann during April and May 1970.

18B.2.2. *Silver eel runs*

The effects of a number of environmental variables on the movements of migrating silver eels have been investigated on the Bann. For example, it has been found that there is a close correlation between the downstream migration of silver eels and water discharge (Anon, 1970) (see Fig. 18B.10). Since the amounts of water going down the Bann can be controlled by sluices, the catch of silver eels can be regulated to some extent. Frost (1950) also noted this relationship between eel migration and water flows, but she further noted that some of the best catches are made on nights when there is little or no moonlight (see Fig. 18B.11) i.e. light tends to inhibit the migratory movements of eels. There is also some observational evidence that a strong southerly wind improves catches of silver eels at Toome, possibly by stimulating directional migration by disturbing the lough in the direction of the outflowing River Bann.

An example of the problem of managing the resource against the conflicting demands of fisheries, drainage and water supply in the 1983 situation is taken from Wood (1984) and illustrated in Fig. 18B.12. Lough levels and discharge were below the statutory minimum until October and the freshets recommended by the Lough Neagh Working Group to coincide with the dark phase of the moon were not released. In August and September virtually no eels were caught. A slight increase in discharge in early October moved

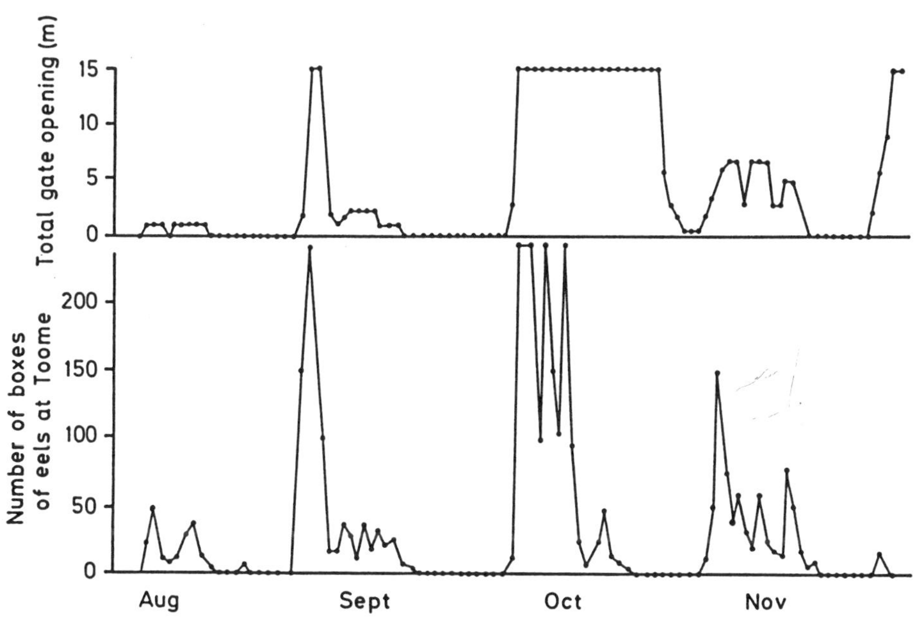

Figure 18B.10. Relationship between daily silver eel catch and total gate opening at Toome.

eels in the River Bann below Toomebridge but the greatly increased discharge in mid-October over-rode any moon-phased emigration and resulted in over 40 tonnes being caught on one night. No freshet and poor catch occurred in early November and, by chance, in December heavy rain, high discharge and the phase of the moon coincided to produce a further catch of 15 tonnes on one night. The erratic monthly harvest added up to a reasonable annual total but the fishery was exposed to lost early season markets, over-taxed storage and transport of live eels and low prices during the mid-October glut.

18B.2.3. *Sex ratios in the silver eel catch*

The total annual weight of silver eels caught in the lower Bann has been recorded since 1914, but not until 1943 was any attempt made to break the catch down into age, sex and size range (Frost, 1950). Early records from a 1927 public inquiry indicate that at that time the silver eel catch was probably entirely female (Anon, 1965). However, by 1943 27.1% of a catch sample at Toome were male, and subsequent samples from Portna during the period 1944–1946 were 39.5% male (Frost, 1950). Jones (unpublished) found that this trend for increasing numbers of males in the catch was evident in samples taken in 1956 and 1957, when 61.1% and 62.7% of commercially caught

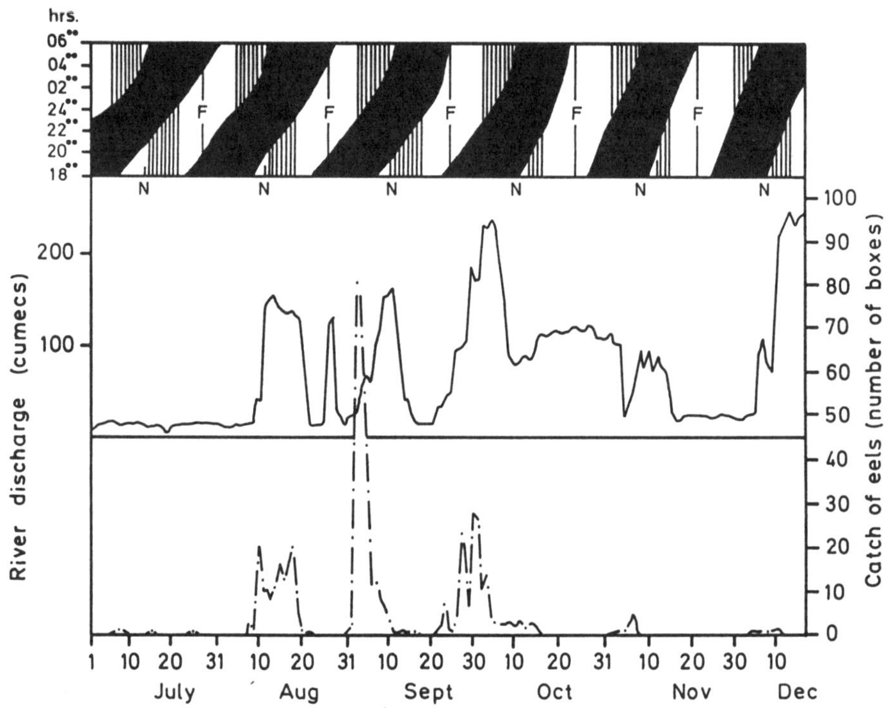

Figure 18B.11. Nightly catch of silver eels at Toome in 1942 plotted in relation to river flow and phases of the moon and their nightly duration; moon shown as F – Full, N – New, striations first and last quarters, and black, no moon.

silver eels at Toome were male. Parsons, Vickers and Warden (1977) investigated these and subsequent changes in the silver eel fishery with particular reference to the possible influence of elver transport. Frost (1950) had previously suggested that elver transport would increase the total population of eels, and since evidence from ageing studies on Windermere indicated that male eels migrated younger then females, this would result in a temporary increase in the percentage of males. Parsons *et al.* (1977) confirmed that male silver eels from Lough Neagh are younger and smaller than females, giving rise to a pronounced bimodality in the lengths of the commercial catch (see Fig. 18B.13). These authors also recorded seasonal variation in the sex ratio of migrating silver eels, and from 1965–1969 there was a significant seasonal increase in the percentage of female silver eels in the catch (Fig. 18B.14a). However, from 1970 onwards, when male eels represented an increasingly larger proportion of the run, this relationship was not apparent (Fig. 18B.14b). After making allowance for possible bias in the sex ratio due to seasonal variation in sampling by various authors, Parsons *et al.* (1977) confirmed that the two periods of elver transport to Lough Neagh have each

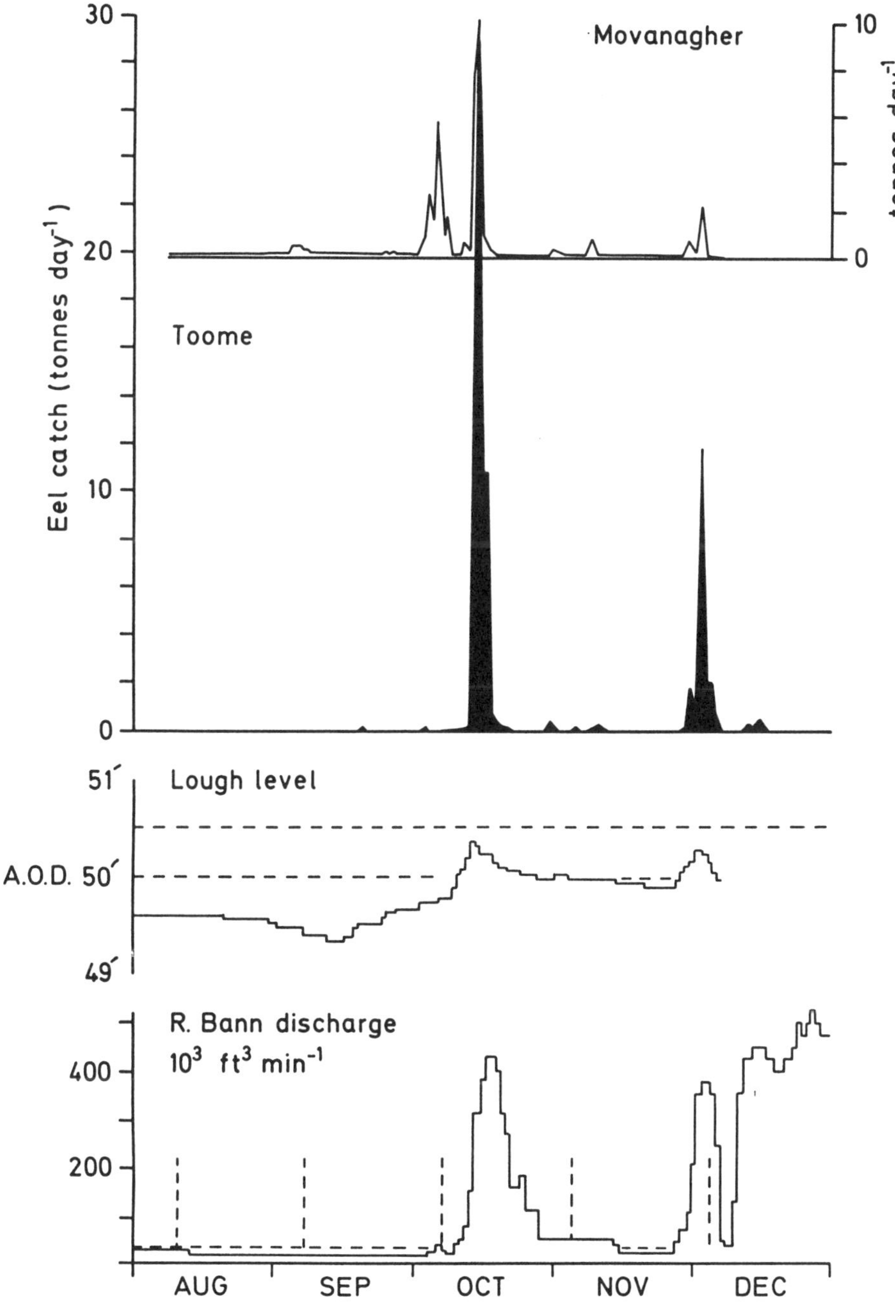

Figure 18B.12. Silver eel catches at Movanagher and Toome in relation to the recommended and actual discharge of the River Lower Bann in August to December 1983. (Data from the Lough Neagh Fishermen's Cooperative and DoE(NI).)

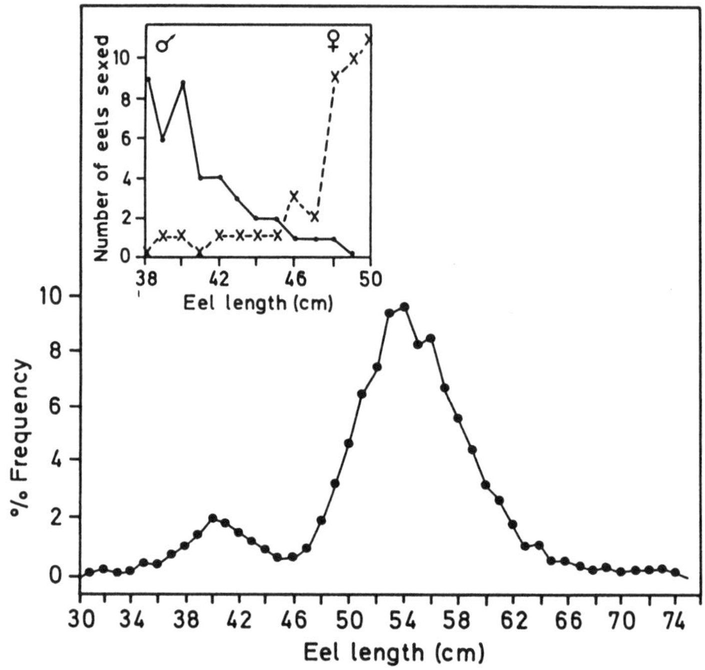

Figure 18B.13. Length frequency of 4906 silver eels measured from 1965 to 1969. Inset shows the length of a sample of silver eels that was sexed within the size range 38–50 cm.

been followed by a significant change in the sex ratio of silver eels trapped at Toome (Fig. 18B.15). A large proportion of the elvers caught in the freshwater traps at the Cutts are therefore eventually differentiating into males. This is contrary to the hypothesis of D'Ancona (1960), who suggested that male elvers have an affinity for brackish water and remain at high densities in estuaries while the females move upstream.

Parsons *et al.* (1977) discuss several possible reasons for the relationship between elver transport and silver eel sex ratio. Firstly they point out that Frost's (1950) suggestion that the change might be temporary due to the different ages at which males and females migrate to the sea cannot be valid since the proportion of males continued to increase after 1950 instead of reverting to normal. Similarly the high proportion of males has been maintained in the catches up to 1978 when sampling was discontinued (Anon, 1978).

It has also been suggested that elvers reaching the Cutts are genotypically sexually determined, but that differentiation only takes place in freshwater and results in a geographic variation in sex ratio. For this hypothesis a behavioural change would have to occur between potentially male and female elvers at some stage between the sea and Lough Neagh, and the males would

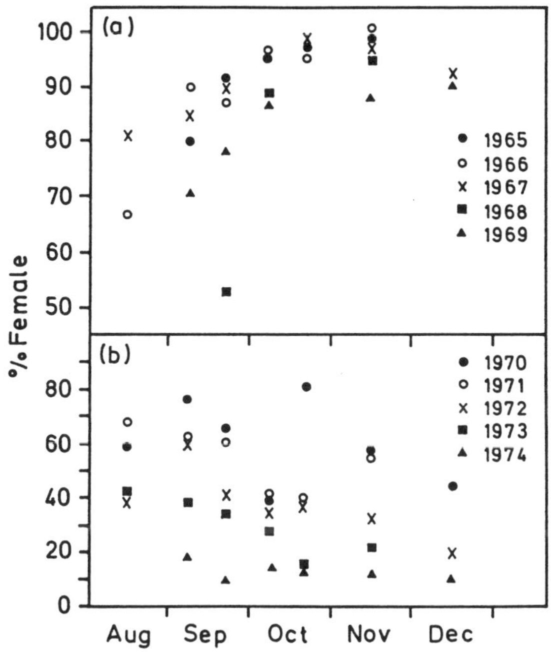

Figure 18B.14. The percentage of female eels in the silver eel catch as the season progresses from August to December 1965–1974.

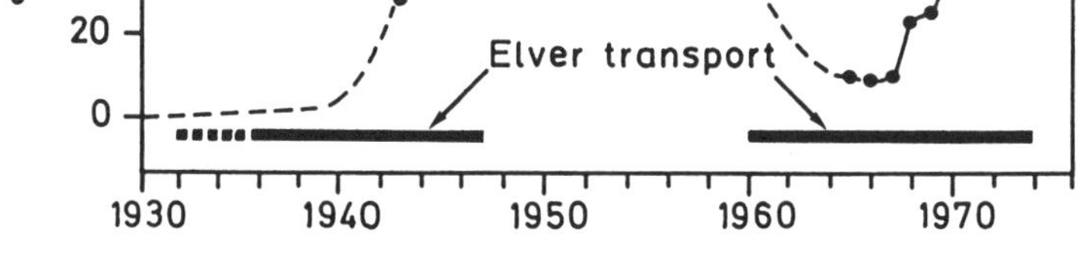

Figure 18B.15. Percentage of males in the samples of silver eels measured at Toome during September and October. Figures for 1943–1945 from Frost (1950) and for 1956–1957 from Jones (unpublished). The dotted line shows the estimated values in the absence of any sampling, and the periods of elver transport are also indicated.

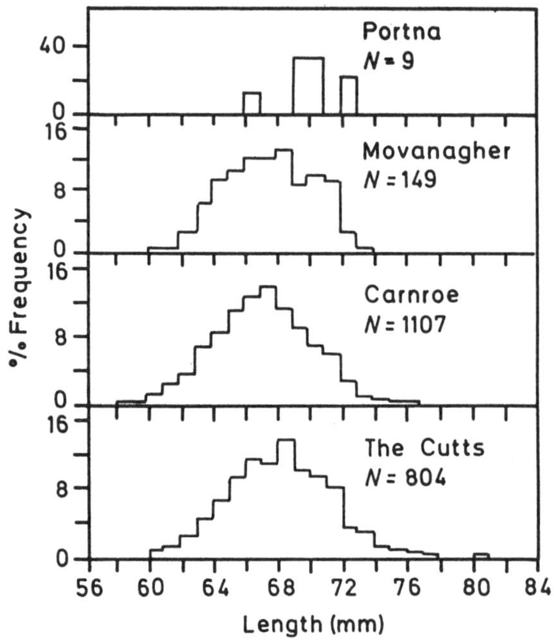

Figure 18B.16. Histograms of length frequency distribution for elvers trapped at various sites on the Lower River Bann during 1974.

not normally progress up-river as far as the lough. However, at various upstream sites Parsons *et al.* (1977) sampled elvers which had migrated past the traps, and found that within three months a few had travelled as far as Portna. This is about half the distance to Lough Neagh, and larger samples at Carnroe and Movanagher showed no bimodality in elver length distribution which might have indicated the onset of sexual differentiation (Fig. 18B.16).

It is also possible that an increased brown (also referred to as yellow) eel fishing mortality would alter the sex ratio since the smaller male eels would be returned to the lough as undersized while the larger females would be retained for market. However, as far as is known, the change in sex ratio between 1940 and 1956 did not accompany a period of increased fishing mortality, and the catch per effort for all sizes of brown eels had not altered significantly up to 1974 when it was assessed by Parsons *et al.* (1977).

The final hypothesis is that sex determination in eels is phenotypic, and elver transport has produced overstocking of Lough Neagh with the result that increased competition for food has caused small eels to differentiate into males. There is some evidence for this in that a significant decrease in the mean lengths and condition factors of male silver eels coincides with the years when male eels derived from the first year of elver transport were

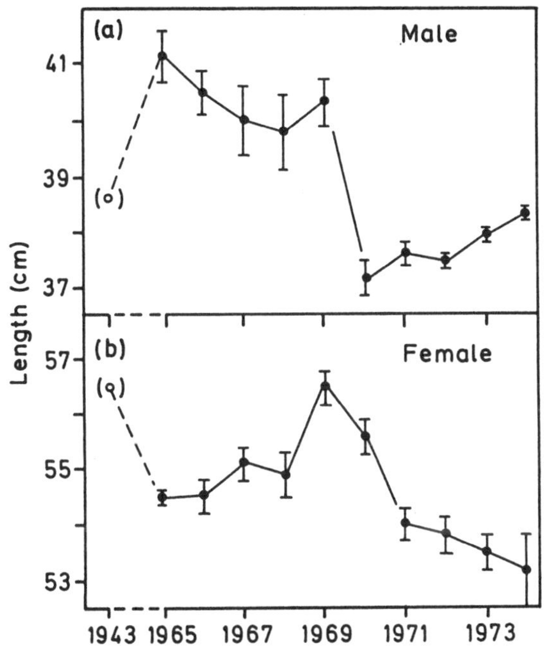

Figure 18B.17. Mean length ± 2 standard errors of male and female silver eels trapped at Toome from 1965 to 1974. Mean length in 1943 also shown (Frost, 1950).

migrating i.e. 1969–1970 (see Fig.18b.17a and 18B.18a). Similarly, the female silver eels, which do not migrtate until they are a few years older than males had significantly reduced lengths and condition factors from 1971–1974 (Fig. 18B.17b and 18B.18b). However, the aging studies from which these conclusions were derived were determined from the otoliths using the technique described by Frost (1945), and a subsequent investigation on Lough Neagh eels suggested that this method tended to overestimate the growth rate (Anon, 1966). This controversy surrounding the accuracy of ageing studies on Lough Neagh eels has not yet been satisfactorily resolved. On a more subjective level, the hugh swarms of chironomids which annually emerge (see Chapter 16) argue against severe competition for food.

18B.2.4. *The food of eels*

Other aspects of the biology of Lough Neagh eels have been studied as part of an investigation into the commercial fishery during the 1960s (Anon, 1965, 1966). The significant reduction in the condition factors of eels from 1965 to 1966 noted by Parsons *et al.* (1977) coincided with a marked change in diet at this time (Table 18B.3). Chironomidae made up the bulk of the diet in each year. However, *Mysis* changed from being a primary food organism in

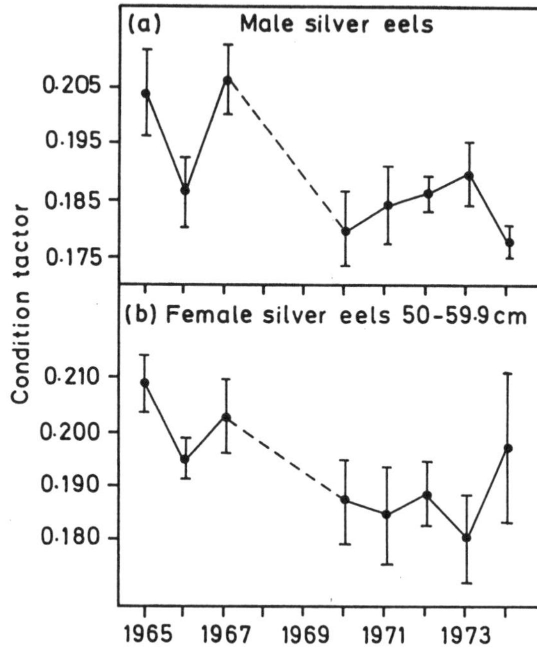

Figure 18B.18. Mean condition factor of samples of male and female silver eels caught at Toome from 1965 to 1974. Analysis of female eels restricted to eels of 50–59.9 cm length.

Table 18B.3. Percentage composition of the diet of Lough Neagh eels sampled from June to October during 1965 and 1966

Year	Substrate type	Percentage composition of diet								
		Chironomidae	*Asellus*	*Valvata*	*Mysis*	*Gammarus*	Leeches	Cadddis	Fish	Others
1965	Mud	35.8	3.0	5.6	40.2	6.6	0.6	0.5	1.7	6.0
1966	Mud	59.4	2.4	23.9	2.1	5.8	0.1	1.3	2.9	2.1
	Rocky	43.3	43.7	1.8	0.2	8.6	0.2	0.4	1.2	0.6

1965 to being virtually insignificant in 1966, when *Valvata* and *Asellus* became more important. Whether this is a perhaps common year to year variation, or a long term trend remains uncertain, although a brief study by Marrion (1986) identified chironomids as the clear preference food of eels in the summer of 1985. A difference in the diet of eels from different areas was also demonstrated in 1966, with *Asellus* being dominant in eels taken over hard ground and *Valvata* being more abundant in eels taken over muddy ground.

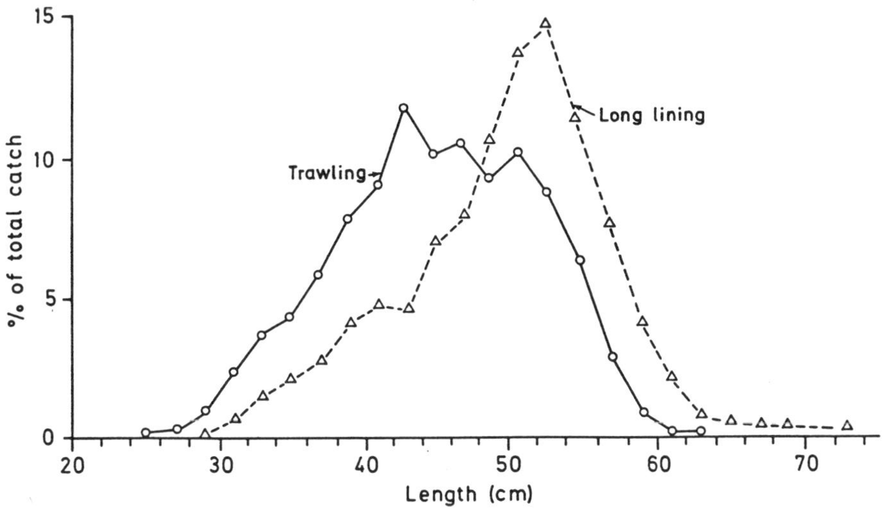

Figure 18B.19. Size distribution of eels captured in Lough Neagh by long lining (△--△) and trawling (○—○) during 1965 and 1966.

18B.2.5. *Trawling*

Much of the investigation into the commercial fishery in the 1960s arose from concern over the advent of trawling of eels at this time. The illegal use of the otter trawl from this purpose in Lough Neagh is reported to have begun on a trial and error basis early in the 1960 season, and by the middle of the season had become established on a large scale with about fifty out of one hundred and fifty boats resorting to the method. No satisfactory assessment of the damage to the benthic fauna due to trawling was completed, although it was calculated that about 1000 m^2 of the substratum would be disturbed by the otter boards per hour of trawling by each boat. Investigations did indicate that up to 35% of the catch in the commercial trawl mesh of 1 cm knot to knot were undersized, compared to a mean of only 15% in the longline fishery (see Fig. 18B.19). (There were also considerable differences in the length frequencies of eels taken on the three types of bait most commonly used by long liners on the lough, although this did not appear to be consistent between years (see Table 18B.4).) There was no evidence of mortality in undersized trawled eels which were returned to the water whereas 23.7% of eels captured by longline died after release.

Other species of fish, and particularly juveniles, were also captured by trawling e.g. pollan, perch, trout. It was felt that this could have a cumulative adverse effect on the stocks of these, but this study was not conclusive.

Table 18B.4. Details of eel catches in 1965 and 1966 using three types of bait on longlines in Lough Neagh

Bait	Catch per effort in 1966 (No. of eels per 100 hooks)	Percentage undersized	
		1965	1966
Worm	17.5	16	31.1
Perch	17.6	21	8.2
Pollan	22.7	4	6.7

However, the major objections to trawling were confirmed in that it was too efficient and could result in overfishing of the eel stocks, and also that it was incompatible with the traditional longline fishery due to destruction of the lines. For these reasons, the Lough Neagh Fishermen's co-operative, which now runs the fishery, successfully enforced a ban on trawling for eels in the early 1970s.

18B.3. References

Anonymous (1965) Lough Neagh eel investigation – report for 1965. Biology of Lough Neagh eels. Dept. Agric. for N.I., unpbl. rep., 30 pp.

Anonymous (1966) Lough Neagh eel investigation – report for 1966. Biology of Lough Neagh eels (*Anguilla anguilla* L.). Dept. Agric. for N.I., unpbl. rep., 51 pp.

Anonymous (1970) Department of Agriculture for Northern Ireland, Annual Report on Research and Technical Work. HMSO, pp. 77–91.

Anonymous (1971) Department of Agriculture for Northern Ireland, Annual Report on Research and Technical Work. HMSO, pp. 97–119.

Anonymous (1973) Department of Agriculture for Northern Ireland, Annual Report on Research and Technical Work. HMSO, p. 101–128.

Anonymous (1975) Department of Agriculture for Northern Ireland, Annual Report on Research and Technical Work. HMSO, pp. 102–127.

Anonymous (1978) Department of Agriculture for Northern Ireland, Annual Report on Research and Technical Work. HMSO, pp. 119–139.

Anonymous (1980) Report on the Sea and Inland Fisheries of Northern Ireland. HMSO, 26 pp.

Bell, P. (ed.) (1971) Lough Neagh Working Group Advisory Report. Vols I and II, Government of Northern Ireland.

Crozier, W.W. and Kennedy G.J.A. (1987) Marine survival and exploitation of R. Bush hatchery salmon (*Salmo salar* L.) as assessed by microtag returns to 1986. Int. Con. for the Explor. of the Sea. Working Paper CM 1987, 8 pp.

D'Ancona, U. (1960) The life cycle of the Atlantic eel. Symp. Zool. Soc. Lond. 1: 61–77.

Frost, W.E. (1945) The age and growth of eels (*Anguilla anguilla*) from the Windermere catchment area. J. Anim. Ecol. 14: 26–36, 106–124.

Frost, W.E. (1950) The eel fisheries of the River Bann, Northern Ireland, and observations on the age of the silver eels. J. Cons. perm. Int. Explor. Mer. 16: 358–383.

Hillas, A.B.E. (1904) Record of salmon marking experiments in Ireland 1902–1905. Fish. Ir. Sc. Invest. 7: 315–353.

Holt, E.W.L. (1903) Record of salmon marking experiments in Ireland, 1898–1902. Fish. Ir. Sc. Invest. 13: 165–196.

Lowe, R.H. (1951) Factors influencing the runs of elvers in the River Bann, Northern Ireland. J. Cons. perm. Int. Explor. Mer. 17: 299–315.

Lowe, R.H. (1952) Letter to the editor on 'Factors influencing the runs of elvers in the River Bann, Northern Ireland. J. Cons. perm. Int. Explor. Mer. 18: 241–244.

Marrion A.G. (1986) A study of the feeding ecology of the European eel, *Anguilla anguilla* L. in Lough Neagh. Unpublished MSc. Thesis, University of Ulster, 76 pp.

Menzies, W.J.M. (1936) The run of elvers in the River Bann, Northern Ireland. J. Cons. perm. Int. Explor. Mer. 11: 249–259.

Parsons, J. (1971) The relationship between salmon catches on the River Bann and various measurements of water flow. In: P. Bell (ed.) Lough Neagh Working Group, Advisory Report, Vol. II, Appendix III, HMSO, pp. 13–16.

Parsons, J., Vickers, K.U. and Warden, Y. (1977) Relationship between elver recruitment and changes in the sex ratio of silver eels *Anguilla anguilla* L. migrating from Lough Neagh, Northern Ireland. J. Fish. Biol. 10: 211–229.

Vickers, K.U. and Ferris, G.D. (1970) Movement of salmon around Ireland. XI. From Carnlough and Torr Head (1969). Proc. R. Ir. Acad. 70: 171–178.

Vickers, K.U. and Parsons, J. (1971) Interim report on the River Bann salmon tagging experiment, June and July 1970. In: P. Bell (ed.) Lough Neagh Working Group, Advisory Report, Vol. II, Appendix III. HMSO, pp. 17–26.

Went, A.E.J. (1947) Salmon marking in Ireland. Proc. R. Ir. Acad. 51: 173–195.

Went, A.E.J. (1958) Salmon movements around Ireland. VIII. From drift nets along the coast of County Donegal (1953–1957). Proc. R. Ir. Acad. 59: 205–212.

Went, A.E.J. (1963) Tagging experiments, 1962. Foyle Fish. Comm. Ann. Rep., App. IV: 43–49.

Went, A.E.J. (1964a) Irish salmon – a review of investigations up to 1963. Sc. Proc. R. Dub. Soc. A1(15): 365–412.

Went, A.E.J. (1964b) The pursuit of salmon in Ireland. Proc. R. Ir. Acad. 63: 191–244.

Went, A.E.J. and Vickers, K.U. (1953) Salmon movements around Ireland – V. From North Count Antrim (1951 and 1952). Proc. R. Ir. Acad. 56: 13–28.

Went, A.E.J. and Vickers, K.U. (1956) Salmon movements around Ireland – VI. From Portballintrae, Co. Antrim (1953 and 1954). Proc. R. Ir. Acad. 58: 17–21.

Went, A.E.J. and Vickers, K.U. (1971) A comparison between the salmon and grilse of the Foyle and Bann systems Foyle Fish. Comm. Ann. Rep., App. IV, 4 pp.

Wood, R.B. (1984) The present status of Lough Neagh as a fishery. *In* Proc. Inst. Fish. Mgmt. Conference (N. Ireland), pp. 61–81.

Wood, R.B. and Smith R.V. (1986) Lough Neagh: 100 years of man's influence on the water quality of a major inland fishery. In: W.W. Crozier and P.M. Johnston (eds.) Proc. 17th Inst. Fish. Mgmt. Conference, N. Ireland, pp. 111–126.

18. The fish of Lough Neagh
Part C. Investigations on the brown trout (*Salmo trutta* L.)

W.W. CROZIER and A. FERGUSON

18C.1. Introduction

18C.1.1. *Historical references*

In common with other large Irish lakes, there are many historical accounts of several types or races of brown trout *Salmo trutta* occurring in Lough Neagh. These accounts range from detailed scientific descriptions to anecdotal information of questionable reliability.

The type encountered most frequently in the literature is the great lake trout or ferox. Dubourdeau (1812) and Sampson (1814) refer to this as *Salmo lacustris* and Yarrell (1841) and Day (1887) as *Salmo ferox*. The local name for this type is usually budagh (Harris, 1744; Pennant, 1812; Couch, 1865; Thompson, 1856) but budagh (Boate,1652) and breddagh (Doyle, 1854) have also been used. These names apparently mean 'big fat fellow' (Day, 1887) a reflection of the large size which this type can attain. Buddagh have been reported as inhabiting deep water in the Lough (Day, 1887), having a piscivorous diet (Yarrell, 1841) and growing to 30 lbs (Thompson, 1856) or even 50 lbs (= 23 kg) (Sampson, 1814). According to Thompson (1856) they spawn in the inflowing rivers but not as far upstream as other trout.

The second name given to certain Lough Neagh trout is the dollaghan (sometimes dolochan or dollaghen). Harris (1744) refers to this as being 'a trout peculiar to Lough Neagh, never exceeding 18 inches (460 mm) in length, and always spawning in the inflowing rivers'. Sampson (1814) mentions the dollaghen but was uncertain as to how it should be classified noting merely that it was different from the *lacustris* type.

Another variety of brown trout described from several Irish lakes including Lough Neagh is the gillaroo. This type aroused considerable interest among 19th century naturalists. Although most often regarded as a variety of *Salmo trutta* (Pennant, 1812; Thompson, 1856), Günther (1866) gave it specific status as *Salmo stomachius*. The term gillaroo or 'red fellow' (Maxwell, 1904) refers to the colour of this type which has a yellowish background upon which are distinctive large red spots, thought to be a reflection of its mollusc

R. B. Wood and R. V. Smith (eds.), Lough Neagh, 419–437.

feeding habit (Pennant, 1812; Couch, 1865). It has been reported as preferring the rocky-bottomed parts and spawning in the shallower regions of lakes and not in the rivers (Thompson, 1856; Day, 1887). The gillaroo is mentioned as occurring in Lough Neagh by Yarrell, (1841) and Thompson (1856), the latter noting that they were taken up to 12 lbs.

Thompson (1856) records the sea trout (anadromous brown trout) as being present in Lough Neagh. However he notes that 'in Lough Neagh, the sea trout is less brilliant in colour than the sea trout from around the coast'. Recent information (Fahy, 1977; Kennedy and Vickers, Chapter 18, Part A) suggests that the sea trout does not penetrate as far upstream as Lough Neagh. It seems likely that Thompson was not referring to sea trout as such but to what is currently referred to by local fishermen as salmon trout which feed in the Lough as adults. Bassett (1888) mentions the salmon trout of Lough Neagh but does not discuss its status.

Frost and Brown (1967) refer to sonaghen of Lough Neagh but as this is the only reference to this type it seems likely that it is a case of confusion with dollaghan.

18C.1.2. *Recent studies*

Very little information is available on the brown trout of Lough Neagh during the first half of this century. Cragg-Hine (1972) carried out an investigation of population density, biomass and growth of juvenile brown trout at 19 river stations in the Lough Neagh catchment. Population density was found to be low in the Blackwater and Upper Bann systems with contrastingly high values in the Moyola and Main systems. Growth rates varied minimally among the rivers examined, however significant length at age differences were present within the Moyola system. Kennedy (1976) presented a short survey of the growth characteristics of brown trout in the Lough based on an examination of 442 fish from commercial catches. Pitcher and Murray (1977) in a report commissioned by the Main Angling Club examined feeding growth and population density of brown trout from the River Main.

A study primarily concerned with the population genetics of brown trout in the Lough and inlet rivers was carried out in the period October 1979–September 1982 by the first author. A summary of the information obtained is presented below and full details are available elsewhere (Crozier, 1983; Crozier and Ferguson, 1986).

18C.2. Population genetics

Much discussion has taken place among fish biologists as to the relative contributions of genetic and environmental influences in producing the range of phenotypic variation observed in brown trout (e.g. Frost and Brown, 1967; Ferguson and Mason, 1981). In traditional studies involving examination of

Table 18C.1. Occurrence and distribution of enzyme polymorphism among all samples examined (comprising 32 river samples, dollaghan and salmon trout)

Enzyme	Locus	Allele[a]	No.[b]	No. polymorphic[c]	Max. freq.
Aspartate aminotransferase	$_s$Aat-1,2	140	29	22	0.359
		45		1	0.022
Aspartate aminotransferase	$_s$Aat-4	74	33	33	0.556
Creatine kinase	Ck-2	115	34	16	0.444
		75		5	0.071
Diaphorase	Dia	90	29	26	0.234
Glycerol-3-phosphate dehydrogenase	Gpd-2	50	34	34	0.722
Isocitrate dehydrogenase	$_s$Idh-1	160	34	29	0.213
Isocitrate dehydrogenase	$_s$Idh-2	130	34	2	0.062
Lactate dehydrogenase	Ldh-5	105	34	9	0.216
Malate dehydrogenase	$_s$Mdh-2	152	34	34	0.692
Malate dehydrogenase	$_s$Mdh-3,4	125	29	26	0.348
Phosphoglucose isomerase	Pgi-2	135	34	19	0.304
Phosphoglucose isomerase	Pgi-3	110	34	5	0.020

[a] Only variant alleles given as 100 allele was universally present.
[b] Number of samples screened for that locus.
[c] 99% criterion used to define polymorphism.

morphological characters it is often difficult to separate these two interacting forces. In recent years electrophoretic studies of enzymes and other proteins, the primary gene products have enabled studies to be made of genetic variation within and between species, minimising the effects of environmental modulation (Ferguson, 1980). These studies have made use of the fact that separate electrophoretic phenotypes of a given protein, which are usually the products of alternative alleles segregating at the controlling gene locus, provide useful markers for the investigation of reproductive isolation and other aspects of the population genetics of fish (Allendorf and Utter, 1979).

In the present study electrophoretic examination of trout genetic variation was carried out using methods nomenclature and genetic interpretations modified from Taggart *et al.* (1981). The analysis presented below is based on a detailed study of allelic frequencies at polymorphic loci (Table 18C.1) in 32 river samples comprising a total of 1678 individuals and from 351 Lough-caught trout which were obtained from commercial netsmen.

In any lake or river system there are opportunities for genetic isolation of fish populations arising purely as a result of the physical separation of the spawning areas. In the Lough Neagh system the presence of migratory lake trout will inevitably complicate the picture as they may act either to inflate or deflate the potential for genetic differentiation depending on their genetic status and breeding organisation. Examples of inter-sample genetic variation are shown for three of the polymorphic loci in Table 18C.2 for samples of trout in the six main river systems and in Table 18C.3 for eight samples in the Six Mile Water system. Analysis by means of the genic contingency Chi-

Table 18C.2. Frequencies of '100' allele for samples taken from each of the major inflowing Lough Neagh rivers. Among sample heterogeneity is tested using the genic contingency χ^2 test of Workman and Niswander (1970)

River	n	Pgi-2	Gpd-2	$_s$Mdh-3,4
Upper Bann	52	1.0	0.615	0.865
Blackwater	59	1.0	0.534	0.879
Ballinderry	65	0.914	0.654	0.818
Moyola	58	0.982	0.5	0.784
Main	42	1.0	0.551	1.0
Six Mile Water	151	0.904	0.677	0.914
χ^2_5		35.769	16.538	25.928
p		<0.001	<0.001	<0.001

Table 18C.3. Frequencies of '100' allele for samples taken from several tributaries within the Six Mile Water river catchment, together with Workman and Niswander (1970) heterogeneity values

Location	n	Pgi-2	Gpd-2	$_s$Mdh-3,4
Ballymartin	58	0.974	0.879	0.862
Palantine	38	0.987	0.592	1.0
Rathmore	50	0.95	0.61	0.97
Clady	151	0.904	0.677	0.914
Iskelly	36	1.0	0.278	0.986
Ballyclare	23	0.696	0.986	0.917
Straid inflow	36	0.986	0.986	0.917
Straid outflow	109	0.968	0.708	0.949
χ^2_7		71.529	119.598	67.725
p		<0.001	<0.001	<0.001

squared test for heterogeneity (Workman and Niswander, 1970) and by Wright's (1978) F-statistics showed significant heterogeneity both among river systems and among sample sites within the four river systems (Ballinderry, Blackwater, Main and Six Mile Water) which were investigated in detail. This suggests that a large number of at least partially reproductively isolated populations of brown trout are present in the Lough Neagh river systems. Within the Six Mile Water system alone at least four separate populations may be present, each of which may be further sub-divided. It seems that physical barriers and geographical distance are not necessary to permit genetic differentiation in brown trout, although where these are present differentiation is accentuated. This is further emphasised by an examination of the total variance across all polymorphic loci in all samples analysed. Some 69% of this variance was accounted for by within sample variation, i.e. individual variation. Of the remaining, among sample, variation, 58% was attributable to differentiation among tributaries within river systems 35% among samples within tributaries and only 7% was distributed among river

Table 18C.4. Summary of characteristics used to distinguish dollaghan and salmon trout

Character	Dollaghan	Salmon trout
Head size	Small relative to body	Large relative to body
Spotting	Many black spots, usually X shaped, on, above and below lateral line. Occasionally red spots on posterior of lateral line	Black spots, usually few in number, rarely X shaped. Usually found on and above lateral line only. Never red spots
Colouration	Moderate or dull silvering. Brown dorsal surface	Often brilliantly silver. Steel coloured dorsal surface
Flesh	White or pink, rarely red. Firm when fresh	Always deep red. Soft when fresh

systems. The extent to which artificially stocked trout contribute to the observed heterogeneity is unknown. However, it appears that only very limited stocking has been carried out and is unlikely to be a significant factor in producing genetic heterogeneity

Of the Lough-caught trout, 243 were identified as dollaghan, 89 as salmon trout, six as possible gillaroo, one as a possible ferox and 12 were unclassifiable. The criteria permitting distinction of dollaghan and salmon trout are given in Table 18C.4. The representation of gillaroo and ferox in the sample was insufficient for analysis and these types will not be discussed further. Significant overall genetic heterogeneity was found between dollaghan and salmon trout largely as a result of allelic differences at the Ck-2 and Aat-4 loci. The heterogeneity at the Ck-2 locus is attributable to the possesion of the Ck-2(115) allele by dollaghan and the Ck-2(75) allele by salmon trout (Table 18C.5). The presence of unique alleles in the two types would suggest that interbreeding if occurring at all, is at a very low level. Further evidence of reproductive isolation between dollaghan and salmon trout is provided by the deficit of observed heterozygotes at the Idh-1, Ldh-5 and Mdh-4 loci compared with Hardy–Weinberg expectations, when the total sample is considered. The most common cause of heterozygote deficits in a sample is due to the treatment of two or more fully or partially isolated populations, characterised by different allelic frequencies as a single panmictic population. To achieve this genetic differentiation between dollaghan and salmon trout would require that reproductive isolation be maintained over a considerable period of time. Based on estimation of genetic identity (Nei, 1972) a divergence time of at least 4,500 years is indicated. Examination of the geographical pattern of variation at the Ck-2 locus suggests that spatial segregation of the spawning grounds is important with the allelic variants characteristic of dollaghan and salmon trout being differentially distributed in the river systems and indeed among tributaries within particular systems such as the Ballinderry and Blackwater. The more widespread occurrence of the Ck-2(115) allele, typical of dollaghan, suggests that dollaghan originate from a

Table 18C.5. Allelic frequencies at representative polymorphic loci examined in dollaghan (n = 239) and salmon trout (n = 87), together with heterogeneity χ^2 values (Workman and Niswander, 1970) for individual loci and summed across loci

Locus	Allele	Dollaghan	Salmon trout	χ^2	d.f.	p
$_s$Aat-4	100	0.698	0.792	4.900	1	<0.05
	74	0.302	0.208			
$_s$Ck-2	100	0.842	0.994	34.385	2	<0.001
	115	0.158				
	75		0.006			
$_s$Idh-1	100	0.880	0.882	0.005	1	>0.9
	160	0.120	0.118			
$_s$Idh-2	100	0.998	1.0	0.356	1	>0.5
	130	0.002				
Ldh-5	100	0.982	0.987	0.175	1	>0.5
	105	0.018	0.013			
$_s$Mdh-2	100	0.563	0.494	2.489	1	>0.5
	152	0.437	0.506			
$_s$Mdh-3,4	100	0.860	0.876	0.283	1	>0.5
	125	0.140	0.124			
			Total χ^2	42.593	8	<0.001

greater number of rivers, in keeping with the greater numbers of this type in the Lough.

The available genetic information does not indicate genetic differentiation between river-resident and migratory trout It appears that a number of river populations produce trout which contribute to the Lough stocks and which return to their natal areas to spawn thus maintaining reproductive isolation among populations. Thus each of the dollaghan and salmon trout groups comprise a number of distinct populations originating from seperate rivers.

Of the 2,057 fish identified in he field as brown trout, eight were later found on electrophoretic analysis to be hybrids between brown trout and Atlantic salmon *Salmo salar*. However, hybrids were found at only a few sample locations and represented 3.6% of these samples.

18C.3. Growth

It is well documented that among migratory lake trout stocks, migration from the nursery streams to the lake does not occur at the same age in all individuals (Ball and Jones, 1960; Hunt and Jones, 1972; Thorpe, 1974; Treasurer, 1976). The age at which a particular fish migrates to the lake can be determined by a sudden marked increase in scale growth (Allen, 1938). In brown trout three classes of migrants are normally found, migration occurring during the second, third or fourth year of life. These are denoted as 1+, 2+ and 3+ migrants respectively. In all, age-length information for

Table 18C.6. Migrant group representation in dollaghan and salmon trout, given as numbers and percentages (in parentheses) within each group

Population	Age at migration			
	1+	2+	3+	Total
Dollaghan	79 (33.2)	128 (53.8)	31 (13.0)	238 (100)
Salmon trout	31 (35.6)	47 (54.0)	9 (10.4)	87 (100)

238 dollaghan and 87 salmon trout was available, the numbers in each of the three migrant classes being shown in Table 18C.6. The proportions of the three migrant classes are similar to those reported for dollaghan by Kennedy (1976).

The sex ratio in the two types differed significantly from 1:1, and also from each other, in dollaghan (1:1.5) and salmon trout (1:2.7), females being in excess in each case, as has been found for lake trout elsewhere. This bias has been attributed to differential mortality (Southern, 1935; Munro and Balmain, 1956) males maturing earlier than females and surviving for a shorter time. However, more recent work (Campbell, 1977; Fleming, 1983) has shown that in brown trout which migrate to the sea there is a greater tendency for females to become migratory. The lack of excess males in Lough Neagh river populations (see below) tends to exclude a similar explanation in this case. The ratio of females to males is significantly lower in the present study compared with that reported by Kennedy (1976), which was based on trout taken in 1966 and 1975. The scales of Lough Neagh trout did not generally display clearly identifiable spawning marks, thus preventing determination of age at first spawning of mature fish. Based on gonadal inspection, it was determined that, in both dollaghan and salmon trout, males mature at an earlier age than females. Thus 90% of male and 50% of female 4+ dollaghan were mature. A significantly greater proportion of 3+ dollaghan were mature compared with equivalent salmon trout.

Analysis of the growth characteristics of the trout was carried out based on ages determined by scale reading and from back-calculation of lengths at earlier years. Following the method of Bagenal and Tesch (1978) plots of fish length against anterior scale radius were constructed. These indicated that in dollaghan growth is allometric but in salmon trout it is isometric and so corrected lengths for the former were obtained using a modified direct proportion formula (Le Cren, 1947; Bagenal and Tesch, 1978). Mean lengths at age were calculated separately for each sex within the migrant groups of dollaghan and salmon trout. No significant difference in growth was found between the sexes and the pooled growth data are presented in Table 18C.7. Differences within and between dollaghan and salmon trout groups with respect to growth rate were evaluated using analysis of variance followed by pairwise comparisons.

In both dollaghan and salmon trout at age one (all trout still in the rivers)

Table 18C.7. Mean back-calculated lengths (cm) at age of dollaghan and salmon trout migrant groups. Standard deviations in parentheses

		Age							
Group	n	1	2	3	4	5	6	7	8
Dollaghan 1+	76	11.3 (2.45)	23.9 (4.11)	29.6 (4.61)	38.2 (4.74)	47.5 (0.78)			
Dollaghan 2+	128	9.5 (1.90)	17.6 (2.65)	29.0 (3.32)	33.2 (3.36)	38.7 (3.96)	48.0 (2.95)	55.8 (0)	61.2 (0)
Dollaghan 3+	31	9.2 (1.76)	16.3 (2.38)	23.1 (3.46)	31.1 (3.01)	36.4 (2.97)			
Salmon trout 1+	30	8.8 (1.94)	22.8 (4.35)	31.3 (5.08)	37.4 (5.44)				
Salmon trout 2+	47	7.4 (1.53)	16.6 (3.86)	30.0 (3.55)	34.8 (2.73)	43.6 (3.43)	50.7 (0)		
Salmon trout 3+	9	6.7 (1.79)	14.3 (2.71)	21.2 (3.34)	31.7 (0.59)	36.3 (0)			

the 1+ migrant group is significantly longer than the other migrant groups. The greater size of 1+ migrants has been noted in a number of other lakes (Allen, 1938; Campbell, 1957) and is clearly associated with early migration, though not necessarily causally. Also, dollaghan were significantly larger than salmon trout. By age four (all trout in the Lough) the 2+ and 3+ dollaghan migrant groups are similar in length but significantly shorter than those that migrated at 1+. However, by age five there is no difference in length of the dollaghan migrant groups. The three migrant groups of salmon trout are not significantly different in length at age four or five. There is, therefore a relationship between growth rate in the rivers and age at migration, with faster growing fish leaving the nursery streams sooner. However, longer stream residence is compensated for by disproportionately enhanced growth following migration. Differences in growth rate between dollaghan and salmon trout were found only in the first year, by age two equivalent migrant groups being of similar length. This could reflect either environmental or genetic differences, dollaghan and salmon trout recruiting from separate river tributaries. In all cases the lengths obtained in this study for the first two years were greater than those reported by Kennedy (1976). This is probably explainable by recalling that growth of dollaghan was found to be allometric whereas Kennedy assumed isometry. This suggestion is supported by the observation that better comparability of results is obtained at later ages which are less affected by non-correction of allometry (Kipling, 1962; Treasurer, 1976).

It has been shown that potential migratory trout have greater body length at age one than river resident trout (Allen, 1938; Ball and Jones, 1960; Hunt and Jones, 1972; Treasurer, 1976). An analysis of growth was carried out for trout taken from six locations in the River Main system. Clearly samples of river trout contain an unknown proportion of potentially migratory trout. All samples were in agreement with a 1:1 sex ratio, contrasting the significant excesses of females noted in both dollaghan and salmon trout. A tendency towards earlier sexual maturity of river trout compared with lake (migratory) trout was noted, almost all 3+ river trout being mature compared with 28% of lake trout. Back calculated lengths at age are given in Table 18C.8, where data for males and females have been pooled in the absence of significant differences between sexes. Analysis of variance indicated significant differences in length at ages one and two among localities indicating different growth characteristics in different parts of the river system. However at age three no significant differences were present. A comparison of mean growth rates for the Main system with those obtained above for dollaghan and salmon trout was carried out. All three migrant groups of dollaghan were significantly longer at age one compared with river trout. At age two the 2+ dollaghan migrants were significantly longer than river trout, but there was no difference between potential 3+ migrants and river trout at this age. Similarly at age three, no difference in length was evident between river trout and 3+ migrant dollaghan. Kennedy (1976) compared resident Main

Table 18C.8. Mean back-calculated lengths (cm) at age for six trout samples from the River Main. Standard deviations are given in parentheses

		Age				
Sample	n	1	2	3	4	5
Main – 1	16	8.6 (1.57)	16.1 (2.81)	21.9 (1.35)		
Main – 2	21	7.7 (1.83)	17.9 (3.02)	23.5 (2.61)		
Glenwhirry – 1	13	8.12 (1.51)	14.4 (1.78)	22.3 (4.19)	24.6 (0)	32.1 (0)
Glenwhirry – 2	32	9.0 (1.41)	18.0 (5.39)	33.2 (0)	39.7 (0)	
Aghanure	29	8.8 (1.52)	14.4 (1.87)	19.2 (0)		
Aghanageeragh	30	7.7 (1.73)	14.8 (2.20)	19.4 (2.90)	23.8 (0)	
Pooled	141	8.4 (1.58)	15.9 (2.87)	22.6 (4.20	29.4 (8.96)	32.1 (0)

trout with dollaghan, by back calculation to pre-migratory ages from mature trout returning to spawn, and found similar differences in length at age one. In salmon trout at age one, the 1+ migrants were not significantly different from river trout but the 2+ and 3+ migrants were significantly shorter. At age two the salmon trout 2+ and 3+ migrants were not significantly different in length from river trout, a condition found also at age three in 3+ salmon trout migrants. It seems, therefore, that for Lough Neagh trout the propensity to migrate, as well as the potential age of migration, is determined within the first year of life. However, these observations on growth rate in river and migratory trout do not contribute to a resolution of the question of whether or not such differences in behaviour have a genetic basis, especially since fish from different river systems were involved. The mean lengths at age for River Main trout are greater than those presented by Cragg-Hine (1972) for the same system but are in good agreement with those detailed by Pitcher and Murray (1977).

18C.4. Feeding habits

The stomach contents of 226 dollaghan and 74 salmon trout obtained in August 1980 and from April to August 1981 and 1982 were analysed. Following Nilsson (1955), percentage volume percentage occurrence and percentage dominance of each food item in the stomachs were recorded. A visual assessment of the degree of fullness of each stomach was made using the points system of Ball (1961). Given the suggestion, from the biochemical data, of a degree of genetic segregation between dollaghan and salmon trout the feeding data for these fish was analysed separately. It is of interest to make a comparison of the feeding of dollaghan and salmon trout as dietary differences have been recorded among other sympatric trout populations (e.g. Ferguson and Mason, 1981). However, comparison must be restricted to fish caught at the same time, as trout frequently show temporal variations in diet.

The stomach fullness indices of comparable samples of dollaghan and salmon trout were not significantly different. Both types showed highest indices during May, a trend similar to brown trout of Llyn Tegid (Ball, 1961). Due to the collection of samples over a three year period inter-year differences in food availability must be taken into account. For example, salmon trout stomach contents of 26 May 1981 and 13 May 1982 were not significantly correlated (Spearman rank correlation test). However, this probably resulted largely from rank order difference in minor dietary items. The Schoener (1970) index, which is considered less sensitive to between group differences in minor dietary items (Wallace, 1981), indicated significant overlap and reflects the dominance in both samples of chironomid pupae. Four coincident samples of dollaghan and salmon trout were available for comparison: April, 1981; May, 1982; July, 1981; August, 1981 (Table 18C.9).

During April dollaghan feeding was primarily benthic with the main dietary constituents being the crustaceans *Gammarus* and *Asellus*, together accounting for some 45% by volume of food eaten, the former being more abundant. Fish were of importance to dollaghan at this time forming 26% of the total diet. Fish eaten were 0+ perch *Perca fluviatilis*, roach *Rutilus rutilus* and occasionally pollan *Coregonus autumnalis pollan*, and when taken, tended to be the dominant food item. Also of importance were chironomid larvae and pupae, accounting for 26% of the total diet. Items of minor dietary importance at this time included *Simulium* adults, *Hydrobius* and *Corixa*. Similar food items were taken by salmon trout in April with some quantitative differences. However, no significant difference was found in the rank order of importance of the various food items. The quantitative difference in average volume percentage of these food items was sufficient to produce a Schoener index of 0.59, indicating that segregation was just significant (values of less than 0.6 being regarded as significant (Wallace, 1981)).

Dollaghan taken in May 1982 showed evidence of mid- and surface-water feeding with chironomid pupae and *Simulium* adults constituting 97% by volume of the diet. The remaining food items included chironomid larvae, *Simulium* larvae and pupae, Tricoptera larvae, Crustacea and Mollusca. Again in salmon trout taken at the same time chironomid pupae and *Simulium* adults were dominant. In the salmon trout sample of May 1981 chironomid pupae were the dominant item with *Simulium* being absent in that year. Both the rank correlation test and the Schoener index indicated a high degree of dietary similarity in the May 1987 feeding habits of dollaghan and salmon trout.

In the July dollaghan sample chironomid pupae and larvae accounted for

Table 18C.9. Comparative dietary composition of dollaghan (S) and salmon trout (ST) stomachs. For each sample are given percentage volume (V), percentage frequency of occurrence (F) and dominance (D) values. l = larvae, p = pupae, a = adults

Food item	Date No.	4/4/81 (D) 16			4/4/81 (ST) 15			13/5/82 (D) 100			13/5/82 (ST) 15			26/5/81 (ST) 18		
	Index	V	F	D	V	F	D	V	F	D	V	F	D	V	F	D
Chironomid l		10.4	46.7	6.7	43.1	58.3	41.7	0.6	20.4	0	0.1	26.7	0	<0.1	17.6	0
Chironomid p		15.6	33.3	13.3	23.8	50.0	25.0	61.4	81.6	59.2	66.9	86.7	73.4	89.3	94.1	94.1
Chironomid a														5.9	17.6	5.9
Simulium l								<0.1	1.0	0						
Simulium p								1.0	1.0	1.0						
Simulium a		0.3	6.7	0	<0.1	16.7	0	35.3	44.9	38.8	20.2	66.7	13.3			
Trichoptera l																
Trichoptera p								0.1	2.0	0				0.7	17.6	0
Trichoptera a														0.2	11.8	0
Plecoptera a								1.0	1.0	1.0						
Ceratopogon l																
Corixa		1.3	13.3	0	0.1	8.3	0	<0.1	1.0	0						
Hydrobius		0.1	6.7	0	<0.1	8.3	0	<0.1	4.1	0	0.1	6.7	0			
Coleopteran l																
Gammarus		28.3	46.7	33.3	24.5	50.0	25.0	0.2	7.1	0	0.4	13.3	0	0.5	29.4	0
Asellus		17.4	26.7	20.0	1.1	25.0	0	0.1	1.0	0	0.7	6.7	0			
Mysis																
Fish		26.7	26.7	26.7	7.4	8.3	8.3	0.1	1.0	0	11.5	13.3	13.3	3.2	11.8	0
terr. Coleoptera														0.1	11.8	0
terr. Diptera								<0.1	1.0	0	<0.1	6.7	0	<0.1	5.9	0
Mollusca								<0.1	1.0	0						
Annelida								<0.1	1.0	0						
Plant material								<0.1	1.0	0	<0.1	17.6	0			

Table 18C.9 Continued

Food item	Date No.	1/6/82 (D) 32			3/7/81 (D) 10			3/7/81 (ST) 12			1/8/80 (D) 33			18/8/81 (ST) 35			18/8/81 (ST) 14		
	Index	V	F	D	V	F	D	V	F	D	V	F	D	V	F	D	V	F	D
Chironomid l		0.3	10.7	0	<0.1	11.1	0	3.2	22.2	0	<0.1	3.1	0	0.1	12.5	0	0.1	21.4	0
Chironomid p		82.4	89.3	85.7	33.4	44.4	33.3	31.7	55.5	33.3	93.2	100	100	9.5	43.7	9.4	15.4	50.0	14.3
Chironomid a		0.1	3.6	0	55.5	66.7	55.6	52.5	66.7	55.6	0.4	6.2	0	81.6	87.5	84.4	83.9	85.7	85.7
Simulium l																			
Simulium p											<0.1	3.1	0						
Simulium a		3.6	10.7	3.6							<0.1	3.1	0	<0.1	6.2	0			
Trichoptera l		0.2	3.6	0				0.11	11.1	0									
Trichoptera p											<0.1	3.1	0						
Trichoptera a		<0.1	3.6	0							6.3	68.7	0				<0.1	7.1	0
Plecoptera a											<0.1	3.1	0						
Ceratopogon l											<0.1	3	0						
Corixa																			
Hydrobius																			
Coleopteran l																	0.3	7.1	0
Gammarus		0.7	7.1	0	<0.1	11.1	0	1.2	11.1	0				4.7	34.4	3.1	0.3	42.9	0
Asellus		1.2	3.6	0										0.3	3.1	0			
Mysis		7.8	17.9	7.1	10.7	11.1	11.1							3.5	9.4	3.1			
Fish		3.6	3.6	3.6				11.1	11.1	11.1				<0.1	3.1	0			
terr. Coleoptera		<0.1	3.6	0															
terr. Diptera																			
Mollusca					0.4	11.1	0												
Annelida																			
Plant material								0.2	22.2	0	0.1	3.1	0	0.1	9.4	0	<0.1	7.1	0

89% of the diet with the remalnder being composed of Crustacea, in this case *Mysis relicta* being most important. The diet of July salmon trout was similar to that of dollaghan except that chironomid larvae were present and *Mysis relicta* absent. Fish, when taken, were the dominant food item in individual stomachs and accounted for 11% of the total salmon trout July diet. The rank order for July dollaghan and salmon trout dietary constituents was not significantly correlated ($r = 0.31$) but again this is probably a reflection of minor dietary items, as the Schoener index of 0.82 indicates the closely similar proportions of major items.

In dollaghan caught in mid August chironomid adults constituted the dominant food item (82% by volume). Chironomid pupae occurred in a large proportion of the stomachs but constituted only some 10% by volume of the diet. Thus dollaghan at this time were primarily mid- and surface water feeders with benthic Crustacea and chironomid larvae virtually absent. August salmon trout were also predominantly surface feeding with chironomid adults forming 84% (by volume) of the diet with benthic organisms being of limited importance. As with the July sample the rank correlation of dollaghan and salmon trout was significantly different but the Schoener index was not significant.

Overall, there is no clear-cut feeding segregation between dollaghan and salmon trout although there appear to be differences in minor dietary items. However, it must be taken into account that the fish were obtained from commercial netsmen and place, time and method of capture may have led to bias in the feeding analysis. Both types show a seasonal pattern of feeding habits typical of those reported in other studies of lake trout. The majority of Lough-caught trout examined were found to have fed mainly on benthic organisms in the early spring and at or near the surface during late spring, and summer. Inter-year variation suggets that the seasonal changes are closely related to prey abundance. Chironomids of the various life stages form the single most important food group for Lough Neagh trout. Since this group forms an abundant food resource in the Lough both types may be able to use it without this necessarily resulting in competition. Nilsson (1955) has noted that while food segregation occurs for sympatric populations of trout and charr in Sweden, this segregation breaks down during the summer when food is superabundant. An interesting feature of Lough Neagh trout diet is the absence of zooplankton which forms a substantial part of the diet of lake trout in other Irish lakes (Kennedy and Fitzmaurice, 1971; Ferguson and Mason, 1981).

Many authors have described a change in diet of lake trout with increasing age with older trout often becoming increasingly piscivorous (e.g. Frost and Brown, 1967). Lough Neagh trout with fish in their stomachs ranged in size from 28.5 cm to 66.2 cm (mean 40.4 cm). While this overlapped with the range in length of non-piscivorous trout of 17.02 cm to 46.5 cm, the mean length of piscivorous ones was significantly greater. The mean size of piscivorous trout did not differ significantly between dollaghan and salmon trout.

Some of the larger trout had substantial numbers of adult cestodes (*Eubothrium crassum*) in the pyloric caecae. The largest individual (66.2 cm) had an infestation of these that almost completely obstructed the gut and caused severe deformation of the pyloric region.

18C.5. Evaluation and conclusions

18C.5.1. *Present status of Lough Neagh brown trout*

From the information obtained in the present study, it appears that of the trout types originally present in Lough Neagh, only dollaghan and salmon trout are present to-day in significant numbers. There is considerable evidence that the trout stocks in the Lough have declined considerably during the present century as a result of eutrophication, disease (Kennedy and Vickers, Part A of this chapter), destruction of spawning and nursery areas by river drainage and possibly overexploitation. Large trout have in particular been virtually eliminated and Kennedy (1976) reports that only 0.4% of the commercial catch in 1975 was greater than 50 cm as compared with 10.9% of similar length in 1966. In the present study trout over 50 cm accounted for 0.6% of the Lough-caught fish. Thus while the Lough was primarily salmonid in character last century with in addition to several types of trout, Atlantic salmon, Arctic charr *Salvelinus alpinus* and pollan present, it is now dominated by cyprinids especially roach. This switch from salmonid to cyprinid character is a common feature of many lakes in western Europe as a direct result of eutrophication. Indeed Lough Melvin in north west Ireland may be one of few lakes still remaining in a pristine post-glacial salmonid condition (Ferguson and Mason, 1981).

18C.5.2. *Taxonomic and management implications*

Within the rivers draining into Lough Neagh there exist a large number of brown trout populations as shown by the genetic heterogeneity among and within river systems. This heterogeneity is frequently developed over small distances and indicates that populations can be highly localized. The propensity to form local genetically distinct populations is typical of brown trout and other salmonids (Allendorf and Phelps, 1981; Ryman and Stahl, 1981). Two major populations of migratory trout can be identified, dollaghan and salmon trout, between which morphological dissimilarity is paralleled by genetic differentiation. Although the genetic differentiation is small, their status as separate races is supported by differences in sex ratio, age at maturity, body/scale growth ratio and length at age data. Two opposing points of view have been put forward to explain the existence of sympatric populations of salmonid fishes (Behnke, 1972). Formerly, it was thought that a population in a given lake could split into two or more reproductively isolated populations without the need for geographical isolation. The more widespread current view (e.g. Svärdson, 1979) suggests that all present sympatric populations are the result of multiple invasions by forms which differentiated allopatrically in glacial and inter-glacial refugia. Thus it is generally accepted that salmonid races have their origins associated with the events of the Pleistocene glaciations. Different ecological and behavioural specialisations which arose in allopatry would, if sufficiently well developed, act as barriers to interbreeding when the populations became sympatric. The innate tendency of brown trout and other salmonids to return to their natal areas for spawning would help maintain such isolation once etablished. Within extensive river systems such behaviour would also lead to development of separate populations even in the absence of geographical isolation, as has been seen in thc Lough Neagh catchment. According to Charlesworth (1963) 'Lake Neagh' was only one of a number of glacial lakes, including 'Lake Lagan' and several Antrim lakes. Some of these had connections to Lake Neagh but these were probably intermittent, thus allowing periods of isolation for the fish present in them.

The existence of numerous genetically distinct populations of brown trout presents a problem of classification. The widespread application of the polytypic species concept by biologists in this century has led to the recognition of all trout types as belonging to a single polytypic species of brown trout. However, this approach does not draw attention to the great diversity of forms which are distinct in their life history, feeding behaviour, morphology, growth rate and potential age of maturation, time and place of spawning age of migration etc. The alternative of recognising all sympatric reproductively isolated forms as separate species, as required by the biological species concept, can lead to over-splitting. An adequate classification is necessary to make clear to fisheries biologists the extent of genetic heterogeneity which is present among brown trout populations. Also for the conservation of unique populations such as dollaghan and salmon trout it may be appropriate to draw attention to them through their recognition as distinct taxa of at least sub-specific rank.

In recent years the use of natural strains of salmonid fishes for creative management has received more attention (Behnke, 1981) as has the need for conservation of unique gene pools (Ryman, 1981). As noted by Behnke, (1972), the diversity in post-glacial salmonid communities is 'highly susceptible to disruption or destruction by introductions, eutrophication and exploitation'. To that list we must add all the other environmental perturbations to which Lough Neagh is increasingly subject. The diversity of brown trout populations in the Lough Neagh system, albeit considerably depleted in the past century, is a valuable genetic resource. For example, an extensive study of brown trout populations in Britain and Ireland (Ferguson and Fleming, 1983) has failed to find the CK-2(75) allele other than in the Lough Neagh system. The restriction of salmon trout, which possess this allele to a few south-western tributaries renders this genotype particularly vulnerable to destruction, e.g. by river drainage.

Acknowledgements

We express our thanks to the staff of the Fisheries Conservancy Board and Lough Neagh fish dealers, in particular Mr P. Quinn for assistance in obtaining samples of Lough trout. A postgraduate studentship to W.W.C. from the Department of Agriculture N.I. is gratefully acknowledged.

18C.6. References

Allen, K.R. (1938) Some observations on the biology of the trout (*Salmo trutta*) in Windermere. J. Anim. Ecol. 7: 333–349.

Allendorf, F.W. and Utter, F.M. (1979) Population genetics. In: W.S. Hoar, D.J. Randall and J.R. Brett (eds.), Physiology, Vol. VIII. Academic Press, London, pp. 407–454.

Allendorf, F.W. and Phelps, S.R. (1981) Isozymes and the preservation of genetic variation in salmonid fishes. In: N. Ryman (ed.), Fish Gene Pools. Ecol. Bull. (Stockholm) 35: 37–52.

Bagenal, T. B. and Tesch, F. W. (1978) Age and growth. In: T.B. Bagenal (ed.), Methods for Assessment of Fish Production in Fresh Waters. I.B.P. Handbook No. 3. Blackwell Scientific Publications, Oxford.

Ball, J.N. (1961) On the food of the brown trout of Llyn Tegid. Proc. Zool. Soc. Lond. 137: 599–622.

Ball, J.N. and Jones J.W. (1960) On the growth of brown trout in Llyn Tegid. Proc. Zool. Soc. Lond. 134: 1–141.

Bassett, G.H. (1888) The Book of County Armagh: Manual and Directory. Sealy, Bryers and Walker, Dublin.

Behnke, R.J. (1972) The systematics of salmonid fishes of recently glaciated lakes. J. Fish. Res. Bd Can. 29: 639–671.

Behnke, R.J. (1981) Systematic and zoogeographical interpretation of Great Basin trouts. In: R.J. Naiman and D.L. Soltz (eds.), Fishes in North American Deserts. New York, John Wiley.

Boate, G. (1652) Ireland's Natural History. Wright, London.

Campbell, R.N. (1957) The effect of flooding on the growth rate of brown trout in Lough Tummell. Scient. Invest. Freshwat. Fish. Scotl.: 14.

Campbell, J.S. (1977) Spawning characteristics of brown trout and sea trout *Salmo trutta* L. in Kirk Burn, River Tweed, Scotland. J. Fish. Biol. 11: 217–229.

Charlesworth, J.K. (1963) The Geology of Ireland. Oliver and Boyd, London.

Couch, J. (1865) A History of the Fishes of the British Isles. Groombridge and Sons, London.

Cragg-Hine, D. (1972) The juvenile salmonid populations of some streams in the Lough Neagh and Lower Bann catchment areas. Dept. Agric. N.I. Fish Res. Leaflet No. 4.

Crozier, W.W. (1983) Population biology of Lough Neagh brown trout (*Salmo trutta* L.). PhD Thesis, The Queen's University of Belfast, 477 pp.

Crozier, W.W. and Ferguson, A. (1986) Electrophoretic examination of the population structure of brown trout (*Salmo trutta* L.) from the catchment of Lough Neagh, Northern Ireland. J. Fish. Biol. 28: 459–477.

Day, F. (1887) British and Irish Salmonidae. Williams and Norgate, London.

Doyle, J.B. (1854) Tours in Ulster. Hodges and Smith, Dublin.

Dubourdeau, J. (1812) History of County Antrim. Gravesberry and Campbell, Dublin.

Fahy, E. (1977) Characteristics of the freshwater occurrence of sea trout *Salmo trutta* in Ireland. J. Fish Biol. 11: 635–646.

Ferguson, A. (1980) Biochemical Systematics and Evolution. Blackie, Glasgow.

Ferguson, A. and Mason, F.M. (1981) Allozyme evidence for reproductively isolated sympatric populations of brown trout (*Salmo trutta* L.) in Lough Melvin, Ireland. J. Fish Biol. 18: 629–642.

Ferguson, A. and Fleming, C.C. (1983) Evolutionary and taxonomic significance of protein variation in the brown trout (*Salmo trutta* L.) and other salmonid fishes. In G.S. Oxford and D. Rollinson (eds.), Protein Polymorphism: Adaptive and Taxonomic Significance. Academic Press, London.

Fleming, C.C. (1983) Population biology of anadromous brown trout (*Salmo trutta* L.) in Ireland and Britain. PhD Thesis, The Queen's University of Belfast.

Frost, W.E. and Brown, M.E. (1967) The Trout. Collins, London.

Günther, A. (1866) Catalogue of the Physostomi in the Collection of the British Museum. British Museum, London.

Harris, W. (1744) The Ancient and Present State of the County of Down. Reilly, Dublin.

Hunt, P.C. and Jones, J.W. (1972) Trout in Llyn Alaw, Anglesey, Wales. II. Growth. J. Fish Biol. 4: 409–424.

Kennedy, G.J.A. (1976) Age, growth and feeding studies on Lough Neagh trout *Salmo trutta* (L.). Min. Agric. N.I. Fish Res. Leaflet No. 10.

Kennedy, M. and Fitzmaurice, P. (1971) Growth and food of brown trout, *Salmo trutta* L. in Irish-waters. Proc. R. Ir. Acad. 71B: 269–352.

Kipling, C. (1962) The use of the scales of the brown trout (*Salmo trutta* L.) for the back-calculation of growth. J. Cons. perm. Int. Explor. Mer. 27: 304–315.

Le Cren, E.D. (1947) The determination of the age and growth of the perch (*Perca fluviatilis*) from the opercular bone. J. Anim. Ecol. 16: 188–204.

Maxwell, H. (1904) British Freshwater Fishes. Hutchinson and Co., London.

Munro, W.R. and Balmain, K.H. (1956) Observations on the spawning runs of brown trout in South Queich, Loch Leven. Sci. Invest. Freshwat. Fish Scot. No. 13.

Nei, M. (1972) Genetic distance between populations. Amer. Natur. 106: 283–292.

Nilsson, N.A. (1955) Studies on the feeding habits of trout and char in north Swedish lakes. Rep. Inst. Freshw. Res. Drottningholm 36: 163–225.

Pennant, T. (1812) British Zoology Vol. III, Class IV, Fishes. Wilkie and Robinson, London.

Pitcher, T.J. and Murray, A. (1977) Growth, food and density of brown trout *Salmo trutta* L. in the River Main, Co. Antrim, 1977, with notes on the mitigation of channel alteration. Main Angling Club. unpubl. rep.

Ryman, N. (1981) Conservation of genetic resources: experiences from the brown trout (*Salmo trutta*). In: N. Ryman (ed.), Fish Gene Pools. Ecol. Bull. (Stockholm) 34: 61–74.

Ryman, N. and Ståhl, G. (1981) Genetic perspectives of the identification and conservation of Scandinavian stocks of fish. Can. J. Fish. Aquat. Sci. 38: 1562–1575.

Sampson, G.V. (1814) A Memoir Explanatory of the Chart and Survey of the County of Londonderry, Ireland. Nicol and Nicol, London.

Schoener, T.W. (1970) Non-synchronous spatial overlap of lizards in patchy habitats. Ecology 51: 408–418.

Southern, R. (1935) Reports from the limnological laboratory. III. The food and growth of brown trout from Lough Derg and the River Shannon. Proc. R. Ir. Acad. 42B: 87–172.

Svärdson, G. (1979) Speciation of Scandinavian *Coregonus*. Rep. Inst. Freshw. Res. Drottningholm 57: 3–95.

Taggart, J., Ferguson, A. and Mason, F.M. (1981) Genetic variation in Irish populations of brown trout (*Salmo trutta* L.): Electrophoretic analysis of allozymes. Comp. Biochem. Physiol. 69B: 393–412.

Thompson, W. (1856) The Natural History of Ireland, Vol. IV. Reeve, Benham and Reeve, London.

Thorpe, J.E. (1974) The movements of brown trout. *Salmo trutta* L. in Loch Leven, Kinross, Scotland. J. Fish Biol. 6: 153–180.

Treasurer, J.W. (1976) Age, growth and length-weight relationship of brown trout *Salmo trutta* (L.) in the Loch of Strathbeg, Aberdeenshire. J. Fish Biol. 8: 241–253.

Wallace, R.K. (1981) An assessment of diet overlap indices. Trans. Am. Fish. Soc. 110: 72–76.

Workman, P.L. and Niswander, J. D. (1970) Population studies on southwestern Indian tribes. II. Local genetic differentiation in the Papago. Amer. J. Human Genet. 22: 24–49.

Wright, S. (1978) Evolution and the Genetics of Populations. Vol. IV. Variability Within and Among Populations. The University of Chicago Press, London.

Yarrell, W. (1841) A History of British Fishes. Vol. 2. Van Voorst, London.

18. The fish of Lough Neagh

Part D. Investigations on pollan (*Coregonus autumnalis pollan* Thompson)

J.P.F. WILSON

18D.1. Introduction

Pollan have been fished for food in Lough Neagh for centuries and are still fished commercially, forming an important component of the Lough fishery. There are large fluctuations in annual catches: the catch was worth about £0.25 million in 1976, but only about £11,000 per year in the mid-1980s.

In spite of its considerable historic and commercial importance, the biology of the pollan was little studied until the 1970s. Ferguson (1974) and Ferguson *et al.* (1978) established the pollan's taxonomic position, Dabrowski (1981, 1982a,b) and Dabrowski *et al.* (1984) studied nutrition and energy metabolism and Wilson (1983, 1984) and Wilson and Pitcher (1983, 1984a, b, 1985) described growth, feeding and population dynamics.

18D.2. Taxonomy

The pollan occurs in Ireland in Lough Neagh, Lough Erne and the Shannon lakes. It is classified with the whitefish and ciscoes in the subfamily Coregoninae of the family Salmonidae. It was originally described as a separate species in 1835 (Thompson, 1856). Some authors have regarded it as conspecific with one or other of the British coregonines (Schindler, 1957; Svardson, 1957; Wheeler, 1967; Maitland 1972), while others have recognised its similarity with the Arctic cisco *Coregonus autumnalis* (Regan, 1911; Behnke, 1972).

Ferguson *et al.* (1978), in an electrophoretic comparison of holarctic coregonines, found no difference between the pollan and *C. autumnalis*. They examined 27 enzyme loci and a number of proteins, and from their matrix of similarity among the coregonines examined they concluded that the pollan is conspecific with the arctic cisco and is unique in being the only cisco in Western Europe. They proposed the name *Coregonus autumnalis pollan* Thompson.

Unlike the pollan, the Arctic *C. autumnalis* is mainly anadromous. Fergu-

R. B. Wood and R. V. Smith (eds.), Lough Neagh, 439–450.

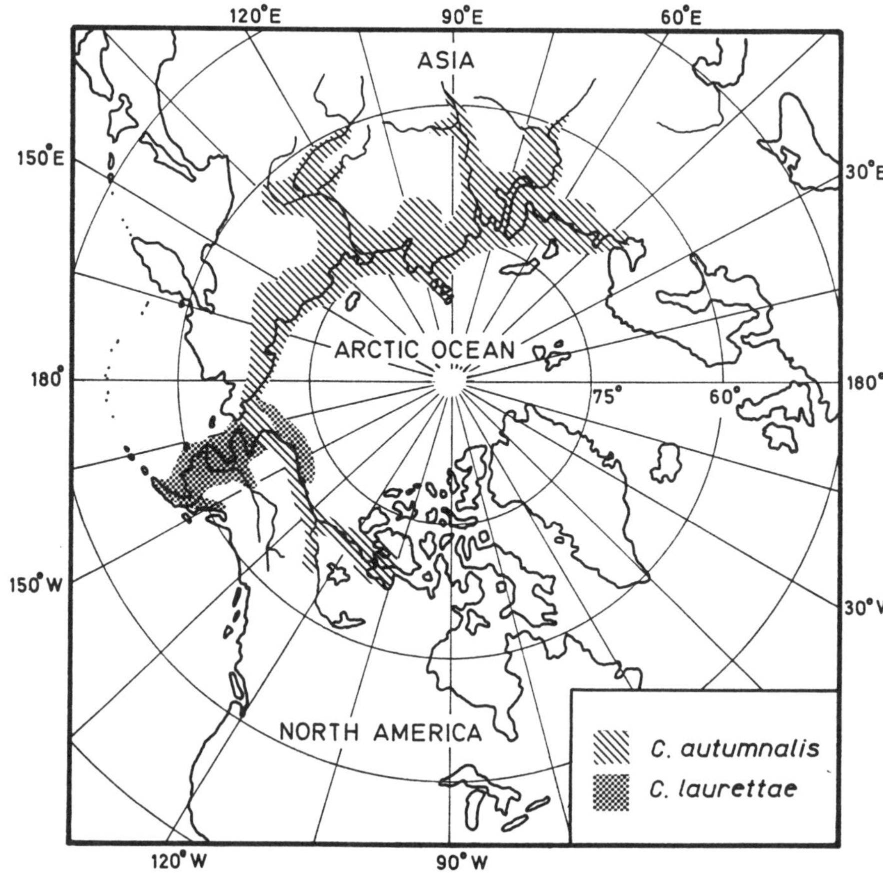

Figure 18D.1. The known geographic ranges of *Coregonus autumnalis* and *C. laurettae*. From McPhail (1966).

son *et al.* suggest that during the last glaciation, *C. autumnalis* was an anadromous fish, living on the edge of the Arctic ice sheet, much as it does at present (Fig. 18D.1). The ice sheet then extended as far as the southern part of the British Isles. When the ice retreated, a fragment of the population remained in the Irish lakes. The British species of whitefish colonised Britain from the south, but were unable to reach Ireland, which had become a separate island.

18D.3. Age and growth

Pollan are quite difficult to age by the methods applied to other salmonids, as they can form more than one check on their scales annually. Wilson and

Pitcher (1984a) developed a consistent method of age determination for pollan, and showed that pollan live to a maximum age of five years attaining a length of about 30 cm and a weight of about 300 g. The seasonal growth of the six year classes of 1972–1977 is shown in Fig. 18D.2.

Dabrowski (1981) found pollan hatchlings had attained 11 mm in length in late February and grown to 70 mm by July. Pollan eggs incubated in lough water in the laboratory took about 416 degree-days to hatch. Dabrowski also described the early development of pollan larvae.

18D.4. Feeding

Pollan exhibit a distinct seasonal cycle in the composition of the diet and in stomach fullness. Stomach fullness is at a maximum in summer, and little food is taken in the spawning season (November–December) (Fig. 18D.3). Adult pollan eat mainly chironomid larvae in winter, and cladocerans (mainly *Daphnia longispina*) in summer (Fig. 18D.4). There is evidence that *Daphnia* spp. are selected by pollan, since copepods, though abundant in Lough Neagh in summer (see Chapter 14), are not an important food source for adults. Copepods were found in the stomachs of O+ fish (Wilson 1984).

Pollan alevins and fry eat mainly nauplii and copepodites. Copepods and *Daphnia longispina* become more important as alevins grow to 20 mg and 45 mg respectively, and fry of more than 1.4 g feed on benthic invertebrates (Dabrowski *et al.* 1984). Pollan fry which were kept in the laboratory and fed zooplankton from the Lough suffered heavy infestation of the cestode *Proteocephalus pollanicola* (Gresson).

Dabrowski (1982c) determined the light thresholds which enabled starved larvae to feed at 50% of their maximum rate. As pollan fry grew from 43 to 90 mg, the light thresholds fell by about one order of magnitude. Lough Neagh water is quite opaque (Chapter 4) and it is possible that this increased visual sensitivity enables older fish to extend the range of their benthic feeding.

18D.5. Reproduction

Pollan in Lough Neagh spawn on stony substrates in the period mid-November to late December. Catches from the spawning grounds in early November consist almost entirely of male pollan with empty stomachs, suggesting that males migrate to the spawning grounds first, while females continue to feed in deeper water.

Pollan become sexually mature in their third year. Each female produces 4,000–12,000 eggs and the mean fecundities of fish from different year-classes is shown in Fig. 18D.5. Fecundity seems to be related to fish length linearly rather than logarithmically. The latter is often the case in coregonines (Healey and Nichol, 1975), but the estimated relationship in pollan may be

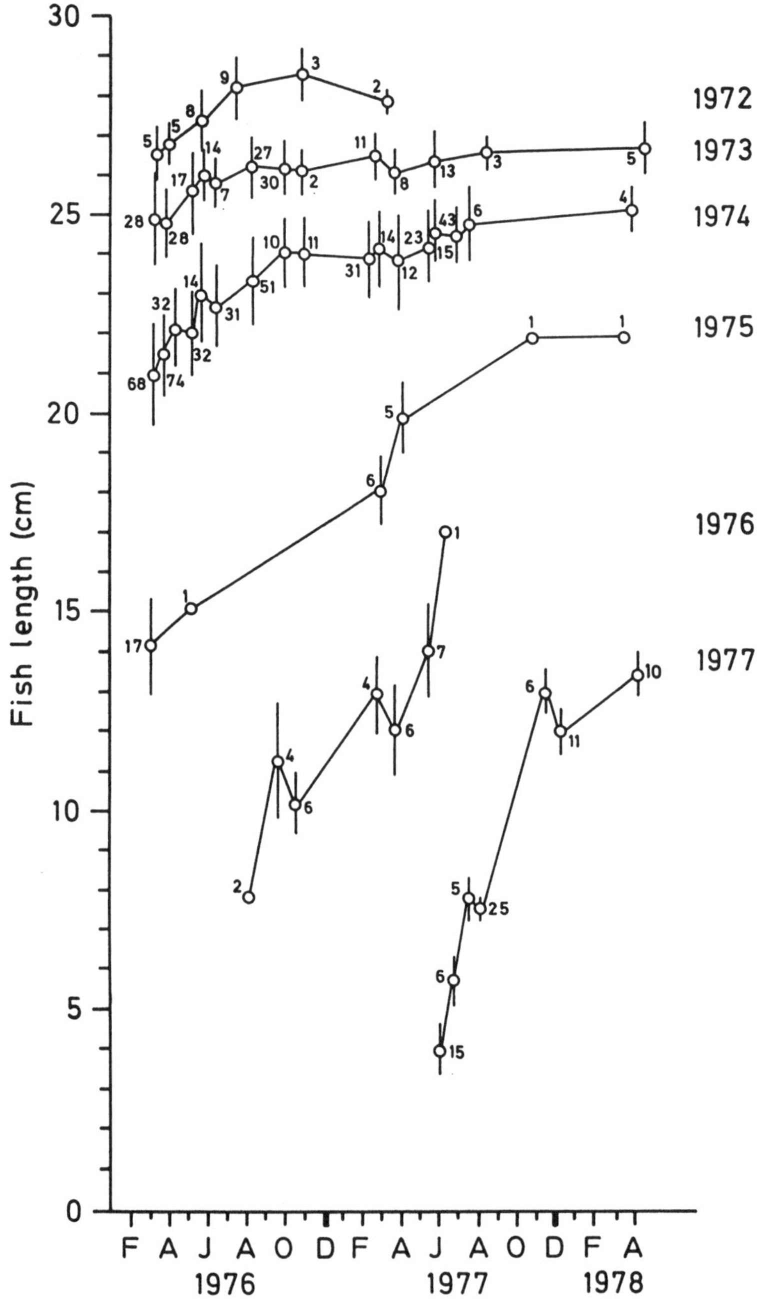

Figure 18D.2. Mean length of pollan of year classes 1972–1977 taken in trawl and draft net samples during 1976–1978. Vertical lines indicate 95% confidence limits; sample size shown for each point. Redrawn from Wilson and Pitcher (1984a).

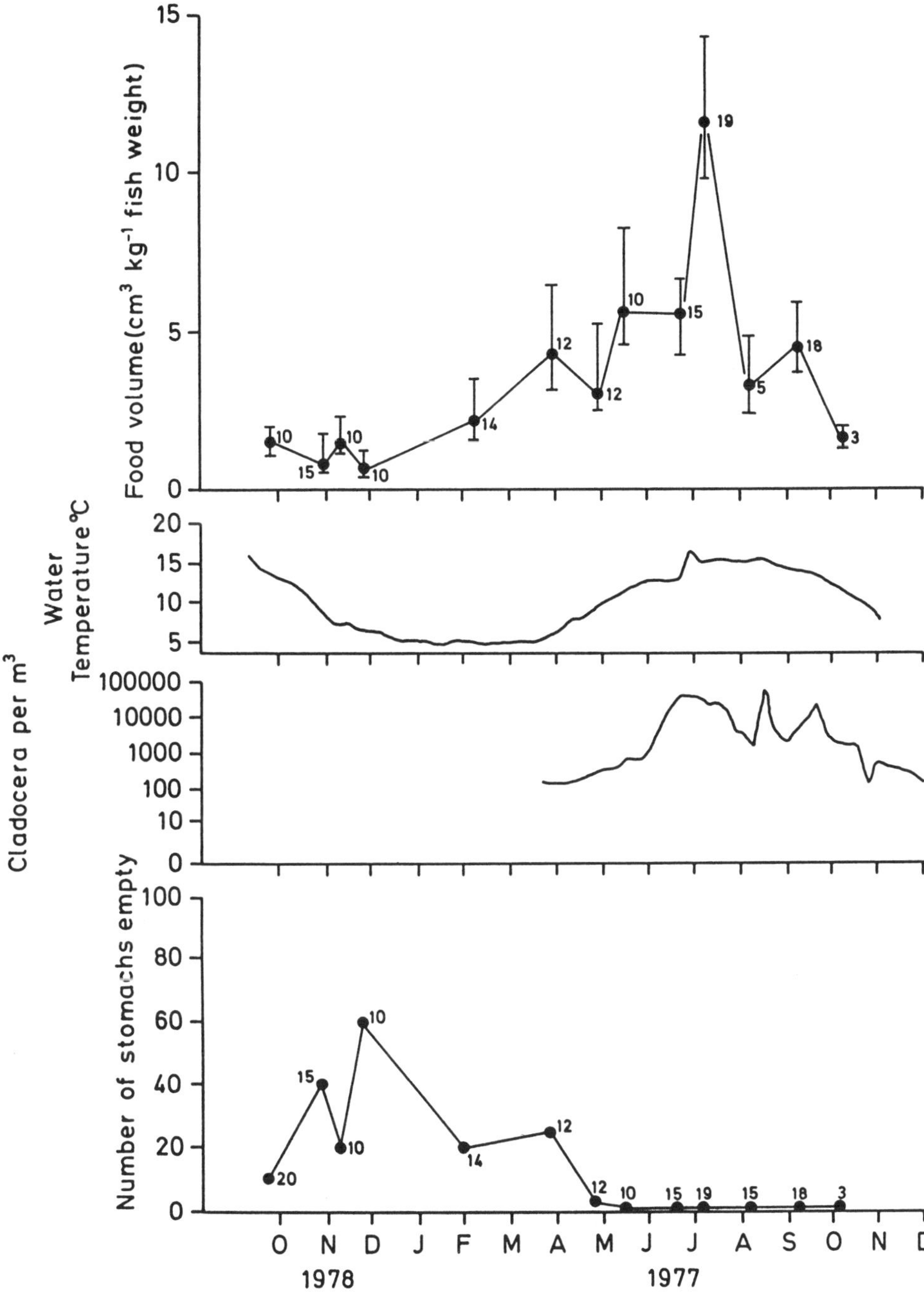

Figure 18D.3. Seasonal variation in mean stomach fullness (quartiles of range indicated) and in percentage of empty stomachs with water termperature and density of Cladocera in Lough Neagh, September 1976 to October 1977. Redrawn from Wilson (1984).

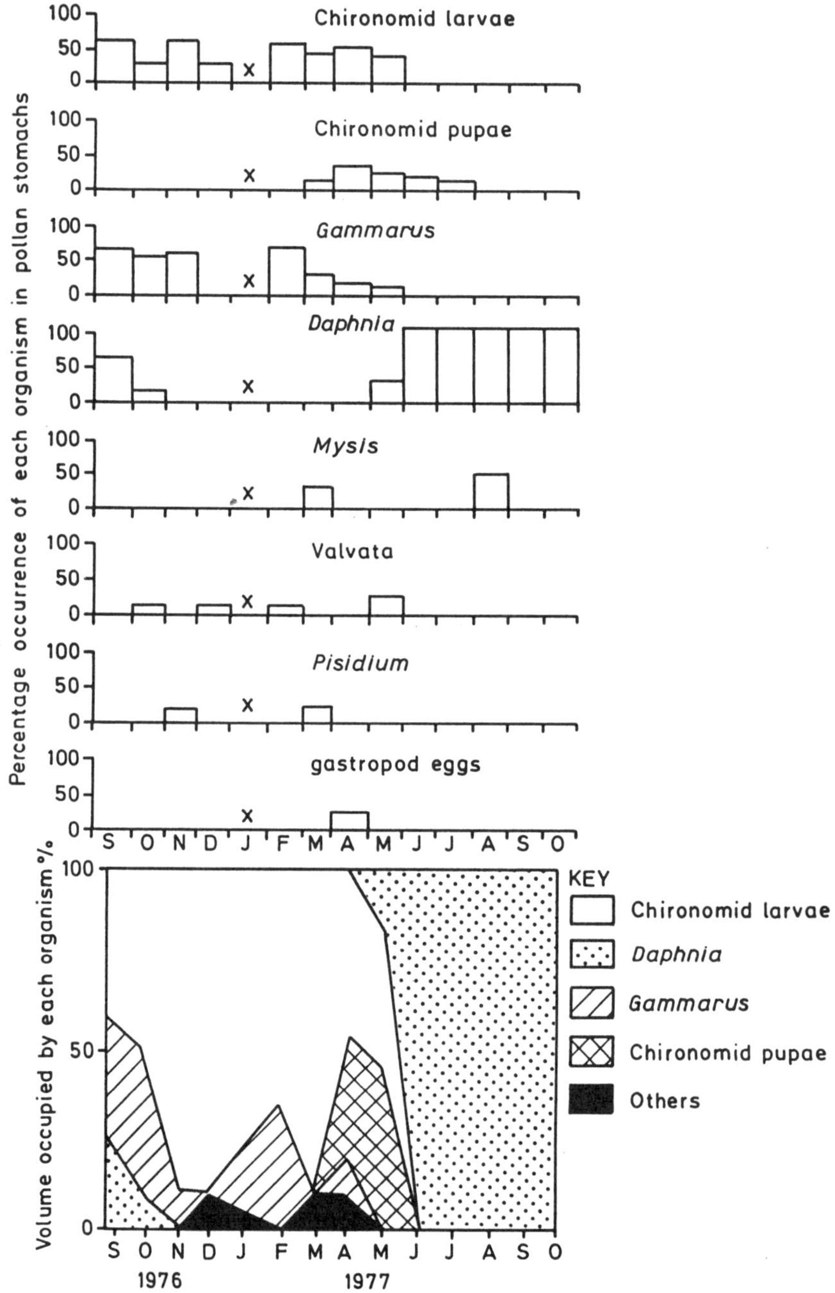

Figure 18D.4. Seasonal variation in food organisms (September 1976 to October 1977) in pollan stomachs by percentage frequency of occurrence (upper histograms) and by percentage volume (lower panel). No samples taken at ×, January 1977. Redrawn from Wilson (1984).

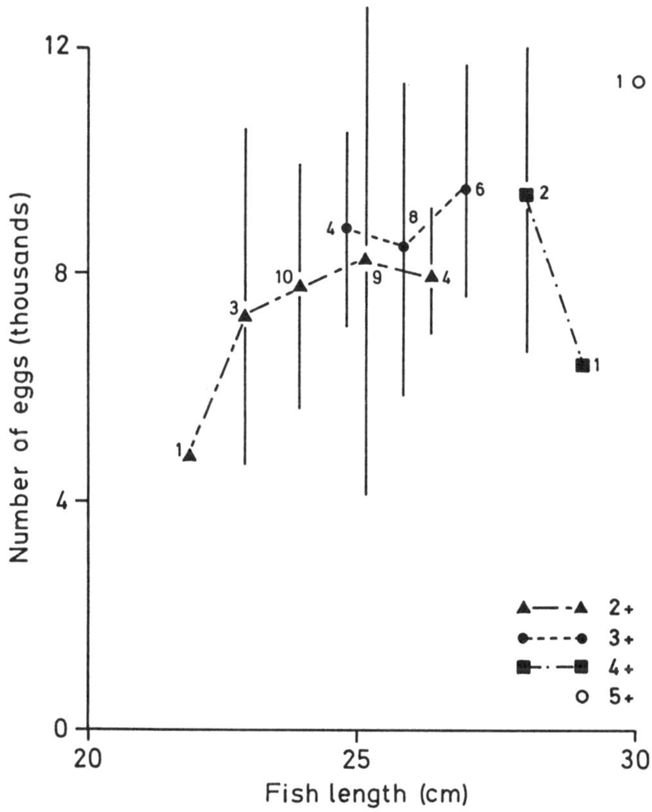

Figure 18D.5. Egg numbers in relation to fish length for 4 year classes of pollan. Vertical lines indicate 95% confidence limits; sample size shown for each point. Redrawn from Wilson and Pitcher.

influenced by the small length range (20–30 cm) of spawning fish. Relative fecundity was significantly lower in 1977 than in 1971 (Wilson and Pitcher, 1984b).

18D.6. Condition

The length-weight relationships calculated for immature pollan, mature males and mature females showed that weight is proportional to length$^{3.2}$ for all three categories. The exponent is significantly different from 3.0, indicating that pollan shape changes during life: the fish become relatively thicker as they grow. Fish caught in 1965 and in 1976 had statistically similar length-weight relationships (Wilson and Pitcher, 1983).

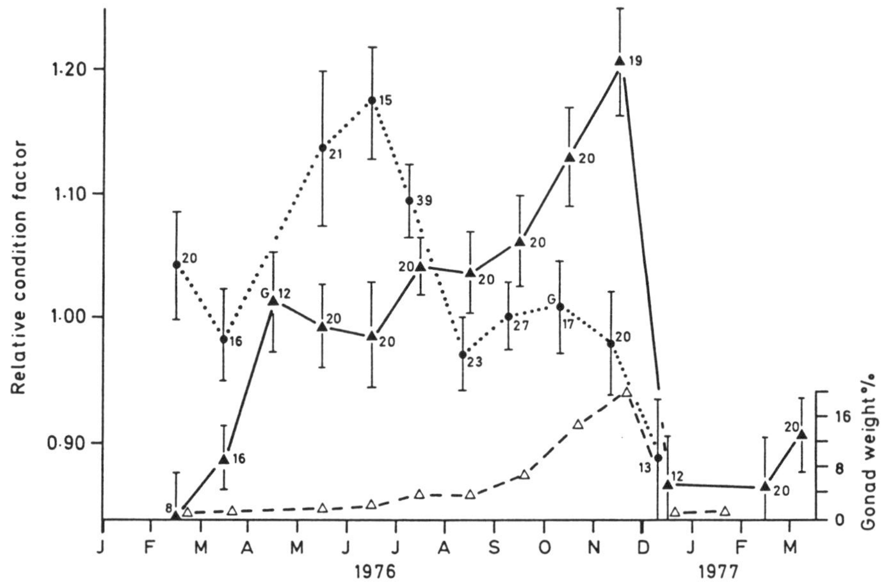

Figure 18D.6. Mean relative condition factors (×), somatic relative condition factors (○) and relative gonad weights (□) of mature female pollan 1976–1977. Sample sizes and 95% confidence limits are indicated. G = gill net samples. Redrawn from Wilson and Pitcher (1983).

A distinct seasonal cycle in condition is apparent in mature fish of both sexes, but not in immature fish. The plot for mature females is shown in Fig. 18D.6. Overall condition k_n reaches a maximum just before spawning. The plot of somatic condition sk_n (based on the relationship of length to body weight minus gonad weight) shows that somatic condition reaches a maximum in June. This is evidence of a decrease in the anabolism of non-reproductive tissue and an increase in the anabolism of ovarian tissue when stomach fullness is at a maximum in mid-summer.

Gonad development seems to be allied to the increase in k_n but not in sk_n (Wilson and Pitcher, 1983). Dabrowski (1982b, 1983) showed that energy in the form of lipids is directed from muscle to gonad growth at this time of year (Figs. 18D.7 and 18D.8). Somatic condition remains high in autumn, while linear growth slows down: pollan often exhibit a secondary check on their scale at this time (Wilson and Pitcher, 1984a), causing the difficulty of using scales to determine age referred to in Section 18D.3.

18D.7. Population dynamics

The total instantaneous mortality rate of pollan was estimated as 1.53 and 2.18 in 1976 and 1977 respectively. Using catch data, the fishing mortality

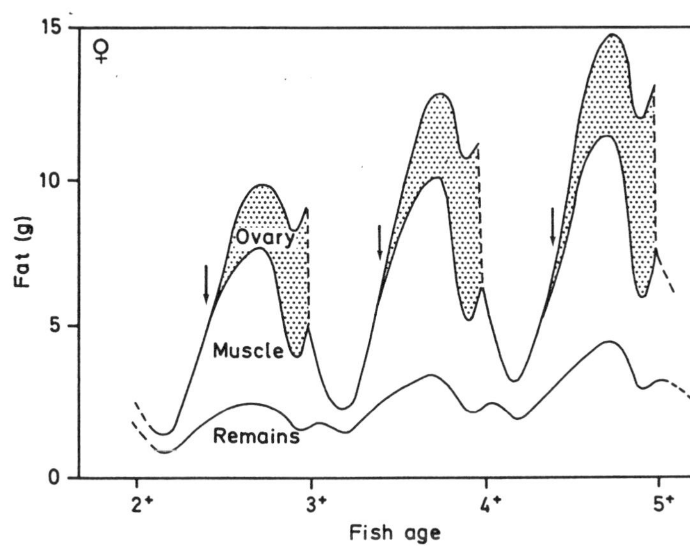

Figure 18D.7. The dynamics of the yearly changes in the total amount of fat in ovary, muscle and remains of mature pollan females. Redrawn from Dabrowski (1982b).

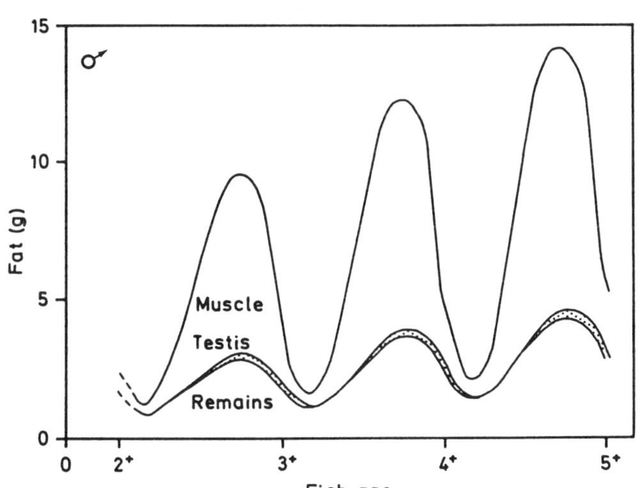

Figure 18D.8. The dynamics of the yearly changes in the total amount of fat in testis, muscle and remains of mature pollan males. Redrawn from Dabrowski (1982b).

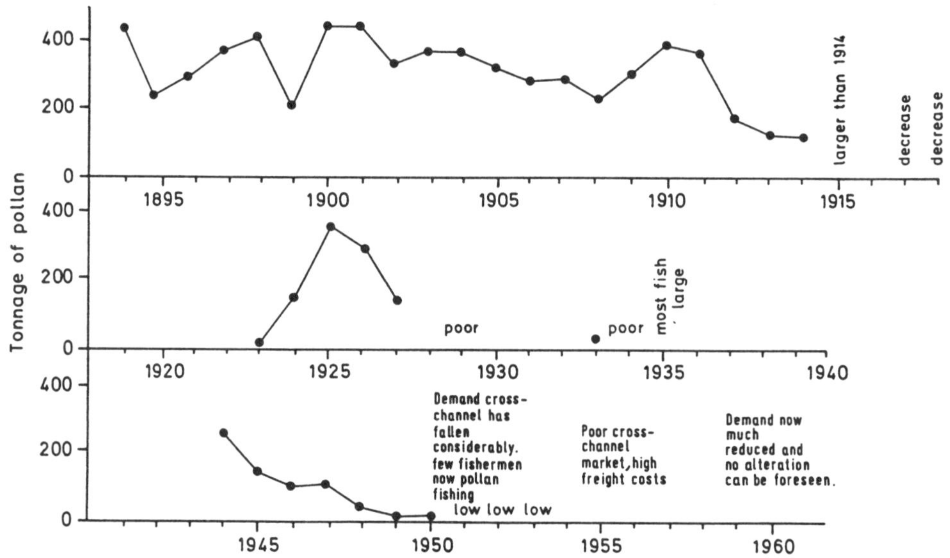

Figure 18D.9. The history of the Lough Neagh pollan fishery, based on the tonnage of fish transported by rail 1894 to 1960 (Data and comments from Reports Sea Inland Fish. Ire. and Reports Sea Inland Fish. N.I.)

and the relative abundance of pollan in 1971, 1976 and 1977 were assessed, and the total mortality of 1.53 was tentatively partitioned into a natural mortality of 0.98 and a fishing mortality of 0.55. Year-class strength in pollan is variable. Pollan are short-lived, so stock size is very dependent on the strength of incoming year-classes, a characteristic of cisco populations (Wilson, 1983).

Pollan become vulnerable to commercial fishing at the beginning of their third year, before they spawn. They are fished with draft nets and gill nets, and some winter trawl fishing was done in the past. Catch records are not kept at present, but the *Reports on the Sea and Inland Fisheries of Ireland* provide records of the tonnage of pollan transported by rail in the first half of this century (Fig. 18.D9). The pollan fishery evidently produced a consistent yield of 100–400 tons in the years 1893–1914. This is a conservative estimate, as large quantities were no doubt consumed locally. The *Reports* frequently comment that the tonnage transported is limited by lack of demand, rather than by lack of fish to catch. In recent decades, the eel fishery became much more valuable than the pollan fishery as demand for eels increased.

Wilson (1979) made two independent estimates of the tonnage of pollan landed in 1976, using commercial catch data and information supplied by fish dealers. The estimates were similar at about 190 tonnes, indicating that the fishery is still capable of producing catches like those documented in the early 1900s. However, in 1977, the commercial catch was much smaller. Catch-effort data showed that the stock size was much reduced, due to the weakness of the 1975 year-class (Wilson, 1983). There is no evidence that pollan have been overfished in recent years. Computer simulation of the population provided evidence that exploitation of the pollan would reduce the population size but not change its year-to-year fluctuation, and indicated that the population could sustain a fishing rate higher than that of 1976 (Wilson and Pitcher, 1985).

18D.8. References

Behnke, R.J. (1972) The systematics of salmonid fishes of recently glaciated lakes. J. Fish. Res. Bd. Can. 29: 639–671.

Dabrowski, K.R. (1981) The spawning and early life history of the pollan (*Coregonus pollan* Thompson) in Lough Neagh, Northern Ireland. Int. Revue ges. Hydrobiol. 66(3): 299–326.

Dabrowski, K.R. (1982a) Proteolytic enzyme activity decline in starving fish alevins and larvae. Env. Biol. Fish. 7(1): 73–76.

Dabrowski, K.R. (1982b) Seasonal changes in the chemical composition of fish body and nutritional value of the muscle of the pollan (*Coregonus pollan* Thompson) from Lough Neagh, Northern Ireland. Hydrobiologia 87: 121–141.

Dabrowski, K.R. (1982c) The influence of light intensity on feeding of fish larvae and fry I. *Coregonus pollan* (Thompson) and *Esox lucious* (L.). Zool. Jb. Physiol. 86: 341–351.

Dabrowksi, K.R. (1983) A note on the energy transformation in body and gonad of coregonid fish. Arch. Hydrobiol. 97: 406–414.

Dabrowski, K., Murawska, E., Terlecki, J. and Wielgosz, S. (1984) Studies on the feeding of *Coregonus pollan* Thompson alevins and fry in Lough Neagh. Int. Rev. ges. Hydrobiol. 69.

Ferguson, A. (1974) The genetic relationships of the coregonid fishes of Britain and Ireland indicated by electrophonetic analysis of tissue proteins. J. Fish. Biol. 6: 311–315.

Ferguson, A., Himberg, K-JM. and Svardson, G. (1978) Systematics of the Irish pollan (*Coregonus pollan* Thompson): an electrophoretic comparison with other Holarctic Coregoninae. J. Fish. Biol. 12: 221–233.

Healey, M.C. and Nichol C.W. (1975) Fecundity comparisons for various stocks of lake whitefish *Coregonus clupeaformis*. J. Fish. Res. B. Can. 32: 404–407.

McPhail, J.D. (1966) The *Coregonus autumnalis* complex in Alaska and northwestern Canada. J. Fish. Res. B. Can. 23: 141–148.

Maitland, P.S. (1972) A key to the freshwater fishes of the British Isles, with notes on their distribution and ecology. Sci. Publ. Freshwat. Biol. Ass. 27: 1–139.

Regan, C.I. (1911) The Freshwater Fishes of the British Isles. London: Methuen.

Schindler, O. (1957) Freshwater Fishes. London, Thames and Hudson.

Svardson, G. (1957) The coregonid problem. VI. The palaearctic species and their intergrades. Rep. Inst. Freshwat. Res. Drottningholm 38: 267–365.

Thompson, W. (1856) The natural history of Ireland IV. Mammalia, reptiles and fishes. H.G. Bohn, London.

Wheeler, A. (1969) The Fishes of the British Isles and North-West Europe. London: Macmillan.

Wilson, J.P.F. (1979) The biology and population ecology of the pollan, *Coregonus autumnalis pollan* Thompson, of Lough Neagh. Unpubl. D. Phil. Thesis, NUU.

Wllson, J.P.F. (1983) Gear selectivity, mortality rate and fluctuation in abundance of the pollan *Coregonus autmnalis pollan* Thompson, of Lough Neagh, N. Ireland. Proc. R. Ir. Acad. Ser. B, 83: 301–307.

Wilson, J.P.F. (1984) The food of the pollan, *Coregonus autumnalis pollan* Thompson, in Lough Neagh, Northern Ireland. J. Fish Biol. 24: 253–261.

Wilson, J.P.F. and Pitcher, T.J. (1983) The seasonal cycle of condition in the pollan, *Coregonus autumnalis pollan* Thompson, of Lough Neagh, Northern Ireland. J. Fish Biol. 23: 365–370.

Wilson, J.P.F. and Pitcher, T.J. (1984a) Age determination and growth of the pollan *Coregonus autumnalis pollan* of Lough Neagh, N. Ireland. J. Fish Biol. 24: 151–163.

Wilson J.P.F. and Pitcher T.J. (1984b) Fecundity of the pollan, *Coregonus autumnalis pollan* Thompson in Lough Neagh, Ireland. J. Life Sci. R. Dubl. Soc. 4, 5: 21–28.

Wilson, J.P.F. and Pitcher T.J. (1985) Predictive modelling of a stock of pollan *Coregonus autmnalis pollan* Thompson, by a Monte Carlo method. Aquaculture and Fisheries Management 1: 103–109.

18. The fish of Lough Neagh
Part E. Ecological studies of the fish community

I.J. WINFIELD, C.M. TOBIN and C.R. MONTGOMERY

18E.1. Introduction

As evidenced by the other chapters of this monograph, Lough Neagh is without parallel in the landscape of the British Isles. With a surface area of approximately 383 km^2 it is by far the largest expanse of freshwater, and its associated principal rivers amount to around 510 km in length. This immense size coupled with a high level of primary productivity (Chapter 11) enables the lough to produce sufficient fish flesh to support several important commercial fisheries, a position which is unique in these Isles. Both detailed considerations and more wide-ranging reviews of these fisheries and associated angling have already been given in this volume (Parts A, B, C, D of this chapter) and elsewhere by Cragg-Hine (1986) and Wood and Smith (1986). It is our intention in the present chapter to review some of the ecological features of the fishes of Lough Neagh fundamental to their commercial and sporting exploitation, activities which are themselves currently worth something in the order of £4,000,000 p.a. to the Province (Wood and Smith, 1986) and, perhaps more importantly, support a fishing community of long standing (Donnelly, 1986).

All parts of the Lough Neagh system are intricately linked with those deep areas of the lough itself where much of the actual fishing takes place. To give but a few of the well-established examples, the eel (*Anguilla anguilla*) uses the Lower Bann on its migrations to and from the lough (Part B), Atlantic salmon (*Salmo salar*) and brown trout (*Salmo trutta*) rely on tributary headwaters for spawning grounds (Parts B, C), and even within the lough the inshore-spawning of the pollan (*Coregonus autumnalis pollan*) (Part D) testifies to the extensive use of the system by its fish inhabitants. Since the early part of 1985 work has been underway at the Freshwater Laboratory investigating lesser-known but equally important aspects of the ecology of the fish populations within Lough Neagh itself. While restricting

This contribution is based on a paper given to a Study Course of the Institute of Fisheries Management (Northern Ireland Branch) in March, 1987.

R. B. Wood and R. V. Smith (eds.), Lough Neagh, 451–471.

our research in spatial terms we have simultaneously attempted to widen our studies to view these animals as part of the Lough Neagh ecosystem, acknowledging their innumerable links with features such as abiotic characteristics, and prey, predator and competitor populations.

It has been argued that such an integrated approach is a necessary next step in the development of our understanding and exploitation of freshwater fish populations (for a brief but excellent exposition of this stance see Andersson, 1984). This philosophy is particularly important with a water body such as Lough Neagh which has seen recent profound changes in its fundamental abiotic and biotic characteristics from mesotrophic beginnings to advanced eutrophy and initial recovery (Wood and Gibson, 1973; Gibson, 1986; Wood and Smith, 1986; see also Chapter 7A, B). As an exaggerated example justifying this approach, it would be of little use calculating maximum sustainable yields for the pollan fishery assuming that natural mortality is constant, as fishery scientists often do, when in reality it may be increasing or decreasing dramatically due to changing oxygen levels associated with changes in the eutrophic state of the lough. Even when eutrophication has not developed sufficiently far to cause lethal oxygen shortages, it can still influence fish communities by shifting their internal competitive balances. Such an effect is at least partly responsible for the often observed succession from salmonid, through percid, to cyprinid dominance as a water body becomes eutrophicated, as happened for example in another large European lake, Lake Constance (Hartmann, 1977). Community studies thus require research effort to be expended on species which, while of no commercial value themselves, are still of commercial relevance due to their adverse effects on fishery target species. A classic example of such a pest species is the commercially-worthless but often competitively-superior roach (*Rutilus rutilus*) which elsewhere has been shown numerous times (see Johansson and Persson, 1986) to depress sympatric perch (*Perca fluviatilis*) populations, a species which in Lough Neagh supports a potentially lucrative fishery (Montgomery, 1988).

In fact, the above hypothetical example involving pollan, a pollution-susceptible salmonid, might not be too far from the truth. In the Lough Neagh of the late 1960's, extreme eutrophication resulted in fish kills which understandably gave rise to fears about the long-term future of fisheries on the lough (see Kennedy, 1978). Phosphorus was subsequently identified as the major factor behind this pollution, and in the early 1980's remedial steps were initiated (for a precis of this detective story see Wood and Smith, 1986). While this has successfully led to a downturn in the maximum algal crops (Gibson, 1986), and thus the probability of further oxygen-related fish kills, the problem still remains that Lough Neagh now provides environmental conditions more suited to cyprinid rather than to salmonid populations. However, Wood and Smith (1986) were able to conclude that the quality of the water in Lough Neagh is fundamentally capable of supporting significant commercial fisheries. The story, of course, does not end here. In order to exploit the fish of Lough Neagh optimally, or even merely competently, it is necessary to manage them rationally. In a water body as unique as Lough Neagh, such management requires a detailed appreciation of the local ecology of the fish populations.

18E.2. Approach and aims

The integrated approach of the present research has been discussed at length above, as has its spatial restriction to the lough proper. However, the study of such functional relationships first requires a basic familiarity with the various factors involved. While Lough Neagh now has a considerable history of research into its abiotic features, primary producers, zooplankton and macroinvertebrates (see Wood and Smith, 1986), before 1985 there had been no studies of its fish *community*, although specific works on the eel, pollan and trout had been carried out (for examples see Parsons *et al.* (1977), Dabrowski (1981), Part D, and Crozier (1984) respectively). However, given the subsequent invasion of roach, discussed in more detail below, even the findings of these autecological studies may not be applicable to the present day populations.

Consequently, much of the research programme to-date has been aimed at establishing a comprehensive description of the Lough Neagh fish populations. In addition to the basic tasks of determining the fish species present and their relative abundance, patterns of within-lough distribution have also been investigated. Only when these aims have been fulfilled can the study progress to investigate more functional relationships such as interactions within the fish community, its effects on the rest of the ecosystem, and ultimately realistic appraisals of the fisheries themselves.

18E.3. An initial description: the Lough Neagh fish community of May 1985

One of the earliest steps in the present research programme was an initial extensive sampling of Lough Neagh in May, 1985. Five trawls, each of five minutes duration, were carried out at each of six sites around the lough (in the general areas of Ballyronan Bay, Toome Bay, Antrim Bay, Rams Island, off Washing Bay, and Ardboe (Irish Grid Refs. are given in Table 18E.1)) using a bottom trawl with a 6 mm bar mesh cod-end. One factor which must be borne in mind in the interpretation of all subsequent trawl data is that of the lower limit of the size-selectivity of the trawl mesh. While the lower limit of efficient scale-fish retention is only 40–50 mm in length, a size which they attain in Lough Neagh by their second year of life, for the relatively much thinner eel this critical length is of the order of 300 mm. Thus the true abundance of the eel population, which contains many individuals several

Table 18E.1. Site details and catch compositions (by numbers) of the initial extensive sampling of Lough Neagh in May, 1985

Site	Irish grid ref (approximate)	Depth (m)	Species composition (%)					Sample size
			Perch	Roach	Eel	Pollan	Other	
Ballyronan Bay	H 955 865	3	2.89	95.22	0.00	0.14	1.73	691
Toome Bay	H 975 865	10	99.10	0.00	0.85	0.00	0.05	2112
Antrim Bay	J 100 845	11	91.16	0.70	5.81	2.09	0.23	430
Rams Island	J 065 714	11	94.51	0.00	5.04	0.44	0.00	674
Washing Bay	H 935 665	5	3.42	88.78	0.83	0.12	6.85	847
Ardboe	H 995 755	10	80.39	0.00	16.85	2.76	0.00	362

years old but still below 300 mm in length, will be consistently and significantly underestimated.

The following eight species were taken in this initial sampling: eel (*Anguilla anguilla*), roach (*Rutilus rutilus*), bream (*Abramis brama*), gudgeon (*Gobio gobio*), brown trout (*Salmo trutta*), pollan (*Coregonus autumnalis pollan*), three-spined stickleback (*Gasterosteus aculeatus*), and perch (*Perca fluviatilis*). Subsequent research has revealed the additional six species: river lamprey (*Lampetra fluviatilis*), Atlantic salmon (*Salmo salar*), pike (*Esox lucius*), stoneloach (*Noemacheilus barbatulus*), rudd (*Scardinius erythrophthalmus*), and tench (*Tinca tinca*).

Quantitative results of this initial sampling in terms of the relative abundance of species are given in Table 18E.1, together with approximate site locations and depths. While a total of eight species of fish were taken, in numerical terms the community was dominated by perch and roach which together always constituted at least 80% of the catch (although it must be remembered that the true abundance of eels will have been underestimated due to the reasons mentioned above). Furthermore, there is a clear relationship between site depth and catch composition with perch and roach largely restricted to the deep and shallow sites respectively. Thus at first sight there is little overlap in spatial distribution between the perch and newly-arrived roach populations, although the subsequent section will show this conclusion to be premature.

18E.4. Distribution patterns: heterogeneity in the horizontal and vertical planes

While the above initial sampling revealed the existence of fundamental interspecific differences in distributions around the lough, further work has defined in greater detail these gradients in horizontal and vertical distribution, and also revealed additional temporal and intraspecific components.

Much of this subsequent fieldwork has concentrated on two convenient sites in the north-west part of the lough, i.e. Ballyronan Bay, taken as a

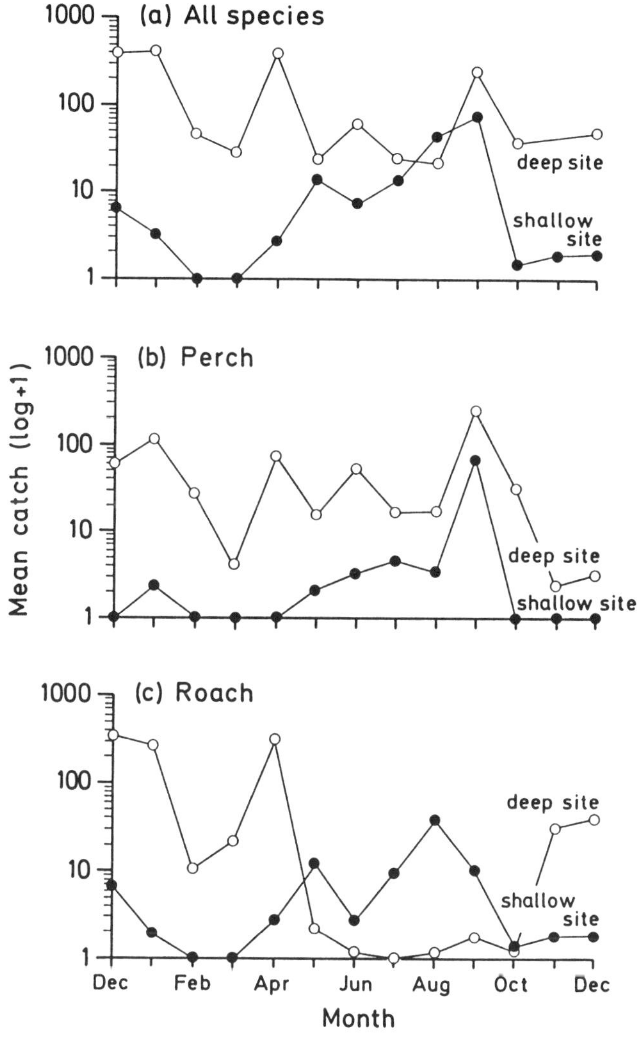

Figure 18E.1. Seasonal changes in the abundance of (a) all species, (b) perch, and (c) roach, at a shallow (Ballyronan Bay) and a deep (Toome Bay) site during 1986.

typical inshore, shallow area, and Toome Bay, representative of the offshore, deep areas. The abundance of fish at these two sites over the whole of 1986 is shown in Fig. 18E.1. When all species are viewed together (Fig. 18E.1a) there is a clear seasonal pattern of distribution with fish being consistently present at the deep site, but being greatly reduced in numbers (and at times completely absent) from the shallow site during the late autumn, winter, and early spring months.

On a species basis, perch follow this general pattern of continuous presence in the deep water with an essentially summertime visit of a small minority of the population to the inshore region (Fig. 18E.1b, note that the apparent great increase in abundance at the shallow site in September is largely due to young-of-the-year (0+) becoming large enough to be retained by the trawl mesh). The pattern of distribution of the roach, the lough's second most abundant species (Fig. 18E.1c), shows a summertime visit to the shallow site and a corresponding vacation of the deep water at this time of year. Thus while for the perch only a small component of the population moves into the shallow area during the summer, in the roach this migration is carried out by almost all of the population. This year-long investigation has shown that perch and roach are indeed spatially separated during the summer, as was evidenced in the initial May sampling, but that this aspect of resource partitioning breaks down during the colder parts of the year.

In addition to the wintertime overlapping of the perch and roach populations described above, limited investigations have shown that there is also a component of overlap during the summer months among the newly-hatched larvae of these two species. These investigations have been restricted because larval sampling is notoriously difficult and requires specialised, expensive and labour-intensive equipment (Bagenal, 1978), and so remains a problem both here in Lough Neagh and in lakes and rivers elsewhere. The fact that common diets during this stage of the life cycle may lead to significant competitive interactions further emphasises the importance of this gap in our knowledge. Furthermore, while sampling scale-fish larvae presents considerable problems to say the least, catching elvers and slightly older eels in anything other than small streams is even more difficult.

In addition to the above complexities in horizontal distribution patterns, the three-dimensional nature of the aquatic habitat also allows extensive movements in the vertical plane. Studies of such complex spatial distributions are fundamentally important to fish ecology as resource partitioning on a spatial or habitat basis is widespread in freshwater fish communities (Werner, 1984). The vertical distributions of perch and roach at the shallow and deep sites are shown in a grossly simplified form in Fig. 18E.2. These data were obtained by the use of fine-mesh gill nets stretching from the water surface to the lough bottom in May, 1987. Again, further sampling has documented the existence of seasonal, diel and intraspecific components in this aspect of the fish populations' distribution ecology.

Roach are essentially confined to the shallow site at this time of year where, over the 24 hour period, they inhabit the entire water column. The same vertical pattern is true for the perch of the inshore area, but the much larger component of their population found in the deeper water is consistently restricted to the deepest, darkest layers of the water column. Pollan were present at both sites but showed a consistent restriction to the upper 3 m of the water column which corresponds with the approximate penetration of significant levels of light into Lough Neagh (Jewson, 1977; Dabrowski and Jewson, 1984).

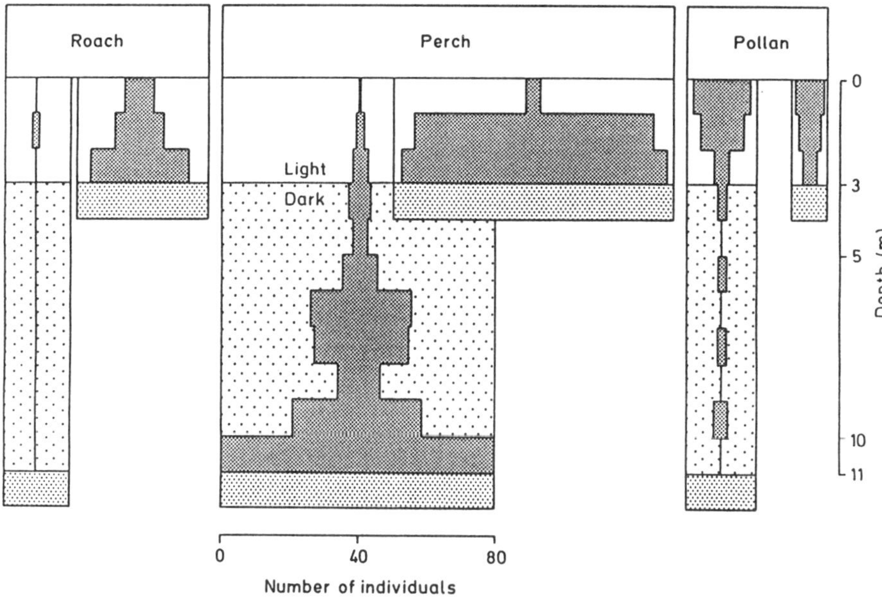

Figure 18E.2. The depth distributions of roach, perch and pollan at a shallow (Ballyronan Bay) and a deep (Toome Bay) site in May, 1987. The lough bottom is represented by a darkly-stippled bar while that part of the water column in perpetual darkness is indicated by light stippling.

While a knowledge of these precise distributions is obviously important for an accurate interpretation of feeding ecology, and hence the roles of these fish in the Lough Neagh ecosystem, it is also fundamentally important for the interpretation of trawl samples. The trawl employed in this study exclusively samples the bottom 3–4 m of the water column and thus any fish above this level will not be caught. While the daytime perch of the deep water, and roach when present, are distributed towards the bottom of the lough and hence will be sampled efficiently by the trawl, this is clearly not always the case for the pollan. This limitation will be acknowledged in the discussion of population trends in the next section. At the shallow site the entire water column is sampled by the trawl and so no equivalent problems arise.

18E.5. Population trends of the major species

The preceeding sections have described the fish populations of Lough Neagh as they were in early 1985, and reviewed their patterns of intra-lough distribution which were elucidated during the subsequent two years. However, during this time some of the populations have not remained static but have

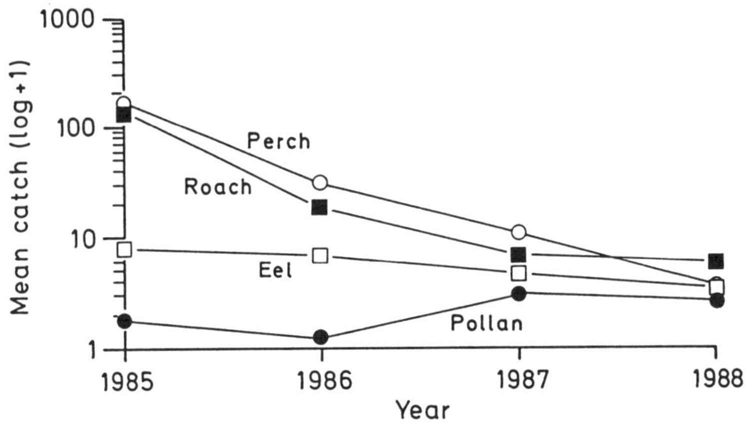

Figure 18E.3. Trends in the abundance of perch, roach, eel and pollan over the period 1985–1987. Data are from extensive sampling around the lough in the early summer of each year.

changed significantly in a number of ways. The population trends have been monitored by extensive sampling around the lough identical in methodology to the initial extensive sampling, during the early summer and early winter of each year. The following accounts of the population trends of the major species are based on data collected during the summer surveys. As sampling effort and character have been held constant, the catches can be taken to reflect changes in the abundances of the various species and thus avoid the need to make difficult extrapolations to absolute abundancies (Fig. 18E.3).

The pollan is probably the lough's most eutrophication-susceptible fish species and so it is encouraging to see that its recent trend in abundance is slightly upwards (note that the fewer numbers of pollan caught compared to the other scale-fish largely reflect its vertical distribution pattern). Oscillations in abundance are the norm rather than the exception in the coregonids (see Hamrin, 1986), and the fishery on Lough Neagh has historically shown a widely fluctuating yield (Part D of this chapter). Such patterns are largely attributable to intraspecific competition for planktonic food, rather than to abiotic features of the environment or overfishing. Indeed, the basic ecology of the coregonids is such that they make ideal species for exploitation in fisheries as they show excellent growth rates and their high post-spawning natural mortality renders them less susceptible than other fishes to overfishing (Hamrin, 1986). However, this very dynamic nature of coregonid populations also means that astute, interactive management is necessary for optimal exploitation. This point will be returned to later.

Although environmental conditions are certainly getting no worse and probably improving, as evidenced by both direct water quality sampling (Gibson, 1986) and the current performance of pollan, the perch population has declined markedly over the three years of this study. As the potentially-competing roach is also decreasing, for reasons of its own discussed below, it is unlikely that this decrease is due solely to interactions within the fish community. Current perch fishing practices on Lough Neagh, including the taking of very small, immature fish (Anon., 1986), are a more likely reason for the observed dramatic decline in stocks. This issue was considered in detail by Montgomery (1988) who pointed out that the small individual size of the present catch (*c*. 120 mm fork length) results in many fish being removed before they are able to reproduce and thus contribute to future stocks. Until such management problems are rectified the Lough Neagh perch fishery will never realise its considerable potential.

While the above decline is lamented by all who have any connections with the fish of the Lough Neagh system, an even more striking decline observed over the same time period in the roach population (Fig. 18E.3) is more controversial. While Lough Neagh fishermen would not miss this fish at all, and would be glad to avoid the unpleasant task of removing several thousand worthless roach from several hundred yards of gillnet intended for a more valuable catch, local and visiting coarse anglers are not so happy with their declining sport. Although the roach has not become extinct in Lough Neagh, and gillnets set in the lough may still eventually become choked with them, it is clear that their numbers have declined significantly from previous high levels.

The absence of any quantitative sampling of the roach soon after its arrival in the Lough Neagh system in 1970 (Cragg-Hine, 1973) means that its local history will always remain something of a mystery, but records kept by roach-seeking anglers do provide additional evidence of changes in the roach populations of the system's rivers (Victor Refausse, Ulster Coarse Fishing Federation, pers. comm.). Although in general there are many problems in the interpretation of angling returns, and in this specific case there was in addition little organised angling during the early period of the roach invasion, it is clear that catches and by implication roach stocks have declined markedly over the last few years. As in the lough itself, this decline is not shown by all members of the fish community, again indicating that basic environmental factors are not involved. It must be acknowledged that these angling activities, while extensive, are nevertheless restricted to the rivers of the Lough Neagh and other systems and have never been practised on the lough itself. However, extensive migrations of roach between running and standing waters have been documented elswhere (Jordan and Wortley, 1985) and it is unlikely that the stocks of roach in Lough Neagh and its tributaries are discrete.

The reason behind this dramatic decline in the roach stocks of Lough Neagh, and possibly those of its rivers and elsewhere, provides a striking example of the need to look at fish as part of the freshwater ecosystem. Cyprinids such as the roach arc amongst thc most cnvironmentally-tolerant of all freshwater fishes, and hence the decline cannot be attributable to poor water quality. Furthermore, while large amounts of roach are removed from the lough as a by-catch of the perch and pollan fisheries, these amounts cannot by themselves account for the magnitude and character of the observed reduction. To understand the decline it is necessary to consider some aspects of the biology of the roach in relation to the peculiar environmental conditions of Lough Neagh.

Laboratory examinations of samples of Lough Neagh roach taken in 1984 and subsequent years have revealed that the local population is extensively infested by the tapeworm *Ligula intestinalis* (Tobin, 1986), a widespread parasite of members of the Cyprinidae. For reasons attributable to the other vectors in this parasite's life cycle, the tapeworm is largely restricted to the younger roach where it infests up to 50% of individuals and can account for up to 33% of the combined weight of parasite and host.

Studies elsewhere of this parasite, which constitutes an important pest in some Eastern European freshwater fisheries, have shown that the tapeworm enters its fish host through a specific group of its zooplanktonic prey, the *Cyclops* genus of copepods (Dubinina, 1966). The parasite's life cycle is completed when the infested fish is consumed by a bird, typically a gull (*Larus* spp.) or a grebe (*Podiceps* spp.), in which the tapeworm rapidly matures and produces eggs which are then passed back into the environment where they may be taken up by cyclopoid copepods.

Clearly, the abundance of this parasite in a freshwater fish community depends not only on the abundance of its cyprinid hosts, but also of *Cyclops* and piscivorous birds, and the adequate provision of shallow areas where the predator-prey interactions can take place. In this context, Lough Neagh is an ideal habitat for *Ligula* because, in addition to containing several species of cyprinids, its zooplankton is dominated by *Cyclops abyssorum* (Chapter 14), and piscivorous birds including great crested grebes (*Podiceps cristatus*), cormorants (*Phalacrocorax carbo*), grey herons (*Ardea cinerea*), lesser black-backed gulls (*Larus fuscus*), and black-headed gulls (*Larus ridibundus*) are common (I.J.W., pers. obs.). Furthermore, during the summer months the roach population inhabits the lough's extensive shallow areas.

The effects of *Ligula* infestations on roach populations elsewhere have been well documented. The presence of the tapeworm in the fish's body cavity can influence its bouyancy making it more susceptible to predation by birds (Sweeting, 1976), and thus causing a decrease in population numbers directly. In this context, Bean and Winfield (1989) demonstrated a clear effect of *Ligula* on the horizontal and vertical distribution of another cyprinid in Lough Neagh, the gudgeon *Gobio gobio*. However, there is also a second, more indirect way in which this parasite exerts a depressing effect on roach and other cyprinids. The parasitic drain on the host's resources leads to a general decline in body condition, and more particularly to a dramatic reduction in the fecundity of females (Arme and Owen, 1968; Arme and Owen, 1970; Wilson, 1971) which eventually leads to a population decline through repeatedly poor recruitment (Wilson, 1971; Sweeting, 1976; Burrough *et al.*, 1979; Kennedy and Burrough, 1981).

The existence of such a *Ligula*-induced effect can be detected by examin-

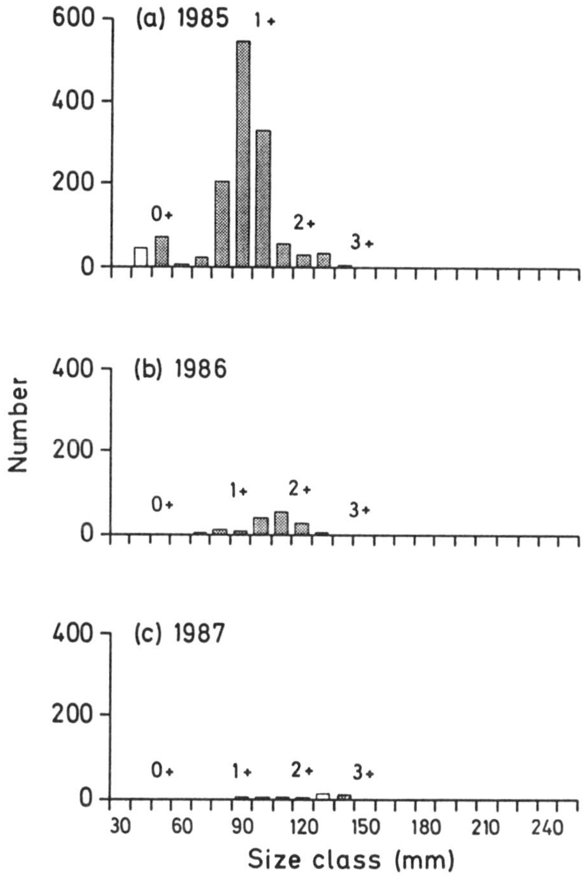

Figure 18E.4. The population structure (expressed as length frequency distributions) of roach from the shallow sites of the early summer extensive sampling of (a) 1985, (b) 1986, and (c) 1987. Approximate ages of the size groups are indicated.

ation of the host's population structure in successive years where continued poor recruitment will lead to an under-representation of young, small fish. This is precisely what has happened in the roach of Lough Neagh over the period 1985–1987 (Fig. 18E.4). While catches in 1985 were dominated by 1+ fish (0+ are too small to be sampled reliably by the trawl mesh), such fish are now rare and the population is dominated by older, larger individuals. Such declines in roach populations observed elsewhere have been followed by oscillations in the abundance of both host and parasite continuing over numerous years (Kennedy and Burrough, 1981). It is therefore unlikely that *Ligula* will cause the roach to become extinct in Lough Neagh, but any more precise predictions of future trends are complicated by the presence in the lough of at least one other infested cyprinid, the gudgeon (Bean and Winfield, 1989), which may act as a reserve pool from which *Ligula* can continually re-infest any recovering roach population. Similar examinations of roach from the declining populations outside of Lough Neagh have never been made, although reports of a marked reduction in the abundance of smaller fish (Victor Refausse, Ulster Coarse Fishing Federation, pers. comm.) suggest that similar infestations may be prevalent elsewhere in the Province.

Despite the considerable problems involved in their quantitative sampling, any survey of fish population trends in Lough Neagh would be incomplete without a consideration of the eel. Trends in the abundance of the larger size classes of this species (recall the limitations of trawling) are shown in Fig. 18E.3. In contrast to the marked changes in the scale-fish populations, the eel has shown only a slight downward trend, the significance of which is difficult to assess because, for reasons given earlier, these are only the largest individuals of the population. Trends in the much more abundant smaller individuals are unknown. Although our sampling of the eel has been admittedly inadequate for any purposes other than the most superficial monitoring, its commercial importance on Lough Neagh has generated an additional source of information in the form of the records of the Lough Neagh Fishermen's Co-operative.

Figure 18E.5 illustrates long-term trends in the incoming elver run, and subsequent brown and silver eel catches made by the fishery. These graphs are essentially updates of earlier figures published by Parsons *et al.* (1977), Wood (1984) and Wood and Smith (1986), but their importance is such that they deserve continued dissemination. The post-1980 steep decline in the elver run observed in Lough Neagh has also been recorded in other parts of Europe (Moriarty, 1986), although a similar reduction noticed elsewhere in 1945–1950 clearly did not occur locally. Stock-recruitment relationships in the eel have been understandably little researched given the practical problems involved with a spawning site several thousand miles away. However, the relationship between the numbers of elvers entering a local freshwater body and the subsequent number of adults they produce is more amenable to investigation and must be marked as a high priority of future eel research. In view of our ignorance in such matters, it is not even certain that the declining numbers of elvers reaching Lough Neagh should be a cause for concern. While this state of ignorance is annoying in scientific terms, it is alarming when the future of a multi-million pound fishery may be at stake.

In seeming contradiction to the decline in the elver run, catches of brown eels have shown an increase over the last 50 years (Fig. 18E.5). However, it is certain that fishing effort has not been constant during this time, but has increased significantly with changing fishing practices. In the absence of detailed information on fishing effort expended, which is probably unobtainable in a fishery the size and nature of Lough Neagh's, such information can tell us nothing of the long-term population trend of eels in the lough. For this we must consider the silver eel catch.

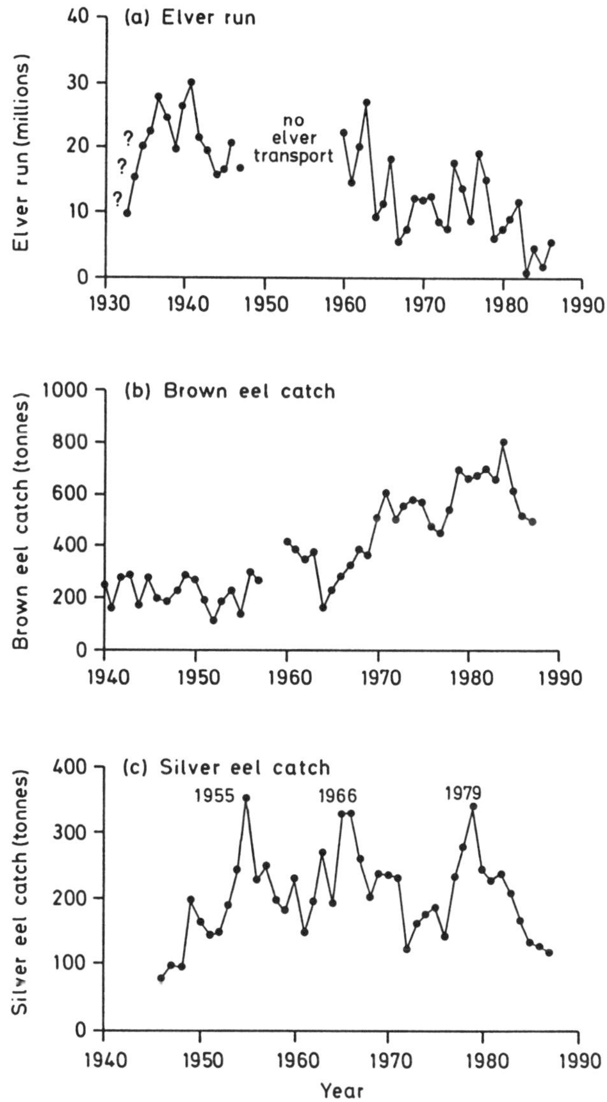

Figure 18E.5. Long-term trends in the sizes of (a) the elver run, (b) the brown eel catch, and (c) the silver eel catch. Data are taken from Parsons *et al.* (1977), with more recent values provided by the Lough Neagh Fishermen's Co-operative. Note that the earliest three points on the elver run are probably significant underestimates of the true values.

While brown eels are fished for during the summer months by a large number of independent fishing boats, the silver eels are caught exclusively by fixed traps during their autumn seaward migration. Consequently, in the absence of fundamental changes to the operation of the traps, fishing effort is less variable and thus catches can be taken to reflect the abundance of mature eels in Lough Neagh. When viewed over a sufficiently long timescale, the catches of silver eels show an apparently cyclical pattern with a period of 11–13 years and peak catches in 1955, 1966 and 1979 (Fig. 18E.5c). While with only three observed peaks it is difficult to be sure mathematically that this is a cycle, and being unable to offer any biological reason for its existence, it is still worth reporting because in this context the post-1980 decline in silver eel catches does not look so worrying. Perhaps the fastest way to test the validity of this model is to wait until 1991 when catches should be at the next peak.

Having been able to offer some optimism for the future of the silver eel fishery on Lough Neagh, it is unfortunate that another potentially serious problem has appeared on the horizon. A swim-bladder nematode of eels, *Anguillicola crassa*, has recently been introduced to Europe from its native Japan. In the European eel this parasite is very pathogenic and evidence is accumulating that it may prevent migration to the Sargasso Sea (Deborah Fitch, University of Exeter, pers. comm.). In association with the Ministry of Agriculture, Fisheries and Food, researchers at the University of Exeter are documenting the occurence and spread of this parasite. Fortunately, samples taken during the summer of 1988 revealed that so far this parasite has not been introduced to Lough Neagh.

18E.6. Post-description research: interactions within and without the fish populations

Having established an extensive description of the fish populations of Lough Neagh and a basic monitoring routine, research is now being directed more towards interactions within the fish community and its position in the Lough Neagh ecosystem generally. Such work requires extremely intricate studies of the fish covering, amongst many other topics, the broad fields of feeding ecology, and growth and mortality rates. Much of this work is still in progress and all that can be given in the present chapter is a very brief overview of the kinds of issues being addressed.

Interspecific competition is acknowledged as an extremely important factor in the structuring of lacustrine fish communities (Werner, 1984). The arrival of the often competitively-dominant roach in Lough Neagh is therefore the subject of much of our work. Johannson and Persson (1986) review much of the relevant literature and show that in water bodies elsewhere the roach routinely and comprehensively outcompetes perch populations during the early zooplanktivorous stages of their life cycles. It is not yet clear to what extent such interactions occur locally given such factors as the recent *Ligula*-induced decline in roach, and peculiar characteristics of the physical conditions and zooplankton populations of Lough Neagh. We remain more ignorant of the possible competitive interactions between roach and the under-sampled young eels. While Tesch (1977) reports that stream-dwelling eels may compete with salmonids, their position in lake communities generally is unknown.

The interactions of fish populations amongst themselves have long been investigated by fisheries scientists and animal ecologists, but their effects on other lake components have only recently begun to be studied. Nevertheless, it is now clear that fish commonly depress prey populations and a relevant example involving percids and cyprinids in a large eutrophic lake, Lake Trummen in Sweden, is given by Anderson *et al.* (1978). Both zooplankton and benthic macroinvertebrates were reduced by the experimental manipulation of fish relative to a fish-free control, and furthermore these effects themselves influenced several aspects of water quality including algal abundance and nutrient availability. Similar work in another large eutrophic lake, Lake Balaton in Hungary (Tatrai and Istvanovics, 1986), has studied further the influence of fish on the water chemistry of lakes and has showed that in Lake Balaton the nitrogen excretion of bream is equal to 15% of the lake's annual external load. Similar effects may occur in Lough Neagh, giving another dimension to the recent invasion of roach. Not surprisingly, work is now in progress around the world investigating the possibility of combating eutrophication by the selective removal of fish populations, and Van Densen *et al.* (1986) review the potential of this new science of biomanipulation in European waters.

If food supplies are limiting, fish may compete and hence interact with other kinds of animals outside of their own community if they share a common diet. On Lough Neagh, one such group is the waterfowl which overwinters in large numbers making the lough internationally important for these birds (Chapter 20). Some work, as yet unpublished, investigating the diet of these ducks has been carried out on Lough Neagh which, together with studies undertaken elsewhere (collated in Cramp and Simmons, 1977) has shown that species such as the tufted duck (*Aythya fuligula*) feed extensively on benthic macroinvertebrates, as do adult roach. Figure 18E.6a shows the trend in overwintering numbers of tufted duck on Lough Neagh over the last 20 years. It can be seen that a marked decline in numbers since 1980, which contrasts with a general upward trend elsewhere in the British Isles, co-incides with the local establishment of the roach and is possibly due to food competition. We await with interest the future trend in these overwintering numbers as the recent poor recruitment of roach works its way through to a reduction in the numbers of benthivorous adults.

Fish do not only interact with birds as potential competitors, they also form the prey of a number of piscivorous species. One such bird is the great crested grebe (*Podiceps cristatus*), and Fig. 18E.6b illustrates its local

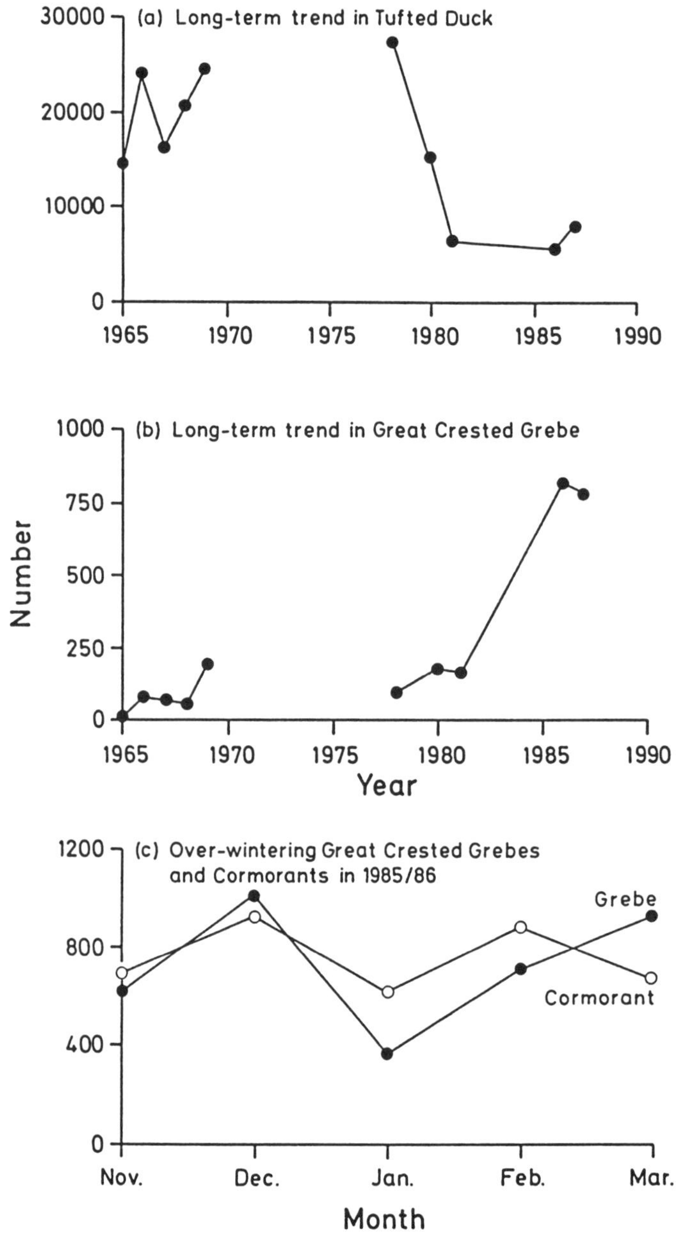

Figure 18E.6. Long-term trends in the overwintering populations of (a) tufted duck, and (b) great crested grebe (each point is the mean of that winter's three highest monthly counts, data provided by the Northern Ireland Ornithologists' Club and the Department of the Environment (Northem Ireland)), and (c) the numbers of great crested grebes and cormorants overwintering in 1985/86 (data provided by D.O.E. (N.I.) and Julian Greenwood (Stranmillis College, Belfast) respectively).

overwintering history. Again there is a significant change in abundance after 1980 when, in contrast to tufted duck, its numbers increased. With a diet consistingly almost exclusively of small fish, it is more than likely that this increase is due to an increase in prey availability resulting from the arrival of the roach.

One final example from the bird world demonstrates the complexity of community and ecosystem ecology. As Fig. 18E.6c shows, the great crested grebe is not the only piscivorous bird that overwinters in large numbers on Lough Neagh. During the winter of 1985/86, in addition to several hundred grebes, around 600–900 cormorants (*Phalacrocorax carbo*) were consistently present on the lough. Assuming daily consumption rates of 200 g and 560 g respectively (approximate mid-points of the ranges quoted by Cramp and Simmons, 1977), during the period between November and March these birds can be estimated to consume around 92 tonnes of fish. To put this figure into context, the fraction taken by the cormorants (70 tonnes) in just this restricted period is equivalent to over half of the previous autumn's silver eel catch (135 tonnes). Of course, not all of these consumed fish will be commercially valuable species, indeed the great majority are probably roach, but these figures do illustrate the importance of piscivorous birds to the fish community of the lough. Recall also the role of such birds in the *Ligula* story.

There is one further twist to the bird story. Although there are unfortunately no long-term counts of cormorants overwintering on Lough Neagh over the last 20 years, casual observations do suggest that these numbers have increased significantly. Furthermore, it has been suggested (Kennedy and Greer, 1988) that one reason for this increase is that the lough's newly-established roach population has provided a new source of food for cormorants overwintering in the Province. One apparent ramification of this proposed change in cormorant dispersal patterns is manifested outside of the Lough Neagh system. Having overwintered on a diet of roach in Lough Neagh, it has been suggested that some of these cormorants subsequently move on to the nearby River Bush in the early spring to exploit the seaward migrations of young salmonids, where they consume a significant proportion of the runs (Kennedy and Greer, 1988). While it is understandable that coarse anglers busily aiding the spread of roach around Ireland could not have anticipated such complex and damaging effects of their activities, such indirect results would not surprise anyone familiar with the complexity of ecosystems.

18E.7. Concluding remarks

The fish community of Lough Neagh is not very diverse, largely as a result of the last glaciation, but it is operating in a large and unique environment. Complex patterns of distribution are a major feature of the ecology of fish in the lough making their sampling and study a difficult task. This already convoluted situation has been complicated further still in recent years by the anthropogenic introduction of the roach. The full implications of this negligent act remain to be resolved. As might be expected given the lough's history of man-induced changes, including widespread and extreme eutrophication, the fish community is not in equilibrium. Fundamental changes are occurring in several fish populations, and will probably always continue to do so.

We are presently investigating several aspects of the interactions between the fish populations of Lough Neagh, and interpreting them in the light of knowledge gained from other lakes. Many of these aspects are now well understood generally, and our problem is one of transferring the principles discovered to the unique environment of Lough Neagh. Wider, less understood, interactions involving the fish populations are also being investigated. Questions involving interactions with waterfowl and, more in the future, water chemistry are particularly exciting and promise to make significant contributions to limnology generally.

We have now dealt at some length with the fundamental ecology of the fish community of Lough Neagh and perhaps it is fitting to close this chapter on an applied note with a few remarks on how our research may contribute to the better management of the local fisheries. The state of flux referred to above necessitates that the fish populations continue to be routinely monitored in the future, and furthermore the complex spatial distributions discovered mean that this sampling will have to be extensive. Only in this way will the scale-fisheries of Lough Neagh ever be managed as efficiently as similar fisheries are quite routinely on the continent, fisheries with which they compete for a common market.

Such studies should cover basic features such as population abundance and individual sizes. The pollan in particular can show great changes in growth rates over a short period, and thus great resulting differences in adult lengths. Hamrin (1986) has shown that in Swedish coregonid fisheries such changes in adult size can mean that gill net meshes varying by just 2 mm can increase or decrease catches by as much as 50%. A strategic investment in scientific fisheries management may also transform the present day economics of formerly important fishery species such as the pike and bream.

The roach must also be included in such fisheries research even though it is commercially worthless on the lough. Its potential competitive effects are obvious and have been discussed in some detail, and its numbers seem sure to increase again over the next decade. In addition, the importance of this species to the tourist trade in the system's rivers is considerable. Apparently widespread declines in riverine stocks may be attributable to *Ligula* infestation, and equally possibly they may not. We simply do not know.

The management of an eel fishery is another kettle of fish. While in contrast to the scale-fisheries there are no worries over the management of local spawning stocks, this species is still part of the lough's fish community.

Again we cannot even guess at present at possible interactions between the eel and, amongst other species, the introduced roach.

The Northern Ireland Economic Council has identified the fisheries of Lough Neagh as one area of the Province's industries that is in acute need of development (Anon., 1987). While there are admittedly considerable problems to be overcome, we now have a sound descriptive base on which to build the applied, scientific sampling which is an essential prerequisite for any future economic development.

Acknowledgements

We are indebted to a variety of people and bodies for their considerable help in the research reported in this paper. Matt Quinn, Colin Bean and others too numerous to mention helped with many long hours of work on the lough and in the laboratory. Sampling operations on the lough were facilitated by the co-operation of the Department of Agriculture (Northern Ireland), the Fisheries Conservancy Board, the Honourable the Irish Society, and the Lough Neagh Fishermen's Co-operative. Denise Winfield collated and analysed many of the waterfowl data which were themselves kindly provided by the Northern Ireland Ornithologists' Club, the Department of the Environment (Northern Ireland), and by Julian Greenwood of Stranmillis College (Belfast) from the results of the 1985/86 B.T.O./S.O.C. Cormorant Survey. Father Oliver Kennedy of The Lough Neagh Fishermen's Co-operative and Victor Refausse of the Ulster Coarse Fishing Federation provided valuable data on eel and coarse fish catches respectively, while Tony Fitzsimons of the Freshwater Biological Investigations Unit (Department of Agriculture, Northern Ireland) gave free access to unpublished zooplankton data. Finally, Brian Wood and Dave Jewson made indirect but fundamental contributions to this work through their involvement in many hours of discussion.

18E.8. References

Andersson, G. (1984) The role of fish in lake ecosystems – and in Limnology. Nordisk Limnologsymposium, Oslo, pp, 189–197.

Andersson, G., Bergren, H., Cronberg, G. and Gelin, C. (1978) Effects of planktivorous and benthivorous fish on organisms and water chemistry in eutrophic lakes. Hydrobiologia 59: 9–15.

Anonymous (1986) Annual Report of the Fisheries Conservancy Board for Northern Ireland, 1986.

Anonymous (1987) Economic strategy: the fishing industry. Northern Ireland Economic Council, Report No. 66.

Arme, C. and Owen, R. (1968) Occurrence and pathology of *Ligula intestinalis* (L.) infections in British fishes. J. Parasitology 54: 272–280.

Arme, C. and Owen, R. (1970) Observations on the tissue response within the body cavity of fish infected with the plerocercoid larvae of *Ligula intestinalis* (L.) (Cestoda: Pseudophyllidea). J. Fish Biol. 2: 35–37.

Bagenal, T.B. (1978) Methods for the assessment of fish production in freshwaters. 3rd edition. Blackwell Scientific Publications, Oxford, 365 pp.

Bean, C.W. and Winfield, I.J. (1989) Biological and ecological effects of a *Ligula intestinalis* (L.) infestation of the gudgeon, *Gobio gobio* (L.), in Lough Neagh, Northern Ireland. J. Fish Biol. 34: 135–147.

Burrough, R.J., Bregazzi, P.R. and Kennedy, C.R. (1979) Interspecific dominance amongst three species of coarse fish in Slapton Ley, Devon. J. Fish Biol. 15: 535–544.

Cragg-Hine, D. (1973) Coarse fish and fishery management in Northern Ireland. Proc. 6th Brit. Coarse Fish. Conf., University of Liverpool, pp. 52–59.

Cragg-Hine, D. (1986) Coarse fish populations and their exploitation in Britain and Northern Ireland. Proceedings of the 17th Annual Study Course of the Institute of Fisheries Management, University of Ulster at Coleraine, pp. 173–182.

Cramp, S. and Simmons, K.E.L. (1977) Handbook of the Birds of Europe, the Middle East and North Africa: the Birds of the Western Palearctic, Volume I. Oxford University Press, Oxford, 720 pp.

Crozier, W.W. (1984) Population structure and biology of brown trout from the Lough Neagh system. Proc. Inst. Fish. Mgmt (N. Ireland Br.) Study Course, Univ. Ulster., pp. 53–60.

Dabrowski, K.R. (1981) The spawning and early life history of the pollan (*Coregonus pollan* Thompson) in Lough Neagh, Northern Ireland. Int. Revue ges. Hydrobiol. 66: 299–326.

Dabrowski, K.R. and Jewson, D.H. (1984) The influence of light environment on depth of visual feeding by larvae and fry of *Coregonus pollan* (Thompson) in Lough Neagh. J. Fish Biol. 25: 173–181.

Donnelly, D.J. (1986) On Lough Neagh's shores. The Donnelly Family, Galbally, Co. Tyrone, 333 pp.

Dubinina, M.N. (1980) Tapeworms (*Cestoda, Ligulidae*) of the fauna of the U.S.S.R. Amerind Publishing Co. Pvt. Ltd., New Delhi, 320 pp.

Gibson, C.E. (1986) Preliminary results on phosphorous reduction in Lough Neagh – assessing the effect against a background of change. Hydrobiol. Bull. 20: 173–182.

Hamrin, S.F. (1986) Ecology of vendace, *Coregonus albula*, with special reference to factors important to the commercial fishery. Arch. Hydrobiol 22: 51–72.

Hartmann, J. (1977) Fischerliche veranderungun in kulturbedingt eutrophierenden seen. Schweiz Z. Hydrol. 39: 243–254.

Jewson, D.H. (1977) Light penetration in relation to phytoplankton content of the euphotic zone of Lough Neagh, N. Ireland. Oikos 28: 74–83.

Johansson, L. and Persson, L. (1986) Planktivorous fish. In Riemann, B. and Sondergaard, M. (eds.), Carbon Dynamics in Eutrophic Temperate Lakes. Elsevier Science Publishers, Amsterdam, pp. 237–266.

Jordan, D.R. and Wortley, J.S. (1985) Sampling strategy related to fish distribution, with particular reference to the Norfolk Broads. J. Fish Biol. 27, Suppl. A: 163–174.

Kennedy, C.R. and Burrough, R.J. (1981) The establishment and subsequent history of a population of *Ligula intestinalis* in roach (*Rutilus rutilus* (L.)). J. Fish Biol. 19: 105–126.

Kennedy, G.J.A. (1978) Impact on fisheries. Proceedings of a Symposium on Lough Neagh – the Algal Problem and its Solution, Institute of Biology (Northern Ireland Branch) and Department of Extra-mural Studies (Queen's University), pp. 22–27.

Kennedy, G.J.A. and Greer, J.E. (1988) Predation by cormorants, *Phalacrocorax carbo* (L.), on the salmonid populations of an Irish river. Aquaculture and Fisheries Management 19: 159–170.

Montgomery, C.R. (1988) The perch fishery of Lough Neagh, Northern Ireland. Proc. Inst. Fish. Mgmt (N. Ireland Br.) Study Course, Univ. Ulster.

Moriarty, C. (1986) The European eel – discoveries and developments. Proceedings of the 17th Annual Study Course of the Institute of Fisheries Management, University of Ulster at Coleraine, pp. 100–110.

Parsons, J., Vickers, K.U. and Warden, Y. (1977) Relationship between elver recruitment and changes in the sex ratio of silver eels *Anguilla anguilla* L. migrating from Lough Neagh, Northern Ireland. J. Fish Biol. 10: 211–229.

Sweeting, R.A. (1976) Studies on *Ligula intestinalis* (L.) effects on a roach population in a gravel pit. J. Fish Biol. 9: 515–522.

Tatrai, I. and Istvanovics, V. (1986) The role of fish in the regulation of nutrient cycling in Lake Balaton, Hungary. Freshw. Biol. 16: 417–424.

Tesch, F.-W. (1977) The Eel. Chapman & Hall, London, 434 pp.

Tobin, C.M. (1986) A record of *Ligula intestinalis* (L.) (Cestoda) from roach (*Rutilus rutilus* (L.)) in Lough Neagh. Ir. Nat. J. 22: 78.

Van Densen, W.L.T., Dijkers C. and Veerman, R. (1986) The fish community of the Loosdrecht lakes and the perspective for biomanipulation. Hydrobiol. Bull. 20: 147–163.

Werner, E.E. (1984) The mechanisms of species interactions and community organization in fish. In: Strong, D.R., Simberloff, D., Abele, L.G. and Thistle, A.B. (eds.), Ecological Communities: Conceptual Issues and the Evidence. Princeton University Press, Princeton, pp. 360–382.

Wilson, J.P.F. (1979) The biology and population ecology of the pollan, *Coregonus autumnalis pollan* Thompson, of Lough Neagh. Unpublished PhD thesis, University of Ulster, 123 pp.

Wilson, R.F. (1971) The decline of a roach *Rutilus rutilus* L. population in Chew Valley Lake. J. Fish Biol. 3: 129–137.

Wood, R.B. (1984) The present status of Lough Neagh as a fishery. Proc. Inst. Fish. Mgmt (N. Ireland Br.) Study Course, Univ. Ulster., pp. 61–81.

Wood, R.B., and Gibson, C.E. (1973) Eutrophication and Lough Neagh. Water Research, 7: 173–187.

Wood, R.B. and Smith, R.V. (1986) Lough Neagh: 100 years of man's influence on the water quality of a major inland fishery. Proceedings of the 17th Annual Study Course of the Institute of Fisheries Management, University of Ulster at Coleraine, pp. 111–126.

19. The vegetation of Lough Neagh wetlands

R.D. DAVIDSON

19.1. Introduction

The shorelines of Lough Neagh and the surrounding lowlands support extensive areas of wetland vegetation. In recent times these wetlands have been greatly modified by the impact of man. Drainage schemes that lowered the lake level have had the most dramatic impact, forcing entire plant communities to shift their ground while others have been altered significantly by the new water regimes. This has led to the reduction or extinction of some species and the increase and spread of others. For example, Harron and Rushton (1986), in reviewing the effects of successive lough lowerings on the wetland flora, use the term 'devastating' and report severe reductions of willow, reeds, bulrush, reedmace, sedges and yellow loosestrife. More extremely, they found that *Thelypteris thelypteroides* (*T. palustris*), *Teesdalia nudicaulis*, *Sium latifolium*, *Cladium mariscus*, *Carex lasiocarpa* and *Carex buxbaumii*, previously found in but one or two stations in the Lough Neagh area had disappeared. *Lathyrus palustris* and *Carex acuta* were in particular thought to have suffered from the effects of drainage, while the macrophyte-covered Lough Beg has lost some 30% of its pre-drainage area. Other less dramatic and sudden, but nonetheless influential, impacts include woodland clearance, peat cutting, grazing, eutrophication and agricultural reclamation.

More generally, Harron and Rushton (1986) briefly outline the characteristics of the Lough Neagh wetland flora (some 700 species) in terms of its significance for rare Irish species and for the general distribution of the Irish flora.

It is the aim in this chapter to discuss the vegetation not only of the open waters of the Lough itself but also the vegetation that is found today on ground that has had at least an historical association with its floodwaters. The lough level today is fairly stable but it used to be both higher and more variable before the various drainage schemes of the last 100 years or so. The influence of these floodwaters can still be seen on low-lying ground around the lough margin but is particularly evident on the extensive flat land to the south of the lough. The vegetation will be discussed in terms of plant

R.B. Wood and R.V. Smith (eds.), Lough Neagh, 473–490.

communities of open water, peatlands, woodlands and grasslands. No recent work has concentrated effort to examine the vegetation in terms of plant communities alone but considerable effort has gone into the examination of the occurrence and distribution of individual species. Most of this work was carried out by Harron in the 1970s and this account draws extensively on his *Flora of Lough Neagh* (Harron and Rushton, 1986) and various reports compiled for the Conservation Branch of the Department of the Environment for Northern Ireland (Boddie and Porter, 1971; Harron, 1977). Particular sites referred to in the text can be located in Fig. 19.1.

19.2. Physical and historical background

Most of the lough shoreline is composed of drift deposits which include boulder clays, sands and gravels with the underlying basalt only outcropping along some sections of the north shore. Landward the shore is gently shelving but is in most cases marked by low cliff sections cut in slightly elevated boulder clay drift. These steep cliff sections are all now a short distance from the present lake margin as they mark the pre-drainage shoreline established before the first lake lowering of the mid-19th century. On the lake bed, offshore sand bars occur in many places to produce local sheltered conditions suitable for aquatic plant life.

To the south and south west of the lough the drift is overlain by extensive areas and in places considerable depths of peat. These extensive flat lands bordering most of the shore and rivers in the area were regularly flooded before drainage and are thus composed of fen peat deposits while areas further inland are more often raised bog peatlands. The peatlands are frequently broken by slightly elevated boulder clay drumlins.

Three major drainage schemes detailed in Chapter 2, Table 2.6, have reduced the level of the lough and restricted flooding. The first scheme was carried out between 1846 and 1858 with further schemes in 1942 and 1959. The combined result of the drainage schemes has been the lowering of the level of the lough by an estimated 2.3 m and the reduction of fluctuations from an estimated 3.5 m to notionally <0.5 m today.

Other changes have taken place which were not directly influenced by the drainage schemes. Forest clearance from the better soils was more or less complete by the 17th century (McCracken, 1971). Over many centuries peat extraction, with controlled drainage, has reduced the extent and level of most of the raised bogs. Drainage of the raised bogs means that their surfaces are no longer actively growing. The water of the lough is only slightly alkaline (~pH 8.2) but a recent development affecting the aquatic plant communities has been the rapid increase in the eutrophic status of the lough which may have led to the extinction of some species and the spread of others (Stewart and Corry, 1938). During the last century the canal system linking Belfast and Newry with the lough facilitates the spread of several species.

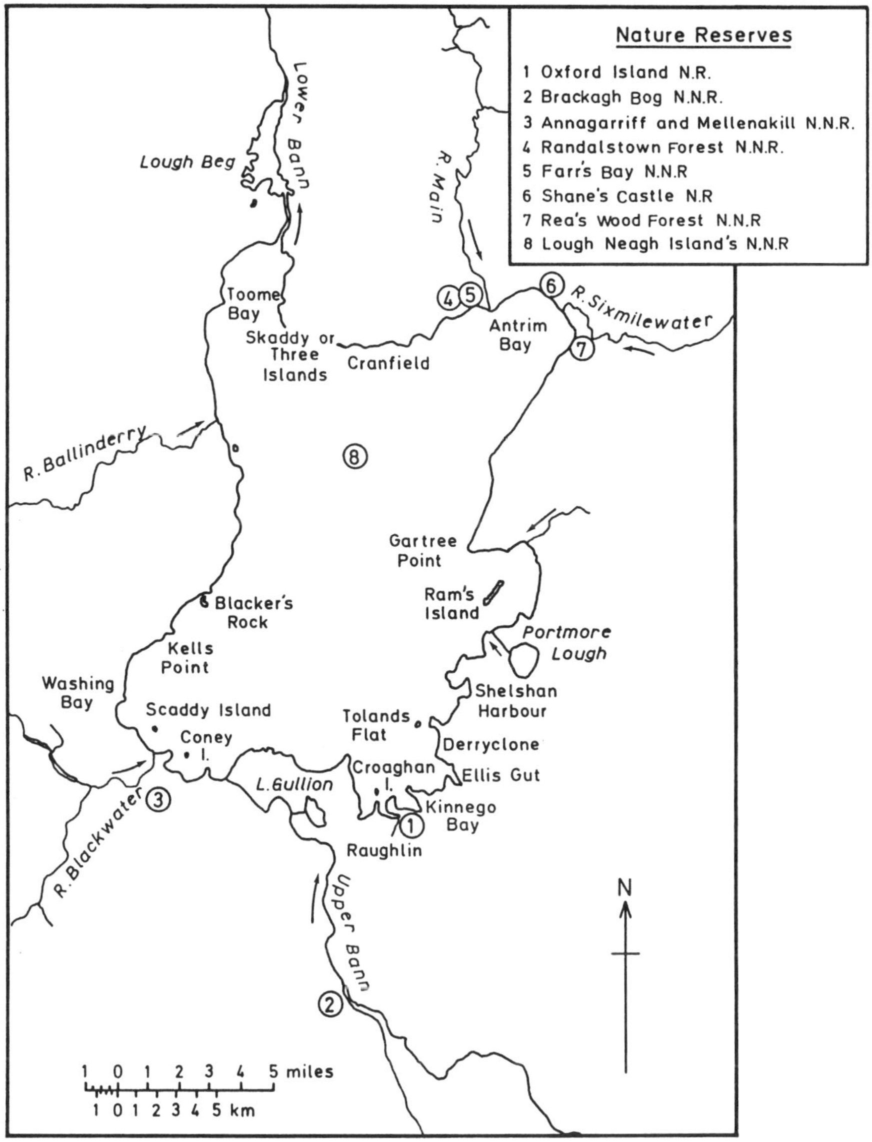

Figure 19.1. Map of Lough Neagh showing Nature Reserves and other places mentioned in the text.

Some aspects of the history of the Lough Neagh vegetation are illustrated in the pollen analyses of Lough Neagh sediments reported in Battarbee (1978).

19.3. Plant communities

The association between the plant communities of open water, reedswamp, fen, raised bog, grasslands, alder and willow carr, birch woodlands and mature woodlands are outlined diagrammatically in Fig. 19.2.

Shallow waters not subject to violent wave action support submerged plant communities. If the water is sheltered enough an emergent reedswamp community can establish itself. The tendency then is for the sheltered water within the reedswamp to trap silt and organic debris to build up an organically-rich sediment layer. As the accumulation proceeds, the landward side of the reedswamp tends to progress to a more terrestrial plant community. The resultant species-rich fen community, influenced by floodwaters but not permanently submerged, can, in suitable local hydrological conditions of poor drainage, pass to a more oligotrophic plant community dominanted by sphagnum mosses. The water-logged acid conditions greatly inhibit plant decomposition so the accumulation of peat within raised bogs tends to be relatively rapid. Deep peat cutting on raised bogs down to the level of eutrophic groundwater can re-establish the succession of the open water stage. 'Improved' drainage within raised bogs causes them to lose their formative plant communities which are altered by an increase in species better adapted to drier conditions and a decrease in hydrophilous species such as sphagnum mosses. A common consequence of improved drainage on raised bogs is the development of birch (*Betula pubescens*, *Betula pendula*) woodland. Fen communities not overtaken by raised bogs may pass to alder (*Alnus glutinosa*) and willow (*Salix* spp.) carr in the absence of grazing or to wet grassland with heavy grazing. Agricultural development has, in many areas, completely ousted fen plant communities and replaced them with commercial grassland mixtures. Alder and willow carr will eventually pass to more mature woodland but commonly reverts to grassland due to clearance. The birch woodlands of the raised bogs are similarly often cleared to be replaced by grassland or open moorland.

The natural progression of plant communities from submerged aquatics to mature woodland is a very slow process but has been violently disrupted each time the lough level has been lowered and the flood levels reduced. New plant communities have developed very quickly on freshly exposed lake shore while other communities have shifted their ground or altered their composition.

Several 'typical' cross-sections of the lake shore and a river have been drawn to outline the physical relationship between the water margin and some of the major plant communities already discussed.

Figure 19.3 shows a section at Rea's Wood Forest NNR., Co Antrim, on an ungrazed site. Reedswamp, alder/willow and mature woodland have all developed on ground that was formerly lake bed. The trees, by fairly sharply distinct successive stages of growth, still show quite clearly the position of the pre-drainage shorelines. A series of low sand ridges running parallel with

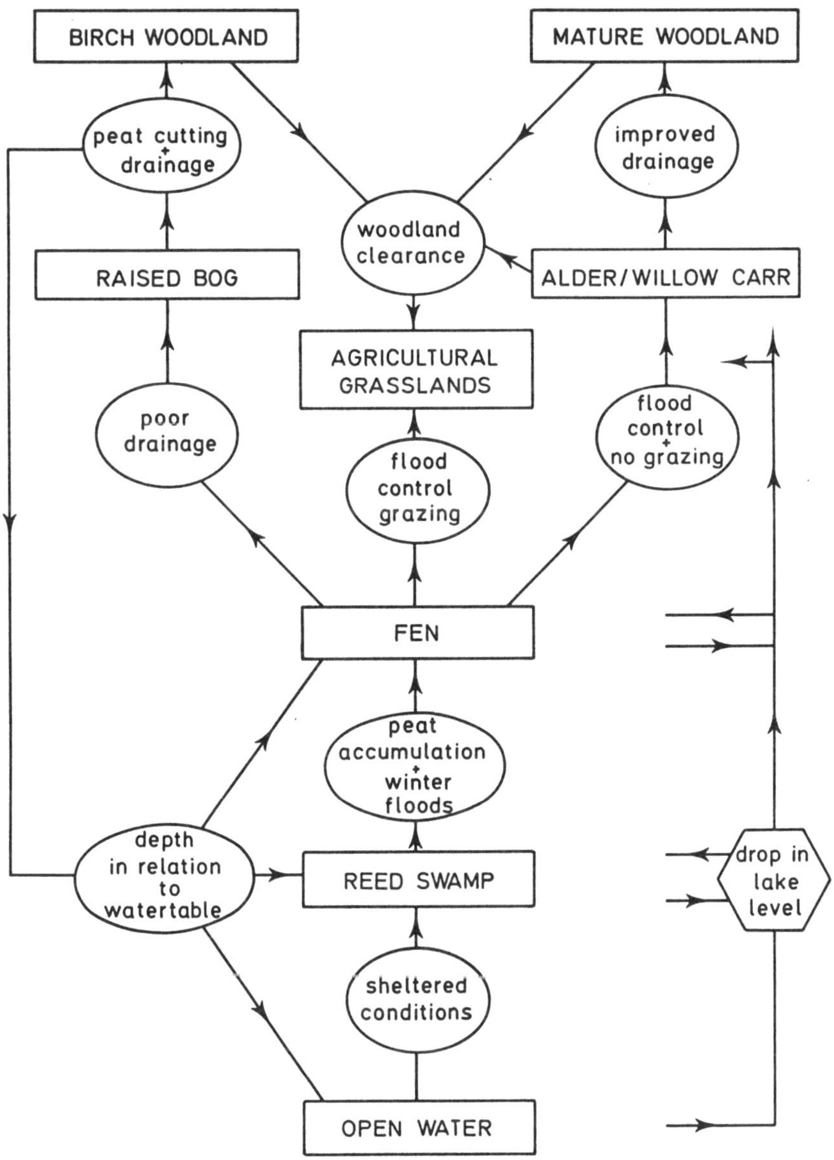

Figure 19.2. Diagram of the relationships between plant communities of of Lough Neagh.

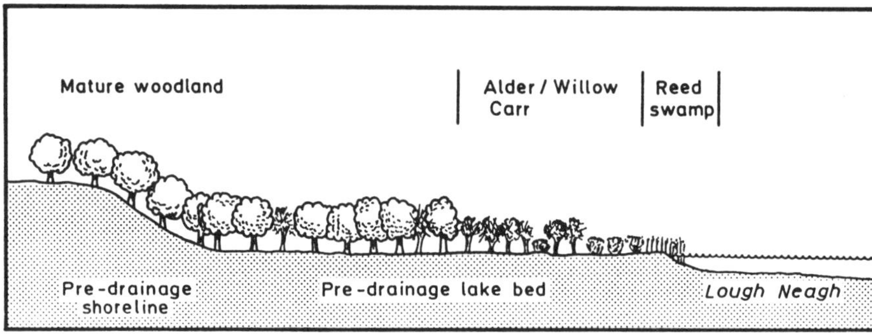

Figure 19.3. Cross-section of shoreline at Rea's Wood.

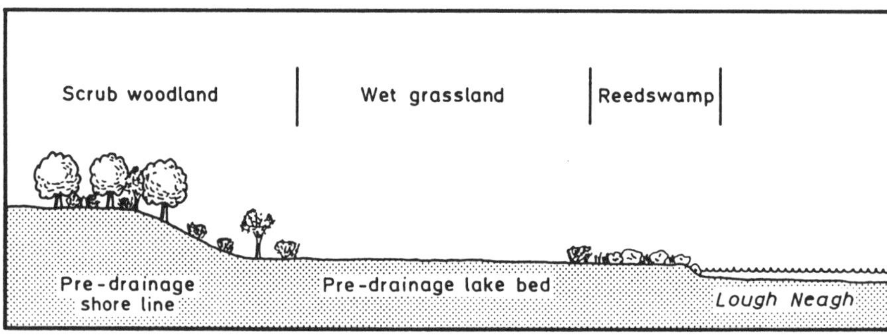

Figure 19.4. Cross-section of shoreline at Derryclone.

the shore within the wood confuse this picture slightly by creating wet and dry sections, with their associated plant communities. In contrast, a similar section through grazed ground lacks the alder/willow carr. The example given in Fig. 19.4 is taken from Derryclone, Co Antrim, where the alder/willow carr is replaced by wet grassland. On grazed shores such as these, a narrow band of alder occasionally occurs close to the water's edge while the steep face of the old shoreline fairly consistently supports mixed scrub.

A shore with a very different character is found at Blacker's Rock, Co. Tyrone (Fig. 19.5). Here the water was very shallow and broken by several islands before the first drainage scheme. The drop in water level has left the landward side of extensive reedbeds dry, broken by stands of more mature trees on the old island sites. The raised bog comes quite close to the shore in this area and is clothed in birch woodland. Along sections of the North Shore the shoreline is rocky (Fig. 19.6). Peatlands are absent here and reedswamp is rare but an unusual 'rock garden' type flora grows in the rock crevices and in places the shore is well wooded. Offshore, submerged aquatics are found where the water is shallow.

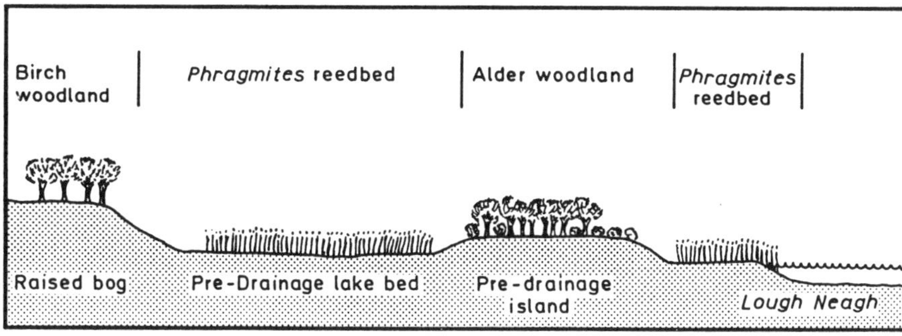

Figure 19.5. Cross-section of shoreline at Blacker's Rock.

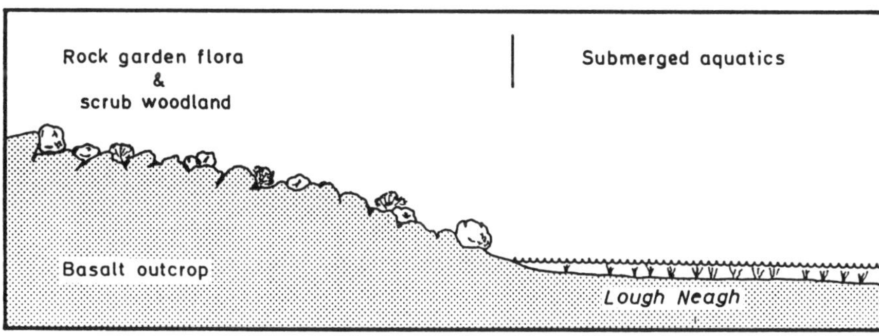

Figure 19.6. Cross-section of shoreline at Cranfield.

Figure 19.7 is a typical section through a river and its margins to the south of the lough, the floodplain of which once supported fen vegetation. The fen communities have been replaced by large fields of re-seeded grassland. The margins of these rivers generally support submerged, floating and emergent aquatic plants.

19.3.1. *Plant communities of open water*

Plants growing in open water are either completely submerged, floating attached, freely floating or emergent. These communities have been strongly influenced by man in recent times through lowered water levels, increasing eutrophication and the introduction and spread of new species. Two formerly abundant species were awlwort (*Subularia aquatica*) and quillwort (*Isoetes lacustris*). These are now described by Harron and Rushton (1986) as very rare. This may have have arisen because of the eutrophic state of the lough as they are essentially plants of mesotrophic waters.

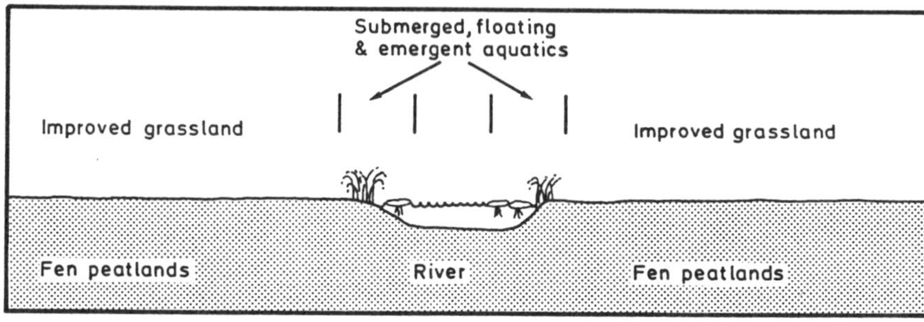

Figure 19.7. Cross-section through River Bann.

19.3.1.1. *Submerged plants*

Completely submerged plants can photosynthesise only where sufficient light penetrates and as the euphotic zone rarely exceeds 3 m in summer (see Chapter 4) these plants survive only in the shallower waters of the lough. They are also excluded from open, exposed shores by turbulent wave action and unstable substrate. As a result, they occur only where they are protected from strong wave action by the configuration of the shoreline or by the presence of off-shore sand bars or islands, or in inlets, pools and drains around the shores of the lough. Lough Beg is shallower than Lough Neagh and is particularly rich in submerged aquatics. The group includes many pondweeds (*Potamogeton* spp.). The commonest are *Potamogeton pectinatus* and *P. filiformis*, the latter growing in shallower water (10–70 cm) and quite often in more exposed conditions. Other fairly frequent pondweeds include *P. lucens*, *P. gramineus*, *P. perfoliatus*, *P. berchtoldii* and *P. crispus*. Of these *P. lucens* occurs in the deepest waters (2 m in Kinnegoe Bay). Less common pondweeds include *P. zizii*, *P. alpinus*, *P. praelongus*, *P. pusillus* and *P. obtusifolius*. *Zannichellia palustris* is frequent in the lough and two fairly abundant species in suitable sheltered waters are spiked water milfoil (*Myriophyllum spicatum*) and canadian pondweed (*Elodea canadensis*), the latter having been introduced *ca.* 1836. A rare water milfoil from N. America (*Myriophyllum alterniflorum* var. *americanum*) occurs on Lough Neagh and Lough Beg. It occurs in other lakes in Ireland but in Britain it is only found in the west of Scotland. *Ranunculus fluitans* has its only Irish station on the lower reaches of the Six Mile Water in Co. Antrim.

Several species of starwort (*Callitriche* spp.) belong to this group, the most widespread being *Callitriche stagnalis* although it is found in pools and drains, rather than in more open situations where *C. hermaphroditica* is more frequently found.

An interesting plant which has a scattered and local distribution elsewhere in the British Isles but which has become locally common in Lough Neagh is *Elatine hyropiper*. It was first noted last century in the Newry Canal (Stewart and Corry, 1938). It is a low creeping plant which grows on muddy substrate in shallow water (0.5 m) where the bottom is not shaded by mats of filamentous algae or larger aquatics.

19.3.1.2. *Floating attached plants*

This group is less tolerant of exposed conditions than the submerged group and is therefore even more confined to sheltered waters. Many plants of the group grow in inlets and drains and find the margins of slow flowing rivers particularly suitable (River Upper Bann and River Blackwater). The species are cosmopolitan and include the yellow water lily (*Nuphar lutea*) amphibious bistort (*Polygonum amphibium*), unbranched bur-reed (*Sparganium emersum*), broad-leaved pondweed (*Potamogeton natans*) and flote-grass (*Glyceria fluitans*).

19.3.1.3. *Free floating plants*

Free floating plants are often found growing in association with floating attached plants; they tend to shade out submerged aquatics. Four species of duckweed (*Lemna*) occur on Lough Neagh. Common duckweed (*Lemna minor*) and greater duckweed (*L. polyrrhiza*) are the most widely distributed and frequently occur together, usually scattered throughout emergent plant communities. *L. polyrrhiza* is worthy of note as it is a rapidly increasing species first recorded on the Lagan Canal in 1880 (Stewart and Corry, 1938) and now very common on Lough Neagh. The other duckweeds are ivy leaved duckweed (*L. trisulca*) which is fairly frequent although it can be easily overlooked as it has a submerged growth habit; and fat duckweed (*L. gibba*) which is described as rare and local.

The frogbit (*Hydrocharis morsus-ranae*) is rare, occurring only in drains at Lough Gullion, and it shares a feature with *Lemna polyrrhiza*, that it reflects the northern extension of the flora of the Central Plain of Ireland. In Britain there are many plants associated with wetlands that are confined to lowland lakes in southern regions; many of these plants recur on the Central Plain of Ireland where they find similar conditions in the numerous, shallow, limestone lakes. In turn many of these plants have extended their range northwards along a narrow tongue of limestone that extends up the corridor of the Blackwater valley to south-east Lough Neagh. *Hydrocharis morsus-ranae* and *Lemna polyrrhiza* are two examples from this group of species with a southern distribution in Britain; others will be referred to later.

19.3.4. *Emergent plants*

This is the reedswamp group composed of plants that are permanently rooted underwater but send up emergent shoots. The water between the plant stems is often sheltered enough to trap not only silt but also dead plant material; this leads to the accumulation of organic matter and the gradual elevation of the rooted surface. It is therefore a transitional community between open water and various terrestrial communities.

Reedswamp communities were much reduced immediately after the 1959 lowering when considerable sections were left dry; most of the shoreline is grazed by cattle so most of these species were quickly lost on the landward side to better adapted marsh and grassland species, while the reedswamp plants re-established themselves at the new shoreline. Where grazing was excluded the freshly exposed areas were invaded by willows (*Salix* spp.) and alder (*Alnus glutinosa*) giving rise to swamp and fen carr; this scrub development is well represented at Farr's Bay, Antrim Bay, Ram's Island, Raughlin Peninsula and Coney Island. Most of the shoreline is very exposed so contrary to expectations for such a shallow body of water reedswamp communities on Lough Neagh are by no means extensive. Abandoned gravel pits and the peatlands where turf cutting has reached down to the level of nutrient rich groundwater provide additional sites.

A very common plant on the shores of Lough Neagh is the reed canary grass (*Phalaris arundinacea*). This is considered a reedswamp species as it tolerates flooding but in fact it often establishes itself just above the mean water level where it is very efficient at binding the soil and withstanding strong wave action. Where the water is shallow, common spike rush (*Eleocharis palustris*) establishes itself in even the most exposed conditions, given a suitable substrate. More sheltered shores support other reedswamp plants, often in pure stands; characteristic are common reed (*Phragmites australis*), reed-mace (*Typha latifolia*), bulrush (*Scirpus lacustris*), branched burreed (*Sparganium erectum*) and bottle sedge (*Carex rostrata*). Lesser reed-mace (*Typha angustifolia*) is local and rare in Ireland but occurs in significant stands on the outer margins of the *Phragmites* beds of Portmore Lough and Lough Gullion. Common emergent plants of shallow water, though not usually in pure stands, include watercress (*Nasturtium officinale*), fool's watercress (*Apium nodiflorum*), water speedwell (*Veronica anagallis-aquatica*), yellow flag (*Iris pseudacorus*), water plantain (*Alisma plantago-aquabca*) and lesser water plantain (*Baldellia ranunculoides*).

Plants of particular interest in this group are the flowering rush (*Butomus umbellatus*) and arrowhead (*Sagittaria sagittifolia*), both of which are frequent in the area but are of definite southern distribution in Britain and have their principal Irish distribution in the Central Plain. *Butomus* is common in drains and inlets while *Sagittaria* is more frequent along the sides of slow moving rivers.

Emergent plants are found in a narrow fringe around the entire lough except where the shore is too exposed. In such cases plant communities are replaced by shifting accumulations of sand and gravel often intermixed with flotsam of tree branches and leaves. Where the shore is rocky, many plants of the emergent group establish themselves in the relative shelter between boulders. The most significant stands of *Phragmites*, with their associated emergent plants, are found at Shane's Castle, Shelshan Harbour, Ellis Gut, Kinnegoe Bay, Raughlin Peninsula, Croghan Island and Scaddy Island. The, smaller lakes are all fringed with *Phragmites* but the most extensive reedbed in the area stretches from Blacker's Rock to Kells Point on the west shore and is in places several hundred metres wide.

19.3.2. *Plant communities of peatlands*

The tendency of reedswamp communities to accumulate partially decayed organic deposits has already been referred to above. In such waterlogged anaerobic conditions the organic layer (peat) can build up above mean water level. The reedswamp plant community then progresses to one of fen; a plant community of peatlands subject to more or less continuous waterlogged conditions but not constantly under standing water. In these situations the ground-water is similar in chemistry to that of the lough and the local rivers, i.e. neutral or slightly alkaline in reaction but rich in nutrients. This suits a wide range of plants and the fen community is an extremely rich one.

Raised bog communities supersede fen communities where local conditions of poor drainage cause the fen peat to accumulate to such a level that the surface vegetation is influenced more by rainwater than by more nutrient-rich groundwater. The leaching of the surface by rainwater leads to the replacement of fen vegetation by more acidophilous vegetation dominated by *Sphagnum* mosses, which quickly accumulate as acid peat and spread to form low, domed, raised bogs. In Ireland the initiation of these raised bogs took place in Atlantic times, about 7,000 years ago. They have developed extensively on low lying ground in some places to a depth of 10 m, especially to the south of the lough in the north of County Armagh, but also along the west shore as far as Toome Bay.

Before the drainage scheme of 1847–1858, large tracts of land were seasonally flooded along the shores of the lough and along the lower courses of many rivers. These areas must have supported well developed fen communities but the extent of the fen has been reduced to almost nothing by various drainage schemes which have progressively eliminated the winter floods. With improved drainage the lowland fens make excellent agricultural land so the natural plant communities were quickly replaced by artificial ones. Where agricultural development has not taken place the restriction of flooding has allowed the ecological succession of herbaceous fen communities to progress to woodland.

As late as 1930, before the final lowering, Duff (1930) records well developed fen communities still surviving over large areas. At this time the mean water level was about 1 m higher than today and fluctuated considerably from summer to winter. The extent of the fen was described by Small (1931) and included the Three Islands area, Shane's Castle, Shelshan Harbour to Portmore Lough, from Oxford Island shore to Maghery, all along the Upper Bann and Blackwater rivers, around Annagarriff and Derryadd Lakes, numerous small patches to the south-west and south-east, along the entire margins of the lough and at the mouth of all rivers.

Today, with improved drainage and greater control of water levels, these areas are no longer subject to winter flooding and fen communities are restricted to narrow margins adjacent to open water and deeply cut over sections within the raised bog areas. An especially interesting mosaic of fen communities in cut over sections within remnants of raised bog is well illustrated at Brackagh Moss and is documented by White (1930). This site has been slightly damaged by improved drainage from the surrounding agricultural land but is now protected as a National Nature Reserve and efforts are being made to maintain the water table at the highest practical level.

The findings of Duff (1930), White (1930, 1932) and Small (1931) formed the basis for Tansley's account of the vegetation of Lough Neagh (Tansley, 1948).

19.3.2.1. *Fens*

The fen community is species-rich and includes many plants which are also found in the reedswamp community. Other plants include marsh marigold (*Caltha palustris*), meadowsweet (*Filipendula ulmaria*), purple loosestrife (*Lythrum salicaria*), great willowherb (*Epilobium hirsutum*), cowbane (*Cicuta virosa*), nodding bur-marigold (*Bidens cernua*), tripartite bur-marigold (*B. tripartita*), marsh ragwort (*Senecio aquaticus*), yellow loosestrife (*Lysimachia vulgaris*), marsh forget-me-not (*Myosotis scorpioides*), bittersweet (*Solanum dulcamara*), water mint (*Mentha aquatica*), gypsy-wort (*Lycopus europaeus*), marsh woundwort (*Stachys palustris*), common spotted orchid (*Dactylorhiza fuchsii*) and northern fen orchid (*D. majalis*). Common sedges include *Carex disticha*, *C. paniculata*, *C. nigra*, *C. flacca*, *C. rostrata* and *C. vesicaria*.

Not all the fen plants have survived the new water regime. Some were especially dependent on seasonal flooding and have now disappeared from the area. Notable amongst these are the marsh fern (*Thelypteris palustris*) and saw sedge (*Cladium mariscus*). There have been more recent single records for marsh pea (*Lathyrus palustris*) and northern bedstraw (*Galium boreale*) but these too may now have disappeared completely. Fen plants that are very rare or have recently become extinct in the Lough Neagh area but are much more frequent in the more base-rich East Anglian fens and wetlands of central Ireland are the marsh fern (*Thelypteris palustris*), buckthorn (*Rhamnus catharticus*), alder buckthorn (*Frangula alnus*), marsh pea (*Lathyrus palustris*), saw sedge (*Cladium mariscus*) and the tufted sedge (*Carex elata*).

An orchid worthy of note associated with the fragments of fen that remain is *Spiranthes romanzoffiana*. It is an American plant known to occur in the Lough Neagh basin and in a few other sites in the British Isles but not elsewhere in Europe. It was first discovered at Brackagh Bog by Praeger (1892). Unfortunately it is near extinction at this site but it is still thriving at several other sites around the lough and at Lough Beg. Its occurrence here is difficult to explain but Praeger (1934) points out that there are several North American plants in NW Europe and similarly several European plants on the NE American coast. He suggests a possible early migration via Iceland and Greenland when the two continents were closer together.

Two extremely rare grasses of fen communities in the British Isles have been found in the Lough Neagh area; holy grass (*Hierochloë odorata*) occurs at Shelshan Harbour and northern small reed (*Calamagrostis stricta*) which has not been found recently at a large number of its former sites, but has been recorded from nearby Lough Beg (Kertland and Lambert, 1972), and from some areas in Lough Neagh (see Harron and Rushton, 1986).

Given the general story of the impoverishment of Lough Neagh's fen flora, one doubts whether if Tansley (1939) were writing today he would still observe that Lough Neagh's fenlands are 'probably the most extensive in the British Isles still remaining comparatively unspoiled'.

19.3.2.2. *Raised bogs*

All of the raised bogs have been altered by various degrees of cutting, draining and burning. As a result there is very little active growth of peat; the *Sphagnum* moss communities that developed most of the peat have been much reduced or eliminated locally. Hamilton (1982) and Hammond (1979) record the alarming rate at which, overall, Ireland's peatlands are being modified and Lough Neagh's are no exception.

Typical plants of the raised bog surfaces today include: sundew (*Drosera rotundiflora*), cross-leaved heath (*Erica tetralix*), heather (*Calluna vulgaris*), cotton grass (*Eriophorum angustifolium*), bog asphodel (*Narthecium ossifragum*) and royal fern (*Osmunda regalis*). Cranberry (*Vaccinium oxycoccus*) and bog rosemary (*Andromeda polifolia*) are found only locally in the area within the Peatlands Park, County Armagh, but are interesting in that they are typical constituents of lowland raised bogs in the British Isles. Andromeda occurs here as an extension of its range from the Central Plain of Ireland where it is very common. All three species of Drosera: *D. rotundifolia*, *D. anglica* and *D. intermedia* occur within the Peatlands Park.

It has already been pointed out that the drained peat surfaces support fewer *Sphagnum* species than a raised bog in its natural condition. Instead they often have a more or less continuous cover of *Calluna* with a ground cover of non-*Sphagnum* mosses. Burning encourages purple moor grass (*Molinia caerulea*), bracken (*Pteridium aquilinum*) or rosebay willowherb (*Epilobium angustifolium*), and, where burning is less effective, vigorous stands of hairy birch (*Betula pubescens*) develop. Areas that are cut over to a fairly low level and regularly burnt develop a poor fen community dominated by *Molinia* associated with devil's bit scabious (*Succisa pratensis*), marsh pennywort (*Hydrocotyle vulgaris*) and sweet vernal grass (*Anthoxanthum odoratum*). This poor fen community is well represented over a large area at Derrymore south of Portmore Lough. The drains in this area support bogbean (*Menyanthes trifoliata*), greater bladderwort (*Utricularia vulgaris*), bog pondweed (*Potamogeton polygonifolius*) and bottle sedge (*Carex rostrata*).

19.3.3. *Plant communities of woodlands*

According to McCracken (1971) oak woodlands were extensive around the shores of the Lough during the early 17th century; the only areas devoid of woodland cover at this time were a small section in the south west at Washing Bay and a larger section on the east shore between Portmore Lough and Antrim. To the south of the lough, in north Armagh, alder and willow grew on the low-lying fens and oak was restricted to the protruding pockets of glacial drift. By the beginning of the 18th century these oak woodlands had all been cleared except for a few remnants on the north shore which were finally cleared at the end of that century. The land that once supported these oakwoods is now almost entirely converted to agricultural use. Many of the townland names on the southern shores of the lake contain the word 'derry' which strongly indicates earlier extensive occurrence of oak woods.

Oak woodlands have been lost but other semi-natural woodlands have been established. The scarp of the old lake shore supports a band of semi-natural woodland dominated by blackthorn (*Prunus spinosa*) and ash (*Fraxinus excelsior*) with occasional spindle (*Euonymus europaeus*), crab apple (*Malus sylvestris*), guelder rose (*Viburnum opulus*) and hazel (*Corylus avellana*). Buckthorn (*Rhamnus catharticus*) was never common but may now have disappeared, Harron and Rushton (1986) describing it as rare, sparse and decreasing.

The scarp is generally cut in boulder clay, and wooded sections support a fragmented woodland ground flora. The Cranfield shore in the north of the Lough is something of an exception as here the basalt outcrops reach the lake shore and the scarp is marked by large boulders and bare rock. The ash woodland on this site is interesting in that it supports native oak (*Quercus petraea*) and wych-elm (*Ulmus glabra*). The long history of woodland cover on the site is reflected in the richness of the ground flora which includes many woodland grasses, namely *Festuca gigantea*, *Melica uniflora*, *Bromus ramosus*, *Brachypodium sylvaticum*, *Elymus caninus* and *Milium effusum*. A curious occurrence is the fern, the giant large-leaved polypody (*Polypodium australe*) which is otherwise rare in Northern Ireland and not normally found away from chalk or limestone.

There are several large estates on the shores of the lough. The owners of these lands set aside sections for woodland management that today support mature beech (*Fagus sylvatica*) and sycamore (*Acer pseudoplatanus*) with a mixture of native trees. Woodland of this type is best represented along the Antrim Bay shore (Rea's Wood Fig. 19.1) but other examples occur at Gartree Point, Rams Island, Raughlin Peninsula and on Coney Island.

As a result of the various lough lowerings freshly exposed margins of the lake bed have been colonized by vegetation. The areas best protected from grazing have generally been those adjacent to existing woodland and it is these areas that have developed the best alder and willow carr. There is typically a marked break in the character of the canopy between the planted hardwoods and the naturally regenerated alder and willow carr which in itself reflects, by stage of growth, successive lake lowerings. Additional carr woodland has also developed on numerous small scattered sites along the shore of Lough Neagh and in association with the other small lakes notably on the north east shore of Portmore Lough. Alder, perhaps aided by its ability to fix nitrogen, quickly colonises freshly exposed sands and gravels; the peatier and wetter sites support a wide range of willow species, the most frequent of which are *Salix fragilis*, *S. alba*, *S. atrocinerea*, *S. caprea* and *S. viminalis*. Blackcurrant (*Ribes nigrum*) and guelder rose (*Viburnum opulus*) are frequent but alder buckthorn (*Frangula alnus*) a common constituent of East Anglian carr is very rare. On drier ground, as on old sand and gravel bars, ash (*Fraxinus excelsior*) and hairy birch (*Betula pubescens*) are common and planted sycamore (*Acer pseudoplatanus*) spreads easily.

The herbaceous vegetation of drier sections is locally dominated by bluebell (*Hyacinthoides non-scriptus*) but the wetter areas support a rich and luxuriant flora with several rarities. Tall plants typical of fen are found here such as meadowsweet (*Filipendula ulmaria*), hemlock water drop-wort (*Oenanthe crocata*) and iris (*Iris pseudacorus*). Growing in great sheets in damp sites in the Antrim Bay area is the greater bittercress (*Cardamine amara*) which is otherwise rare in Ireland. Characteristic sedges are *Carex remota*, *C. vesicaria*, *C. paniculata* and *Scirpus sylvaticus*. A sedge thought to be extinct in Ireland, *C. elongata*, was re-found at several damp woodland sites by Harron (Harron, 1974). The Himalayan balsam (*Impatiens glandulifera*) first recorded in 1937 (Stewart and Corry, 1938) has established itself in several localities and as a very succesful introduced plant may eventually have detrimental effects on the native vegetation locally.

Birch (*Betula pubescens*) forms vigorous pure stands throughout much of the cut-over bog in the area. It is especially characteristic of overgrown ramparts that were once used as paths to traverse the bog but now retained only as land boundaries. Its distribution is therefore closely associated with that of the raised bog.

19.3.4. *Plant communities of grasslands*

All the grasslands in the immediate vicinity of the lough are man-made and are mostly used for summer grazing of domestic livestock. In the absence of management the natural transition from open water to dry land passes through reedswamp to willow and alder carr to oak woodland. Where seasonal flooding is more persistent the tree species do not colonise and a tall fen vegetation develops. The reclamation of the fen areas has already been mentioned; they have, for the most part, been re-seeded with agricultural grassland mixtures to produce a sward which is, incidently, much favoured by flocks of wintering swans (see Chapter 20).

The narrow strip of exposed lake bed that fringes the lough is grazed throughout its length apart from the reedswamp and woodland sections already mentioned and other small and inaccessible sites. The type of grassland that develops is controlled by the soil (which is neutral, but varies from sands and gravels to boulder clay and peat), the water regime of the site and the management. All these factors vary on a very local scale and as no systematic work on the grassland communities has been carried out it is possible to make only very general comments about their species composition and their distribution.

On the lake shore, poorly drained ground that is not continuously saturated supports a wet meadow flora which includes meadow buttercup (*Ranunculus acris*), creeping buttercup (*R. repens*), ladies smock (*Cardamine pratensis*), moneywort (*Lysimachia nummularia*), common rush (*Juncus effusus*) and the grasses *Deschampsia cespitosa*, *Holcus lanatus*, *Agrostis canina*, *A. stolonifera*, and *Poa trivialis*. Heavy clay soils often support hard rush (*Juncus inflexus*) while peat soils yield the sedge *Carex disticha*. Very wet grasslands on mineral soils are termed marshes and typical species of this group include many that are found in both reedswamp and fen. Examples are the marsh marigold (*Caltha palustris*), ragged robin (*Lychnis flos-cuculi*), marsh cinquefoil (*Potentilla palustris*), marsh pennywort (*Hydrocotyle vulgaris*), marsh bedstraw (*Galium palustre*), the rush *Juncus acutiflorus* and the grass *Agrostis stolonifera*.

The distribution of the re-seeded grasslands corresponds with that of former fen and to a lesser extent to reclaimed raised bog.

Other local sections of former lake bed have also been re-seeded; the reclamation is often done by bulldozing higher ground onto lower ground occupied by reedswamp or marsh communities on the margin of the lough to produce a uniform, efficiently drained surface.

19.3.5. *Conservation*

With each fall in water level the plant communities of Lough Neagh have been subjected to severe and rapid changes. All have been altered in character by the loss of some species and the addition of others. Some plant communities have almost disappeared completely, as was the fate of the once extensive fen communities. Further lowerings are unlikely, but other pressures are mounting which will threaten the continued survival of the natural and semi-natural areas that remain. Agricultural development is the main threat, and many sites have been destroyed in recent years.

The Conservation Branch of the Department of the Environment through the statutory powers vested in it by the Amenity Lands Act NI (1965) and, more recently, by the Nature Conservation and Amenity Lands (N.I) Order 1985, have taken certain measures to control damaging development. This has been done in two ways, firstly, by declaring the entire shoreline and much of the low lying ground to the south of the lough an Area of Scientific Interest. This is not a very effective system in controlling land management as carried out by farmers but it does give the Conservation Branch a say in developments that require planning permission or in cases where farmers wish to use agricultural improvement grants. Secondly, a more effective but a very local measure is the establishment of National Nature Reserves. These have been set up on several prime sites of scientific value by acquiring the land or drawing up agreements with the landowners for the reserves to be managed primarily for the protection of the flora and fauna.

Eight nature reserves have been established in the Lough Neagh area and although they are small in extent they encompass a wide range of vegetation types.

Three different agencies are involved in nature reserve management. Craigavon Borough Council owns and manages Oxford Island Nature Reserve. The reserve vegetation is representative of the Lough Neagh shoreline, wet grasslands on former lake bed and scrub woodland on the scarp of the old shoreline.

The Royal Society for the Protection of Birds lease the woodland along the shore of the Shane's Castle Estate primarily as a reserve for birds but the shoreline vegetation with alder and willow carr on the former lake bed is also afforded protection. The Countryside and Wildlife Branch of the Department of the Environment controls the remaining six reserves. Brackagh Bog National Nature Reserve (NNR) is a raised bog site that has been cut over to such an extent that a mosaic of open water and fen vegetation types has developed. Annagarriff and Mullenakill NNR contain two plots of uncut raised bog sites which have been partially drained and support areas of semi-natural woodland. Randalstown Forest, Farr's Bay and Rea's Wood Forest NNRs are all sites on the former lake bed which support developing alder and willow carr, open fen and shoreline vegetation. The Lough Neagh Islands NNR was established primarily for the benefit of breeding birds but the islands support a wide range of shoreline vegetation communities.

19.4. References

Battarbee, R.W. (1978) Observations on the recent history of Lough Neagh and its drainage basin. Phil. Trans. Soc. London. Series B. 281: 303–345.

Boddie, G.F. and Porter, H.T. (1971) Vegetation survey of the Lough Neagh islands. Report to Conservation Branch, DoE(NI).

Duff, M. (1930) The ecology of the Moss Lane Region, Lough Neagh. Proc. R. Ir. Acad. 39B: 477–96.

Hamilton, A.C. (1982) Peatland, Chapter 9, in: Cruckshank, J.G. and Wilcock, D.N. (eds.), Northern Ireland Environment and Natural Resources. The Queen's University of Belfast and the New University of Ulster, 294 pp.

Hammond, R.F. (1979) The peatlands of Ireland. Soil Survey Bull. 35. An Foras Taluntais. Dublin.

Harron, J. (1974) *Carex elongata* refound on Lough Neagh shores. Ir. Nat. J. 18: 91–92.

Harron, J.E. (1977) A preliminary study of areas of special ecological interest occurring in the Lough Neagh Basin. Report to Conservation Branch, DoE(NI).

Harron, J. and Rushton, B.S. (1986) Flora of Lough Neagh. Publ. Ir. Nat. J. Committee and the University of Ulster, 270 pp.
Kertland, M.P.H. and Lambert, D.S. (1972) Supplement to the vascular section of the 2nd edition (eds. Praeger, R.L. and Megaw, W.R. 1938) of A Flora of the North-East of Ireland by Stewart S.A. and Corry T., Belfast Naturalists' Field Club and Quota Press, Belfast.
McCracken, E. (1971) The Irish woods since tudor times. David and Charles, Newton Abbot.
Praeger, R.L. (1892) Botanical rambles in County Armagh. In: Proceedings Report of the Belfast Natural History and Philosophical Society for 1892–3. Also in Belfast. Nat. Hist. Phil. Soc., Selections from 150 years of Proceedings 1831–1981, pp. 135–137; chosen and edited by M. McNeilly. Publ. The Society, Belfast, 1981.
Praeger, R.L. (1934) The botanist in Ireland. E.P. Publishing Wakefield (1974 reprint).
Small, J. (1931) The fenlands of Lough Neagh. J. Ecol. 19: 383–388.
Stewart, S.A. and Corry, T. (1938) A flora of the North-East of Ireland. Quota Press, Belfast.
Tansley, A.G. (1939) The British Isles and their vegetation. Cambridge U.P.
Tansley, A.G. (1948) Britain's green mantle (2nd Edition). George Allen and Unwin, 327 pp.
White, J.M. (1930) Recolonisation after peat cutting. Proc. R. Ir. Acad. 39B: 453–476.
White J.M. (1932) The fens of North Armagh. Proc. R. Ir. Acad. 40B: 233–283.

20. Birds of Lough Neagh and Lough Beg

J.S. FURPHY

20.1. Introduction

It is a matter of regret that so little scientific study of the avifauna of Loughs Neagh and Beg has yet been undertaken, which is perhaps surprising in view of the known international significance of the area for wildfowl. It is, as discussed in Section 20.7, a Ramsar site. The amateur Northern Ireland Ornithologists' Club has carried out counts of the wildfowl at different periods, and these represent the principal effort to assemble population data. Wardening staff of the Department of the Environment, Craigavon Borough Council and the Royal Society for the Protection of Birds have also undertaken counts of wildfowl in selected areas or for specific purposes. However, all these counts have severe limitations – there is normally no record of the activities of the birds at the time of the count; they take no account of the sometimes extensive movements which take place at dusk; local movements due to weather conditions can be obscured; except for swans, no attempt is made to differentiate numbers by sex or by age class; no account is taken of the numbers of wildfowl on other waters within easy flying distance; there is no attempt to assess mortality due to wildfowlers. Thus there are several beliefs about the wildfowl which cannot be verified or denied – for example, it is assumed that birds rarely occur more than one mile out from the shore, and that tufted duck *Aythya fuligula* have strong specific preferences when feeding off hatches of Chironomid midges.

No systematic ringing of the wildfowl has been undertaken, so little is known of their movements, particularly on a global scale although D'Arcy (1978) illustrates local migratory routes e.g. between Lough Beg and Belfast Lough. Only about six sites are regularly used for the trapping of passerines, and waders have never been systematically ringed. However, even as this chapter is being written, changes in this lamentable lack of study are in hand, a study on the population dynamics of one gull population is being carried out by an amateur worker, Dr Ian and Mrs Denise Winfield of the University of Ulster, funded by the DoE Wildlife and Conservation Branch, are undertaking a thorough analysis of all previous bird counts, and more vigorous

R. B. Wood and R. V. Smith (eds.), Lough Neagh, 491–506.

and regular bird counting is being co-ordinated again by DoE Conservation Branch. However, written at this time, this chapter has to be based very much on personal observations and on wildfowl counts made by the Northern Ireland Ornithologists' Club. In this chapter the birds are considered within broad habitat groupings. Areas mentioned in the text are shown in Fig. 20.1.

20.2. Birds of open water

Although Loughs Neagh and Beg have long been renowned for their large wildfowl populations, no reliable estimates of numbers were available until the Northern Ireland Ornithologists' Club undertook counts of wildfowl, coots and grebes at monthly intervals from 1964 to 1969. Six further complete counts were carried out in the winters of 1979, 1980 and 1981. All data quoted here are from these counts, kindly made available by the Club. Comparisons with the overall Irish situation refer to information published in Hutchinson (1979).

Mallard – *Anas platyrynchos*. This is probably the most familiar duck to the layman, being the most widespread species in the region. It is also the most numerous of the surface-feeding or dabbling ducks on Loughs Neagh and Beg with a maximum count of 10,700 in August 1966 and a peak midwinter (November to January) count of 5,400 in November 1965. The seasonal variations in population are shown in Fig. 20.2a as monthly means of the 1965–1969 data. The more recent counts indicated a maximum midwinter population in excess of 2,500, a little less than the average of the earlier period. The distribution of mallard for the 1965–1969 period is shown in Table 20.1.

Clearly, the largest refuge areas are Lough Beg and the north-east of Lough Neagh, both of which are shallow areas, the former with especially rich macrophytic vegetation (see Chapter 19). It may be that because of the prevailing south-west winds the north-east area has much vegetable material drifting ashore, providing better feeding than might be expected from the vegetation growing *in situ*. The west and south coasts have many shallow stretches but because of disturbance, primarily from wildfowlers, populations representing around 15% of the total are found there only in the nesting season.

An unknown number of birds are reared by wildfowlers and released into the wild in August and September of each year, and it is also unknown if they remain in the area of release or migrate south-westwards as has been recorded elsewhere.

Teal – *Anas crecca*. The distribution pattern of this species differs from that of mallard principally in that the bulk of the population is at Lough Beg. The average monthly populations during the 1965–1969 period are shown in Fig. 20.2a.

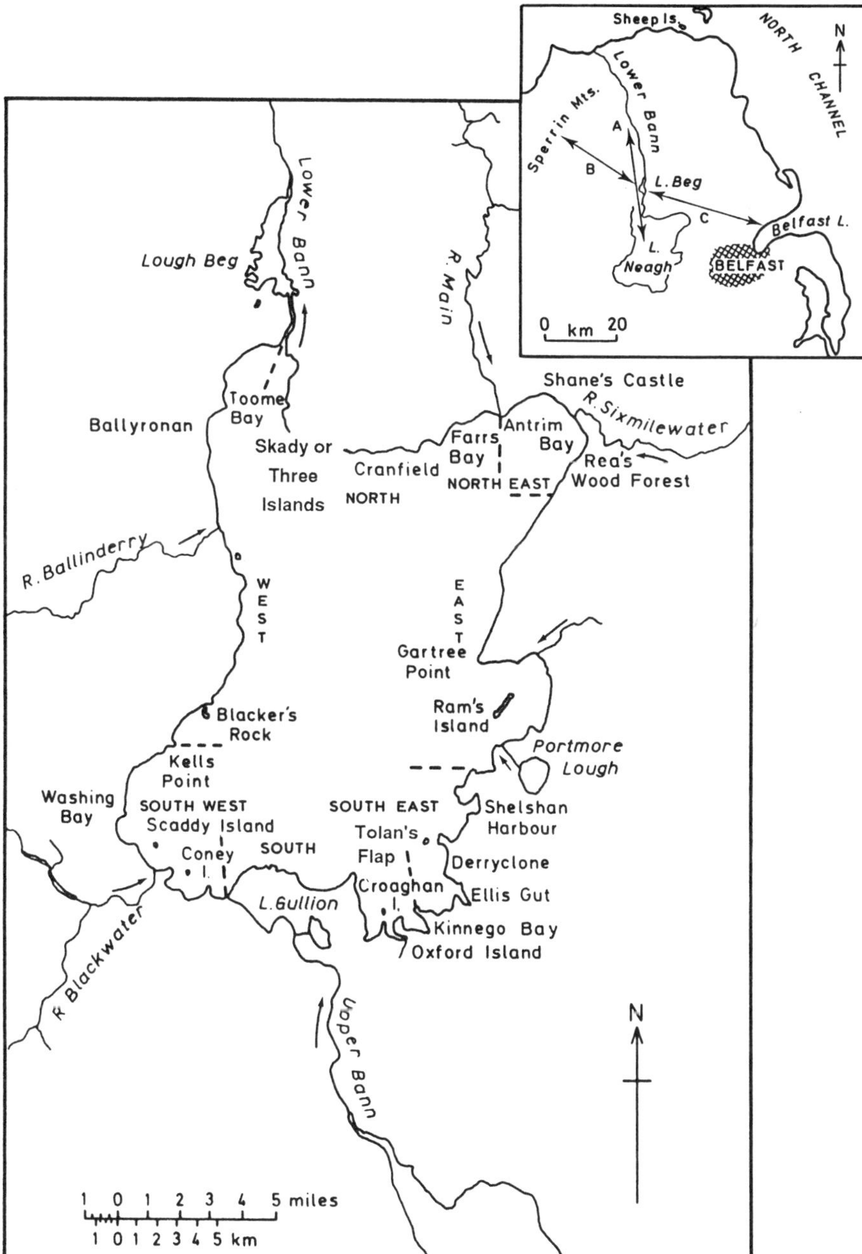

Figure 20.1. Map of Lough Neagh and Lough Beg showing the areas used to determine the distribution of birds and areas of special interest referred to in the text. Inset shows the local route of some bird migration given by D'Arcy (1978).

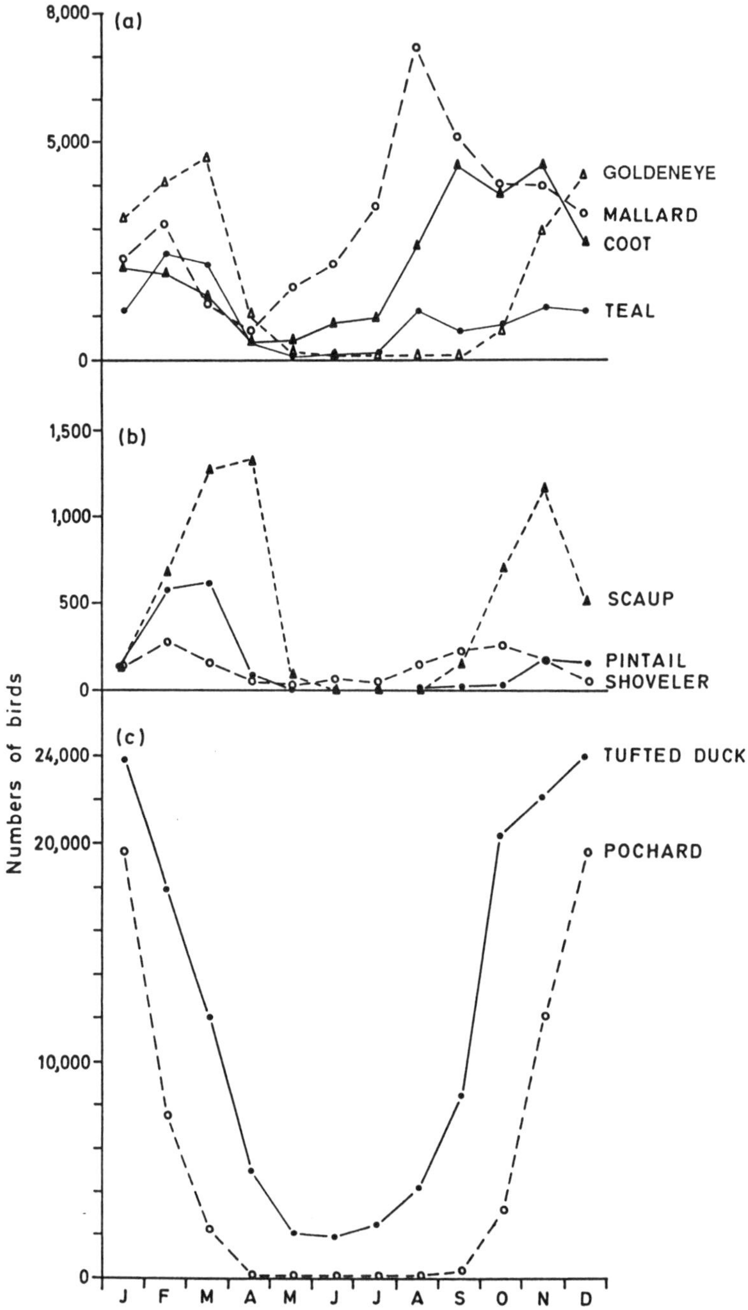

Figure 20.2. The seasonal variation in populations of wildfowl on Lough Neagh and Lough Beg. Data are the means of counts made each month from 1965 to 1969 by the Northern Ireland Ornithologists' Club.

Table 20.1. The distribution of mallard on Lough Neagh and Lough Beg. The data are expressed as the mean percentages of the total counts at monthly intervals from 1965–1969, and the areas are shown in Fig. 20.1

Month	Beg	Neagh						
		W	SW	S	SE	E	NE	N
Jan	13.1	3.6	2.9	17.7	2.2	6.7	47.6	6.1
Feb	24.5	8.8	6.2	8.4	2.0	7.2	37.5	5.1
Mar	23.2	14.7	11.2	9.5	2.0	7.2	27.3	5.2
Apr	17.3	14.3	5.1	15.4	2.1	11.3	23.9	9.6
May	17.9	10.7	9.5	16.2	2.6	7.9	26.0	9.7
June	8.0	13.2	3.4	16.1	2.2	12.6	36.5	8.0
July	13.6	15.5	3.1	13.4	2.0	11.0	37.0	4.0
Aug	28.4	5.5	7.6	18.1	0.8	8.3	27.8	3.2
Sep	22.3	8.9	4.8	8.8	1.1	7.5	43.6	3.1
Oct	5.4	4.7	2.1	8.9	1.7	4.6	67.5	5.3
Nov	7.5	5.5	4.6	11.6	4.9	5.3	56.1	4.4
Dec	3.2	4.0	0.5	7.1	0.6	4.3	78.9	1.6

The average of the three midwinter counts in the 1979–1981 period in which Lough Beg was counted was 1,134.

Gadwall – *Anas strepera*. Very little is known about this species, the highest count of which has been 89; it is however believed that a significant part of the winter population is in an inaccessible area.

Wigeon – *Anas penelope*. With a total Irish population in winter of over 100,000 birds, the maximum Lough Neagh/Lough Beg count recorded of 3,565 represents a very small proportion of the Irish total. Lough Beg has over 60% of the local birds at most stages of the winter. Very few birds are seen during most summers, and these are probably wounded birds unable to attempt migration.

Pintail – *Anas acuta*. This is the only duck with a maximum recorded in spring (see Fig. 20.2b).

It can be assumed that Lough Beg, where almost all local birds are seen, is an important staging post for more than 10% of the birds overwintering in Ireland.

Shoveler – *Anas spatula*. A few breed, and fluctuating numbers occur throughout the year (see Fig. 20.2b), with Lough Beg again holding the bulk of the population.

Scaup – *Aythya marila*. There is a substantial population in the Lough Neagh basin (Fig. 20.2b) although this is usually a marine species. The maximum recorded during the 1979–1981 counts was 1,200 representing 18% of the total United Kingdom population of a fluctuating species. On Lough Neagh the vast majority of birds are concentrated in one flock which feeds in a deep trench between Ballyronan and Skady Island.

Tufted duck – *Aythya fuligula*. Among the diving ducks, this is the most familiar, being a resident with very large numbers in winter (see Fig. 20.2c)

Table 20.2. The distribution of tufted duck on Lough Neagh and Lough Beg. The data are expressed as the mean percentages of the total counts made at monthly intervals from 1965–1969, and the areas are shown in Fig. 20.1

Month	Beg	Neagh						
		W	SW	S	SE	E	NE	N
Jan	4.2	11.2	14.8	13.2	6.6	32.6	9.9	7.7
Feb	9.5	6.3	23.7	13.6	5.3	21.6	9.0	10.7
Mar	11.0	7.8	12.6	24.1	8.5	14.8	9.1	12.1
Apr	17.8	18.0	11.2	22.2	9.4	8.6	6.0	7.2
May	9.9	12.0	22.1	20.7	5.9	3.9	12.2	13.3
June	5.3	10.3	16.3	28.9	12.4	4.5	11.0	11.3
July	2.9	41.8	5.9	15.3	4.7	3.8	20.1	5.6
Aug	2.5	35.1	11.4	13.3	3.6	7.3	13.5	13.1
Sep	7.1	19.1	36.2	13.0	2.7	9.4	5.7	7.0
Oct	5.6	15.4	14.6	38.0	6.6	7.8	5.1	6.6
Nov	8.0	14.4	15.1	22.2	9.9	10.9	9.7	10.9
Dec	9.6	13.0	9.3	35.3	5.3	8.6	6.5	12.0

and taking readily to feeding by humans at public parks. The distribution of tufted duck in the 1965–1969 period (Table 20.2) shows that deeper waters are the most favoured and that Lough Beg is relatively unimportant. The numbers which occur make the area the most important for this species in Europe west of the Netherlands, with over 50% of Irish birds (and equal to 45% of the total British population). Some notable monthly counts have been: 30,445 (December 1965), 28,856 (November 1967), 27,859 (February 1968) and 34,176 (January 1969). Although a count of over 19,000 was made in November 1979, it is believed that numbers have declined during the last ten years. Despite the vast numbers which occur, there is insufficient evidence to indicate the source of migrants.

Pochard – *Aythya ferina*. Lough Neagh is most important for this species with the numbers present at times exceeding those in the whole of Great Britain. Counts approximating to, or exceeding, the British total have been: 28,161 (December 1965), 37,648 (January 1966), 41,144 (January 1980); on two of these occasions individual flocks of over 19,000 were noted.

Even allowing for the fluctuations in numbers due to movements of birds to Ireland from Britain or Europe in hard weather, the monthly averages for the 1965–1969 counts shown in Fig. 20.2c) are impressive. A small, but apparently increasing, number breed principally on the south shore of Lough Neagh.

Goldeneye – *Bucephala clangula*. Over 60% of the Irish population over-winters here (see Fig. 20.2a) and perhaps 25% of the total British Isles population. There appear to be significant differences between the distribution of the pochard and goldeneye which are demonstrated in Table 20.3. Although small numbers of birds are seen through the summer months, no case of breeding has ever been proved.

Table 20.3. The winter distribution of pochard and goldeneye on Lough Neagh and Lough Beg. The data are expressed as the mean percentages of the total counts made at monthly intervals from 1965–1969, and the areas are shown in Fig. 20.1

Month		Beg	Neagh						
			W	SW	S	SE	E	NE	N
Nov	Pochard	55.0	18.5	8.6	9.3	2.4	0.1	2.1	4.1
	Goldeneye	0.2	23.5	4.1	39.1	5.8	3.5	21.6	2.0
Dec	Pochard	42.3	9.8	19.1	18.6	2.5	1.3	2.1	4.2
	Goldeneye	0.2	16.4	15.2	29.3	10.2	12.1	12.5	4.2
Jan	Pochard	27.7	13.8	17.9	28.3	3.8	1.3	0.6	6.8
	Goldeneye	0.7	15.6	15.3	36.8	2.5	10.2	15.0	3.8

Table 20.4. The distribution of coot in Lough Neagh and Lough Beg at four times of the year. The data are expressed as the mean percentages of the total counts made at monthly intervals from 1965–1969, and the areas are shown in Fig. 20.1

Month	Beg	Neagh						
		W	SW	S	SE	E	NE	N
Feb	18.7	0.4	0.7	8.0	3.3	26.3	7.1	30.9
May	4.4	9.5	9.7	19.5	7.2	10.4	11.4	28.1
Aug	11.2	1.0	4.3	31.1	0.8	24.2	18.0	9.1
Nov	18.3	0.0	0.0	4.0	18.3	9.5	4.9	48.1

Coot – *Fulica atra.* The breeding population of somewhere between 200 and 400 pairs is augmented by migrants in autumn, and the overall seasonal pattern is shown in Fig. 20.2a. Distribution of the population varies considerably throughout the year (see Table 20.4) although no explanation can be offered.

In addition to the species listed the cormorant (*Phalacrocorax carbo*) is also found on Lough Neagh but no quantitative estimates of its considerable population size are yet available. Although present all year round, the cormorant does not breed on the lough; ringing studies suggest that the population is sustained by the arrival of juveniles originating from colonies on the north coast, particularly Sheep Island, Co. Antrim.

20.2.1. *Comment on birds of open water*

While it has been pointed out earlier that scientific information on the wildfowl of Lough Neagh is scarce, it is immediately apparent, from a re-examination of Fig. 20.2, the extent to which Lough Neagh and Lough Beg, with their rich feeding grounds and usually ice-free waters, are important over-wintering stations. Of the species illustrated, only mallard, coot and tufted duck are present in significant numbers from May to August and even then, the populations of the last named are only about 10% of the peak winter values. Scaup and, to a lesser extent, goldeneye and shoveler, appear to have two annual peaks of population, one in early winter and another in late spring. In the absence of more information both from this and other areas it must remain an interesting speculation that species such as Scaup are in fact using Lough Neagh as a staging post en route to and from more southerly winter sites. Mallard, with peak populations in late summer, show a marked contrast with the other species, but it is known to breed in large numbers here (see Section 20.3 of this Chapter).

20.3. Birds of Lough Neagh islands

The many islands of Lough Neagh are of considerable value to birds at most times of the year and some species whose populations have been dealt with in Section 20.2 will also be referred to here and in Section 20.4 because of their preferred breeding sites. The oldest islands (i.e. those existing before recent lough lowerings (see Chapter 2)) with patches of mature woodland, have breeding populations of small passerines including blackcap *Sylvia atricapilla*, wren *Troglodytes troglodytes* and chaffinch *Fringilla coelebs*, as well as providing secure nest sites for the large hooded crow *Corvus corone* and magpie *Pica pica* which are often molested as pests. Ram's Island has a thriving colony of grey heron *Ardea cinerea*.

Most of the ornithological interest is concentrated on islands which were exposed by various, recent lowerings of Lough Neagh; some of these have bouldery or shingle areas utilised by under 100 pairs of common terns *Sterna hirundo*. This species has been forced to abandon a number of sites by increasing numbers of gulls or by thickening vegetation.

By far the most numerous gull is the black-headed gull *Larus ridibundus* of which several thousands of nests are made on islands, in large colonies of up to 20,000 pairs as at Scaddy Island. This species has been an inland nesting species only during this century, and the expansion has been most marked since 1960. A slower rate of increase has been noted in the case of the lesser black-backed gull which is confined to relatively few colonies, the total population being of the order of 300 pairs. This species was traditionally migratory, but during the last five years over 100 have spent the winter in the area. Since the late 1960s a small number of herring gulls, *Larus argentatus*, have been nesting among the lesser black-backs; their total breeding population is less than 20 pairs. In most years one or two pairs of great black-backed gulls *Larus marinus* attempt to nest, and since 1980 common gulls *Larus canus* have been attempting to colonise islands in the south of Lough Neagh.

Where there is adequate cover on islands (as also on the shores of the Loughs) several duck species nest. Those species which are ill-equipped to walk, build their well-concealed nests close to the water's edge e.g. tufted duck *Aythya fuligula* pochard *Aythya ferina* and red-breasted merganser *Mergus serrator*; their young are able to swim soon after hatching. Most surface-feeding ducks can nest well away from the water's edge, and indeed mallard *Anas platyrynchos* may choose sites up to 1 mile inland. However, gadwall, *Anas strepera*, of which there are less than 10 pairs, choose sites adjacent to water. It can be assumed that most teal, *Anas crecca*, breed away from the loughs as very few are observed in May and June, whereas by August the average total for both Loughs Neagh and Beg will have risen to 1,100. A very few pairs of shoveler, *Anas spatula*, breed regularly, but pintail *Anas acuta* is very infrequently proved to do so. Garganey, *Anas querquedula*, may also breed, but proof is lacking, despite regular early summer sightings. These small numbers contrast with those of mallard, with over 1,000 pairs assumed to be breeding each year.

Shelduck *Tadorna tadorna* are typically found nesting in the concealment of rabbit burrows, but on the islands of Lough Neagh they nest on the ground, usually, but not always, under vegetation. This species leaves the area soon after the young are fit to fly, moving to one of the vast moulting sites of Bridgewater Bay, England, or the Heligoland Bight; it is not known to which sites Lough Neagh birds move. From September to early March very few are seen but by the end of March the full breeding population of 80 to 100 pairs has returned.

Surprisingly small numbers of wading birds nest on the islands, the most widespread being redshank *Totanus totanus* and common sandpiper *Tringa hypoleucos*; a mere handful of pairs of ringed plover *Charadrius hiaticula* use sandy or gravelly shores on which to lay their eggs. Occasionally a pair of oystercatchers *Haematopus ostralegus* succeed in raising a brood.

Some islands have stone structures such as walls or towers, the cavities in which provide nesting sites for pied wagtail *Motacilla alba* and grey wagtail *Motacilla cinerea*. Both species feed extensively on shore invertebrates.

20.4. Birds of reed beds

The expanses of reedswamp which occur in several areas of Lough Neagh and around some of the smaller adjacent waters (see Chapter 19) provide nesting sites for species which because of their special adaptations are unable to move easily on land and therefore require to nest close to or on water, but in well-protected locations. They also need to build structures which will float in waters with fluctuating levels and which are subject to varying wind conditions.

The most significant of these birds is the great crested grebe *Podiceps cristatus*, several colonies of which exist; the total population during the 1975 breeding census was just under 300 birds, but severe gales had damaged many of the reed-beds causing the desertion of probably in excess of 40% of the birds. Since then there has been a good recovery but increased disturbance, particularly as the result of fishing activities, has been largely responsible for the birds deserting some traditional sites and resorting to new ones, some of which are unsuitable such as rocky promontories without cover. The little grebe *Podiceps ruficollis* will also nest in reeds but prefers to nest under trees; its population in summer is also increasing. Unlike the great crested grebe, the little grebe appears to be sedentary; most of the larger species move away in autumn, and there is circumstantial evidence to suggest that these birds move to the salt water Belfast Lough in winter.

The only other widespread reed-bed nesting species which builds a floating nest is coot *Fulica atra*, which will also nest on dry land. It is found either colonially or solitarily, and shows considerable aggression both inter- and intra-specifically. A few kingfishers (*Alcedo atthis*) are frequently seen in autumn and winter but little or nothing is known of their numbers or distribution. Whereas the above species generally feed in open water, the water rail *Rallus aquaticus* is most secretive and rarely feeds either in open water or on open ground; consequently its distribution is poorly known, despite the quality of its song and the fact that its most favoured singing time is a late autumn evening.

In contrast to the situation prevailing over much of mainland Europe, Ireland has only one widely distributed warbler which is confined to reedbeds; the inappropriately named sedge warbler *Acrocephalus schoenobaenus*. The nest is built in reeds or waterside vegetation. A migrant species, there is evidence of considerable autumnal passage through the area, with particularly large numbers being recorded at Portmore Lough, which appears to be a strategic link in a series of sites where birds accumulate large deposits of body fat prior to long-distance migration towards Africa.

In 1983 two reed warblers *Acrocephalus scirpaceus* were trapped at Portmore Lough, and as this is a species apparently in the process of colonising Ireland, it is possible that breeding on a very small scale is occurring around Lough Neagh as it is in the neighbouring County of Down.

Since 1974 a small population of ruddy duck *Oxyura jamaicensis*, a North American species now established in a feral state in Britain, has been breeding on the south and west shores of Lough Neagh. The origin of the first birds is obscure; it is less likely that they were genuine transatlantic vagrants than birds which had been released locally or had flown from Britain.

20.5. Birds of grassland

For a distance of some miles inland from both Loughs Neagh and Beg a high proportion of the land is down to grass, either as a permanent or rotational crop. Over the past four centuries, and more especially during the last twenty years, this proportion has been increasing as shoreline scrub is reclaimed and as rough pastures are agriculturally improved.

These grasslands, of whatever type, are important to grazing birds, the

Table 20.5. The distribution of mute swan and whooper swan during the four seasons on Lough Neagh and Lough Beg. The data are expressed as the mean percentages of the total counts made at monthly intervals from 1965–1969, and the areas are shown in Fig. 20.1

	Beg	Neagh						
		W	SW	S	SE	E	NE	N
Mute swan								
NDJ	32.7	7.3	9.9	9.6	3.1	11.3	3.8	22.1
FMA	30.5	9.2	9.8	9.0	3.6	13.6	4.7	19.6
MJJ	17.7	12.7	14.1	17.0	3.6	15.2	5.8	14.0
ASO	17.6	13.4	18.5	12.6	1.6	17.0	4.6	14.7
Whooper swan								
NDJ	57.4	8.9	9.6	4.5	11.2	4.1	1.1	3.0
FMA	63.3	7.6	2.0	2.1	9.1	15.0	0.2	0.7

largest of which, with obvious aquatic affinities are the swans. The mute swan *Cygnus olor* is the only resident species, and being omnivorous, grazes least. Its numbers have considerably declined in the last ten years; counts taken from 1965 to 1969 indicated that the maximum winter population averaged approximately 820, whereas midwinter counts from 1979 to 1981 showed that this number had dropped to 265. No reason for the decline is suggested, but it must be pointed out that non-breeding flocks are not necessarily faithful to any one area; on the negative side, one occurrence of lead poisoning involving several birds which fed close to a clay-pigeon shooting range has been recorded.

Whooper swans *Cygnus cygnus* feed about equally on vegetable and animal matter; several flocks appear capable of obtaining all their green food from macrophytic vegetation and so never appear on fields. As with the previous species there has been a reduction in population over the same period; counts on comparable dates indicate a fall in population from 378 to 206. However, it is difficult to assess the reasons for such a decline, as this species has fluctuated in numbers in the past.

Table 20.5 shows the distribution of the two swan species and illustrates the considerable importance of Lough Beg in winter; the more widespread distribution of mute swans in the breeding season is clear. Bewick's swan *Cygnus columbianus* is primarily a grazing species; it is small and more goose-like in appearance and behaviour than the other species. Flocks, at times in excess of 100 birds, can be regularly seen on favoured fields to which there is considerable annual loyalty.

Formerly flocks of grey geese would occupy grass or stubble fields in winter, but in recent years it is only to the north-east of Lough Neagh that birds come with any regularity. Up to 250 grey-lag geese *Anser anser*, probably including some feral birds, and less than 40 greenland white-fronted geese *Anser albifrons* occur. The duck species with the greatest fondness for grazing on short swards is wigeon *Anas penelope*. To avoid disturbance, feeding in these areas is often at night.

Most of the other species are exploiting the populations of invertebrates in the soil or on the vegetation, with the exception of the finch flocks which in winter eat seeds of many species. Several wading birds occur, their differing, specially adapted, bills allowing a variety of food sources to be taken. Many flocks of the short-billed lapwing *Vanellus vanellus* and golden plover *Charadrius apricaria* occur throughout the area, feeding off the ground surface. The less common curlew *Numenius arquata* uses its long curved bill for probing, and the redshank *Totanus totanus*, while using its moderately long bill in grassland is most frequently met with singly on the shores. Of these species, only lapwing nests in this habitat, and the only other wader doing so is dunlin *Calidris alpina* in very small numbers on the west shores of Lough Beg which are subject to inundation, causing frequent loss of nests.

Nesting in thick tufts of longer permanent grass, or in silage swards are skylarks *Alauda arvensis* and meadow pipit *Anthus pratensis*. These species are superficially similar, with brown, black and buff plumage to add concealment to the bird on the nest. Skylarks are quite omnivorous, while pipits are highly insectivorous. Although most skylarks leave in autumn, at least one flock of over 250 birds has been seen in winter. Pipits are also migratory, and there is a substantial emmigration in autumn.

Now confined to a very few localities is the corncrake *Crex crex*. Although its decline is due to a variety of factors, the change in agricultural practice whereby grass is cut for silage early in the summer undoubtedly has had a drastic effect on the successful nesting of this species, which nests on the ground and of which the female is most reluctant to leave the clutch when disturbed.

Most passerines also feed on grassland, including very large flocks of starling *Sturnus vulgaris* (some of which roost in reed-beds and others in urban situations), rook *Corvus frugilegus*, jackdaw *Corvus monedula*, redwing *Turdus musicus*, fieldfare *Turdus pilaris* and song thrush *Turdus philomelos*. Where there are seeds as in unimproved grassland, cardueline finches may occur in flocks which may number over two thousand birds of many species including chaffinch *Fringilla coelebs*, greenfinch *Chloris chloris*, linnet *Carduelis cannabina*, redpoll *Carduelis flammea* and yellowhammer *Emberiza citrinella*.

20.6. Birds of woodland

There are many different types of woodland around both loughs, but none is sufficiently extensive to be completely dominant in any one area. As a result, few of the bird species are confined to one type of woodland, but most range freely through several woodland habitats in search of food, nest sites or shelter. Some species, such as grey heron *Ardea cinerea*, rook *Corvus frugilegus* and jackdaw *Corvus monedula*, require tall sturdy mature trees in which to nest colonially; older trees provide hole-nesting sites for the last-named species as well as for stock doves *Columba oenas*. Wood pigeon *Columba palumbus* often uses mature trees, especially when covered with ivy *Hedera helix*. Both kestrel *Falco tinnunculus* and long-eared owl *Asio otus* prefer to utilise old crow *Corvus corone* or magpie *Pica pica* nests, which can be found in several woodland types; the other and most generally distributed small raptor, the sparrow hawk *Accipter nisus*, prefers the seclusion of older thick woods. The largest breeding raptor in the area is buzzard *Buteo buteo* of which one pair successfully bred in 1983 in mature estate woodland. One ground-nesting game species, woodcock *Scolopax rusticola*, usually nests at the base of a mature tree, but is not infrequently located in coniferous woodland.

The passerines with the strongest affinity to mature woodland are blackcap *Sylvia atricapilla* and chiffchaff *Phylloscopus collybita*; the former requires the presence of a considerable shrub layer and has increased as a breeding species during the last fifteen years. There has also been a similar increase in the number of blackcaps which overwinter in the area; in one orchard alone 68 different individuals were trapped and ringed in one winter; it is believed that these birds are migrants from northern Europe and not residents. The past ten years has witnessed the virtual disappearance of another woodland bird – the garden warbler *Sylvia borin* which has unaccountably left its known sites in the north-east corner of Lough Neagh where it was confined to mature estate woodland. The other leaf warbler of the area, the willow warbler *Phylloscopus trochilus*, occurs more frequently in the lower alder and willow scrub of the Lough Neagh shore-lands.

Outside the breeding season most of the passerine species move through the woodlands, often non-selectively, in flocks usually of mixed species. The four tit species great *Parus major*, blue *Parus caeruleus*, coal *Parus ater* and long-tailed *Aegithalos caudatus* associate with tree creeper *Certhia familiaris* and goldcrest *Regulus regulus*. Occasionally siskin *Carduelis spinus* may accompany them, particularly when they are feeding on alder *Alnus glutinosa*; a more frequent feeding companion is redpoll *Carduelis flammea*. Greenfinch *Chloris chloris*, goldfinch *Carduelis carduelis* and linnet *Carduelis cannabina* occur more in scrub than in woodland, though they prefer hedges and grassland. Open glades in mature woodland, especially of oak *Quercus* spp. or beech *Fagus sylvatica* attract flocks of chaffinch *Fringilla coelebs* sometimes of several hundreds, and occasionally including a handful of brambling *Fringilla montefringilla*; both are migrant species in addition to chaffinch being the commonest breeding woodland species. Bullfinch *Pyrrhula pyrrhula* is seldom found in parties of more than eight or ten birds; although more properly a woodland species, it is most frequently met with in hedgerows. The only other woodland bird which appears to be solitary is the wren *Troglodytes troglodytes* which, however, will roost communally, particularly in hard weather.

The majority of the remaining woodland species are usually met with in the extensive hedgerows away from the loughs; of the thrush group the resident mistle thrush *Turdus viscivorus* probably is the species with the least winter augmentation by European migrants; the others in this group are song thrush *Turdus philomelos* and blackbird *Turdus merula*; fieldfare *Turdus pilaris* and redwing *Turdus musicus* are winter visitors. Robins *Erithacus rubecula* are generally distributed, but the similarly-sized stonechat *Saxicola torquata* is confined to a few areas of open scrub including whin *Ulex europaeus*.

Yellowhammer *Emberiza citrinella* and reed bunting *Emberiza schoeniclus* occur in both hedgerow and scrub habitats; the former has declined in recent years perhaps due in part to a loss of the invertebrate food supply required for its nestlings.

Two other migrant warbler species occur principally in scrub: the grasshopper warbler *Locustella naevia* and the whitethroat *Sylvia communis*. Found throughout all woodland habitats is the spotted flycatcher *Muscipapa striata*, the latest arrival of all summer migrants. Also widespread, but resident, is the dunnock *Prunella modularis*. The house sparrow *Passer domesticus* is not only a familiar of man, but occurs widely in hedgerows and ruins. Since 1960 the tree sparrow *Passer montanus* has colonised the Lough Neagh Basin; in Ireland the distribution of this species shows a very strong bias towards waterside environments unlike the situation elsewhere in Europe.

Many of the woodland and scrub-loving species listed above will also be found on the cut-over peat bogs which have been recolonised by birch *Betula pubescens*, but one species is confined to the more heathery parts of this habitat – the nightjar *Caprimulgus europaeus*. Formerly occurring on several areas, this species has declined to such an extent that only one bird was reported singing in 1983. The factors responsible for the decline are unknown.

20.7. Nature conservation

At the international level, Loughs Neagh and Beg appear as one site on the lists of sites appended to the Ramsar Convention – the Convention on the Conservation of Wetlands of International Importance to Wildfowl and Waders. Under this Convention, contracting governments undertake to ensure the conservation of the sites, to ensure that they do not deteriorate in terms of habitat quality, and in the case of irreversible damage to replace a site of similar quality on the list.

The principal legislative instrument relating to nature conservation until very recently was the Amenity Lands (Northern Ireland) Act 1965, and the highest designation possible under it was that of nature reserve. This is defined as 'any area reserved, managed and used for the protection and maintenance therein of wild life, wild vegetation and objects of geological, physiographical or other scientific or special interest, in the public interest and in order to provide opportunities for study or research for students of such life, vegetation and objects'. Such nature reserves are acquired by the Department of the Environment for Northern Ireland by outright purchase, lease or by agreement with the landowner, of all or some of the interests in the land. Four such reserves have been declared on Lough Neagh, all with the prefix 'National' indicating that they are deemed to be leading examples of their particular habitats.

Three of the National Nature Reserves consist of woodland, shoreline scrub and pools in the north-east of Lough Neagh at Randalstown Forest, Farr's Bay and Rea's Wood Forest. The fourth NNR comprises most of the islands in Lough Neagh, and its declaration is an indication of the importance of the islands to nesting wildfowl, terns and gulls. During the winter months wildfowling is undertaken by arrangement with the owners of the shooting rights, the Shaftesbury Estates of Lough Neagh, Ltd.

The nature reserve at Oxford Island is owned and managed by the Craigavon Borough Council; it has several shoreline habitats, a range of grassland types, hedges and woodland and is currently proposed as NNR. The Department of the Environment also leases the shooting rights over two areas in Lough Neagh which are managed as Wildfowl Refuges; one is managed in association with the Shaftesbury Estates of Lough Neagh Ltd., and the other in addition with the Shane's Castle Estates and the Royal Society for the Protection of Birds. This latter arrangement is linked to the non-statutory nature reserve at Shane's Castle, wherein addition to large numbers of wildfowl, the site affords opportunities for the observation and study of passerines in the woodland.

Loughs Neagh and Beg are also Areas of Scientific Interest. Under the Amenity Lands Act such areas are notified because of their importance in nature conservation terms. The land remains in the possession of the landowners, and the conservation interests are safeguarded by an exercise of a measure of planning control whereby all planning applications are subject to consideration of the implications on the scientific interest.

At the voluntary level, many of the wildfowling clubs and syndicates maintain portions of the areas for which they have leased the shooting rights as refuge areas.

While, clearly, effort is being made to provide protected areas it must be highly regretted that we know so little about the bird life assocated with Lough Neagh and Lough Beg. Movements and migrations and the interactions between the birds and the rest of the ecosystem will take many years to clarify at the present rate of studying but, as hinted earlier, more structured research is beginning and legislative changes are improving the prospects that this important but neglected aspect of the ecology of Lough Neagh will receive due attention.

Acknowledgements

I am very grateful to the Northern Ireland Ornithologists' Club for permission to use data from its wildfowl counts. I have relied on information from the wardening staff of the Department of the Environment (NI), Craigavon Borough Council and the Royal Society for the Protection of Birds, R.W. Culbert, Mrs. P. Davidson, R.D. Davidson and D. Allen; J.P. Dutton and H.C.A. McBride also provided useful information; I am also bound to acknowledge a debt to all those with whom I have watched the birds of this area with pleasure.

20.8. References

D'Arcy, G. (1978) Birds at Lough Beg. Blackstaff Press, Belfast.
Hutchinson, C. (1979) Ireland's Wetlands and their Birds. Irish Wildbird Conservancy, Dublin.

21. Lough Neagh into the twenty-first century

R.B. WOOD

21.1. Taking stock

There never can be an ideal time to publish a review of ongoing studies such as those which Lough Neagh continues to enjoy twenty years after the great upsurge in scientific interest in 1967. Much as the accountant draws a line at what he chooses to be the end of a financial year, so our choice of giving our account now is to an extent arbitrary. With the possible exception of geological and early settlement history almost all chapters could be marginally or significantly updated given the continuing and indeed expanding volume of work currently being carried out. Even in those historical accounts interpretations may in time be more finely tuned and in disciplines such as palaeolimnology (Chapter 7), sediments are being analysed for other constituents which amplify and clarify if not substantively change current thinking.

With other aspects of the work, the reader will have noted that several of the chapters dealing with long term data such as those on the seasonality of nutrient cycles, phytoplankton and zooplankton (Chapters 9, 10 and 14) and the experimental studies which illuminated their interpretation (Chapters 4, 11, and 13) cover, in the main, only the first decade or so of what might be called Lough Neagh's modern era. Long term studies of that sort continue to this day because it is felt that more profound ecological understanding is obtainable from long runs than from the much more common 3 year PhD runs which afford so little scope for interpreting year-to-year variation. Clearly we anticipate that interpretations will improve but believe that the 10 year runs given here are, in their own right, a major contribution.

In Chapter 1, Lough Neagh was introduced under the shadow of a widespread concern at its advanced state of eutrophication, and it was pointed out that much of the work carried out in Chapters 8, 9, 10, 11 and 16 owed its origin to that concern. Indeed, the FBIU was set up specifically to seek out the causes and possible remedies for this unacceptable condition and Smith in Chapter 8 reported on the first steps in 1981–1983 to make use of the FBIU's findings for management. Sewage-borne available phosphate had

R.B. Wood and R.V. Smith (eds.), Lough Neagh, 507–512.

by then been clearly identified as the critical agent and a process for removing significant quantities of phosphate from sewage effluent developed (Gray, 1985). The sustained response of Lough Neagh may take years to emerge (it will be remembered from Chapter 7 that Lough Neagh had been increasingly enriched over about 100 years) but there is already cause to believe that Smith's optimism in Chapter 8 is justified. Gibson (1986) and Smith *et al.* (1986) have produced reassuring evidence that the rise in levels of phosphate recorded between 1968 and 1981, and the corresponding higher crops of phytoplankton, have been halted perhaps even reversed. Figs. 21.1 and 21.2 illustrate the increasing phosphate and chlorophyll *a* concentrations up to the time of phosphate removal and the reversal since, together with the welcome lessening of the blue-green algal concentration.

Perhaps the most encouraging aspect of Lough Neagh's initial response to phosphate removal is the precision with which the FBIU's carefully developed model predicted it. In the first couple of years, some 60 tonnes yr^{-1} of sewage derived phosphate-P was kept from the lough. The measured response in terms of chlorophyll a through its strong correlation with phosphate was within a tonne or two of that. Of course it is likely that year-to-year variables will to an extent make such a neat signal somewhat more noisy but, so far, confidence in the model and by inference the precision of the equations, co-efficients and parameters fed into it must remain very high.

Those privileged to go on studying Lough Neagh for the next couple of decades will doubtless determine with interest whether and how closely other facets of the ecology, for example, the chironomid and zooplankton populations, respond to the changed nutrient loading. While the immediate management problem of reducing algal nuisance may, let us hope, be in hand, the real value of following any recovery will surely lay in its contribution to understanding how lakes work, how parts interact, how new equilibria are established. We now have a massive, whole-lake experiment available, confused no doubt with other, ever-changing inputs, not merely phosphate.

From its modest beginnings in 1968, research on Lough Neagh could not be expected to advance on all fronts with equal vigour or success. Where most effort was placed great progress was madc and by now some of those chapters, although providing the essential platform for later ideas and progress, seem somewhat dated. It would be wrong to assume however that this volume somehow marks the end of creative and imaginative research; if anything it marks the beginning. In some areas we are conceited enough to believe that recent and current work, both in concept and technique, bears favourable comparison with research anywhere else. In other areas, of which macrophytes and wildfowl are obvious examples, much baseline work needs to be done before they can be integrated into the whole-system ecology which remains our long term, and probably unattainable, goal. Without pre-empting what later writers might include in a volume Lough Neagh Mark II, it is worth briefly looking at what is in hand and what the future may bring. Continuing the long term studies of phyto- and zooplankton has

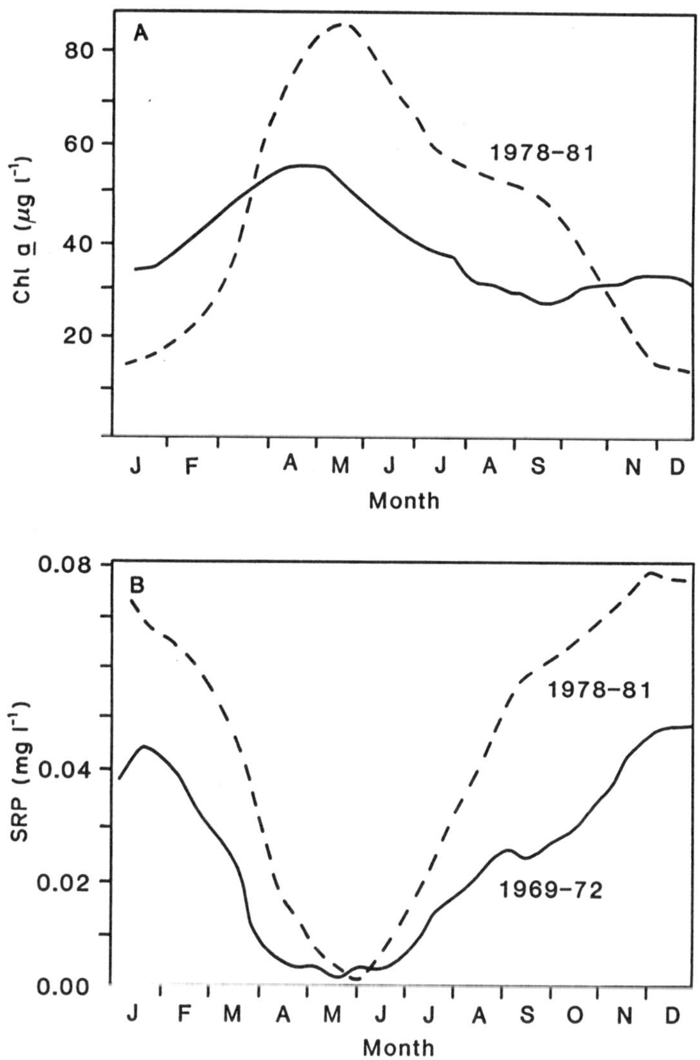

Figure 21.1. Mean weekly values of (A) chlorophyll *a* and (B) soluble reactive phosphate in Lough Neagh in 1969–72 (——) and in 1978–81 (- - -). Data from Smith *et al.* (1986).

already been mentioned, not only in relation to nutrient inputs but with increasing concern for their links to higher levels in the food web and for the underlying factors which control their performance. Nutrient budget studies are far from over and Gibson is moving towards a much more profound understanding of processes than were needed merely to incriminate phosphate and exonerate nitrate from the management of eutrophication.

Already applying modern techniques, the FBIU is using an aircraft-

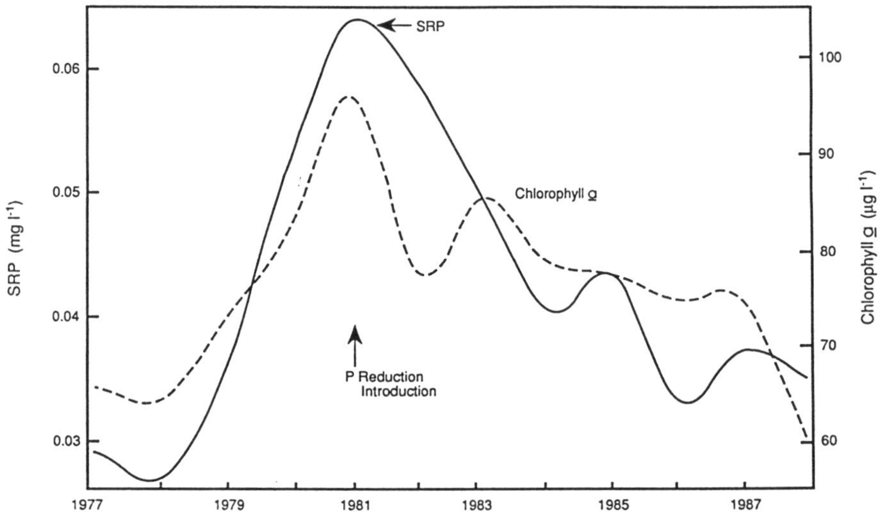

Figure 21.2. Mean annual soluble reactive phosphorus and mean April–June chlorophyll *a* for the years 1977–1988. Phosphorus reduction at sewage treatment works was first introduced in 1981.

mounted Daldalus Scanner as a remote sensor to develop an algorithm to estimate algal concentrations. Calibrated against *in situ* measurements the technique will be used to interpret satellite data. Acid rain and the environmental impact of nitrate losses from grassland figure prominently in current work and Smith and Stewart's recent analysis of Lough Neagh's nitrate inputs may have significance for nitrates in drinking water far beyond this catchment.

In Chapter 3, Carter briefly referred to the work of Soliman and Jewson on wave properties and water movements in Lough Neagh. On to that preliminary study (and on such as those of Carter and Murphy, Chapter 16 and Quinn, Chapter 17) Jewson and his group are vigorously building not merely to provide information on how effluents discharged into Lough Neagh might be dispersed but as a significant contribution to the ecology of the lough, particularly, though not exclusively, in the inshore areas where physical and chemical processes so powerfully interact with benthic and planktonic algae, invertebrates, bacteria, fish and birds to produce a particularly challenging study of food web inter-connections. Radiotelemetry of water current data and sophisticated isotope tracer techniques are being brought to bear.

On the palaeoecological front, Rippey and his group are nearing completion of the first phase of a comprehensive record of trace metal flux to Lough Neagh (coupled with contemporary inputs) which not only describes aspects of Lough Neagh's industrial history but contributes fundamentally to our understanding of the sedimentary processes, of atmospheric pollution on this upwind edge of Europe and, through international links, to the analysis of Europe's acid rain problem. Given Lough Neagh's role in water supply and the burgeoning legislation covering it, trace organics and pesticides already figure in Lough Neagh research thinking.

One area of study which, given its economic and social importance is surprisingly only now emerging from the baseline stage, is that of fish research. Whilst this book can devote some 90 pages to extant studies, it is fair to say that those concerned feel strongly that the level of sustained work is inadequate, given not only the economic but also the ecological importance of fish. Winfield has rapidly gathered around him a vigorous group but three of the four scientists are postgraduates for which replacements may or may not be forthcoming. Even if they are, comprehensive population studies covering many species, including introductions such as the roach, and aimed both at sensible exploitation and ecological wisdom, do not necessarily fall easily into PhD-sized packages. It is almost incomprehensible that eels remain almost completely unstudied between elver recruitment and brown eel catch. With salmon and trout, upon which much leisure and tourist potential depends, the story is almost the same. There is no comprehensive analysis of catch, and escapement into nursery areas and spawning records are by the FCB's own admission, inadequate. Such work as is done is often in reaction to anglers' complaints which are not always totally objective nor a sound basis for good research.

In the main, as reported in Chapter 1, Lough Neagh's rivers are in pretty good condition but in significant danger of deterioration, partly from drainage activities but mostly from agricultural effluents to which intensive animal rearing and rapidly increasing silage production makes a very significant contribution. Further work on the ecology of rivers and on the application of farming methods (such as those proposed by Gornall and Wood (1988) for treating farm wastes) may be anticipated. It is unlikely that slurry processing will have the equivalent environmental impact on rivers as phosphate removal appears to be having on the open lough but it may play a valuable role.

Rivers are of course not merely water-filled channels in which fish live. With their riparian flora and fauna, they represent a habitat of considerable conservation, amenity and landscape value and it is good to be able to report that the Conservation and Wildlife Branch of DoE have funded a survey of the Six Mile Water with regard to macrophytes, riparian flora and associated amenity as part of their role in contributing to river management. Other surveys may follow and the Winfields have completed for the same government agency a thorough re-analysis of all previous bird counts (Winfield and Winfield, 1987a, and b). A more systematic bird census is in hand and some research has begun.

Part of the changing pattern of study of Lough Neagh which we can anticipate for the future springs from the evolving political and public awareness of environmental problems and the legislation, local, national, European and world wide which is being enacted. The Nature Conservation and Amenity Lands (NI) Order 1985 was Northern Ireland's positive response to the Balfour Report and the greatly enhanced status of the Conservation and Wildlife Branch, the establishment of the Environmental Protection Division within DoE(NI) and the designation, in 1992, of L. Neagh as an Area of Special Scientific Interest (ASSI) give clear hope for pro-active as well as re-active effort.

At the European level a plethora of directives and proposals is being brought to bear on most things aquatic and, more generally, the directive requiring environmental impact analysis as part of planning proposals for major developments raises expectations that underlying environmental science will play its proper role alongside economic and political decision making. This may come into force none too soon. Reference was made in Chapter 2 to the lignite deposits close to the lough shore. Planning seemed to reach a fairly advanced stage, with loughside power generation commonly held to be the only economic means of exploiting it. Environmental consequences could have been profound and although lignite exploitation has been put off for the time being it was disconcerting to note that that decision was made without any evidence known to us of a *comprehensive* environmental impact analysis, only small isolated bits of study. Perhaps, when the question next arises, as it surely will, Lough Neagh's capacity to cope will be better understood and argued.

Chapter 1 began with a quaint analogy equating scientific knowledge of Lough Neagh to Aesop's bundle of sticks. We bring the analogy full-circle with the message that new saplings for the next bundle have already been planted, some are well grown, some are still seedlings and some await freshly prepared planting grounds.

21.2. References

Gibson, C.E. (1986) Preliminary results on phosphorus reduction – assessing the effect against a background of change. Hydrobiol. Bull. 20: 173–182.

Gornall, L.K. and Wood, R.B. (1988) Potential environmental benefits from processing farm effluents. Ecol. Bull. 39: 143–145.

Gray, A.V. (1985) Removal of phosphate at sewage treatment works and the implications on phosphate loadings into Lough Neagh. J. Inst. Wat. Eng. Sci. 39: 137–154.

Smith, R.V., Jordan, C. and Stewart, D.A. (1986) The application of empirical modelling to eutrophication management in Lough Neagh. In: Water quality modelling in the inland natural environment. BHRA Conference, pp. 309–324.

Winfield, D.K. and Winfield, I.J. (1987a) Lough Neagh Waterfowl Ecology Report commissioned by the Countryside and Wildlife Branch of DoE(NI).

Winfield, D.K. and Winfield, I.J. (1987b) Initial studies of the wintering wildfowl of Lough Neagh, Northern Ireland. Paper read at the Br. Ecol. Soc. Conference, Cambridge, December 1987.

TAXONOMIC / BIOLOGICAL INDEX

GENERAL INDEX

MONOGRAPHIAE BIOLOGICAE

1. *Physiologia Comparata et Oecologia*, Vol. I. 1948 out of print
2. *Physiologia Comparata et Oecologia*, Vol. II. 1952 out of print
3. *Physiologia Comparata et Oecologia*, Vol. III. 1954 out of print
4. *Physiologia Comparata et Oecologia*, Vol. IV. 1957 out of print
5. (1) R. A. Kennedy: *The Metzner Theory of Urine Formation.* 1957
   (2) M. D. Grmek: *On Ageing and Old Age.* Basic Problems and Historic Aspects of Gerontology and Geriatrics. 1958
   (3–4) O. Bassir: *Biochemical Aspects of Human Malnutrition in the Tropics.* 1962 out of print
6. F. S. Bodenheimer: *Animal Ecology Today.* 1958 out of print
7. R. R. Gates: *Taxonomy and Genetics of Oenothera.* 1958 ISBN 90-6193-061-8
8. A. Keast, R. L. Crocker and C. S. Christian (eds.): *Biogeography and Ecology in Australia.* 1959 out of print
9. S. Stanković: *The Balkan Lake Ohrid and Its Living World.* 1960 out of print
10. E. Rivnay: *Field Crop Pests in the Near East.* 1962 ISBN 90-6193-063-4
11. M. Kozhov: *Lake Baikal and Its Life.* 1962 ISBN 90-6193-064-2
12. M. S. Mani: *The Ecology of Plant Galls.* 1964 out of print
13. E. van Oye (ed.): *The World Problem of Salmonellosis.* 1964 ISBN 90-6193-066-9
14. D. H. S. Davis (ed.): *Ecological Studies in Southern Africa.* 1964 out of print
15. J. van Mieghem and P. van Oye (eds.): *Biogeography and Ecology in Antarctica.* 1965 ISBN 90-6193-067-7
16. H. Boyko (ed.): *Salinity and Aridity.* New Approaches to Old Problems. 1966 out of print
17. M. J. Coe: *The Ecology of the Alpine Zone of Mount Kenya.* 1967 ISBN 90-6193-068-5
18. E. J. Fittkau, J. Illies, H. Klinge, G. H. Schwabe and H. Sioli (eds.): *Biogeography and Ecology in South America*, Vol. I. 1968 out of print
19. E. J. Fittkau, J. Illies, H. Klinge, G. H. Schwabe and H. Stioli (eds.): *Biogeography and Ecology in South America*, Vol. II. 1969 ISBN 90-6193-071-5
20. R. N. Kaul (ed.): *Afforestation in Arid Zones.* 1970 out of print
21. R. Battistini and G. Richard-Vindard (eds.): *Biogeogaphy and Ecology of Madagascar.* 1972 ISBN 90-6193-073-1
22. J. E. Bishop: *Limnology of a Small Malayan River Sungai Gombak.* 1973 ISBN 90-6193-074-X
23. M. S. Mani (ed.): *Ecology and Biogeography in India.* 1974 ISBN 90-6193-075-8
24. E. K. Balon and A. G. Coche (eds.): *Lake Kariba.* A Man-made Tropical Ecosystem in Central Africa. 1974 ISBN 90-6193-076-6
25. W. D. Williams (ed.): *Biogeography and Ecology in Tasmania.* 1974 ISBN 90-6193-077-4
26. A. de Vos: *Africa, the Devastated Continent?* Man's Impact on the Ecology of Africa. 1975 ISBN 90-6193-078-2
27. G. Kuschel (ed.): *Biogeography and Ecology in New Zealand.* 1975 ISBN 90-6193-079-0
28. I. Prakash and P. K. Ghosh (eds.): *Rodents in Desert Environments.* 1975 ISBN 90-6193-080-4
29. J. Rzóska (ed.): *The Nile.* Biology of an Ancient River. 1976 ISBN 90-6193-081-2
30. G. Kunkel (ed.): *Biogeography and Ecology in the Canary Islands.* 1976 ISBN 90-6193-082-0

MONOGRAPHIAE BIOLOGICAE

31. M. J. A. Werger (ed.): *Biogeography and Ecology of Southern Africa*, 2 vols. 1978 ISBN 90-6193-083-9
32. C. Serruya (ed.): *Lake Kinneret* [Lake of Tiberias / Sea of Galilee]. 1978 ISBN 90-6193-085-5
33. Ph. D. Mordukhaí-Boltovskoi (ed.): *The River Volga and Its Life.* 1979 ISBN 90-6193-084-7
34. Ch. W. Heckman: *Rice Field Ecology in Northeastern Thailand.* The Effect of Wet and Dry Seasons on a Cultivated Aquatic Ecosystem. 1979 ISBN 90-6193-086-3
35. M. Kalk, A. J. McLachlan and C. Howard-Williams (eds.): *Lake Chilwa.* Studies of Change in a Tropical Ecosystem. 1979 ISBN 90-6193-087-1
36. B. R. Allanson (ed.): *Lake Sibaya* [South Africa]. 1979 ISBN 90-6193-088-X
37. H. Löffler (ed.): *Neusiedlersee.* The Limnology of a Shallow Lake in Central Europe. 1979 ISBN 90-6193-089-8
38. J. Rzóska: *Euphrates and Tigris.* Mesopotamian Ecology and Destiny. With Contributions by J. F. Talling and K. E. Banister. 1980 ISBN 90-6193-090-1
39. K. A. Brodsky: *Mountain Torrent of the Tien Shan.* A Faunistic-Ecology Essay. Translated from Russian by V. V. Golosov. 1980 ISBN 90-6193-091-X
40. M. S. Mani and L. E. Giddings (eds.): *Ecology of Highlands.* 1980 ISBN 90-6193-093-6
41. A. Keast (ed.): *Ecological Biogeography of Australia*, 3 vols. 1981 ISBN 90-6193-092-8
42. J. L. Gressit (ed.): *Biogeography and Ecology of New Guinea*, 2 vols. 1982 ISBN 90-6193-094-4
43. H. Heathwole, T. Done and E. Cameron: *Community Ecology of a Coral Cay.* A Study of One-Tree Island, Great Barrier Reef, Australia. 1981 ISBN 90-6193-096-0
44. P. S. Maitland (ed.): *The Ecology of Scotland's Largest Lochs.* Lomond, Awe, Ness, Morar and Shiel. 1981 ISBN 90-6193-097-9
45. K. Müller (ed.): *Coastal Research in the Gulf of Bothnia.* 1982 ISBN 90-6193-098-7
46. G. K. Rutherford (ed.): *The Physical Environment of the Faeroe Islands.* 1982 ISBN 90-6193-099-5
47. J. I. Furtado and S. Mori (eds.): *Tasek Bera* [Malaysia]. The Ecology of a Feshwater Swamp. 1982 ISBN 90-6193-100-2
48. G. R. South (ed.): *Biogeography and Ecology of the Island of Newfoundland.* 1983 ISBN 90-6193-101-0
49. J. A. Thornton (ed.): *Lake McIlwaine.* The Eutrophication and Recovery of a Tropical African Man-made Lake. 1982 ISBN 90-6193-102-9
50. R. W. Edwards and M. P. Brooker: *The Ecology of the Wye* [U.K.]. 1982 ISBN 90-6193-103-7
51. T. Petr (ed.): *The Purari* [Papua New Guinea]. Tropical Environment of a High Rainfall River Basin. 1983 ISBN 90-6193-104-5
52. H. Kuhbier, J. A. Alcover and C. Guerau d'Arellano Tur (eds.): *Biogeography and Ecology of the Pityusic Islands.* 1984 ISBN 90-6193-105-3
53. J.-P. Carmouze, J.-R. Durand and C. Lévêque (eds.): *Lake Chad.* Ecology and Productivity of a Shallow Tropical Ecosystem. 1983 ISBN 90-6193-106-1
54. S. Horie (ed.): *Lake Biwa* [Japan]. 1984 ISBN 90-6193-095-2
55. D. R. Stoddart (ed.): *Biogeography and Ecology of the Seychelles Islands.* 1984 ISBN 90-6193-107-X
56. H. Sioli (ed.): *The Amazon.* Limnology and Landscape Ecology of a Mighty Tropical River and Its Basin. 1984 ISBN 90-6193-108-8

MONOGRAPHIAE BIOLOGICAE

57. C. H. Fernando (ed.): *Ecology and Biogeography in Sri Lanka.* 1984 ISBN 90-6193-109-6
58. S. J. Casper (ed.): *Lake Stechlin* [Germany]. A Temperate Oligotrophic Lake. 1985 ISBN 90-6193-512-1
59. U. Th. Hammer: *Saline Lake Ecosystems of the World.* 1985 ISBN 90-6193-535-0
60. B. R. Davies and K. F. Walker (eds.): *The Ecology of River Systems.* 1986 out of print
61. P. de Deckker and W. D. Williams (eds.): *Limnology in Australia.* 1986 ISBN 90-6193-578-4
62. Y. Yom-Tov and E. Tchernov (eds.): *The Zoogeography of Israel.* The Distribution and Abundance at a Zoogeographical Crossroad. 1988 ISBN 90-6193-650-0
63. F. D. Por: *The Legacy of Tethys.* An Aquatic Biogeography of the Levant. 1989 ISBN 0-7923-0189-7
64. B. R. Allanson, R. C. Hart, J. H. O'Keeffe and R. D. Robarts: *Inland Waters of Southern Africa.* An Ecological Perspective. 1990 ISBN 0-7923-0266-4
65. F. di Castri, A. J. Hansen and M. Debussche (eds.): *Biological Invasions in Europe and the Mediterranean Basin.* 1990 ISBN 0-7923-0411-X
66. R. W. Edwards, A. S. Gee and J. H. Stoner (eds.): *Acid Waters in Wales.* 1990 ISBN 0-7923-0493-4
67. S. K. Jain and L. W. Botsford (eds.): *Applied Populations Biology.* 1992 ISBN 0-7923-1425-5
68. C. Dejoux and A. Iltis (eds.): *Lake Titicaca* [Peru/Bolivia]. A Synthesis of Limnological Knowledge. 1992 ISBN 0-7923-1663-0
69. R.B. Wood and R.V. Smith (eds.): *Lough Neagh.* The Ecology of a Multipurpose Water Resource. 1993 ISBN 0-7923-2112-X

KLUWER ACADEMIC PUBLISHERS – DORDRECHT / BOSTON / LONDON